MOLECULAR EPIDEMIOLOGY OF CHRONIC DISEASES

A Note on the Cover–Artist Statement

My body is the primary tool in my process of application and removal of colour to create paintings. I use it as a paintbrush to create gestural works within the spacial limitations of the body, acknowledging the marks it makes whilst establishing control over my palette, creating a balance between chance and intention.

This process allows me to explore how closely interlinked the body is to the mind, realising the effect of emotional strains, shocks and experiences on the body, and translating them into a visual piece of work. My subject matter is drawn from my own challenges and relationships, whilst remaining visually obscure.

My work is a cross-section of my life, a mesh of emotions and an honest expression of identity.

GISELLE WILD

MOLECULAR EPIDEMIOLOGY OF CHRONIC DISEASES

Editors

Chris Wild
University of Leeds, UK

Paolo Vineis
Imperial College, London, UK

Seymour Garte
University of Pittsburgh Cancer Institute, Pittsburgh, PA, USA

John Wiley & Sons, Ltd

Other Wiley Editorial Offices

John Wiley & Sons Inc., 111 River Street, Hoboken, NJ 07030, USA

Jossey-Bass, 989 Market Street, San Francisco, CA 94103-1741, USA

Wiley-VCH Verlag GmbH, Boschstr. 12, D-69469 Weinheim, Germany

John Wiley & Sons Australia Ltd, 42 McDougall Street, Milton, Queensland 4064, Australia

John Wiley & Sons (Asia) Pte Ltd, 2 Clementi Loop #02-01, Jin Xing Distripark, Singapore 129809

John Wiley & Sons Canada Ltd, 6045 Freemont Blvd, Mississauga, ONT, L5R 4J3

Wiley also publishes its books in a variety of electronic formats. Some content that appears in print may not be available in
electronic books.

Library of Congress Cataloging-in-Publication Data

Molecular epidemiology of chronic diseases / edited by Chris Wild, Paolo
Vineis, and Seymour Garte.
    p. ; cm.
  Includes bibliographical references and index.
   ISBN 978-0-470-02743-1 (cloth : alk. paper)
 1. Molecular epidemiology. 2. Chronic diseases–Epidemiology.   I. Wild,
Chris, 1959- II. Vineis, Paolo. III. Garte, Seymour J.
  [DNLM: 1. Chronic Disease–epidemiology. 2. Biological Markers.
  3. Epidemiologic Methods. 4. Epidemiology, Molecular. WT 500 M719 2008]
   RA652.5.M648 2008
   614.4–dc22
                                    2008003725

British Library Cataloguing in Publication Data

A catalogue record for this book is available from the British Library

ISBN 978-0-470-02743-1 (H/B)

Typeset in 10/12 pt Times by Thomson Digital, Noida, India
Printed and bound in Great Britain by Antony Rowe Ltd., Chippenham, Wiltshire
This book is printed on acid-free paper
The cover image shows a painting by Giselle Wild. The photograph of the painting was kindly provided
by the artist and is used with her permission.

Contents

Contributors   **xi**

Acknowledgements   **xv**

1 Introduction: why molecular epidemiology?   **1**
Chris Wild, Seymour Garte and Paolo Vineis

2 Study design   **7**
Paolo Vineis

   2.1.   Introduction: study design at square one   7
   2.2.   Epidemiological measures   12
   2.3.   Bias   12
   2.4.   More on confounding   14
   2.5.   Specificities of molecular epidemiology design   15
   2.6.   Conclusions   20
         References   21
         Essential reading   22

3 Molecular epidemiological studies that can be nested within cohorts   **23**
Andrew Rundle and Habibul Ahsan

   3.1.   Introduction   23
   3.2.   Case–cohort studies   26
   3.3.   Nested case–control studies   27
   3.4.   Considerations regarding biomarker analyses in case–cohort
         and nested case–control studies   29
   3.5.   Conclusion   34
         References   35

4 Family studies, haplotypes and gene association studies   **39**
Jennifer H. Barrett, D. Timothy Bishop and Mark M. Iles

   4.1.   Introduction   39
   4.2.   Family studies   39
   4.3.   Genetic association studies   45
   4.4.   Discussion   51
         References   52

5 Individual susceptibility and gene–environment interaction   **55**
Seymour Garte

    5.1.   Individual susceptibility   55
    5.2.   Genetic susceptibility   55
    5.3.   Metabolic susceptibility genes   56
    5.4.   Study designs   59
    5.5.   Gene–environment interaction   60
    5.6.   Exposure dose effects in gene–environment interactions   62
    5.7.   Mutational effects of gene–environment interactions   63
    5.8.   Conclusions   63
           References   64

6 Biomarker validation   **71**
Paolo Vineis and Seymour Garte

    6.1.   Validity and reliability   71
    6.2.   Biomarker variability   72
    6.3.   Measurement of variation   73
    6.4.   Other issues of validation   75
    6.5.   Measurement error   75
    6.6.   Blood collection for biomarkers   79
    6.7.   Validation of high-throughput techniques   79
           References   80

7 Exposure assessment   **83**
Mark J. Nieuwenhuijsen

    7.1.   Introduction   83
    7.2.   Initial considerations of an exposure assessment strategy   83
    7.3.   Exposure pathways and routes   85
    7.4.   Exposure dimensions   86
    7.5.   Exposure classification, measurement or modelling   87
    7.6.   Retrospective exposure assessment   93
    7.7.   Validation studies   94
    7.8.   Quality control issues   94
           References   95

8 Carcinogen metabolites as biomarkers   **97**
Stephen S. Hecht

    8.1.   Introduction   97
    8.2.   Overview of carcinogen metabolism   98
    8.3.   Examples of carcinogen metabolite biomarkers   99
    8.4.   Summary   106
           References   107

9 Biomarkers of exposure: adducts   **111**
David H. Phillips

    9.1.   Introduction   111
    9.2.   Methods for adduct detection   114
    9.3.   Adducts identified in human tissue   116

9.4. Adducts as biomarkers of occupational and environmental exposure to carcinogens   117
9.5. Smoking-related adducts   119
9.6. DNA adducts in prospective studies   119
9.7. Conclusions   121
      References   122

10 Biomarkers of mutation and DNA repair capacity   **127**
      Marianne Berwick and Richard J. Albertini

10.1. Introduction   127
10.2. Classification of mutations   128
10.3. Mutations in molecular epidemiology   128
10.4. DNA repair   130
10.5. Classes of DNA repair   130
10.6. Common assays to measure DNA repair capacity   132
10.7. Integration of DNA repair assays into epidemiological studies   134
10.8. Genetic markers for DNA repair capacity   135
       References   137

11 High-throughput techniques – genotyping and genomics   **141**
      Alison M. Dunning and Craig Luccarini

11.1. Introduction   141
11.2. Background   141
11.3. SNP databases   144
11.4. Study types   144
11.5. Study design   147
11.6. Genotyping technologies   149
11.7. Sample and study management and QC   149
11.8. After the association has been proved – what next?   153
       References   154

12 Proteomics and molecular epidemiology   **155**
      Jeff N. Keen and John B. C. Findlay

12.1. Introduction   155
12.2. General considerations   155
12.3. Sample selection   156
12.4. Proteomics technologies   157
12.5. Illustrative applications   164
12.6. Final considerations   165
       References   165

13 Exploring the contribution of metabolic profiling to epidemiological studies   **167**
      M. Bictash, Elaine Holmes, H. Keun, P. Elliott and J. K. Nicholson

13.1. Background   167
13.2. Cancer   172
13.3. Cardiovascular disease   174
13.4. Neurodegenerative disorders   176
13.5. The way forward   177
       References   177

14   Univariate and multivariate data analysis   181
Yu-Kang Tu and Mark S. Gilthorpe

    14.1.   Introduction   181
    14.2.   Univariate analysis   183
    14.3.   Generalized linear models   186
    14.4.   Multivariate methods   187
    14.5.   Conclusions   196
           References   197

15   Meta-analysis and pooled analysis – genetic and environmental data   199
Camille Ragin and Emanuela Taioli

    15.1.   Introduction   199
    15.2.   Meta-analysis   199
    15.3.   Pooled analysis   201
    15.4.   Issues in pooled analysis of epidemiological studies involving
           molecular markers   201
           References   205

16   Analysis of complex datasets   207
Jason H. Moore, Margaret R. Karagas and Angeline S. Andrew

    16.1.   Introduction   207
    16.2.   Gene–environment interaction   207
    16.3.   Gene–gene interaction   208
    16.4.   Statistical interaction   208
    16.5.   Case study: bladder cancer   213
    16.6.   Genome-wide analysis   215
    16.7.   Summary   218
           References   218

**17   Some implications of random exposure measurement errors
     in occupational and environmental epidemiology**   223
S. M. Rappaport and L. L. Kupper

    17.1.   Introduction   223
    17.2.   Individual-based study   224
    17.3.   Group-based studies   225
    17.4.   Comparing biases for individual-based and group-based studies   228
    17.5.   Conclusions   230
           References   230

18   Bioinformatics   233
Jason H. Moore

    18.1.   Introduction   233
    18.2.   Database resources   234
    18.3.   Data analysis   236
    18.4.   The future   240
           References   240

19 Biomarkers, disease mechanisms and their role in regulatory decisions   **243**
Pier Alberto Bertazzi and Antonio Mutti

    19.1.   Introduction   243
    19.2.   Hazard identification and standard setting   243
    19.3.   Risk characterization: individuals and populations   246
    19.4.   Monitoring and surveillance   249
    19.5.   What to regulate: exposures or people's access to them?   251
    19.6.   Conclusion   252
           References   252

20 Biomarkers as endpoints in intervention studies   **255**
Lynnette R. Ferguson

    20.1.   Introduction: why are biomarkers needed in intervention studies?   255
    20.2.   Identification and validation of biomarkers   257
    20.3.   Use of biomarkers in making health claims   257
    20.4.   Biomarkers of study compliance   258
    20.5.   Biomarkers that predict the risk of disease   259
    20.6.   Biomarkers relevant to more than one disease   259
    20.7.   Biomarkers that predict the optimization of health or performance   264
    20.8.   Conclusions   264
           References   264

**21 Biological resource centres in molecular epidemiology:
collecting, storing and analysing biospecimens**   **267**
Elodie Caboux, Pierre Hainaut and Emmanuelle Gormally

    21.1.   Introduction   267
    21.2.   Obtaining and collecting biospecimens   268
    21.3.   Annotating, storing and processing biospecimens   271
    21.4.   Analysing biomarkers   277
    21.5.   Conclusions   278
           References   278

22 Molecular epidemiology and ethics: biomarkers for disease susceptibility   **281**
Kirsi Vähäkangas

    22.1.   Introduction   281
    22.2.   Ethical aspects in biomarker development for disease susceptibility   283
    22.3.   Ethical aspects of biobanking   286
    22.4.   Molecular epidemiology and society   290
    22.5.   Conclusions   292
           References   293

**23 Biomarkers for dietary carcinogens: the example
of heterocyclic amines in epidemiological studies**   **299**
Rashmi Sinha, Amanda Cross and Robert J. Turesky

    23.1.   Introduction   299
    23.2.   Intake assessment of HCAs   299
    23.3.   HCA metabolism   300

23.4.   Conclusions and future research   304
         References   305

24   Practical examples: hormones   309
       Sabina Rinaldi and Rudolf Kaaks

       24.1.   Introduction   309
       24.2.   Hormone measurements for large-scale epidemiological studies 309
       24.3.   Laboratory methods   313
       24.4.   Validation and reproducibility of hormone measurements   315
       24.5.   Sample collection and long-time storage   316
       24.6.   Does a single hormone measurement represent long-term exposure? 316
       24.7.   Interpretation of measurements of circulating hormones   317
       24.8.   Conclusions   319
               References   319

**25   Aflatoxin, hepatitis B virus and liver cancer: a paradigm
       for molecular epidemiology   323**
       John D. Groopman, Thomas. W. Kensler and Chris Wild

       25.1.   Introduction   323
       25.2.   Defining molecular biomarkers   324
       25.3.   Validation strategy for molecular biomarkers   324
       25.4.   Development and validation of biomarkers for human
                 hepatocellular carcinoma   324
       25.5.   Susceptibility   334
       25.6.   Biomarkers to elucidate mechanisms of interaction   334
       25.7.   Early detection biomarkers for HCC   337
       25.8.   Summary and perspectives for the future   338
               References   338

26   Complex exposures – air pollution   343
       *Steffen Loft, Elvira Vaclavik Bräuner, Lykke Forchhammer,
       Marie Pedersen, Lisbeth E. Knudsen and Peter Møller*

       26.1.   Introduction   343
       26.2.   Personal monitoring of external dose   344
       26.3.   Biomarkers of internal dose and air pollutants   344
       26.4.   Biomarkers of biologically effective dose   345
       26.5.   Biomarkers of biological effects   350
       26.6.   Genetic susceptibility and oxidative stress related to air pollution 352
       26.7.   Conclusion   352
               References   353

Index   359

List of Contributors

Habibul Ahsan
Center for Genetics in Epidemiology
Columbia University
Mailman School of Public Health
New York, NY, USA

Richard J. Albertini
Cell and Molecular Biology
University of Vermont
Burlington, VT, USA

Angeline S. Andrew
Department of Genetics
Norris Cotton Cancer Center
Dartmouth Medical School
Lebanon, NH, USA

Jennifer H. Barrett
Section of Epidemiology and Biostatistics
Leeds Institute of Molecular Medicine
St James's University Hospital
Leeds, UK

Pier Alberto Bertazzi
EPOCA, Epidemiology Research Center
Department of Occupational and
    Environmental Health
University of Milan
Milano, Italy

Marianne Berwick
Division of Epidemiology
UNM School of Medicine
University of New Mexico
Albuquerque, NM, USA

M. Bictash
Biological Chemistry
Division of Biomedical Sciences
Imperial College London
London, UK

Timothy D. Bishop
Section of Epidemiology and Biostatistics
Leeds Institute of Molecular Medicine
St James's University Hospital
Leeds, UK

Elodie Caboux
Department of Biostatistics
Université Catholique de Lyon
Lyon, France

Amanda Cross
Division of Cancer Epidemiology and Genetics
National Cancer Institute
Bethesda, MD, USA

Alison M. Dunning
Cancer Research UK
Department of Oncology
Strangeway's Research Laboratory
Cambridge, UK

P. Elliott
Biological Chemistry
Division of Biomedical Sciences
Imperial College London
London, UK

Lynnette R. Ferguson
Discipline of Nutrition
Faculty of Medical & Health Sciences
The University of Auckland
Auckland, New Zealand

John B.C. Findlay
Institute of Membrane and Systems Biology
LIGHT Laboratories
University of Leeds
Leeds, UK

Lykke Forchhammer
Institute of Public Health
University of Copenhagen
Department of Occupational
    and Environmental Health
Copenhagen, Denmark

Seymour Garte
Graduate School of Public Health
University of Pittsburgh Cancer Institute
Pittsburgh, PA, USA

Mark S. Gilthorpe
Centre for Epidemiology and Biostatistics
University of Leeds
Leeds, UK

Emmanuelle Gormally
Department of Biostatistics
Université Catholique de Lyon
Lyon, France

John D. Groopman
Department of Environmental Health Sciences
Johns Hopkins Bloomberg School of
    Public Health
Baltimore, MD, USA

Pierre Hainaut
Department of Biostatistics
Université Catholique de Lyon
Lyon, France

Stephens S. Hecht
University of Minnesota Cancer Center
Minneapolis, MN, USA

Elaine Holmes
Biological Chemistry
Division of Biomedical Sciences
Imperial College London
London, UK

Mark M. Iles
Section of Epidemiology and Biostatistics
Leeds Institute of Molecular Medicine
St James's University Hospital
Leeds, UK

Rudolf Kaaks
Division of Cancer Epidemiology
German Cancer Research Center
Heidelberg, Germany

Margaret R. Karagas
Department of Genetics
Norris Cotton Cancer Center
Dartmouth Medical School
Lebanon, NH, USA

Jeff N. Keen
Institute of Membrane and Systems Biology
LIGHT Laboratories
University of Leeds
Leeds, UK

Thomas N. Kensler
Department of Environmental Health Sciences
Johns Hopkins Bloomberg School of
    Public Health
Baltimore MD, USA

H. Keun
Biological Chemistry
Division of Biomedical Sciences
Imperial College London
London, UK

Lisbeth E. Knudsen
Institute of Public Health
University of Copenhagen
Department of Occupational
    and Environmental Health
Copenhagen, Denmark

Lawrence Kupper
School of Public Health
University of North Carolina at Chapel Hill
Chapel Hill, NC, USA

Steffen Loft
Institute of Public Health
University of Copenhagen
Department of Occupational
    and Environmental Health
Copenhagen, Denmark

Craig Luccarini
Cancer Research UK
Department of Oncology
Strangeway's Research Laboratory
Cambridge, UK

Peter Møller
Institute of Public Health
University of Copenhagen
Department of Occupational
    and Environmental Health
Copenhagen, Denmark

Jason H. Moore
Department of Genetics
Norris Cotton Cancer Center
Dartmouth Medical School
Lebanon, NH, USA

Antonio Mutti
University of Parma
Department of Clinical Medicine
Nephrology and Health Sciences
Parma, Italy

Mark J. Nieuwenhuijsen
Department of Environmental Science and
Technology
Health and Environment
Imperial College London
London, UK

J. K. Nicholson
Biological Chemistry
Division of Biomedical Sciences
Imperial College London, UK

Marie Pedersen
Institute of Public Health
University of Copenhagen
Department of Occupational
    and Environmental Health
Copenhagen, Denmark

David H. Philips
Institute of Cancer Research
Brookes Lawley Building
Sutton, UK

Camille Ragin
University of Pittsburgh Cancer Institute
UPMC Cancer Pavillion
Pittsburgh, PA, USA

Stephen Rappaport
School of Public Health
University of North Carolina at Chapel Hill
Chapel Hill, NC, USA

Sabina Rinaldi
International Agency for Research on Cancer
Lyon, France

Andrew Rundle
Department of Epidemiology
Mailman School of Public Health
Columbia University
New York, NY, USA

Rashmi Sinha
Division of Cancer Epidemiology and Genetics
National Cancer Institute
Bethesda, MD, USA

Emanuela Taioli
University of Pittsburgh Cancer Institute
UPMC Cancer Pavillion
Pittsburgh, PA, USA

Yu-Kang Tu
Biostatistics Unit
Centre for Epidemiology and Biostatistics
Leeds Institute of Genetics and
    Health Therapeutics
University of Leeds
Leeds, UK

Robert J. Turesky
Department of Health
Albany, New York, USA

Elvira Vaclavik Bräuner
Institute of Public Health
University of Copenhagen
Department of Occupational
    and Environmental Health
Copenhagen, Denmark

Kirsi Vähäkangas
Department of Pharmacology and Toxicology
Kuopio, Finland

Paolo Vineis
Division Epidemiology Public Health
    and Primary Care
Imperial College London
London, UK

Chris Wild
Molecular Epidemiology Unit
Centre for Epidemiology and Biostatistics
The LIGHT Laboratories
University of Leeds
Leeds, UK

Acknowledgements

The Editors would like to thank all the authors for providing excellent manuscripts in a timely fashion and to the following: David Forman, Rudolph Kaaks, Soterios Kyrtopoulos, Stephen Rappaport, Alan Boobis, Julie Fisher, Tony Fletcher, John Molitor, Duccio Cavalieri, Ann Daly, David Phillips, Miriam Poirier, Giuseppe Matullo, Paul Schulte, Thomas Kensler, Emanuella Taioli, Marja Sorsa, Lenore Arab, Regina Santella, and Peter Farmer reviewers who played a critical role in ensuring the quality of the manuscripts:

Special thanks are due to Margaret Jones for collating all the manuscripts and keeping everyone on schedule and for doing all of this with tremendous commitment and good grace.

PV was supported by ECNIS (Environmental Cancer Risk, Nutrition and Individual Susceptibility), a network of excellence operating within the European Union 6th Framework Program, Priority 5: "Food Quality and Safety" (Contract No 513943) and by Compagnia di San Paolo, Torino.

CPW was supported by the NIEHS, USA ES06052 and by the EU NewGeneris (Newborns and Genotoxic Exposure Risks), an integrated project operating within the European Union 6th Framework Program, Priority 5: "Food Quality and Safety" (Contract No 016320-2).

1

Introduction: Why Molecular Epidemiology?

Chris Wild,[1] Seymour Garte[2] and Paolo Vineis[3]

[1]University of Leeds, UK, [2]University of Pittsburgh Cancer Institute, USA,
and [3]Imperial College London, UK

Physicians, public health workers, the press and the public at large are increasingly preoccupied with 'environmental risks' of disease. What are the causes of Alzheimer's disease? Are the causes environmental or genetic, or a mixture of the two? Does exposure to particles from incinerators or traffic exhausts cause cancer? Epidemiology has traditionally tried to answer such important questions. For example, the associations between tobacco smoking and lung cancer, between chronic hepatitis B virus infection and liver cancer, and between aromatic amines and bladder cancer are now considered to be 'causal', i.e. there is no doubt about the causal nature of these relationships. This is because the same positive observations have been made in a large number of settings, the association is very strong, it is biologically plausible, there is a dose–response relationship, and we cannot explain away the association in terms of bias or confounding. In the case of aromatic amines, strong animal evidence was available before observations in humans.

But not all issues of causality in human disease from environmental exposures are so clear. Consider two examples. Is there a 'causal' association between dietary exposure to acrylamide and cancer in humans? Are polycyclic aromatic hydrocarbons (PAHs) a cause of lung cancer?

These examples are obviously much more difficult to resolve than the previous ones. In the case of cigarettes, a simple questionnaire proved accurate enough to allow a reasonable estimation of exposure, while in the case of aromatic amines there were rosters in the chemical industries that allowed unequivocal identification of exposure to single agents for all workers. In contrast, estimation of the intake of acrylamide from French fries and other sources, using dietary questionnaires, is an almost desperate enterprise. For PAHs, the sources of exposure are multiple (e.g. diet, air pollution from car exhaust, heating, industrial pollution, specific occupations), making it almost impossible to have an estimate of total PAH exposure based on a questionnaire. One can measure PAHs in ambient air through air sampling (in the work environment or with a personal monitor); however, the level of PAHs in the air is only indirectly associated to the amount of PAHs that actually enter the body and end up binding to DNA. The ability of PAHs to reach DNA and bind to it depends on individual metabolic capabilities (which are in part genetically determined), involving a number of different enzymes and pathways. Finally, the ability of PAHs to lead to heritable changes in DNA (mutations) is additionally related to the individual's ability to repair DNA damage.

Molecular Epidemiology of Chronic Diseases, Edited by C. P. Wild, P. Vineis, and S. Garte
© 2008 John Wiley & Sons, Ltd

For these reasons, starting at least in 1982 with a paper by Perera and Weinstein but probably before with a paper by Lower (Vineis 2007), 'molecular epidemiology' was introduced into the practice of cancer research. A simple definition is that:

> '... it entails the inclusion in epidemiologic research of biologic measurements made at the molecular level – and is thus an extension of the increasing use of biologically based measures in epidemiologic research' (McMichael 1994).

This corresponds to one of the first (if not the first) definition:

> Advanced laboratory methods are used in combination with analytic epidemiology to identify at the biochemical or molecular level specific exogenous and/or host factors that play a role in human cancer causation (Perera and Weinstein 1982).

However, the terminology has been criticized by some:

> The term 'molecular epidemiology' may suggest the existence of a sub-discipline with substantive new research content. Molecular techniques, however, are directed principally at enhancing the measurement of exposure, effect, or susceptibility, and not at formulating new etiologic hypotheses. As techniques of refinement and elaboration, the integration of molecular measures into mainstream epidemiologic research can offer higher resolution answers in relation to disease causation' (McMichael 1994).

In this book, many examples are drawn from cancer epidemiology but there is also reference to other chronic degenerative conditions, including vascular disease, diabetes and neurodegenerative disease, which share some of the challenges with cancer in terms of establishing causality. In contrast, we have not included specific emphasis on infectious disease, where the term 'molecular epidemiology' was introduced many years earlier than in the cancer epidemiology field.

The goals of the new discipline of molecular epidemiology are the same as those suggested by the few examples above. First of all, to contribute to better estimation of exposure, including 'internal' exposure, through the measurement of end-points, such as chemical metabolites and adducts (e.g. haemoglobin adducts for acrylamide, DNA adducts for PAHs). Second, genetic susceptibility, emerged as became an important subject for enquiry, since it became clear that between exposure and effect there was a layer of metabolic reactions, including activation, deactivation and DNA repair, which affected the dose–response relationship in a fashion analogous to other susceptibility factors, such as age, sex and nutritional status. A further goal of epidemiology is to reduce disease burden by identification of risk factors for disease. It took a long time to discover the association between aromatic amines and bladder cancer, and thus hundreds of workers died from causes that could have been avoided. One limitation therefore of cancer epidemiology is the long latency period (decades) after exposure starts and before the disease is clinically diagnosed. For this reason, epidemiologists have been searching for early lesions that could be reasonably used as surrogates of the risk of cancer. Chromosome aberrations, gene mutations and, more recently, gene expression and epigenetics have been introduced as intermediate markers in the pathway that leads from exposure to overt disease, thus adding to the categories of exposure and susceptibility biomarkers mentioned above. The goal of all these efforts in biomarker development is to allow faster and earlier detection of disease in individuals, as well as to shorten the time needed to identify possible human carcinogens.

Molecular epidemiology studies will normally employ a number of tools for the measurement of exposure, susceptibility and disease, e.g. questionnaires, job-exposure matrices, data from environmental monitoring, routinely collected health data and biomarkers. The biomarkers have certain properties which influence their application. For example, some biomarkers of exposure may only relate to the recent past, whilst the level of others may be affected by the presence of disease (reverse causation). Consequently, careful consideration of the properties of the biomarkers is needed in relation to how they are to be applied from the point of study design through to data analysis. This book therefore begins (Chapters 2–5) by discussing study design, particularly in light of the complex

Table 1.1 Discoveries that support the original model of molecular epidemiology[1]

Marker linked to exposure or disease	Exposure	Reference
Exposure/biologically effective dose		
DNA adducts	PAHs, aromatic compounds	Tang *et al.* 2001
	AFB$_1$	Ross *et al.* 1992
Albumin adducts	AFB$_1$	Wang *et al.* 1996
		Gong *et al.* 2002
Haemoglobin adducts	Acrylamide	Hagmar *et al.* 2005
	Styrene	Vodicka *et al.* 2003
	1,3-Butadiene	Albertini *et al.* 2001
Preclinical effect (exposure and/or cancer)		
Chromosome aberrations	Lung	Bonassi *et al.* 2004
	Leukaemia	Smith *et al.* 2005
	Benzene	Holeckova *et al.* 2004
HPRT	PAHs	Perera *et al.* 2002
	1,3-Butadiene	Ammenheuser *et al.* 2001
Glycophorin A	PAHs	Lee *et al.* 2002
Gene expression	Cisplatin	Gwosdz *et al.* 2005
Genetic susceptibility		
Phenotypic markers	Several cancers	Berwick and Vineis 2000
		Wei, Spitz *et al.* 2000
SNPs		
NAT2, GSTM1	Bladder	Gracia Closas 2005
CYP1A1	Lung	Vineis *et al.* 2003

[1]Perera and Weinstein 1982; NRC 1987.

interplay between the environment and genes, thus laying a foundation for the subsequent parts of the book.

Molecular epidemiology has had several success stories (see Table 1.1) in all three categories of biomarkers – exposure, susceptibility and early response. Biomarkers of exposure can make a significant contribution to establishing disease aetiology. For example, aflatoxins were structurally identified in 1963 and shown to be liver carcinogens in animals shortly afterwards. However, despite evidence from ecological studies, it was only with the development of biomarkers of individual exposure that convincing evidence of the association with human liver cancer risk was obtained in a prospective cohort study in China, published in 1992 (IARC 1993). In the case of *Helicobacter pylori* and gastric cancer, the time between identification of the pathogen and evaluation as a human carcinogen by IARC was around 10 years; this was in no small part due to the early availability of

serum markers (antibodies to bacterial antigens) to establish exposure status (IARC 1994). The latter example also demonstrates the value of being able to establish long-term past exposure to putative risk factors, something which has been easier with infectious agents than with chemical exposures. The development and validation of biomarkers of exposure to environmental risk factors therefore remains an outstanding challenge to molecular epidemiology (Vineis 2004; Wild 2005). The principles of biomarker development, validation and application are discussed by Vineis and Garte and by Nieuwenhuijsen in Chapters 6 and 7, followed by a number of examples of categories of biomarker described in chapters by Hecht, Phillips, and Berwick and Albertini (Chapters 8–10).

Identification of genetic susceptibility to environmental exposures due to low penetrance alleles continues to occupy the attention of many researchers in the field of molecular epidemiology. Indeed, with the arrival of the polymerase chain reaction

(PCR), the facile genotyping for single nucleotide polymorphisms (SNPs) and the ease of application to case–control studies, there has at times appeared to be an imbalance of effort between the development of biomarkers for environmental exposure assessment and genotyping (Wild 2005). The problems of genetic association studies comprising small subject numbers and focusing on single polymorphisms in genes involved in complex pathways has been discussed (Rebbeck *et al.* 2004). These limitations are partially overcome by systematic reviews and meta-analyses (see Ragin and Taioli, Chapter 15). However, the Human Genome Project and subsequent programmes to comprehensively catalogue SNPs have spawned a new generation of large genome-wide association studies with the statistical power to identify at-risk genotypes in a more reliable manner (Scott *et al.* 2007; Zeggini *et al.* 2007). The ability to study the role of genetic susceptibility to environmental risk factors may help to reveal those underlying environmental factors that only entail a risk in a subset of the population.

Approaches to large-scale genotyping, together with the potential pitfalls, are addressed by Dunning and Luccarini in Chapter 11. At the same time, other '-omics' technologies are beginning to find application in population studies and to generate novel hypotheses about disease mechanisms as well as candidate biomarkers. Bictash and colleagues in Chapter 13 describe the contribution of metabonomics in elucidating key pathways affected by both exposure and disease development, whilst Keen and Findlay in Chapter 12 provide a similar introduction to the application of proteomics to epidemiological studies.

The introduction of genomics, proteomics and metabonomics represents one dimension in which the volume of data from molecular epidemiology studies is growing at a startling pace. In addition, in order to unravel the complex interplay between environmental and genetic factors in disease aetiology, often involving modest elevated risks (1.5-fold or less) associated with a particular risk factor, epidemiologists are turning to studies encompassing larger and larger numbers of subjects. This may be, for example, through consortium-led multicentre case–control studies or large prospective cohort studies entailing biological

banks of samples, such as EPIC or UK Biobank. The complexity of these datasets demands increasingly sophisticated biostatistical approaches to derive meaningful conclusions. Several chapters in this book therefore address the analysis of molecular epidemiological studies (Chapters 14–17). An equally important element to the future exploitation of these large datasets is the ability to access and process information through bioinformatics (Moore, Chapter 18).

Epidemiology ultimately aims to lead to a reduced health burden through disease prevention. In Chapters 19 (Bertazzi and Mutti) and 20 (Ferguson) the application of molecular epidemiology to regulatory decision making and in the evaluation of intervention strategies are considered. Some practical issues of sample collection and processing are covered in Chapter 21 (Caboux and colleagues). The collection of biological specimens and the derived genetic and other molecular or biochemical measurements represent sensitive information, with concerns over exploitation of genetic data, e.g. in terms of access to insurance or employment. These issues become even more complex when studies are multicentre and involve collaborations across international borders, where ethical regulations may differ significantly. Some of the ethical considerations are discussed in this volume by Vähäkangas (Chapter 22).

The purpose of this handbook is both to show the potential applications of laboratory techniques to tackle epidemiological problems and to consider the limitations of laboratory work, including the sources of uncertainty and inaccuracy. Issues such as reliability (compared to traditional epidemiological methods, involving questionnaires) and the timing of exposures are also explored. The latter part of the book comprises four chapters (Chapters 23–26) which take individual examples of research areas where molecular epidemiology has led to advances in understanding as well as serving to illustrate some of the limitations of the approach. These final chapters emphasise the main focus of the book, which is on practical and applied aspects of molecular epidemiology, as actually used in the study of human exposure and disease.

References

Albertini RJ, Sram RJ, Vacek PM *et al*. 2001. Biomarkers for assessing occupational exposures to 1,3-butadiene. *Chem Biol Interact* **135**: 429–453.

Ammenheuser MM, Bechtold WE, Abdel-Rahman SZ *et al*. 2001. Assessment of 1,3-butadiene exposure in polymer production workers using *HPRT* mutations in lymphocytes as a biomarker. *Environ Health Perspect* **109**: 1249–1255.

Berwick M, Vineis P. 2000. Markers of DNA repair and susceptibility to cancer in humans: an epidemiologic review. *J Natl Cancer Inst* **92**: 874–897.

Bonassi S, Znaor A, Norppa H *et al*. 2004. Chromosomal aberrations and risk of cancer in humans: an epidemiologic perspective. *Cytogenet Genome Res* **4**: 376–382.

Garcia-Closas M, Malats N, Silverman D *et al*. 2005. NAT2 slow acetylation, *GSTM1* null genotype, and risk of bladder cancer: results from the Spanish Bladder Cancer Study and meta-analyses. *Lancet* **366**: 649–659.

Gong YY, Cardwell K, Hounsa A *et al*. 2002. Dietary aflatoxin exposure and impaired growth in young children from Benin and Togo: cross-sectional study. *Br Med J* **325**: 20–21.

Hagmar L, Wirfalt E, Paulsson B *et al*. 2005. Differences in haemoglobin adduct levels of acrylamide in the general population with respect to dietary intake, smoking habits and gender. *Mutat Res* **580**: 157–165.

Holeckova B, Piesova E, Sivikova K *et al*. 2004. Chromosomal aberrations in humans induced by benzene. *Ann Agric Environ Med* **11**: 175–179.

IARC. 1993. *Some Naturally Occurring Substances: Food Items and Constituents, Heterocyclic Aromatic Amines and Mycotoxins*. Monographs on the Evaluation of Carcinogenic Risks to Humans, vol 56. IARC: Lyon.IARC. 1994. *Schistosomes, Liver Flukes and Helicobacter pylori*. Monographs on the Evaluation of Carcinogenic Risks to Humans, vol 61. IARC: Lyon.

Lee KH, Lee J, Ha M *et al*. 2002. Influence of polymorphism of *GSTM1* gene on association between glycophorin a mutant frequency and urinary PAH metabolites in incineration workers. *J Toxicol Environ Health A* **65**: 355–363.

McMichael AJ. 1994. Invited commentary – 'molecular epidemiology': new pathway or new travelling companion? *Am J Epidemiol* **140**: 1–11.

Perera FP, Mooney LA, Stampfer M *et al*. 2002. Associations between carcinogen–DNA damage, glutathione S–transferase genotypes, and risk of lung cancer in the prospective Physicians' Health Cohort Study. *Carcinogenesis* **23**: 1641–1646.

Perera FP, Weinstein IB. 1982. Molecular epidemiology and carcinogen–DNA adduct detection: new approaches to studies of human cancer causation. *J Chron Dis* **35**: 581–600.

Rebbeck TR, Martinez ME, Sellers TA *et al*. 2004. Genetic variation and cancer: improving the environment for publication of associated studies. *Cancer Epidemiol Biomarkers Prev* **13**: 1985–1986.

Ross RK, Yuan JM, Yu MC *et al*. 1992. Urinary aflatoxin biomarkers and risk of hepatocellular carcinoma. *Lancet* **339**: 943–946.

Scott LJ, Mohlke KL, Bonnycastle LL *et al*. 2007. A genome-wide association study of type 2 diabetes in Finns detects multiple susceptibility variants. *Science* **316:** 1341–1345.

Smith MT, McHale CM, Wiemels JL *et al*. 2005. Molecular biomarkers for the study of childhood leukaemia. *Toxicol Appl Pharmacol* **206**: 237–245.

Tang DL, Phillips DH, Stampfer M *et al*. 2001. Association between carcinogen–DNA adducts in white blood cells and lung cancer risk in the Physicians Health Study. *Cancer Res* **61**: 6708–6712.

Vineis P. 2007. Commentary: first steps in molecular epidemiology: Lower *et al*. 1979. *Int J Epidemiol* **36**: 20–22.

Vineis P. 2004. A self-fulfilling prophecy: are we underestimating the role of the environment in gene–environment interaction research? *Int J Epidemiol* **33**: 945–946.

Vineis P, Veglia F, Benhamou S *et al*. 2003. CYP1A1 (TC)-C-3801 polymorphism and lung cancer: a pooled analysis of 2451 cases and 3358 controls. *Int J Cancer* **104**: 650–657.

Vodicka P, Koskinen M, Stetina R *et al*. 2003. The role of various biomarkers in the evaluation of styrene genotoxicity. *Cancer Detect Prev* **27**: 275–284.

Wang LY, Hatch M, Chen CJ *et al*. 1996. Aflatoxin exposure and risk of hepatocellular carcinoma in Taiwan. *Int J Cancer* **67**: 620–625.

Wei QY, Cheng L, Amos CI *et al*. 2000. Repair of tobacco carcinogen-induced DNA adducts and lung cancer risk: a molecular epidemiologic study. *J Natl Cancer Inst* **92**: 1764–1772.

Wild CP. 2005. Complementing the genome with an 'exposome': the outstanding challenge of environmental exposure measurement in molecular epidemiology. *Cancer Epidemiol Biomarkers Prev* **14**: 1847–1850.

Zeggini E, Weedon MN, Lindgren CM *et al*. 2007. Replication of genome-wide association signals in UK samples reveals risk loci for type 2 diabetes. *Science* **316:** 1336–1341.

2

Study Design

Paolo Vineis

Imperial College of Science, Technology and Medicine, London, and University of Torino, Italy

2.1. INTRODUCTION: STUDY DESIGN AT SQUARE ONE[1]

Usually when we design an epidemiological study, we start with an aetiological hypothesis. The hypothesis can come from the observation of a cluster of disease, or a trend, or a peculiar geographic distribution, or observations made by clinicians, or experimental data in animals. For example, case series documenting 'larger-than-expected' numbers of haematopoietic cancers among shoe workers in Italy (Vigliani and Saita 1964) and Turkey (Aksoy 1974) provided important clues on the role of benzene in leukaemogenesis. However, due to the lack of information on the population at risk and disease rates in a comparison population, the magnitude of association between benzene exposure and haematopoietic cancers could not be assessed on this basis only.

The study design aims essentially at answering two types of questions:

1. Contributing to the identification of *cause–effect* relationships between exposure to putative risk/preventive factors and disease (although causality can be inferred only with a complex reasoning that usually involves also other types of evidence).
2. *Measuring* the exposure–disease association (strength, dose–response, population impact).

A first crucial distinction is between *observational studies*, in which the exposure is not manipulated by the investigator, and *experimental studies*, in which it is. It is usually thought that experimental studies (e.g. randomized controlled trials) are a better study design than simply observational studies, because they allow better control of confounding, thanks to random allocation of potential confounders. The philosopher Popper thought that only experiments allow us to make scientific causal statements, and therefore astronomy would not be a science, certainly an extreme point of view.

Confounders are variables other than the one under study that can provide an alternative explanation for the observed association. For example, exposure to benzene of shoe workers might be associated with other chemical exposures that could turn out to be the real cause of leukaemia. Although the randomized trial may be the strongest design to show causal links, for ethical reasons we cannot randomize exposure to suspect carcinogens, and we have to live with observational epidemiology.

The choice of a study design depends on several elements and one has several options:

1. *Which exposure(s) to study*
   (a) DETERMINANT (e.g. benzene). By 'determinant' we mean the main exposure that is of interest to the researcher.

---

[1] This introduction to study design is elementary. For those who know the basics of study design already, we suggest moving on to Section 2.5, 'Specificities of molecular epidemiology design'.

Molecular Epidemiology of Chronic Diseases, Edited by C. P. Wild, P. Vineis, and S. Garte
© 2008 John Wiley & Sons, Ltd

(b) CONFOUNDERS (extraneous variables providing an alternative explanation). Confounders, in the simplest sense, are defined as other variables that are associated with the determinant under investigation and are risk factors for the disease. Due to these properties, confounders can lead to an alternative explanation to the one under study (e.g. other solvents, or smoking habits, instead of benzene).

(c) EFFECT MODIFIERS (variables which quantitatively influence the cause–effect relationship, e.g. exposure to toluene and xylene). Usually several solvents are used in the same jobs, such as in shoe manufacture. Solvents other than benzene can be confounders, but they can also modify the effect of benzene, in particular by enhancing it.

2. *Which population to study, with which design.* We have several options that depend on availability, convenience, feasibility, costs or ethical issues.

(a) A COHORT STUDY

(i) *Chemical workers in Pliofilm production.* We can enroll a cohort of subjects exposed to the hypothesized determinant, e.g. one of the first cohort studies on benzene included workers involved in Pliofilm production, exposed to benzene (see Example A). Advantages of cohort studies are usually a good level of exposure assessment and the possibility of studying all diseases. Disadvantages are the frequent lack of information on confounders (at least in occupational cohorts), the

---

EXAMPLE A   Cohort studies on benzene exposure and leukaemia

One cohort consisted of 1212 white men employed during 1936–1975 in the production of Pliofilm, a covering material made from rubber hydrochloride, whose production involved the use of large volumes of benzene as a solvent. Various estimates have been made of the exposure levels in this cohort with varying proportions of person–years accumulated by men exposed to different benzene levels. The latest follow-up showed 481 deaths (SMR = 1.0), 111 cancer deaths (SMR = 1.1), 14 leukaemia deaths (SMR = 3.6, CI 2.0–6.0), of which six were from acute myeloid leukaemia (SMR = 5.0, CI 1.8–11.0) and showing a strong association with cumulative exposure (Paxton *et al.* 1994).

A second cohort of 74 828 workers were recruited from 1427 benzene-exposed work units in 672 factories in 12 cities in China. For deceased subjects, cause of death was obtained from medical records, other written factory records or death certificates at local police stations. The subject's physician or next of kin was contacted only if written records listing cause of death could not be obtained. For all subjects whose cause of death was suspected to be cancer or a haematological disorder, hospital and other medical records were sought to verify the diagnosis. The specific histopathology, date of diagnosis, date of death, hospital/place of death, source of medical information and other important diagnosis-related information were abstracted on a standard diagnostic validation form and reviewed by the field centre director and the investigators. For cases of haematological disorder, medical records and all available pathology and peripheral blood smear slides were reviewed by haematopathologists from the Beijing Union Medical College Hospital, the NCI and the Mayo Clinic in Rochester, MN, USA.

There were 1369 deaths, which gave a low overall mortality compared with the general population (SMR = 0.5). A cohort of 35 805 unexposed workers had the same overall mortality (RR = 1.1). The exposed cohort had 42 incident leukaemia cases (RR = 2.6 compared with unexposed, CI = 1.3–5.7). The largest group was acute myelogenous leukaemia (Obs = 23, RR = 3.1, CI = 1.2–10.7), but risk estimates of the same size were seen for chronic myelogenous leukaemia (Obs = 2, RR = 2.6) and acute lymphocytic leukaemia (Obs = 5, RR = 2.8) (Yin *et al.* 1996).

frequent poor quality of information on outcomes (with significant exceptions), the usually small numbers of events for rare diseases, and the long time required to achieve the results if the cohort is prospective. In Example A it should be noticed that only six deaths from acute myeloid leukaemia occurred in the exposed workers.

(ii) *A multicentre cohort study in 12 Chinese cities.* Example A also describes a much larger investigation in China, which identified cases of leukaemia with great accuracy, including histological type. Forty-two cases of leukaemia (including 23 of acute myeloid leukaemia) were registered out of 74 000 workers enrolled.

Blood samples were collected, and this enabled the molecular studies described below.

(b) A CASE–CONTROL STUDY. The logic of the case–control study is not straightforward for beginners. The case–control study is based on the concept of sampling. Rather than studying the whole population, we include all the cases with a certain disease, and only a sample of the general population for comparison with the cases. One has to imagine the general population, in which some people are exposed to benzene (particularly in some areas, e.g. heavy industrialized areas with a high prevalence of exposure: see Example B). In the next 5 years a number of cases of leukaemias will arise

---

EXAMPLE B   Case–control study on benzene exposure and leukaemia

A hospital-based multicentre case–control study, stratified on centre, age and sex, with 280 childhood leukaemia cases and 285 controls, was carried out in France. Data were collected by a standardized interview of the mothers. No clear association was seen between maternal occupational exposure to hydrocarbons during pregnancy and leukaemia, or between residential traffic density and leukaemia. There was an association between dwellings neighbouring a petrol station or a repair garage during childhood and the risk of childhood leukaemia (OR 4.0, 95% CI 1.5–10.3), with a duration trend. The association, which appeared particularly strong for acute non-lymphocytic leukaemia (OR 7.7, 95% CI 1.7–34.3), was not altered by adjustment for potential confounding factors. Results showed an association between acute childhood leukaemia and dwellings neighbouring auto repair garages and petrol stations, which are benzene-emitting sources (Steffen *et al.* 2004).

The most difficult part in case–control studies is exposure assessment. In this particular study, data were collected by specifically trained medical doctors through a face-to-face interview of the mothers of the children (cases and controls), after having given written informed consent. Case and control mothers were interviewed under strictly similar conditions, at the same period, using a standardized questionnaire. For all children, a history of exposure to hydrocarbons at each of their different dwelling places was taken. The period investigated spanned from date of conception to date of diagnosis or interview. *In utero* exposure and childhood exposure were considered separately. Among the different types of exposure investigated was the existence of a neighbouring business, and especially an automobile repair garage or a petrol station. By 'neighbouring business' was meant a property adjoining the child's dwelling. Questions were also asked to determine the possible exposure to heavy traffic roads in the vicinity (< 50 m) of the children's homes. The questionnaire also included questions on parental occupations, which were coded using international codes for job titles and industrial activities. For the categorization of jobs with respect to potential exposure to benzene and other solvents, the authors used a basic job-exposure matrix they had developed with industrial hygienists. A checklist of closed questions on mothers' tasks during their pregnancy was also included in the questionnaire, with explicit items on solvent exposure.

in this population, say 50 cases out of 1 million people. We could of course have a census of the whole population (n = 1 million), identify those exposed to benzene and count leukaemias among the exposed and the unexposed, but this would be an infinite and inefficient job. It is much better to collect all the cases (since they are not many) and only a sample of the general population, say 200 individuals randomly sampled from the whole population base (= controls). Then we can count how many cases and how many controls have been exposed to benzene through occupational exposures or in other ways. While in cohort studies the question is 'how many exposed subjects developed the disease in comparison with the unexposed', in case–control studies the question is 'how many cases with the disease have been exposed compared with healthy subjects (non-cases)'?

Case–control studies have problems that are in some way complementary to those of cohort studies. In particular, while it is possible to collect information on confounders, exposure assessment is usually a problem, particularly if retrospective collection of information for a long time preceding disease is needed. As Example B suggests, exposure to benzene was only inferred through interviews and residential data.

(c) Studies of molecular epidemiology on benzene exposure. Several studies on biomarkers have been conducted among workers exposed to benzene, particularly in the Chinese cohort. Specific chromosomal aberrations have been observed in both preleukaemia and leukaemia patients exposed to benzene, as well as in otherwise healthy benzene-exposed workers (Zhang *et al.* 2005). Using fluorescent *in situ* hybridization (FISH) and the polymerase chain reaction (PCR), Zhang and colleagues (2005) found that high occupational benzene exposure increased the frequencies of

aberrations in chromosomes 5, 7, 9, 8 and 11, aberrations that are frequently seen in acute myeloid leukaemias and in preleukemic myelodysplastic syndrome. In the same studies on Chinese workers, protein-expression patterns were detected by surface-enhanced laser desorption/ionization (SELDI)–time-of-flight (TOF) analysis, a technique used to address proteomics. SELDI–TOF analysis of exposed and unexposed subjects revealed that lowered expression of PF4 and CTAP-III proteins is a potential biomarker of benzene's early biological effects and may play a role in the immunosuppressive effects of benzene (Vermeulen *et al.* 2005). Finally, Lan *et al.* (2006) investigated 20 candidate genes in the same Chinese cohort. After accounting for multiple comparisons, SNPs in five genes were associated with a statistically significant decrease in total WBC counts among exposed workers [IL-1A (–889C.T), IL-4 (–1098T.G), IL-10 (–819T.C), IL-12A (8685G.A) and VCAM1 (–1591T.C). This report provides evidence that single nucleotide polymorphisms (SNPs) in genes that regulate haematopoiesis influence benzene-induced haematotoxicity.

(d) SOME METHODOLOGICAL ISSUES

(i) *Choice of controls* is a difficult issue and has been considered in a number of papers and books (see e.g. Breslow, in Ahrens and Pigeot 2004; Rothman and Greenland 1998). In population-based case–control studies, controls are meant to be a random sample of the population from which the cases arise (study base; see below). Controls should be selected at the time of case ascertainment. This is conceptually clear, but in practical terms representativeness is rarely achieved because of low response rates among eligible population controls. Hospital-based controls have higher response rates but their choice is not straightforward. They should include diseases that are not associated with the risk factor under study, e.g. in a study on lung cancer, chronic onstructive pulmonary disease (COPD) patients would not be a suitable control group because of the strong association with tobacco smoke. As a rule of thumb, controls should not include more than 10% of a given disease.

In more rigorous terms: (a) controls are meant to represent the null hypothesis, i.e. the distribution of exposures that the cases would have experienced had they not become cases; (b) therefore, the controls have to be a sample of the 'source population' of the cases (study base). In the hospital-based design the control series is expected to be a sample of the referral area of one or more hospitals where the cases where recruited. However, this is conceptually and practically difficult.

(ii) *Other study designs* in addition to case–control and cohort studies are worth mentioning (see Box 2.1). The *ecological study* is a study design in which information about exposure and/or disease is aggregated at the population level, i.e. it is not available at the individual level. The main problem with this design is the 'ecological fallacy', i.e. the risk of attributing to the putative risk factor a role that in fact is played by other variables that are associated with it in the population. This is essentially a problem of confounding and is dealt with later when tackling the 'population stratification' problem in genetic research.

A *case series* is just a collection of cases without a comparison group. This design has obvious limitations, since it does not allow the researcher to compare an observed with an expected figure (e.g. proportion of exposed to a putative causal agent). Only in extreme circumstances, i.e. when the observed proportion of exposed cases is clearly not compatible with any expected value, can this design be meaningful (to use an old metaphor from Sir D. Acheson, if I observe 40 instead of the usual 20 sparrows out of my window in London, I will not notice; but if I observe an eagle, I will infer that something unusual is happening).

A *case–case* (or case only) study is a more sophisticated design that has been suggested for at least two purposes: (a) to compare environmental exposures between cases with and cases without certain inherited genetic characteristics (e.g. SNPs), i.e. to study gene–environment interactions; or (b) to compare environmental exposures between cases with and cases without certain acquired pathogenetic

---

Box 2.1 Different types of study design in epidemiology (modified from Ahrens and Pigeot (eds) (2005), *Handbook of Epidemiology*, p.165, Springer with kind permission of Springer Science and Business Media.)

Study type	Reasoning
Ecological	Descriptive association at a group level may be used for development of broad hypotheses
Cross-sectional	Descriptive; individual association may be used for development of hypotheses
Case-control	Increased prevalence of risk factor among diseased may indicate a causal relationship
Cohort	Increased risk of disease among exposed may indicate a causal relationship
Intervention	Modification (reduction) of the incidence rate of the disease confirms a causal relationship

changes, such as gene mutations or other alterations in intermediate markers (Kass and Gold, in Ahrens and Pigeot 2004, p.338; see also example below by Porta *et al.*).

Cross-sectional studies are those in which cases and controls are sampled not according to a longitudinal time frame but in a restricted time window. Cases are 'prevalent', i.e. they have an already diagnosed disease and are still alive in the given time window. The main problem with prevalent cases is that they are survivors, i.e. they may have several features – including genetic characteristics and environmental exposures – that differentiate them from 'incident' cases.

2.2. EPIDEMIOLOGICAL MEASURES

The basic measure of epidemiology is the incidence rate, i.e. the number of new occurrences divided by the *population–time*, the latter being the number of subjects observed multiplied by the time of observation. For example, the observation of 10 cases of leukaemia among 1000 exposed subjects observed for 10 years gives an incidence rate of 10/10 000 person-years, i.e. 1/10 000 *per year*. Let us suppose that this rate is observed among those exposed to benzene, while the rate in the unexposed is 0.2/10,000 per year, based on the observation of two cases out of 1000 subjects also observed for 10 years. Then the *rate ratio* (RR) is $1:0.2 = 5$.

A simpler alternative to the rate is the 'risk', i.e. the proportion of the subjects originally enrolled who developed the disease in a certain time period. In the example above, the risk in the exposed is simply 10/1000 in 10 years of observation among the exposed and 2/1000 in the unexposed. The difference with the rate is that the denominator does not incorporate time.

If you draw a sample from the base of the study, within a population-based case–control study, you obtain the following situation:

	Exposed	Unexposed
Cases	a	b
Controls	c	d
Total	n_1	n_2

If there is no bias, i.e. the controls are sampled irrespective of exposure, c and d, respectively, are random samples of the total exposed (n_1) and unexposed (n_2) populations in the study base. Therefore, the measure $(a/c)/(b/d)$ is a faithful estimate of $(a/n_1)/(b/n_2)$. Since $(a/c)/(b/d)$ can be rewritten as ad/bc, which is called the 'odds ratio' (OR), the latter is a good estimate of the risk ratio. The odds ratio is also a good estimate of the rate ratio if the disease is rare and/or if the controls are sampled at the time cases occur (to know more about the 'rare disease assumption' you should refer to a textbook of epidemiology; see Essential Reading).

2.3. BIAS

The validity of the conclusion above, that is the equivalence of the OR and the RR, holds true only if sampling of the controls is unbiased, i.e. we do not over- or underestimate the exposed in the control sample. In general, if the distribution of the exposure is influenced by the disease status as a consequence of the study design or sampling, we find a spurious (biased) association between exposure and disease. For example, if most cases accept an interview and the donation of a blood sample, while only 50–60% of controls do (as often happens), this will introduce a 'selection bias' due to the fact that responders are likely to be different from non-responders. If we investigate alcohol drinking habits in relation to hepatocarcinoma, it is likely that responding controls include fewer drinkers than a truly random sample of the general population, and this will lead to overestimation of the association. As another example, mothers who gave birth to malformed children (cases) will remember their exposure to drugs during pregnancy in greater detail than will mothers of healthy babies (controls). This 'recall' bias will also distort the estimate of the association in favour of a relationship between certain drugs and birth defects that may not exist in reality.

One of the first attempts to classify bias was by Sackett (1979) in a seminal paper. The catalogue of 35 types of bias proposed by Sackett was very detailed, but in fact can be simplified into six categories of bias in:

1. 'Reading-up' on the field.
2. Specifying and selecting the study sample.
3. Executing the experimental manoeuvre.
4. Measuring exposures and outcomes.
5. Analysing the data.
6. Interpreting the analysis.

Of the 35 biases, nine are discussed in more detail by Sackett. These deserve some discussion, since they are particularly relevant to the development of epidemiology. The first is the *prevalence–incidence bias*: 'a late look at those exposed (or affected) early will miss fatal and other short episodes, plus mild or silent cases and cases in which evidence of exposure disappears with disease onset'. *Admission rate bias* refers to the fact that exposed and unexposed cases have different hospital admission rates, so that their relative odds of exposure to the putative cause will be distorted in hospital-based studies. *Unmasking bias* means that 'an innocent exposure may become suspect if, rather than causing a disease, it causes a sign or symptom which precipitates a search for the disease'. *'Non-respondent bias'* (i.e. 'non-respondents from a specified sample may exhibit exposures or outcomes which differ from those of respondents') is obviously crucial for all social research. *Selection bias* seems to be considered by Sackett as a subgroup of *'membership bias'*, i.e. 'membership in a group may imply a degree of health which differs systematically from that of the general population'. *Diagnostic suspicion bias* is described as follows: 'a knowledge of the subject's prior exposure to a putative cause may influence both the intensity and the outcome of the diagnostic process'. *Exposure suspicion bias* is defined in this way: 'a knowledge of the patient's disease status may influence both the intensity and outcome of a search for exposure'. In more modern definitions, exposure suspicion bias and recall bias ('questions about specific exposures may be asked several times of cases but only once of controls') are considered within the same category of 'information bias'.

More modern definitions of bias have followed. According to Miettinen (1985), bias refers to the validity of contrasts we make within epidemiological studies: 'the key to successful design of a non-experimental study in this area, as in general, is

the emulation of experimentation'. The validity of a randomized clinical trial, considered as the gold standard, rests on three main features:

1. The use of randomization, i.e. *comparability of populations*.
2. The use of blinding, i.e. *comparability of information*.
3. The use of a placebo, i.e. *comparability of effects*.

The key point, in fact, is the first, randomization. Hence, Miettinen suggests a classification of bias into comparison, selection and information bias (according to the prevailing type of design error). Such classification is present in other texts of epidemiology, such as Rothman's (1986) and Hennekens and Buring's textbooks (1987). According to Rothman, *selection bias* is 'a distortion of the effect measured, resulting from procedures used to select subjects that lead to an effect estimate among subjects included in the study different from the estimate obtainable from the entire population theoretically targeted for the study' (Rothman 1986). Selection bias, in fact, depends on selection of the exposed/unexposed subjects on the basis of (i.e. not independently of) the outcome, or on the selection of the diseased/healthy subjects on the basis of exposure status. Similarly, we have *information bias* when the error of classification on one axis (exposure or outcome) is not independent of the classification on the other axis.

Bias in screening practices

Screening for cancer is a field in which biomarkers have been applied extensively and several forms of bias have been described. In 1928 two independent investigators, Papanicolaou in the USA and Babes in Romania, reported that cancer of the cervix could be diagnosed by examining exfoliated cells from the cervical epithelium. Only after the war, however, was their technique systematically introduced in order to detect cervical cancer at early stages and improve survival of the patients. X-ray examination of the breast in asymptomatic women was already advocated in the 1930s by Gershon-Cohen as a means to reduce breast cancer mortality. The first mammography technique was introduced by Egan in the 1960s, and in the same period the first randomized trial (HIP) was started. Among the different methodological issues that were raised concerning mass cancer screening, two peculiar types of bias have been described in early seminal papers by Hutchinson and Shapiro (1968) and Feinleib and Zelen (1969). 'Length bias sampling' concerns the fact that individuals who develop a rapidly progressive disease, and who are thus more likely to die than the majority of individuals with disease, are unlikely to be found in a population that presents for screening. In other words, the screening programme is likely to select subjects with long-lasting diseases, so that the effectiveness of screening in terms of survival is overstated. This phenomenon is certainly relevant to studies of molecular epidemology, since the cross-sectional measurement of a biomarker is likely to select the lesions that have a longer duration, and thus the biomarker characterizes only some part of the natural history and some types of lesions.

A second peculiar kind of bias in screening is the 'lead time' effect. Lead time is defined as the interval between the time of detection by screening and the time at which the disease would have been diagnosed in the absence of screening. It is the time by which screening advances diagnosis of the disease, which does not correspond to the time by which death is postponed. This same concept is likely to be highly relevant to biomarker measurement when this is done for clinical/early diagnosis purposes.

2.4. MORE ON CONFOUNDING

Confounding arises when the exposed and unexposed subpopulations we are investigating would have different risks even if exposure had been absent from these subpopulations (Pearce and Greenland, in Ahrens and Pigeot 2004, p. 372). Confounding is probably the main problem with observational studies. In experiments confounding is tackled with randomization, which allocates potentially confounding variables randomly to the groups that are compared. On average, and depending on the study size and lack of selection bias, after randomization the potential confounders

will have the same distribution among the different groups.

Conditions that need to be met to say that an exposure is a confounder are:

1. The putative confounder must predict disease risk in the absence of the exposure under study (based on prior knowledge).
2. The confounder has to be associated with the exposure under investigation in the source population.
3. Variables that are intermediate in the pathway between exposure and disease should not be treated as confounders; to do so could introduce serious bias.

The third condition is highly relevant to molecular epidemiology, because we expect that intermediate markers, such as C-reactive protein, chromosome aberrations or gene mutations, act as intermediate links in the pathway between exposure and disease, and adjustment for them would seriously hamper our understanding of the caual relationship. Box 2.2 shows the differences between situations in which a variable can and situations in which a variable cannot be reasonably considered a confounder.

Confounding is different from bias. Bias has to do with the study design or the way information is collected, and implies a distortion in the estimate of association. Bias can be avoided by design but not by increasing the sample size, which has to do only with precision of the estimate, i.e. the width of the confidence interval, not with lack of bias. Confounding is an objective problem that derives from the structure of reality, not from mistakes in the study design or selective recall of the patients. The problem comes from the fact that variables are related to each other in the population. For example, coffee drinkers also tend to be smokers, and smoking is a known risk factor for pancreatic cancer. Therefore, if I investigate the association between coffee drinking and cancer of the pancreas, this can be 'confounded' by smoking (Box 2.2). The examples in molecular epidemiology are probably countless (see below). Confounding cannot be easily eliminated by design, and certainly not by increasing the study size. Strategies to control confounding variables include:

Box 2.2 Confounding and intermediate variables (courtesy of M-R Jarvelin)

- Situations in which *C confounds R* (risk factor) *–D* (disease) relationship

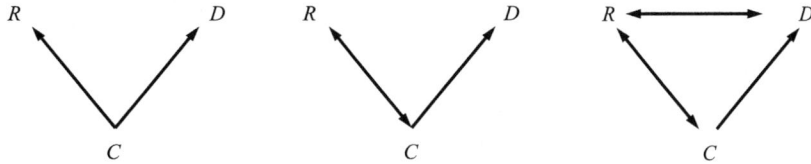

R   D   R   D   R ⟷ D

C   C   C

- Situations in which *C does not* confound *R–D* relationship

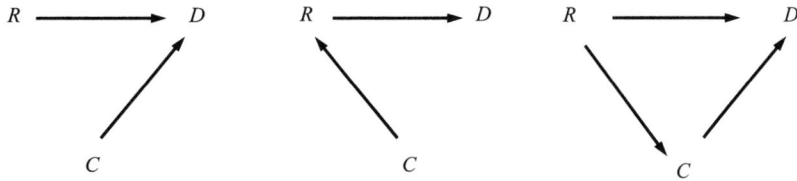

R ⟶ D   R ⟶ D   R ⟶ D

C   C   C

1. *Randomization*, i.e. random assignment of the exposure variable, which can only be done for beneficial, not for hazardous exposures.
2. *Restriction of the study to subcategories of potential confounders*, e.g. non-smokers.
3. *Statistical analysis* (e.g. multivariate analysis, see Chapter 14).

A further elegant way to address confounding is Mendelian randomization (see below). Box 2.3 shows how stratification allows the investigator to identify confounding. The overall OR linking pancreatic cancer to coffee drinking is 4.75. However, when we stratify by smoking, the ORs become 1.0 (no association with coffee) in both smokers and non-smokers. This is an extreme example, due to a very strong association between coffee drinking and smoking in the controls.

Apart from randomization – which is not acceptable with risk factors – the other approaches may have limitations: restrictions may mean drastic reduction in sample size or the inability to investigate effect modification, e.g. if we limit the study to non-smokers. Statistical approaches to confounding – like adjustment through stratification or multivariate analysis – may be limited by the absolute numbers of subjects in the cells after stratification by several potential confounders, with a consequent instability of estimates.

Box 2.3 How to identify a confounder through stratification (see text) Coffee drinking and pancreatic cancer

		Coffee drinking		OR
		Yes	No	
Whole population				
Cancer	Yes	200	50	4.75
	No	800	950	
Smokers				
Cancer	Yes	194	24	1.0
	No	606	76	
Non-smokers				
Cancer	Yes	6	26	1.0
	No	194	874	

OR = odds ratio.

2.5. SPECIFICITIES OF MOLECULAR EPIDEMIOLOGY DESIGN

Molecular epidemiology needs some specific considerations in terms of study design. Very often studies of molecular epidemiology (particularly of the type called 'transitional studies') have a cross-sectional design, i.e. they recruit patients with a specific disease and hospitalized or 'convenience' controls. Cases can be prevalent, i.e. their disease may have been diagnosed some time before recruitment into the study. The implications of such imperfect study designs are vast. Prevalent cases can be survivors, so that the biomarker that is measured can just be a prognostic indicator and not involved in causation. Or, if controls are affected by a disease other than the one under study, the biomarker can be affected by the presence of disease and therefore give a biased representation of the true causal relationship. In one of the pioneering studies on lung cancer and CYP2D6, controls were affected by COPD (Ayesh *et al.* 1984); this study showed a very strong association between a CYP2D6 variant phenotype and lung cancer, but it is likely that this result may have been biased by the choice of controls. Results from subsequent studies were contradictory.

Since metabolic phenotypes are assessed through the administration of a chemical or a mixture (e.g. coffee) and the study of the metabolic fate (metabolites in blood or urine), phenotypes can be affected by the metabolic changes due to the presence of a disease, while genotypes are not. In general, the genotype is stable and thus can be confidently assessed in case–control studies.

The situation is different with intermediate markers, such as mutations or proteomic changes. The logic underlying causal reasoning in epidemiology, i.e. the measurement of intermediate (bio)markers in the context of cohorts in which incidence rates of disease are computed, should not be forgotten. Only prospective studies allow for a proper temporal evaluation of the role of intermediate biomarkers. For example, the analysis of *p53* mutations with a cross-sectional design has been an invaluable tool in the investigation of liver carcinogenesis (Montesano *et al.* 1997), but the cross-sectional design of the early studies did not allow the researchers to exclude that mutations were a consequence of cell selection rather than of the original causal agent, such as aflatoxins (e.g. Aguilar *et al.* 1994). In other words, what was observed was the spectrum of mutations in liver cancers as the consequence of a long and complex process involving the effect of carcinogens, DNA repair and the selection of cells carrying specific mutations conferring a selective advantage to cells.

Table 2.1   Half-life of selected biomarkers

Phenol (metabolite of benzene) in urine: 6 h
Carbon monoxide in blood: 5 h
Haemoglobin aromatic adducts: 120 days
DNA aromatic adducts in lymphocytes: several months

Therefore, in principle, prospective studies are better for the understanding of time relationships between exposure, intermediate biomarkers and disease, although they have the limitation of being usually based on a single spot biological sample, which does not allow the measurement of *intra-individual variation* (see Chapter 6).

Randomized trials with biological samples have been repeatedly performed, e.g. using dietary changes as the intervention and oxidative damage as the outcome. However, though randomized controlled trials are probably the best design to conclude for causality in epidemiology, they also have limitations, particularly the short half-life of most biomarkers and also of most interventions, compared to the long-term exposures that are needed to cause chronic diseases like cancer. Many biomarkers are short-lived, i.e. they represent changes that occurred very close to recruitment/blood drawing/sample collection (Table 2.1).

Another specific question with respect to some biomarkers is that sometimes it is difficult to understand whether the marker is intermediate in the pathway leading from exposure to disease or is just a consequence of exposure, with no role in disease onset, or even an epiphenomenon of disease with no relationship to exposure. Examples of this kind are C-reactive protein (CRP) and micronuclei. CRP seems to be related to inflammation and risk factors for myocardial infarction, but it is not clear whether it is really causally related to the latter (intermediate in the pathway; see below). Micronuclei seem to originate from exposure to clastogens but can lead to cell death and therefore are likely not to be intermediate in the causal pathway.

Special designs in molecular epidemiology

Molecular epidemiology uses the same study designs as other types of epidemiology. Specific features

EXAMPLE C   Nested case–control study on benzene exposure and leukaemia

A nested case–control study was conducted to evaluate the exposure–response relationship between external ionizing radiation exposure and leukaemia mortality among civilian workers at the Portsmouth Naval Shipyard (PNS), Kittery, Maine. The PNS civilian workers received occupational radiation exposure while performing construction, overhaul, repair and refuelling activities on nuclear-powered submarines.

The study population from which the cases and controls were selected consisted of 37 853 civilian workers ever employed in the period 1952–1992, whose vital status and cause of death were obtained through 1996. Of the total cohort, 13 468 workers were monitored for external ionizing radiation. All deaths occurring during 1952–1996 among the entire cohort with an underlying cause of death classified as leukaemia, under the revision of the International Classification of Diseases in effect at the time of death, were selected as cases (n = 115). For each case, four controls were randomly selected from each risk set, using incidence density sampling methods. A risk set consisted of all workers who were under observation at the age of the index case at the time of failure (i.e. were hired at an earlier age than the case's death age and lived longer than the case). Cases and controls were matched on attained age because it is a strong predictor of leukaemia mortality and is therefore an important potential confounder of the associations of interest in these analyses.

Very accurate data were collected on dosimetry and potential exposure to solvents. In addition to radiation doses received in the workplace, a secondary analysis was conducted incorporating doses from work-related medical X-rays and other occupational radiation exposures. A significant positive association was found between leukaemia mortality and external radiation exposure, adjusting for gender, radiation worker status and solvent exposure duration (OR = 1.08 at 10 mSv of exposure; 95% CI 1.01–1.16). Solvent exposure (including to benzene and carbon tetrachloride) was also significantly associated with leukaemia mortality, adjusting for radiation dose, radiation worker status and gender. Incorporating doses from work-related medical X-rays did not change the estimated leukaemia risk per unit of dose (Kubale et al. 2005).

include the nested case–control study, frequently used in cohorts (see Example C and Chapter 3) and special designs conceived for genetic epidemiology. To exemplify one of the most popular among the latter, let us consider the 'case-parent triad' design, in which subjects affected with the disease under investigation are the 'cases', while their parents are the controls. Conceptually, this is a variant of the case–cohort study (Ahsan et al. 2002). The purpose of the study is to investigate whether the distribution of putative at-risk alleles is different among cases and controls. The advantage of the design is that comparison with parents automatically matches for shared environmental exposures and for ethnicity, thus allowing strict control of the 'population admixture' problem (see below). A variant of the case-parent triad design is the one that uses only heterozygous parents and estimates the 'transmission disequilibrium test' (TDT), a χ^2 test for the association between the at-risk allele and the disease. More

on such special designs can be found in Kass and Gold (in Ahrens and Pigeot 2004).

Bias

Molecular epidemiology also has some specific issues when bias and confounding are addressed. Bias can arise because of measurement error, whose extent can be considerable, although often unknown. An example is represented by oxidative damage. When it was originally measured using phenol-based extraction of DNA, samples were exposed to air, which led to gross overestimation of oxidative DNA damage (Collins et al. 2004). If such measurement error is evenly distributed among the groups we are comparing, it simply leads to an underestimation of the association (it is a sort of re-shuffling of the distribution of true biomarker values among the groups). However, if for any reason one group has been exposed to air more than another, this can lead to any direction of bias.

Types of bias that are common in other epidemiological studies may become dramatic when biological samples are collected and biomarkers are measured. Examples are discussed below.

Selection bias related to sample collection

Consider an epidemiological study on orofacial clefts, in which blood samples have been collected but only about 75% of the cases and 50% of controls gave their consent to participation. The point is not in the low response rate but in the different characteristics of respondents and non-respondents among cases and controls. Non-participants were younger and less educated than participants, particularly when fathers were involved (Romitti *et al.* 1998). In general, selective participation exerts its main effect on epidemiological estimates related to environmental or behavioural variables, which tend to be associated to social class and education. However, selection bias can also affect the association with biomarkers. For example, tissue biopsies can be collected to investigate somatic mutations in pancreatic cancer. However, biopsies can be unavailable because: (a) the patient had a diagnosis without a biopsy; (b) the biopsy is too small to extract a sufficient amount of DNA; or (c) the ethical committee or the patient did not allow the use of the tissue.

In a study on pancreas cancer (Hoppin *et al.* 2002), out of more than 1000 eligible patients the investigators were able to extract DNA from only 46 biopsies. The patients with a biopsy available were more frequently white and the tumour size was on average 179 mm, vs. 570 mm among the patients whose biopsy was not made available. This discrepancy is likely to introduce bias if one tries to correlate the prevalence of somatic mutations with exposure characteristics, such as occupation.

A further example is provided by Porta and colleagues (2002). In a case–case study, patients with pancreatic cancer at two general hospitals were retrospectively identified. Their clinical records were abstracted and paraffin-embedded samples retrieved from pathology records. DNA was amplified, and mutations in codon 12 of the *K-ras* gene were detected. Results on the mutations were obtained for 51 of the 149 cases (34.2%). Mutation data were over five times more likely to be available from one of the hospitals. In particular, subjects with mutations were more likely to have received a treatment with curative intent (OR = 11.56, 95% CL 2.88–46.36).

Confounding and population admixture

A confounder is a covariate that changes with exposure and predicts disease. The simple example of smoking, coffee drinking and pancreatic cancer seen above is easy to understand. Unfortunately, often we know too little about biomarkers to hypothesize confounding. We only partially know which are the correlates of DNA adducts, or of metabonomic patterns. Let us consider a couple of examples. The CYP1A2 phenotype has been suspected of being associated with colorectal cancer. However, the enzyme can be induced by cruciferous vegetables, which in turn protect from colorectal cancer; therefore, the observation of an association between colorectal cancer and the phenotype could be confounded by cruciferous vegetable intake. Another example: in a study on the levels of plasma DNA in cancer patients and controls (in the context of a multicentre cohort study) we found that, although the level of plasma DNA seemed to predict the onset of cancer, it was also strongly associated with the recruitment centre. This was due to modalities of blood collection and storage, since a longer time elapsing between blood drawing and storage in liquid nitrogen was associated with higher DNA levels, due to greater white blood cell death. Thus, the association between cancer and plasma DNA levels could be confounded by centre, since cancer rates also differed by centre in this multicentre study (Gormally *et al.* 2004).

Confounding is also possible with genotyping, in a least two ways. First, linkage disequilibrium (LD) can lead to attribute a disease to gene X, while the genuinely causal gene is in LD with X. Second, confounding can be due to the phenomenon of 'population stratification'. Usually the populations

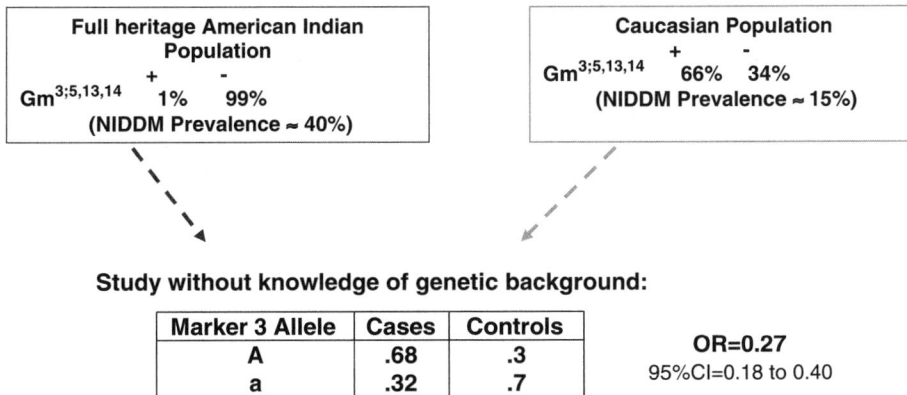

Figure 2.1 Example of population stratification: prevalence of diabetes mellitus according to a genetic marker in two populations. Reproduced with permission from Knowler *et al.* (1998) *The American Journal of Human Genetics*, The University of Chicago Press.

we investigate are mixed, with different subgroups characterized by different disease rates, a different genetic make-up and different prevalences of exposure to relevant risk factors. For example, if a putative gene for hypertension is investigated in a mixed population, it may happen that Afro-Americans, who have higher rates of hypertension and more frequent exposure to risk factors for the disease, also have a different distribution of alleles for the putative gene than Caucasians. This would introduce confounding, leading us to erroneously attributing a causal role to the gene in question. A real example is shown in Figure 2.1, where the haplotype Gm3;5,13,14 (a genetic marker) seems to be associated with type 2 diabetes mellitus. As the figure shows, there is a very strong association between this gene variant and the Pima Indian origin; the latter population, in turn, has one of the highest prevalences of diabetes in the world. If Pima Indians and Caucasians are analysed together (population admixture) a spurious (confounded) association between diabetes and the gene variant is found; the association disappears if the two ethnicities are analysed separately. However, this is an extreme case, since only two ethnic groups are considered and the distribution of the gene variants is extremely unbalanced among them. It has been estimated that the phenomenon of population admixture is unlikely to play a major confounding role in most studies, particularly if several ethnicities are considered in the same investigation, i.e. the population is highly admixed (Wacholder *et al.* 2000).

Mendelian randomization

A recent example of a difficult interpretation of data concerning an intermediate biomarker, as already mentioned, is the role played by C-reactive protein (CRP) in cardiovascular disease (CVD). Is CRP really an intermediate marker, i.e. does it lie in the causal pathway between risk factors and disease, or is it simply an indicator of risk factors? For example, both metabolic changes (such as triglyceride or HDL-cholesterol elevation) and inflammation are risk factors for CVD: is CRP just an epiphenomenon of their alteration, or is it a genuine mechanism that mediates their action on CVD? As Table 2.2A shows, in fact all risk factors for CVD are associated with CRP, and thus confounding is plausible. To answer this question, Mendelian randomization has been proposed. Mendelian randomization consists in assessing the relationship between CVD and a gene variant that is known to influence the levels of CRP. Since a gene variant is randomly assorted from parents to the offspring, the association between the gene variant and CVD would not be confounded by risk factors like those mentioned above, and would thus confirm/falsify an independent role of CRP in causing CVD. Mendelian randomization has three arms: one showing that the exposure (in this case CRP) is associated with the disease (CVD); one showing that CRP is associated with the gene variant; and finally, one showing that the gene variant is associated with CVD. As Table 2.2 suggests, when the CRP genotype 1059G is considered, CRP levels

Table 2.2 (A) Distribution of risk factors for coronary heart disease by quartiles of C-reactive protein (CRP). (B) Levels of CRP and distribution of risk factors for coronary heart disease by CRP genotypes

A

| | C-reactive protein quartile | | | | |
	1	2	3	4	p trend
Hypertension (%)	45.8	49.7	57.5	60.7	< 0.001
BMI	25.2	27.0	28.5	29.7	< 0.001
HDL cholesterol	1.80	1.69	1.63	1.53	< 0.001

B

| | Means by *CRP* genotype 1059G | | |
	GG	GC or CC	
C-reactive protein	1.81	1.39	< 0.001
Hypertension (%)	53.3	53.1	0.95
BMI	27.5	27.8	0.29
HDL cholesterol	1.67	1.65	0.38

Reprinted from Timpson *et al.* 2005 with permission from Elsevier.

vary in the expected way, whereas the risk factors for CVD do not. Worns *et al.* (2006) performed a prospective study in 108 monozygotic (MZ) and 60 same-sex dizygotic (DZ) twin pairs to analyse the genetic and environmental contributions to plasma CRP and IL-6 levels. Heritability of IL-6 was 0.61, indicating that plasma IL-6 levels are to a major part influenced by genetic determinants; however, for CRP, heritability was only 0.22, pointing to a moderate genetic influence.

The third arm of the Mendelian randomization reasoning is the study of the association between the gene variant and CVD, which has not been yet tested extensively in epidemiological studies. In an investigation in the Framingham Heart Study, variance-component linkage analyses of blood levels of four biomarkers of vascular inflammation [C-reactive protein (CRP), interleukin-6 (IL-6), monocyte chemoattractant protein-1 (MCP-1), and soluble intercellular adhesion molecule-1 (sICAM-1)] were conducted in 304 extended families from the Framingham study, using data from a genome scan. Heritability estimates ranged from 14% (IL-6) to 44% (MCP-1) after log transforming and adjusting for covariates. Significant linkage to MCP-1 was found on chromosome 1 (LOD = 4.27 at 186 cM; genome-wide p = 0.005) in a region containing inflammatory candidate

genes, such as *SELE*, *SELP* (E- and P-selectin) and *CRP*. Other hotspots with LOD scores > 2 were found for *MCP-1* on chromosome 1 (LOD = 2.04 at 16 cM; LOD = 2.34 at 70 cM) and chromosome 17 (LOD = 2.44 at 22 cM) and for *sICAM-1* on chromosome 1 at 229 cM (LOD = 2.09), less than 5 cM from the interleukin-10 (IL-10) gene. Apparently, therefore, multiple genes on chromosome 1 may influence inflammatory biomarker levels and may have a potential role in development of CVD, but no clear role for CRP has yet emerged (Dupuis *et al.* 2005) (NB: LOD means 'logarithm of odds', and is a measure of linkage disequilibrium).

2.6. CONCLUSIONS

Epidemiology, based on simple tools such as interviews and questionnaires, has achieved extremely important goals for public health, including the discovery of the causal relationships between salt intake and hypertension, peptic ulcer disease and *Helicobacter pylori*, smoking and lung cancer, and asbestos and mesothelioma. Even a complex issue, such as the relationship between air pollution (a low-level exposure) and chronic disease has been successfully dealt with by time-series analysis and cohort studies not based on the laboratory. The use of molecular epidemiology methods is meant to provide a specific set of new tools to answer specific scientific questions. The following are examples:

1. A better characterization of exposures, particularly when levels of exposure are very low or different sources of exposure should be integrated in a single measure.
2. The study of gene–environment interactions
3. The use of markers of early response, in order to overcome the main limitations of cancer epidemiology, i.e. the relatively low frequency of specific forms of cancer and the long latency period between exposure and the onset of disease.

In addition to the usual sources of bias that need to be addressed in all epidemiological studies, limitations of molecular epidemiology should be acknowledged: the complexity of many laboratory methods, with partially unknown levels of measurement error or inter-laboratory variability; the limited knowledge of the sources of bias and confounding; and the uncertain biological meaning of markers, as in the case of some types of adducts or some early response markers. However, success stories, such as human papilloma virus (HPV) and cervical cancer, aflatoxins and liver cancer, or benzene and leukaemia show that the contribution from molecular epidemiology can be considerable.

Acknowledgements

This chapter was made possible by a grant from the European Community (6th Framework Programme, Grant No. 513943) for the Network of Excellence – Environmental Cancer Risk, Nutrition and Individual Susceptibility (ECNIS) (WP4 and WP8).

References

Aguilar F, Harris CC, Sun T *et al.* 1994. Geographic variation of *p53* mutational profile in nonmalignant human liver. *Science* 1994; **264**: 1317–1319.

Ahsan H, Hodge SE, Heiman GA *et al.* 2002. Relative risk for genetic associations: the case–parent triad as a variant of case–cohort design. *Int J Epidemiol* 2002; **31**: 669–678.

Ayesh R, Idle JR, Ritchie JC *et al.* 1984. Metabolic oxidation phenotypes as markers for susceptibility to lung cancer. *Nature* **312**: 169–170.

Collins AR, Cadet J, Moller L *et al.* 2004. Are we sure we know how to measure 8-oxo-7,8-dihydroguanine in DNA from human cells? *Arch Biochem Biophys* **423**(1): 57–65.

Dupuis J, Larson MG, Vasan RS *et al.* 2005. Genome scan of systemic biomarkers of vascular inflammation in the Framingham Heart Study: evidence for susceptibility loci on 1q. *Atherosclerosis* **182**: 307–314.

Feinleib M, Zelen M. 1969. Some pitfalls in the evaluation of screening programs. *Arch Environ Health* **19**: 412–415.

Hennekens CH, Buring JE. 1987. *Epidemiology in Medicine*. Little, Brown: Boston, MA.

Hoppin JA, Tolbert PE, Taylor JA *et al.* 2002. Potential for selection bias with tumor tissue retrieval in molecular epidemiology studies. *Ann Epidemiol* **12**: 1–6.

Hutchinson GB, Shapiro S. 1968. Lead time gained by diagnostic screening for breast cancer. *J Natl Cancer Inst* **41**: 665–681.

Knowler WC, Williams RC *et al.* 1988. Gm3,5,13,14 and type 2 diabetes mellitus: an association in American-Indians with genetic admixture. *Am J Hum Genet* **43**: 520–526.

Miettinen OS. 1985. *Theoretical Epidemiology*. Wiley: New York.

Montesano R, Hainaut P, Wild CP. 1997. Hepatocellular carcinoma: from gene to public health. *J Natl Cancer Inst* **89**: 1844–1851.

Kubale TL, Daniels RD, Yiin JH *et al.* 2005. A nested case–control study of leukemia mortality and ionizing radiation at the Portsmouth Naval Shipyard. *Radiat Res* **164**: 810–819.

Paxton MB, Chinchilli VM, Brett SM *et al.* 1994. Leukemia risk associated with benzene exposure in the pliofilm cohort. II. Risk estimates. *Risk Anal* **14**: 155–161.

Porta M, Malats N, Corominas JM *et al.* 2002. Pankras I Project Investigators. Generalizing molecular results arising from incomplete biological samples: expected bias and unexpected findings. *Ann Epidemiol* **12**: 7–14.

Romitti PA, Munger RG, Murray JC *et al.* 1998. The effect of follow-up on limiting non-participation bias in genetic epidemiologic investigations. *Eur J Epidemiol* **14**: 129–138.

Rothman KJ. 1986. *Modern Epidemiology*. Little, Brown: Boston, MA.

Sackett DL. 1979. Bias in analytic research. *J Chron Dis* **32**: 51–63.

Steffen C, Auclerc MF, Auvrignon A *et al.* 2004. Acute childhood leukaemia and environmental exposure to potential sources of benzene and other hydrocarbons; a case–control study. *Occup Environ Med* **61**(9): 773–778.

Timpson NJ, Lawlor DA, Harbord RM *et al.* 2005. C-reactive protein and its role in metabolic syndrome: Mendelian randomization study. *Lancet* **366**: 1954–1959.

Vermeulen R, Lan Q, Zhang L *et al.* 2005. Decreased levels of CXC-chemokines in serum of benzene-exposed workers identified by array-based proteomics. *Proc Natl Acad Sci USA* **102**(47): 17041–17046.

Yin SN, Hayes RB, Linet MS *et al.* 1996. A cohort study of cancer among benzene-exposed workers in China: overall results. *Am J Ind Med* **29**(3): 227–235.

Wacholder S, Rothman N, Caporaso N. 2000. Population stratification in epidemiology: studies of common genetic variants and cancer: quantification of bias. *J Natl Cancer Inst* **92**: 1151–1158.

Wörns MA, Victor A, Galle PR, Höhler T. 2006. Genetic and environmental contributions to plasma C-reactive protein and interleukin-6 levels – a study in twins. *Genes Immun* 7(7): 600–5.

Essential reading

Ahrens W, Pigeot I (eds). 2005. *Handbook of Epidemiology*. Springer: Berlin.

Hennekens CH, Buring JE. 1987. *Epidemiology in Medicine*. Little, Brown: Boston, MA.

Miettinen OS. 1985. *Theoretical Epidemiology*. Wiley: New York.

Perera FP. 2000. Molecular epidemiology: on the path to prevention? *J Natl Cancer Inst* **92**(8): 602–612.

Ransohoff DF. 2005. Bias as a threat to the validity of cancer molecular-marker research. *Nat Rev Cancer* **5**(2): 142–149.

Schulte P, Perera F. 1993. *Molecular Epidemiology. Principles and Practice*. Academic Press: San Diego, CA.

Rothman KJ, Greenland S (eds). 1998. *Modern Epidemiology*, 2nd edn. Lippincott-Raven: Philadelphia, PA.

Sackett DL. 1979. Bias in analytic research. *J Chron Dis* **32**: 51–63.

Toniolo P, Boffetta P, Shuker DEG *et al.* 1997. Application of biomarkers in cancer epidemiology. IARC Scientific Publication No. 142. IARC: Lyon.

Vineis P, Malats N, Lang M *et al.* 1999. Metabolic polymorphisms and susceptibility to cancer. IARC Science Publication No. 148. IARC: Lyon.

3

Molecular Epidemiological Studies that can be Nested within Cohorts

Andrew Rundle[1] and Habibul Ahsan[2]

[1]Mailman School of Public Health, Columbia University, NY, and [2]Center for Cancer Epidemiology and Prevention, University of Chicago, IL, USA

3.1. INTRODUCTION

Cohort studies are considered the most rigorous tool in observational epidemiology for making causal inferences (Rothman 1986; Greenland 1987). Other designs, such as case–control studies, derive their validity from their ability to estimate effect measures that otherwise would be calculated from a cohort study (Rothman 1986; Greenland 1987). In a cohort study, a group of individuals free of the disease of interest are identified, exposure status is assessed and the individuals are then followed up through time for disease incidence (Rothman 1986). For rare diseases, such as cancer, cohort studies involve tens or hundreds of thousands of study subjects (Riboli and Kaaks 1997; Ahsan *et al.* 2006). Cohort studies are commonly depicted in illustrations as though all the cohort members were simultaneously enrolled into the cohort. In truth, it may take several years for the cohort to be assembled and for questionnaire data and biological samples to be collected; furthermore, enrollment may occur across multiple sites, with staggered recruitment occurring in different years (Riboli and Kaaks 1997). Two measures of effect relating the exposure to disease outcomes are commonly calculated from cohort studies, the relative risk and incidence rate ratio (Rothman 1986; Greenland 1987). These measures of effect are used to make causal inferences regarding whether or not the exposure increased the risk of disease development (Rothman 1986; Greenland 1987).

The relative risk is calculated as the disease incidence in the exposed divided by the disease incidence in the unexposed (Rothman 1986). Disease incidence is a proportion and, in a closed cohort with complete follow-up, is calculated as the number of cases of disease observed to occur during follow-up, divided by the total number of people under observation. Disease incidence can be calculated for the cohort overall and separately for the exposed and unexposed. Relative risks are calculated by dividing the incidence in the exposed by the incidence in the unexposed, and a relative risk > 1 implies that the exposure increased the risk of developing disease over the course of follow-up. The incidence rate ratio is the ratio of two incidence rates and is often calculated when individuals enter and leave the cohort at different times, such as when enrollment into the cohort occurs over several years and when follow-up is incomplete because of competing causes of death or loss to follow-up (Rothman 1986). The incidence rate for the exposed group is calculated as the number of cases among the exposed divided by the person–years of follow-up accrued by the exposed portion of the cohort. Similarly, the incidence rate is calculated for the unexposed, and a ratio of these two rates that is > 1 implies that the exposure increases the rate of disease development. When the disease incidence is rare and there is minimal loss to follow-up, the relative risk and incidence rate ratio

Molecular Epidemiology of Chronic Diseases, Edited by C. P. Wild, P. Vineis, and S. Garte
© 2008 John Wiley & Sons, Ltd

(A)

Exposed N_1 → Diseased A (D, D, D, D); Healthy B

Unexposed N_0 → Diseased C (D, D); Healthy D

Relative Risk

$I_1 = A/N_1$

$I_0 = C/N_0$

$RR = I_1/I_0$

$$RR = \frac{A/N_1}{C/N_0}$$

(B)

exposed N_1 → $N_1 t$ → Diseased A (D, D, D); Healthy B

Unexposed N_0 → $N_0 t$ → Diseased C (D, D); Healthy D

Incidence Rate Ratio

$I_1 = A/N_1 t$

$I_0 = C/N_0 t$

$IRR = I_1/I_0$

$$IRR = \frac{A/N_1 t}{C/N_0 t}$$

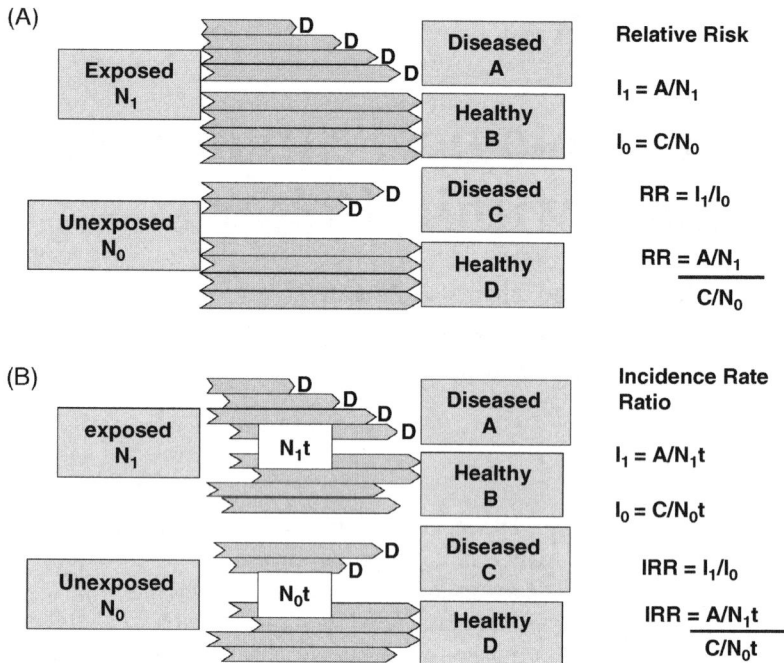

Figure 3.1 Schematic illustration of a cohort study and the calculation of the relative risk and rate ratio. (A) Closed cohort with complete follow-up and calculation of the relative risk. (B) Cohort in which subjects enter and leave the cohort at different times and person–time of follow-up is used to calculate the rate ratio

are very similar in magnitude. Figure 3.1A, B, illustrates these two scenarios and the calculation of the relative risk and the rate ratio.

When the exposure data is based on questionnaire data that are collected at baseline enrollment into the cohort, analyses of the risk of disease associated with exposure are inexpensive to conduct. This is not to say that follow-up and data entry and cleaning are not expensive – these elements of cohort studies can be very expensive – just that, once collected, there are few barriers to analyses of databased information. However, there is a growing number of large cohort studies established over the past decades in which biological samples have been collected and stored for future molecular epidemiology studies (Tang *et al.* 2001; Riboli and Kaaks 1997; Ahsan *et al.* 2006). The stored biological samples are precious and represent a limited resource that must be used as efficiently as possible. Furthermore, it can be very expensive to thaw out samples and analyse biomarker levels in the entire cohort.

Two related yet methodologically distinct study designs, case–cohort and nested case–control studies, are typically used for molecular epidemiological studies within prospective cohort studies because they offer logistical efficiency over full cohort analyses. These designs utilize cases that would be included in standard cohort analyses, but rely on a sample of the underlying non-diseased portion of the cohort to serve as a referent group. The sampling procedures allow analyses to be conducted on far fewer study subjects and yet provide estimates of the relative risk or incidence rate ratio that would have been calculated from the full cohort. In both designs an odds ratio is calculated and, depending on the sampling, can be understood to estimate either the relative risk or the incidence rate ratio. The differences between these two designs and the trade-offs in selecting one design over another have been described previously (Wacholder 1991; Wacholder *et al.* 1991, 1992; Langholz and Thomas 1990; Suissa *et al.* 1998; Barlow *et al.* 1999). However, choosing between

(A)

Relative Risk

$I_1 = A/N_1$

$I_0 = C/N_0$

$RR = I_1/I_0$

$RR = \dfrac{A/N_1}{C/N_0}$

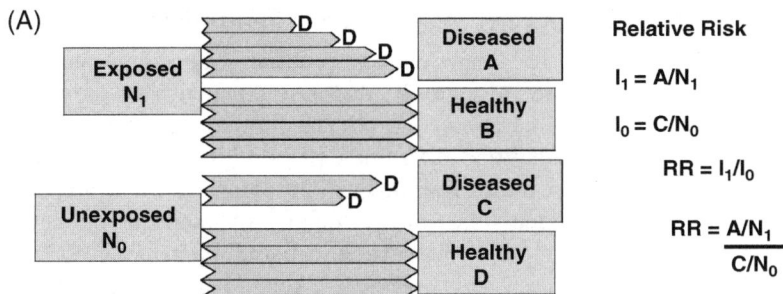

In a case-cohort study in a cohort with complete follow-up, all of the cases and a sample, r, of the cohort are used

	Cases	Cohort
Exposed	A	$N_1\,(r)$
Unexposed	C	$N_0\,(r)$

$$\text{Cross product} = \frac{AN_0\,(r)}{CN_1\,(r)} = \frac{A/N_1}{C/N_0} = RR$$

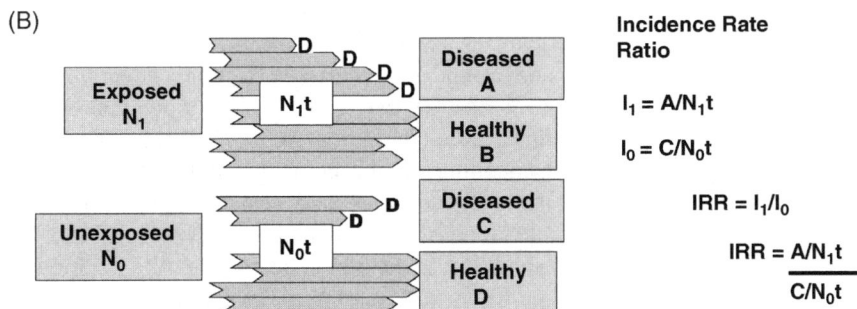

(B)

Incidence Rate Ratio

$I_1 = A/N_1 t$

$I_0 = C/N_0 t$

$IRR = I_1/I_0$

$IRR = \dfrac{A/N_1 t}{C/N_0 t}$

In a case-cohort study in a cohort that has variable follow-up, all of the cases are used and a sample, r, of the cohort is used to estimate the follow-up experience of the cohort

	Cases	Cohort
Exposed	A	$N_1 t\,(r)$
Unexposed	C	$N_0 t\,(r)$

$$\text{Cross product} = \frac{AN_0 t\,(r)}{CN_1 t\,(r)} = \frac{A/N_1 t}{C/N_0 t} = IRR$$

Figure 3.2 Schematic illustration of a case–cohort study conducted within a closed cohort (A) and within a cohort with variable follow-up time (B). Within a closed cohort, a case–cohort study can be conducted that will allow for the calculation of the relative risk. Within a cohort with variable duration of follow-up, a case–cohort study can be designed that allows for the estimation of the rate ratio the two designs and the conduct of the studies becomes more complicated when biomarker analyses are added to the designs (Hunter 1997; Rundle *et al.* 2005). The two designs are described and issues unique to biomarker analyses that impact the conduct of these studies are discussed in the remainder of this chapter.

3.2. CASE–COHORT STUDIES

Design and calculable measures of effect

The case–cohort design was first proposed as the 'case–base' design by Miettinen (1982) and was subsequently refined by Prentice (1986). In this approach, the exposure experience of a random subset (known as 'subcohort') of the total baseline cohort is selected as a 'control' or referent group and compared to the experience of the cases. In some instances the subcohort is chosen as a stratified random sample, with stratification on important confounding factors (Salomaa *et al.* 1999; Ballantyne *et al.* 2004). Apart from random errors, the distributions of exposure (i.e. biomarker) or any other factors in this random subset should represent that of the total cohort at the beginning of the observation period. The ratio of exposed to unexposed individuals (with respect to the biomarker of interest) in the referent group is the same as in the total cohort at baseline. As such, the estimated odds ratio using a case–cohort study represents the risk ratio for the same biomarker–disease relationship had a full cohort analysis involving all cohort members been performed (Sato 1994; Schouten *et al.* 1993; Rothman 1986). Alternatively, the follow-up data for the subcohort (including those who develop disease) may be treated as representative of the person–time experience of the total cohort and case–cohort studies can be analysed to estimate the rate ratio for the biomarker–disease relationship (Prentice 1986; Barlow *et al.* 1999). If a baseline cohort member later develops the disease of interest during the observation period, he/she is also treated as a case and thus also contributes to the exposure distribution among the cases (Prentice 1986; Barlow *et al.* 1999). Figure 3.2A, B, respectively, illustrate case–cohort analyses of a cohort with complete follow-up and of a cohort with variable follow-up.

Case–cohort designs offer flexibility

The major advantage of this design is that the referent group, randomly selected from the baseline cohort, can be used as a referent group for multiple different case groups arising from the cohort (Wacholder 1991). However, this supposition relies on the quality of exposure measurement in the subcohort being the same as for the case series. This is likely to be true for stored questionnaire data (Dirx *et al.* 2001), the use of industrial hygiene records and job histories to model occupational exposures (Zeka *et al.* 2004), or the modelling of ambient exposures using data-based residential histories and air monitoring records (Abrahamowicz *et al.* 2003). However, this assumption may not hold for biomarker analyses.

Additionally in this design, an investigator may select the time scale best suited for the particular set of statistical analyses (Wacholder 1991). A common time scale for cohort and case–cohort analyses is duration of follow-up; however, age and calendar year can both independently influence disease risk and may be more appropriate time scales (Breslow *et al.* 1983; Wacholder 1991). Depending on the context and the biomarker in question, different time scales might be more or less appropriate. For a genetic polymorphism for which risk is thought to accrue over the duration of the life course, age might be the most appropriate time scale rather than duration of follow-up. For a biomarker that might be influenced by a strong generational cohort effect, such as blood lead levels (Pirkle *et al.* 1994), calendar year might be more appropriate. For a biomarker that reflects the onset of exposure, such as a new occupational exposure, time since exposure onset might be most appropriate. In general the case–cohort design can accommodate separate analyses of each of these biomarkers, using different time scales for each analysis.

Analytical complexity

As described above, all of the cases occurring during follow-up of the cohort are included in case–cohort analyses as well as a sample of the

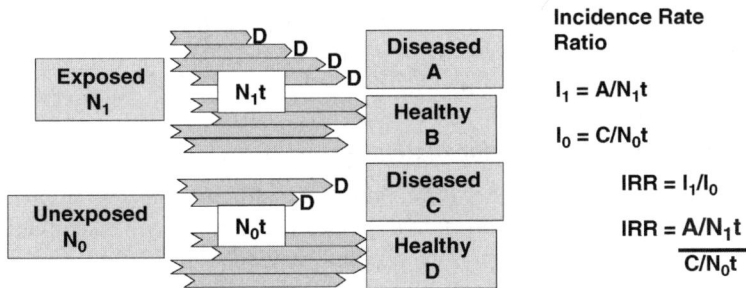

In a nested case-control study all of the cases and a matched sample, r, of the exposed and unexposed person time are used

	Cases	Controls
Exposed	A	$N_1 t (r)$
Unexposed	C	$N_0 t (r)$

$$\text{Cross product} = \frac{A N_0 t (r)}{C N_1 t (r)} = \frac{A/N_1 t}{C/N_0 t} = IRR$$

Figure 3.3 Schematic illustration of a nested case–control study conducted in a cohort with variable duration of follow-up. Note that the diagram is very similar to that of a case–cohort study conducted in a cohort study with variable follow-up. The difference is that controls are selected conditionally to match the person–time follow-up of the cases, whereas in the case–cohort design the subcohort is selected randomly and followed up to provide an estimate of the person–time experience of the overall cohort

total cohort at baseline. Typically, relatively few of the cases arise from the individuals sampled to be part of the subcohort; the rest of the cases arise from non-sampled portion of the cohort. For the cases that arise from outside of the subcohort, there is debate regarding the point at which the person–years for these subjects contribute to the calculation of the rate ratio (Barlow *et al.* 1999). Prentice (1986) and Barlow (1994) describe approaches employing a weighting of 1 in the denominator of the pseudo-likelihood for cases that arise outside the subcohort, while an approach by Self and Prentice (1988) uses a weighting of 0. Additionally, Barlow's approach weights the cases in the subcohort prior to failure and the subcohort controls by the inverse of their sampling fraction. At this time, the optimal weighting scheme for case–cohort analyses is unclear (Barlow *et al.* 1999). However, as will be described later, storage conditions and laboratory variability may differentially affect biomarker assays of samples

from cases arising from within the subcohort as compared to samples from cases arising from outside the cohort. Different weighting schemes may be more or less appropriate for different biomarker analysis scenarios and the most appropriate approaches have not yet been described.

3.3. NESTED CASE–CONTROL STUDIES

Design and calculable measures of effect

In the nested case–control design, matched controls are selected when each case occurs, i.e. they are selected from the cohort members who are still disease-free at the time a case occurs. This sampling procedure is referred to as risk set sampling. The matching on the time at risk (i.e. duration between enrollment/biospecimen collection and disease onset) causes the ratio of exposed to unexposed number of controls to represent, on average, the ratio of exposed to unexposed person–time for the total cohort. Thus, the odds ratio derived from the nested case–control study estimates the incidence rate ratio for the biomarker–disease relationship, had a full cohort analysis involving all cohort members been performed. Figure 3.3 illustrates a nested case–control study and the calculation of the rate ratio.

Matching

In a nested case–control design, the matching of controls to cases on follow-up duration between biological sample collection and disease development is seen as a major advantage. Additionally, it is common for controls to be matched to cases on other likely confounders, such as age and gender. However, additional matching conditions make the control series less representative of the person–time experience of the total overall cohort. Another consideration with matching is that the disease risk associated with a matching factor cannot be studied, although the role of a matching factor as an effect modifier can be investigated. A complaint with nested case–control studies is that the intricate matching used in the design generates a highly selected control series that cannot be used for other studies and does not provide representative data on the entire cohort (Suissa *et al.* 1998). In past work, substantial differences in beta coefficients, around two-fold, have been observed for age and gender as predictors of body mass index (BMI) when cross-sectional analyses were performed in the controls from a nested case–control study and then repeated in the over-all cohort the case–control study was nested in (Rundle *et al.* 2005). The differences in the betas occur because the controls are not representative of the over-all cohort and the matching factors act as effect modifiers of the associations between age, gender and BMI. However, since the primary goal of both designs is to identify prospective exposure–disease associations, the ability to conduct valid cross-sectional studies is of secondary importance in choosing a study design. Another issue is that the matching on duration of follow-up fixes the time scale of the analyses, reducing analytical flexibility. It is common for several biomarkers to be analysed in blood samples from subjects included in a nested case–control study, and the most biologically relevant time scale may be different

for each of the biomarkers, yet all analyses are conducted using the same time scale (Tang *et al.* 1998; Airoldi *et al.* 2005; Peluso *et al.* 2005).

Counter-matching

Another limitation of the individual matching procedure is that only matched sets that are discordant on exposure are informative, which reduces the statistical power of the design. To address this problem, a procedure known as counter-matching has been described (Langholz and Clayton 1994; Cologne 1997; Steenland and Deddens 1997). Controls are inversely matched to cases on a correlate/proxy of the exposure of interest in a manner that seeks to maximize the number of discordant pairs, increasing statistical efficiency (Langholz and Clayton 1994; Cologne 1997; Steenland and Deddens 1997). As an example of counter-matching, imagine an occupational cohort in which a nested-case control study is planned to determine whether an exposure-related biomarker is associated with disease outcome, and there are data available from job records on the occupational exposure that can be used as matching criteria. The distribution of exposure among the cases is used to assign cases to high or low exposure groups, perhaps by using the median exposure level as a cut point. A control would then be matched to each case, with the control randomly selected from among disease-free subjects who are in the exposure category to which the case does not belong. So, if a case has high exposure, a control would be randomly selected from among disease-free subjects in the low exposure group. Biomarker levels would then be assessed and case–control analyses of the biomarker data would be conducted, with a weighting factor representing the sampling probability for each subject. Counter-matching produces increased statistical efficiency over sampling that is random with respect to exposure, and this approach has been extended to studies of gene–environment interactions (Steenland and Deddens 1997; Andrieu *et al.* 2001; Bernstein *et al.* 2004). In contrast to matching on confounders, which yields controls that are similar to cases on those factors and provides for efficient control for confounding, counter-matching on exposure correlates attempts to maximize variation of the exposure of interest within case–control sets (Langholz and Clayton 1994). This approach can yield substantial statistical efficiency in comparison to nested case–control designs, which use random selection of controls from among risk sets (Steenland and Deddens 1997).

Individuals may be included in the analyses multiple times

In a nested case–control study, a control matched to a case at time *A* may subsequently develop the disease and becomes a case at time *B*. In this scenario, the study subject is analysed as both a control (at time *A*) and a case (at time *B*). This might be expected to be an uncommon occurrence for a rare outcome such as cancer, but when one considers matching controls to cases on age, smoking status and cigarettes smoked per day (Perera *et al.* 2002), the probability of a control later becoming a case is not trivial. Additionally, in designs where controls are matched to cases on many factors, there may be relatively few subjects available to be selected as controls within certain strata of matching variables. Thus, a subject may be selected as a control for a case at time *A* and again as a control at time *B*. For questionnaire data, the same data value, e.g. cigarettes smoked per day at enrollment, is used for the subject at time *A* when he/she is a control and at time *B* when he/she is a case, or if selected again as a control. However, as will be described later, things are more complex when biomarker data are used from a subject who contributes information more than once in a study.

3.4. CONSIDERATIONS REGARDING BIOMARKER ANALYSES IN CASE–COHORT AND NESTED CASE–CONTROL STUDIES

Past publications have discussed both practical or logistic concerns and comparisons of statistical efficiency and power (Wacholder 1991; Wacholder *et al.* 1991; Langholz and Thomas 1990). However, the use of biomarkers adds additional complexities to comparing these two designs (Hunter 1997; Rundle *et al.* 2005). The issues with biomarkers that must be considered are analytical batch effects, storage effects and freeze–thaw cycles.

Batch effects

Analytical batches are comprised of groups of biological samples that are analysed together under a particular set of conditions. When multiple batches are used in a study, the laboratory strives to duplicate across batches the conditions of the laboratory analyses (e.g. reagent stocks, equipment calibration) that may affect laboratory results. The goal is to prevent the analytical results derived from any given sample from being influenced by the particular analytical batch in which the sample happened to be analysed. However, this goal is often not met for many laboratory assays (Falk *et al.* 2000; Aziz *et al.* 1999; Tworoger *et al.* 2004; Taioli *et al.* 1994; Falk *et al.* 1999; Rybicki *et al.* 2004). Even commercial ELISA kits from the same manufacturer have shown substantial variation between different lots of kits (Aziz *et al.* 1999).

Batch effects can create random noise or bias, which is expected to have the same consequences as misclassification of exposure in epidemiological studies (Rundle 2000; Schulte and Perera 1993). If the measurement error is evenly distributed among cases and referents, such misclassification usually causes an underestimation of the association between analyte and disease. However, when the errors are unevenly distributed between cases and the referent group, bias arises. Thus, it is important to have the same proportion of cases and controls in different batches to avoid such bias. Batch effects can occur randomly with respect to time or assays can appear to drift with time, either producing lower or higher results over time (Schulte and Perera 1993).

There are many constraints that prevent all of the samples from being analysed together and necessitate the use of batches. These constraints may be caused by the limits of technology, such as the number of wells on an ELISA or PCR plate, or the number of samples that can be loaded into a robotic device. The constraints might be administrative and driven by the need to complete particular milestones for funding agencies or by the availability of reagents. Subject recruitment and transport and shipping limitations may create logistical limitations requiring samples to be analysed in batches. Finally, the limits of endurance of the laboratory technicians or other staffing issues may require samples to be analysed in batches.

Batch effects and case–cohort studies

Two scenarios for the initial case–cohort study being launched within a cohort need to be considered. The first scenario is when an investigator launches a new cohort study with an *a priori* plan for conducting case–cohort analyses at a future date. This will be referred to as a *prospective case–cohort study*. The second is when, after a period of follow-up, a case–cohort study is conducted within an existing cohort. This will be referred to as a *retrospective case–cohort study*. In the prospective scenario, the subcohort is selected as soon as the cohort is fully assembled and the cases are identified later, during the course of cohort follow-up. In the prospective scenario, there are incentives to begin laboratory analyses of samples from subcohort members as soon as the subcohort is identified. The ability to quickly begin laboratory work is similar to the logistical efficiency advantage that has been previously cited for case–cohort studies, when extensive field work is required to gather data from study subjects (Wacholder 1991). Early completion of the bulk of the laboratory work also allows for cross-sectional studies of correlates of biomarkers to begin right away (Ahsan *et al.* 2006). One of the advantages of the case–cohort design is that, since the subcohort is a random sample of the cohort, cross-sectional analyses of biomarkers in the subcohort provide valid information on the overall cohort. In addition, often as a cohort is assembled prevalent cases of the disease of interest are identified and interesting analyses of these prevalent cases and subcohort can be conducted.

In such a prospective case–cohort study, as incident cases accrue during follow-up, the investigators can begin analysing biological samples from cases. However, if subcohort samples were analysed immediately after the subcohort was selected, this approach would very closely align case–cohort status to analytical batch, setting the stage for bias. Biological samples from cases that arose from the subjects selected to be part of the subcohort will already have been analysed, and

thus the same laboratory value for the biomarker is applied to that individual when they appear in the data as a case and as a member of the subcohort. In essence, the same biological sample and laboratory analysis is used when the subject is a case and when the subject is a member of the subcohort. In general, however, the majority of cases arise from the vast bulk of cohort members not selected as part of the subcohort. Since samples from these cases were not analysed in batches of biological samples from the subcohort, the circumstances of analyses for these samples will likely differ from those of the subcohort. If batch effects operate as laboratory drift over time, then biased results are assured, and the direction of the bias depends on the direction of the drift. In a large study with batch effects acting as random noise across many batches, it is possible that, on average, the batch effects will generate a pattern of random noise in the cases that is similar to the pattern in the subcohort, in essence generating random measurement error and bias to the null. However, in a fashion analogous to that seen in studies with small sample sizes, in studies with a small number of analytical batches, sampling variation plays a larger role and the possibility of biased results increases (Hirschhorn *et al.* 2002; Ioannidis *et al.* 2003). Taking this point to an extreme, if batch variability is random, but a study only uses two batches and cases and referents are imbalanced across the batches, bias can occur, essentially as in the case of laboratory drift.

The retrospective scenario is more typical, in which there is an existing cohort that has been followed up for a number of years and accrued enough cases for case–cohort analyses. In this scenario, the cases would be compared to a subcohort retrospectively selected from the baseline members of the cohort. Samples from the subcohort and the case series would be analysed contemporaneously, and samples should be randomized to batches to avoid bias due to batch effects.

One of the often-cited strengths of case–cohort studies is that the subcohort can be used as a referent group for a variety of different case series, including those with different latencies (Wacholder 1991). A case–cohort study could be implemented early in cohort follow-up for a common cancer and then, after much further follow-up, the subcohort could be used as a referent group for a series of rarer or long-latency cancers. Tremendous logistical efficiency is gained in this way, because data from the referent group already exists in study databases. However, if the analyses use biomarkers susceptible to batch effects, this advantage may essentially be outweighed by the potential for bias. It may be of interest to analyse the same biomarkers in subsequent case series that were previously analysed in the subcohort and a prior case series. However, unless biological samples from the subcohort are re-analysed, biological samples from successive case-series will be analysed in analytical batches that contain no samples from the subcohort, essentially guaranteeing batch effects. Again, if the batch effect represents random noise and many batches are used, it is theoretically possible that, overall, the batches of case samples will reflect the same random noise as the subcohort and bias will be minimal. However, this possibility can not be verified in the data, calling into question the results. If the batch effect operates as laboratory drift, bias is guaranteed. As noted above, new aliquots of biological samples from the previously analysed subcohort samples can be analysed. These analyses can take case–control status into consideration during batch allocation or subjects can be randomized to batches. However, having to re-analyse samples from the subcohort nullifies the efficiency advantage of the case–cohort design.

Batch effects and nested case–control studies

The individual matching used in a nested case–control study allows for easy matching on analytical batch, removing batch effects. However, depending on how the study is implemented, there are several nuances that need to be considered. As discussed above, subjects may be included in the data multiple times, either as a control and then later during follow-up as a case, or multiple times as a control. If a nested case–control study is implemented at the end of cohort follow-up, batching can be arranged such that a single biological sample from an individual represented in the dataset multiple times can be placed in one batch, along with sam-

ples from all the subjects to which that individual is matched. In this way, one aliquot of biological sample can be used to provide biomarker data for each of the occurrences of that individual in the dataset. Another scenario is when cases accrued in the past and their retrospectively matched controls are analysed, while follow-up of the cohort and new case identification continues prospectively for another couple of years. Here it is possible for a subject selected to be a control for a past case (time *A*), to become a case during the continued follow-up and be matched to a set of controls at time *B*. In this scenario, a biological sample from the subject may have already been analysed previously when the subject acted as a control at time *A*. However, it would be improper for biomarker data generated when the sample was analysed at time *A* to also be used as the data when the subject is a case at time *B*. This is because, if biomarker data, generated when the subject was a control at time *A*, is used when the subject is a case at time *B*, biomarker data from the controls at time *B* will generated from a different lab batch than the biomarker data from the case at time *B*. In essence, this would break the matching of cases and controls on laboratory batch. It would be most appropriate if a new aliquot of biological material from the case at time *B* were analysed in a batch with biological material from the controls matched to the case at time *B*. Likewise, if a subject serves as a control for multiple cases, multiple samples from that control should be analysed, one for each instance where the subject is represented in the dataset. This is analogous to making sure the quality of exposure data is the same for cases and controls. Clearly, this issue is a larger problem for more common diseases, and may not be impossible to address if the laboratory assay consumes large volumes of biological sample.

Storage effects

The second issue is that for some biomarkers the levels of the biomarker may be influenced by the duration of sample storage, or that storage duration will reduce the accuracy of the assay (Evans *et al.* 1996; Kronenberg *et al.* 1994; Shih *et al.* 2000; Lathey *et al.* 1997; Woodrum *et al.* 1996; Bolelli *et al.* 1995; Phillips *et al.* 1988; Lagging *et al.* 2002; Ng V *et al.* 2003; Risio *et al.* 2003; Fergenbaum JH *et al.* 2004). The typical worry is that biomarker levels may decay and decline over time; however, there are examples of measured biomarker levels increasing over time (Bolelli *et al.* 1995). Changes in biomarker levels during storage can be substantial. Evans and colleagues have reported a 19% decrease in lipoprotein-α in serum samples stored for 3 years at $-70°C$ (Evans *et al.* 1996). Additionally, for samples stored at $-80°C$ for 3 years, Bolelli and colleagues reported a 30% increase in serum-free testosterone and a 40% decrease in progesterone (Bolelli *et al.* 1995).

Another concern is that for mechanical freezers, temperature can vary across freezers, and within a freezer the internal temperature can vary by location and time (Su *et al.* 1996; Helsing *et al.* 2000). After long-term storage in such freezers, it may be appropriate to match samples on storage location within the freezer (Helsing *et al.* 2000). Storing samples in the liquid phase of liquid nitrogen probably provides the most consistent storage conditions. However, in large multi-site cohort studies, samples are often stored centrally in nitrogen and then shipped to investigators' laboratories for analyses, where they may be stored in mechanical freezers for substantial time periods. It may even be advisable to physically arrange the samples in these mechanical freezers in a configuration appropriate for the study design. For a nested case–control study, matched sets of cases and controls could be stored in close proximity, and for case–cohort studies the samples could be stored in a random order.

The impact of storage effects on study validity depends on whether it alters the level of the biomarker or reduces assay precision. If storage effects alter assay precision, random error may be increased for biomarkers stored longer, reducing statistical power, and, depending on how the study is implemented, may cause bias. If storage duration alters the biomarker level, the effect on study validity depends on whether biomarker levels change at an absolute linear rate, in a half-life-like fashion, or at a relative rate. Depending on the mathematical function describing how biomarker levels change and how the study is implemented, bias may be away or towards the null.

For large multi-centre cohorts, it may take many years to assemble the entire cohort and thus storage time may be affected by logistical considerations. If recruitment begins at the various centres at different times, storage duration becomes associated with centre and the demographic and exposure characteristics of the subjects recruited at that site. In European Prospective Investigation of Cancer (EPIC), the Oxford site was one of the first to begin recruiting; subjects recruited at this site tended to have a higher socio-economic status and a large proportion of the subjects were vegetarians. Thus, the length of sample storage is not randomly distributed across lifestyle risk factors. A related issue is that biomarker levels may be influenced if sample handling and processing vary by site.

Storage effects and case–cohort studies

As described in the section on batch effects, in a prospective implementation of the case–cohort study the vast majority of the cases are likely to arise from cohort members not selected as part of the subcohort. If laboratory analyses of samples from the subcohort begin immediately after the study is initiated, there is likely to be substantial time that elapses between when samples from the subcohort are analysed and when the cases are identified and their samples are analysed. In this implementation, storage duration is closely aligned with case-subcohort status and, if storage effects occur, can produce bias. In the retrospective approach to the case–cohort study, cases and subcohort members are identified simultaneously and storage time need not be so starkly associated with case-subcohort status. However, if enrollment into the cohort took place over a number of years, across multiple sites, storage duration could still vary substantially across cohort members and storage duration could be associated with the demographic characteristics of the various sites. In these circumstances, statistical control for storage duration should be included in the statistical analyses.

In circumstances in which biomarkers from a second or third case series with longer latencies are compared with biomarker levels measured in the original subcohort, storage duration becomes increasingly associated with case-subcohort status.

As the duration of time between the laboratory analyses conducted in the subcohort and the analyses conducted for the new case series grows, the difference in the length of time the samples from the cases and subcohort members were stored before being analysed increases. Statistical control for duration of storage becomes more difficult for longer-latency cancers, due to sparse data at the tails of the distribution of storage times. Studies of longer-latency case series may require that biomarker levels in biological samples from the subcohort be re-analysed concurrently with samples from the new case series, so that the effects of storage duration can be removed.

Storage effects and nested case–control studies

Since biological samples are generally collected at baseline entry into the cohort, the matching on duration of follow-up in a nested case–control design effectively matches on sample storage duration. The typical analytical approach of using conditional logistic regression analyses with this design effectively controls for storage effects. In the discussion of batch effects and instances where an individual might be represented in the dataset at multiple times, it was noted that biological samples should be re-analysed at each time point at which the subject is represented in the data. Storage effects provide a further rationale for this recommendation. If biomarker data generated for an individual at the first instance he/she appeared in the study was then re-used for later instances when the individual appeared in the study, the storage duration for the biological samples from subjects matched to that individual at his/her later occurrences in the study would be longer than the storage duration of the biological sample from the individual whose data is re-used from the earlier analyses. This re-use of the data point from the earlier occurrence of the subject in the dataset effectively breaks the matching on storage duration. Again, this is a strong rationale for generating a new biomarker data point for each time a subject appears in the dataset.

Freeze–thaw cycles

Whole blood samples drawn from study subjects are typically separated into constituent parts, such as white blood cells, serum and red blood cells, and then frozen in multiple small aliquots. Many laboratory assays require volumes of sample that are smaller than the stored aliquots, so during a particular study a sample aliquot may be thawed, sampled from and then the remaining biological sample refrozen for future studies. However, components of the biological sample may be altered by thawing and re-freezing cycles (Sgoutas and Tuten 1992; Brey et al. 1994; Bilodeau et al. 2000; Halfon et al. 1996; Groschl et al. 2001; McLeay et al. 1992; Holten-Andersen et al. 2003; Lahiri and Schnabel 1993; Bellete et al. 2003). Thus, when a biological sample is thawed a second time to be assayed for a new biomarker study, the result of the laboratory analyses may differ from the results that would have been generated if the sample was being thawed for only the first time. There are several physical and chemical changes associated with freezing that may alter the chemical properties of a biological sample (Brey et al. 1994). As a sample freezes, there is a concentration of solutes in the residual liquid phase of the sample and increases in ionic strength which may cause protein precipitation and denaturation (van den Berg and Rose 1959; Brey et al. 1994). Furthermore, ice crystals precipitation causes the pH of the remaining liquid phase to change by several pH points, which may cause lipid degradation and protein denaturation (van den Berg and Rose 1959; Williams-Smith et al. 1977; Brey et al. 1994). Freeze–thaw cycles can have a substantial impact on biomarker levels. Sgoutas and colleagues reported a 27% decrease in lipoprotein-α levels, as measured by ELISA, after two freeze–thaw cycles to $-20°C$, and a 23% decrease after four cycles to $-70°C$ (Sgoutas and Tuten 1992). Bellete et al. (2003) reported that three freeze–thaw cycles reduced the content of DNA stored in sterile distilled water by more than half. In these experiments, DNA decay was influenced by the copy number of the target DNA sequence and the volume of distilled water (Bellete et al. 2003).

Thus, the sequence in which hypotheses are tested in a given bank of samples can influence biomarker levels and study results. For a given aliquot of biological material in which an investigator plans to measure several biomarkers, the sequence of assays and required freeze–thaws should be carefully planned. For instance, in a saliva sample, cortisol should be analysed within the first or second round of freeze–thaws, while progesterone can be analysed in later rounds of thawing (Groschl *et al.* 2001).

Freeze–thaw cycles and case–cohort studies

For the first case–cohort study launched in a cohort, implemented either prospectively or retrospectively, it seems unlikely that the effects of freeze–thaw cycles will inherently bias the analyses. It is unlikely that the prevalence of samples that have been thawed will differ between cases and the subcohort members. Even if previous biomarker studies had been conducted within a cohort, multiple aliquots of each sample fraction are likely to be available for analyses, so if one has previously been thawed, another could be used. Should it occur that the number of freeze–thaw cycles differs between the case-series and the subcohort, statistical control for the freeze–thaw cycles could be implemented.

In the case of studies of later, longer-latency case series, depending on the biological material to be analysed, the sample aliquots from the subcohort may have already been thawed and re-frozen in prior rounds of laboratory analyses. Biological samples from the subcohort members will be in higher demand than samples from subjects not selected into the subcohort, because the selected individuals serve as the referent group for multiple studies. For instance, if a serum biomarker was of interest in the original case–cohort study of one case type and a different serum biomarker is of interest in the subsequent case–cohort study for a different case type, serum aliquots from the subcohort may have already gone through a round of thawing and re-freezing. The samples from the new case series are less likely to have been previously thawed, because the samples were not of interest until the subjects were identified as cases. Thus, when new rounds of laboratory analyses are conducted in subsequent case series and the subcohort, storage conditions may impact study validity.

Freeze–thaw cycles and nested case–control studies

The individual matching in the nested case–control design allows for the possibility of including in the design, matching on the number of freeze–thaw cycles to which a subject's biological sample has been subjected. Again, one must consider what to do in circumstances where an individual appears multiple times in the dataset. It has been recommended above that an aliquot of biological sample be analysed for biomarker data at each time point at which the individual appears in the dataset. However, if there is only one or only a few aliquots stored per subject, repeat laboratory analyses may require that the same aliquot be re-used and be subjected to multiple freeze–thaw cycles. In this circumstance, one must weigh the measurement concerns associated with freeze–thaw cycles vs. those arising from batch and storage effects. Again, matching may provide a solution; aliquots that have also be previously thawed could used from the subjects matched to individual in question.

3.5. CONCLUSION

Both case–cohort and nested case–control designs allow for the prospective analysis of risk of disease associated with a biomarker with greater logistical efficiency than full-cohort analyses. Each design allows for the non-biased calculation of the incidence rate ratio that otherwise could have been calculated from a full-cohort analysis. In both designs, all of the cases that would have been included in cohort analyses are used and the experience of the cohort is ascertained through the use of sampling. The main issue to be considered with these designs is whether the quality of the biomarker data is equivalent in the cases and the referent group. The case–cohort design allows for more flexibility in statistical analyses, because the time scale is not fixed by the design, and various time scales that are most biologically relevant for different biomarkers can be used in the analyses. This flexibility in analyses may be important for the particular investigations and this consideration may dictate the choice of the study design. On balance, however, the nested case–control design appears to be a better choice for handling the logistical issues that are particular to biomarker analyses. The strength is that nature of control selection alleviates many of the problems discussed here. The drawback is, however, that the controls are highly selected and have few analytical uses beyond testing the hypotheses of the particular study.

References

Abrahamowicz M, Schopflocher T, Leffondre K et al. 2003. Flexible modeling of exposure-response relationship between long-term average levels of particulate air pollution and mortality in the American Cancer Society study. *Journal of Toxicology & Environmental Health Part A* **66**(16–19): 1625–1654.

Ahsan H, Chen Y, Parvez M et al. 2006. Health Effects of Arsenic Exposure Longitudinal Study (HEALS) – Design and Description of a Multidisciplinary Collaborative Investigation. *J Exp Sci Environ Epidemiol* **16**(2): 191–205.

Airoldi L, Vineis P, Colombi A et al. 2005. 4-Aminobiphenyl-Hemoglobin Adducts and Risk of Smoking-Related Disease in Never Smokers and Former Smokers in the European Prospective Investigation into Cancer and Nutrition Prospective Study. *Cancer Epidemiol Biomarkers Prev* **14**(9): 2118–2124.

Andrieu N, Goldstein AM, Thomas DC et al. 2001. Counter-matching in studies of gene-environment interaction: efficiency and feasibility. *Am J Epidemiol* **153**(3): 265–274.

Aziz N, Nishanian P, Mitsuyasu R et al. 1999. Variables that affect assays for plasma cytokines and soluble activation markers. *Clinical & Diagnostic Laboratory Immunology* **6**(1): 89–95.

Ballantyne CM, Hoogeveen RC, Bang H et al. 2004. Lipoprotein-associated phospholipase A2, high-sensitivity C-reactive protein, and risk for incident coronary heart disease in middle-aged men and women in the Atherosclerosis Risk in Communities (ARIC) study. [see comment]. *Circulation* **109**(7): 837–842.

Barlow W. 1994. Robust variance estimation for the case-cohort studies. *Biometrics* **50**: 1064–1072.

Barlow WE, Ichikawa L, Rosner D et al. 1999. Analysis of case-cohort designs. *Journal of Clinical Epidemiology* **52**(12): 1165–1172.

Bellete B, Flori P, Hafid J et al. 2003. Influence of the quantity of nonspecific DNA and repeated freezing and thawing of samples on the quantification of DNA by the Light Cycler. *Journal of Microbiological Methods* **55**(1): 213–219.

Bernstein JL, Langholz B, Haile RW et al. 2004. Study design: evaluating gene-environment interactions in the etiology of breast cancer – the WECARE study. *Breast Cancer Res* **6**(3): R199–214.

Bilodeau JF, Chatterjee S, Sirard MA et al. 2000. Levels of antioxidant defenses are decreased in bovine spermatozoa after a cycle of freezing and thawing. *Molecular Reproduction & Development* **55**(3): 282–288.

Bolelli G, Muti P, Micheli A et al. 1995. Validity for epidemiological studies of long-term cryoconservation of steroid and protein hormones in serum and plasma. *Cancer Epidemiology, Biomarkers & Prevention* **4**(5): 509–513.

Breslow N, Lubin J, Marek P et al. 1983. Multiplicative models and cohort analysis. *J Am Stat Assoc* **78**: 1–12.

Brey RL, Cote SA, McGlasson DL et al. 1994. Effects of repeated freeze-thaw cycles on anticardiolipin antibody immunoreactivity. *American Journal of Clinical Pathology* **102**(5): 586–588.

Cologne JB. 1997. Counterintuitive matching. *Epidemiology* **8**(3): 227–229.

Dirx MJ, Voorrips LE, Goldbohm RA et al. 2001. Baseline recreational physical activity, history of sports participation, and postmenopausal breast carcinoma risk in the Netherlands Cohort Study. *Cancer* **92**(6): 1638–1649.

Evans RW, Sankey SS, Hauth BA et al. 1996. Effect of sample storage on quantitation of lipoprotein(a) by an enzyme-linked immunosorbent assay. *Lipids* **31**(11): 1197–1203.

Falk RT, Gail MH, Fears TR et al. 1999. Reproducibility and validity of radioimmunoassays for urinary hormones and metabolites in pre- and postmenopausal women. *Cancer Epidemiology, Biomarkers & Prevention* **8**(6): 567–577.

Falk RT, Rossi SC, Fears TR et al. 2000. A new ELISA kit for measuring urinary 2-hydroxyestrone, 16alpha-hydroxyestrone, and their ratio: reproducibility, validity, and assay performance after freeze-thaw cycling and preservation by boric acid. *Cancer Epidemiology, Biomarkers & Prevention* **9**(1): 81–87.

Fergenbaum JH, Garcia-Closas M, Hewitt SM et al. 2004. Loss of antigenicity in stored sections of breast cancer tissue microarrays. *Cancer Epidemiol Biomarkers and Prev* **13**(4): 667–672.

Greenland S. 1987. Interpretation and choice of effect measures in epidemiologic analyses. *Am J Epidemiol* **125**: 761–768.

Groschl M, Wagner R, Rauh M *et al.* 2001. Stability of salivary steroids: the influences of storage, food and dental care. *Steroids* **66**(10): 737–741.

Halfon P, Khiri H, Gerolami V *et al.* 1996. Impact of various handling and storage conditions on quantitative detection of hepatitis C virus RNA. *Journal of Hepatology* **25**(3): 307–311.

Helsing KJ, Hoffman SC, Comstock GW. 2000. Temperature variations in chest-type mechanical freezers. *Clinical Chemistry* **46**(11): 1861.

Hirschhorn JN, Lohmueller K, Byrne E *et al.* 2002. A comprehensive review of genetic association studies. *Genetics in Medicine* **4**(2): 45–61.

Holten-Andersen MN, Schrohl AS, Brunner N *et al.* 2003. Evaluation of sample handling in relation to levels of tissue inhibitor of metalloproteinases-1 measured in blood by immunoassay. *International Journal of Biological Markers* **18**(3): 170–176.

Hunter D. 1997. Methodological issues in the use of biological markers in cancer epidemiology: cohort studies. In: Toniolo P, Boffetta P, Shuker D, Rothman N, Hulka B, Pearce N, editors. Application of Biomarkers in Cancer Epidemiology. *Lyon: International Agency for Research on Cancer.* p 39–42.

Ioannidis JP, Trikalinos TA, Ntzani EE *et al.* 2003. Genetic associations in large versus small studies: an empirical assessment. *Lancet* **361**(9357): 567–571.

Kronenberg F, Lobentanz EM, Konig P *et al.* 1994. Effect of sample storage on the measurement of lipoprotein[a], apolipoproteins B and A-IV, total and high density lipoprotein cholesterol and triglycerides. *Journal of Lipid Research* **35**(7): 1318–1328.

Lagging LM, Garcia CE, Westin J *et al.* 2002. Comparison of serum hepatitis C virus RNA and core antigen concentrations and determination of whether levels are associated with liver histology or affected by specimen storage time. *Journal of Clinical Microbiology* **40**(11): 4224–4229.

Lahiri DK, Schnabel B. 1993. DNA isolation by a rapid method from human blood samples: effects of MgCl2, EDTA, storage time, and temperature on DNA yield and quality. *Biochemical Genetics* **31**(7–8): 321–328.

Langholz B, Clayton D. 1994. Sampling strategies in nested case-control studies. *Environ Health Perspect 102 Suppl* **8**: 47–51.

Langholz B, Thomas DC. 1990. Nested case-control and case-cohort methods of sampling from a cohort: a critical comparison. *American Journal of Epidemiology* **131**(1): 169–176.

Lathey JL, Marschner IC, Kabat B *et al.* 1997. Deterioration of detectable human immunodeficiency virus serum p24 antigen in samples stored for batch testing. *Journal of Clinical Microbiology* **35**(3): 631–635.

McLeay WR, Horsfall DJ, Seshadri R *et al.* 1992. Epidermal growth factor receptor in breast cancer: storage conditions affecting measurement, and relationship to steroid receptors. *Breast Cancer Research & Treatment* **22**(2): 141–151.

Miettinen O. 1982. Design options in epidemiologic research: An update. *Scand J Work Environ Health* **8**(Suppl 1): 7–14.

Ng V, Koh D, Fu Q *et al.* 2003. Effects of storage time on stability of salivary immunoglobin A and lysozyme. *Clin Chim Acta* **338**(1–2): 131–134.

Peluso M, Munnia A, Hoek G *et al.* 2005. DNA Adducts and Lung Cancer Risk: A Prospective Study. *Cancer Res* **65**(17): 8042–8048.

Perera FP, Mooney LA, Stampfer M *et al.* 2002. Associations between carcinogen-DNA damage, glutathione S-transferase genotypes, and risk of lung cancer in the prospective Physicians' Health Cohort Study. *Carcinogenesis* **23**(10): 1641–1646.

Phillips GB, Yano K, Stemmermann GN. 1988. Serum sex hormone levels and myocardial infarction in the Honolulu Heart Program. Pitfalls in prospective studies on sex hormones. *Journal of Clinical Epidemiology* **41**(12): 1151–1156.

Pirkle JL, Brody DJ, Gunter EW, *et al.* 1994. The decline in blood lead levels in the United States. *The National Health and Nutrition Examination Surveys (NHANES)* [see comment]. JAMA **272**(4): 284–291.

Prentice R. 1986. A case-cohort design for epidemiologic cohort studies and disease prevention trials. *Biometrika* **73**: 1–11.

Riboli E, Kaaks R. 1997. The EPIC Project: Rationale and Study Design. European Prospective Investigation into Cancer and Nutrition. *Int J Epidemiol* **26**(suppl. 1): s6–s14.

Risio M, De Rosa G, Sarotto I *et al.* 2003. HER2 testing in gastric cancer: molecular morphology and storage time-related changes in archival samples. *International Journal of Oncology* **23**(5): 1381–1387.

Rothman K. 1986. Modern Epidemiology. Boston, MA: Little, Brown and Company.

Rundle A. 2000. A molecular epidemiologic case-control study of breast cancer [Doctoral]. *New York: Mailman School of Public Health.* p. 152.

Rundle AG, Vineis P, Ahsan H. 2005. Design Options for Molecular Epidemiology Research within Cohort Studies. *Cancer Epidemiol Biomarkers Prev* **14**(8): 1899–1907.

Rybicki BA, Rundle A, Savera AT *et al*. 2004. Polycyclic aromatic hydrocarbon-DNA adducts in prostate cancer. *Cancer Research* **64**(24): 8854–8859.

Salomaa V, Matei C, Aleksic N *et al*. 1999. Soluble thrombomodulin as a predictor of incident coronary heart disease and symptomless carotid artery atherosclerosis in the Atherosclerosis Risk in Communities (ARIC) Study: a case-cohort study. [see comment]. *Lancet* **353**(9166): 1729–1734.

Sato T. 1994. Risk ratio estimation in case-cohort studies. *Environmental Health Perspectives 102 Suppl* **8**: 53–56.

Schouten EG, Dekker JM, Kok FJ *et al*. 1993. Risk ratio and rate ratio estimation in case-cohort designs: hypertension and cardiovascular mortality. [see comment]. *Statistics in Medicine* **12**(18): 1733–1745.

Schulte P, Perera F. 1993. Molecular epidemiology: principles and practices. San Diego: *Academic Press*.

Self S, Prentice R. 1988. Asymptotic distribution theory and efficiency results for case-cohort studies. *Ann Stat* **16**: 64–81.

Sgoutas DS, Tuten T. 1992. Effect of freezing and thawing of serum on the immunoassay of lipoprotein(a). *Clinical Chemistry* **38**(9): 1873–1877.

Shih WJ, Bachorik PS, Haga JA *et al*. 2000. Estimating the long-term effects of storage at -70 degrees C on cholesterol, triglyceride, and HDL-cholesterol measurements in stored sera. *Clinical Chemistry* **46**(3): 351–364.

Steenland K, Deddens JA. 1997. Increased precision using countermatching in nested case-control studies. *Epidemiology* **8**(3): 238–242.

Su SC, Garbers S, Rieper TD *et al*. 1996. Temperature variations in upright mechanical freezers. *Cancer Epidemiology, Biomarkers & Prevention* **5**(2): 139–140.

Suissa S, Edwardes MD, Boivin JF. 1998. External comparisons from nested case-control designs. *Epidemiology* **9**(1): 72–78.

Taioli E, Kinney P, Zhitkovich A *et al*. 1994. Application of reliability models to studies of biomarker validation. *Environmental Health Perspectives* **102**(3): 306–309.

Tang D, Phillips D, Stampfer M *et al*. 2001. Association between carcinogen-DNA adducts in white blood cells and lung cancer risk in the physicians health study. *Cancer Res* **61**(18): 6708–6712.

Tang D, Rundle A, Warburton D *et al*. 1998. Associations between both genetic and environmental biomarkers and lung cancer: evidence of a greater risk of lung cancer in women smokers. *Carcinogenesis*. **19**(11): 1949–1953.

Tworoger SS, Yasui Y, Chang L *et al*. 2004. Specimen allocation in longitudinal biomarker studies: controlling subject-specific effects by design. Cancer Epidemiology, *Biomarkers & Prevention* **13**(7): 1257–1260.

van den Berg L, Rose D. 1959. Effect of freezing on the pH and composition of sodium and potassium phosphate solutions: the reciprocal system KH_2PO_4-Na_2HPO_4-H_2O. *Arch Biochem Biophys* **81**: 319–329.

Wacholder S. 1991. Practical considerations in choosing between the case-cohort and nested case-control designs. *Epidemiology* **2**(2): 155–158.

Wacholder S, Gail M, Pee D. 1991. Selecting an efficient design for assessing exposure-disease relationships in an assembled cohort. *Biometrics* **47**(1): 63–76.

Wacholder S, Silverman DT, McLaughlin JK *et al*. 1992. Selection of controls in case-control studies. III. Design options. *American Journal of Epidemiology* **135**(9): 1042–1050.

Williams-Smith DL, Bray RC, Barber MJ *et al*. 1977. Changes in apparent pH on freezing aqueous buffer solutions and their relevance to biochemical electron-paramagnetic-resonance spectroscopy. *Biochemical Journal* **167**(3): 593–600.

Woodrum D, French C, Shamel LB. 1996. Stability of free prostate-specific antigen in serum samples under a variety of sample collection and sample storage conditions. *Urology* **48**(6A Suppl): 33–39.

Zeka A, Eisen EA, Kriebel D *et al*. 2004. Risk of upper aerodigestive tract cancers in a case-cohort study of autoworkers exposed to metalworking fluids. *Occupational & Environmental Medicine* **61**(5): 426–431.

4

Family Studies, Haplotypes and Gene Association Studies

Jennifer H. Barrett, D. Timothy Bishop and Mark M. Iles

University of Leeds, UK

4.1. INTRODUCTION

While there are many similarities between genetic epidemiology and environmental epidemiology, there are several striking differences. Environmental epidemiology has used the case–control design extensively to quantify the association between environmental exposure and risk of disease. Genetic epidemiology has to date focused on families, taking advantage of the Mendelian laws of inheritance (Khoury *et al.* 1993; Thomas 2004; Morton 1978). The collection of information on sets of relatives (families) allows studies to quantify the familial risk as a measure of the extent that important risk factors aggregate in families, to provide some insight into the relative contributions of genes and environment and to establish whether genes have a potential role in disease susceptibility (Susser 1985).

Any analysis of families should take account of important environmental factors to avoid over-estimating the contribution of genes to disease aetiology. In exceptional circumstances, strong environmental factors that aggregate in families may explain an increased risk to close relatives, and in general environmental factors common to families increase the correlation between relatives (Hopper and Carlin 1992). The clearest way of identifying important environmental factors is through standard epidemiological studies (e.g. case–control or cohort studies), since correlations between related individuals for exposure to the environmental risk factor will reduce the power to identify the environmental

risk factor from family studies. While these correlations would have to be large before there would be a substantial decrease in power, there are other influences that remain unmeasured which will also affect the power.

In this chapter, the various possible sampling units are discussed, along with the types of hypothesis that can be examined using each structure (whether family-based or not). This will also serve as an indication of the research process that would be applied to the study of a disease for which there was little previous information about the role of genetic factors.

4.2. FAMILY STUDIES

Is there an increased risk of disease in relatives of cases?

For a disease which is partially determined by genes, relatives of a case will be at increased risk because they have genetic variants in common with the case (King *et al.* 1984). As a higher proportion of cases will carry the disease-related alleles than the general population, their relatives are also more likely to carry the disease alleles than members of the general population. The level of increased risk will depend upon the strength of the genetic relationship and the specific role of genetic factors in determining risk (Easton and Peto 1990). In general, more distant relatives will have lower risks than close relatives, since closer relatives are more likely to share genetic risk factors with the case ('proband') than are more distant relatives. The simplest form of family study is to examine the frequency of disease among the relatives of probands who are systematically identified, such as through a population-based register (Goldgar *et al.* 1994). Usually attention is restricted to first-degree relatives (parents, siblings and offspring). The relatives define a cohort exposed by their genetic relatedness to the proband; their frequency of disease is compared with that of the general population. There is, however, one twist to the usual epidemiological approach. Classically, the exposure must of course precede the disease; in family studies the exposure should be thought of as being exposure to the genes of the proband and not to the disease in the proband, since the relative will always have some genotypes in common with the proband, but the disease onset could occur in the relative temporally prior to that in the proband. Comparison with the risk of disease in the relatives of controls (unaffected persons also identified systematically) or with incidence rates for the whole population gives a measure of the strength of association of family history with disease occurrence, using standard statistical methods. At this stage, it is also useful to stratify by the type of genetic relationship (e.g. parents vs. siblings) and look for differences in risk between these two groups or to examine the risks by gender of the case (Risch 1990). It is also useful to stratify probands by age group and to determine whether relatives of younger probands have higher risks compared to the general population than do relatives of older probands, indicating that the familial effects are stronger at young ages (Risch 2001). Differences may be due to differences in the importance of genetic factors according to age. For instance, in female breast cancer, the first-degree relatives of early onset cases have a three-fold increased risk of breast cancer, while the relatives of later onset cases have only a 1.5-fold increased risk. The statistical analyses required here are standard to epidemiology.

Is the familial aggregation due to genes or environment?

Disease may aggregate in families because of shared genetic factors or shared environment or both. In the initial stages of studying a disease showing some familial aggregation, attempts to dissect the increased risk to relatives into genetic and environmental components are warranted, although usually this is logistically complex. One direct way of addressing this issue is via the twin study. Twins are divided into two groups: monozygous (MZ) twins, who arise from the splitting of one fertilized egg or zygote and so should be genetically identical, and dizygous twins (DZ), who arise from two ova and two sperm and so are not genetically identical and in fact are related as two siblings. By definition, MZ twin pairs must be of the same sex as each other, while approximately half of the DZ pairs will be of the same sex. Unless gender plays no role in disease expression, the main interest will be in comparing MZ with same-sex (SS) DZ twin pairs (Motulsky and Vogel 1997).

Since the MZ twins are genetically identical to each other, any pair that is discordant for a disease shows that genes are not sufficient to determine disease expression, i.e. environmental factors or chance play some role in determining risk or expression (Khoury *et al.* 1993). The similarity between the two twins of a pair can be measured by an intra-class correlation coefficient for a quantitative trait or by a concordance rate for disease status. The concordance rate is simply the proportion of twin pairs in which the co-twin of an affected twin is also affected. The comparison of most interest is the concordance rate in MZ twin pairs with the concordance rate in SS DZ twin pairs. If genes play no role in determining risk, then the two concordance rates should be identical, while if genes are important the MZ concordance rate should be higher than the DZ concordance rate (Hopper *et al.* 2005; Risch 2001).

Twin studies are designed to compare the effect of having different genetic make-up (as for DZ pairs) with that of having the same genetic make-up (as for MZ pairs) while both twins of a pair are brought up in the same household. Adoption studies are the reverse in the sense that they are designed to compare the effect of differing household environment on individuals with similar genetic backgrounds (Rice and Borecki 2001). Adoption studies are performed by comparing the frequency of disease among the offspring of an affected parent whose children have been adopted out with the frequency in children who have unaffected biological parents but an affected adoptive parent. If the first group have higher frequencies than the second, then this suggests that genetic factors predominate, while if disease is more frequent in the second group, then environmental factors are more important. Such approaches have been much more widely used in behavioural studies than in, for instance, cancer studies.

The 'best' evidence for a genetic effect would come from either a twin study or an adoption study, but both of these are major enterprises, and in many circumstances the number of individuals eligible for the study will be extremely limited. For the twin study, a set of twins is required where at least one has been diagnosed with the disease of interest. Even for a reasonably common disease, the combined frequency of the disease and twinning is sufficiently low to make the study infeasible unless dealing with a huge population base.

What is the genetic mechanism?

The studies described to date have only been able to show that genetic factors appear to be important. To make further progress, this observation must be made more precise by identifying how the genetic factors are transmitted (Elston and Spence 2006; Khoury *et al.* 1993; King *et al.* 1984). The information of interest includes the number of loci involved, the probability that specific combinations of alleles lead to disease and the frequency of susceptible genotypes. With this knowledge, predictions could be made about the risk of subsequent disease in currently unaffected individuals and progress can be made in understanding the disease biology. The only way to understand the details of the clustering is to study families and to examine the pattern of disease within the families, i.e. the specific genetic relationships of affected individuals. Segregation analysis is an aid to describing the characteristics of this clustering. Families can be characterized by the affection status of the parents – both affected, exactly one affected or neither affected – and the frequency with which their offspring are affected can be examined. In the

simplest cases, such as for a rare, high-penetrance, dominant disease, families will rarely be found where both parents are affected, but when one parent is affected, on average half of the children will be affected. For a rare recessive disease, in most families neither parent will be affected and on average one-quarter of the offspring will be affected. The same approach suffices for the complex diseases which are the focus of this chapter, with the segregation patterns essentially characterizing the family aggregation but being less interpretable in terms of precise genetic mechanism. The characteristics of the clustering of disease within families are dependent on the specific genetic mechanism. Resolving the mechanism requires the fitting of probabilistic models representing genetic effects to the information collected on families (Elston and Spence 2006). These models define the mode of inheritance of the disease, including the number of loci involved in determining risk, the frequency of individuals at risk in the general population and the importance of unmeasured factors in disease expression. It is of particular interest to distinguish between the familial aggregation of disease being explained by the effect of a single gene or by the actions of many loci each having small effect (see e.g. Antoniou and Easton 2006).

Where is the gene?

If there is a gene which partially determines the risk for disease, then the locus for this gene must be on a specific chromosome (Morton 1955). If the location can be defined at least approximately, then estimates of the risk of disease in specific family members can be improved, using the information on the location, and molecular biology studies can be initiated aimed at identifying the specific gene defect. In many situations, far more information about genetic factors can be obtained by studying larger pedigrees instead of nuclear families (Cannon-Albright *et al.* 1991). Simply, the more related individuals that are studied, the more accurately the location of the gene(s) should be shown.

There are two basic approaches to the identification of the underlying gene(s): linkage and association. After a brief discussion of linkage analysis, the remainder of this chapter focuses on genetic association studies.

Linkage analysis

If a single gene or multiple genes affect risk of disease, how can such genes be identified? Three facts can be exploited to address this:

1. The rules of genetic inheritance are well known and define how similar, on average, the genotypes of close relatives are to each other on the basis of their genetic relationship.
2. The closer that two loci are located to each other, the less frequently will there be recombination between them during meiosis, i.e. the more closely the loci are linked to each other.
3. Closely related affected persons will tend to share the disease-associated variants more often than would be expected according to (1) (Vogel and Motulsky 1979).

One of the earliest products of the Human Genome Project was the identification of polymorphic sites within the genome for which assays can be produced for efficient identification of a person's genetic variants ('marker' loci; Donis-Keller et al. 1987; Weissenbach et al. 1992). If conclusions about a genetic effect are correct, then it should be possible to find one or more marker loci located close to the gene. When a disease gene is linked to a marker, then other markers can be found that are also close to this gene. The aim is to find markers that flank the disease gene, i.e. markers that define the region in which the disease gene must lie.

Figure 4.1 shows a pedigree containing two siblings with the same disease and the results of genotyping a set of DNA markers on an autosomal chromosome while attempting to find the gene for this disease. The figure shows the genotyping results on a trace as taken from the software on equipment measuring the DNA content (in this case, the genetic variant is scored by the length of the DNA fragments, measured by the highest peaks on the trace) and the tabulated results. Inspection of the pedigree shows that person 102 has inherited the 161 base-pair (bp) allele for marker D9S1749 from her mother and

a 187 bp allele for D9S736. Her brother, 103, has inherited the 141 bp allele for D9S1749 from his mother and a 187 bp allele for D9S736. For this family, then, the two affected siblings have inherited discrepant alleles from the mother and from the father but the information on D9S736 is less clear. Analysis incorporating the known proximity of D9S1749 and D9S736 will assist in the interpretation of the D9S736 inheritance. The interpretation of this information in terms of disease mapping depends upon the accumulated evidence from many such families and from any insight into the mode of disease risk inheritance. For instance, if the disease in Figure 4.1 were a rare recessive disease, then this family would give evidence against a gene in the region of these markers because of the discrepant inheritance for the maternal alleles at D9S1749.

For complex diseases, studies are often based on affected relative (usually sibling) pairs. Consider an affected sibling pair (as in Figure 4.1) where the parental genotypes at an autosomal locus are 1–2 (father) and 3–4 (mother). Each child would have one of the genotypes 1–3, 1–4, 2–3 or 2–4 and, under Mendelian inheritance, each possibility would be equally likely. Consider now two children: they could have exactly the same genotype (e.g. 1–3 and 1–3), entirely different genotypes (e.g. 1–3 and 2–4) or have one allele is common (e.g. 1–3 and 1–4). Under Mendelian inheritance rules, the chance that they have exactly the same genotypes is 0.25, the chance of entirely different genotypes is 0.25 and the chance of having one allele in common is 0.5 (as the shared allele could be the paternal or maternal allele). Overall, over a large number of sibling pairs, on average 25% should be genetically identical at this locus, 50% should have one allele in common, and 25% should be genetically discordant. These proportions are expected simply on the basis of the genetic relationship of the sibling pairs (notice that different alleles are likely to be segregating in each family but each family can be scored in this way). If this locus happened to be in the proximity of a disease-associated gene and only affected sibling pairs were considered, then more than 25% of the sibling pairs would

Person	D9S1749	D9S736
100	141-161	187-187
102	151-161	187-189
103	141-167	187-187

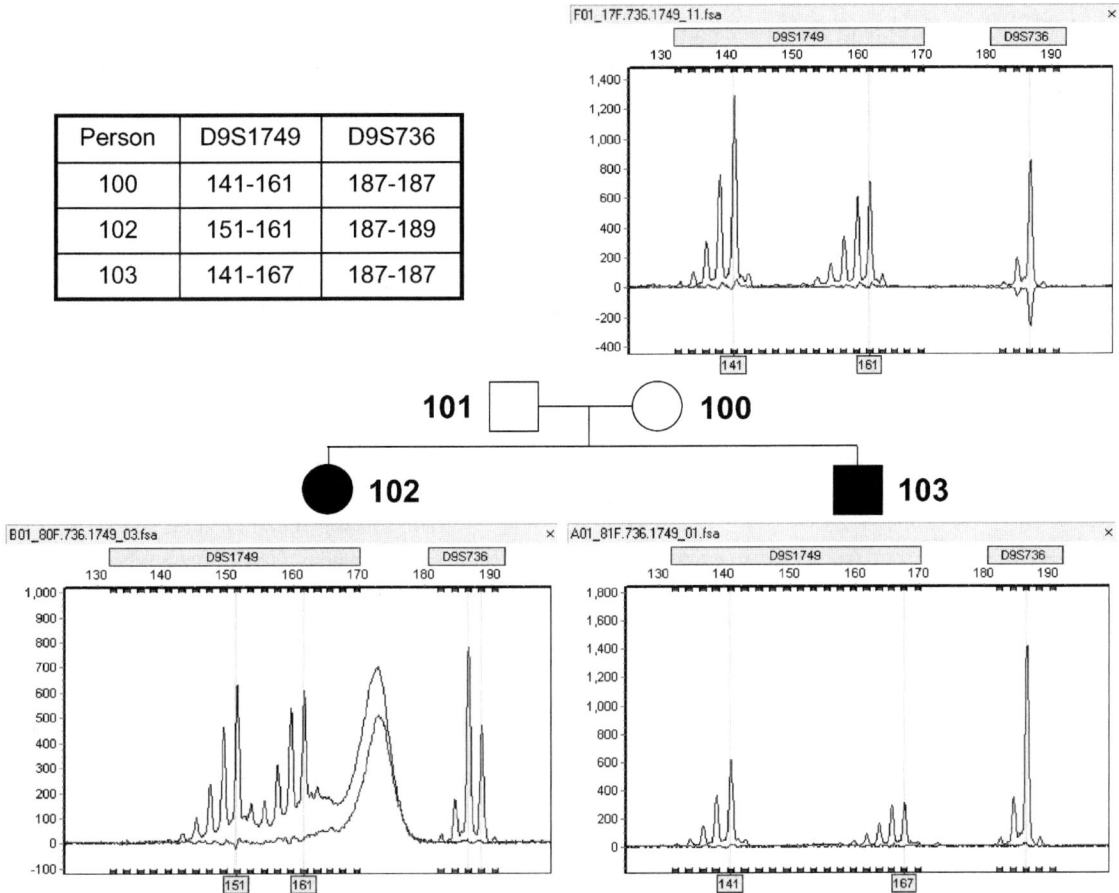

Figure 4.1 The results from a four-person pedigree with genotyping results for the mother (100), and the two offspring (102 and 103). The three genotyping panels show the laboratory-derived results, in which the length of the alleles are measured for two DNA markers (D9S1749 and D9S736; both markers are on chromosome 9). The primers have been chosen so that the two markers can be assessed concurrently. For person 100, the allele lengths are 141 and 161 base pairs (bp), indicated by the peaks of the left-hand patterns. The jagged nature of the trace is due to the nature of the polymorphism and the type of assay. The table at top left shows the results obtained for both markers for all three persons, indicating for the D9S1749 marker that the mother passed on the 161 allele to her daughter and the 141 allele to the son, meaning that the untyped father must have had the 151–167 genotype

be expected to be genetically identical and possibly more than 50% would share one allele, or, equivalently, the discordant pairs should be less than 25% of the pairs. Statistical tests based on these observations can be used to determine whether the marker is in the vicinity of a disease-associated gene (Risch 1990b; Suarez *et al.* 1978). A more detailed introduction to genetic linkage analysis can be found elsewhere (Teare and Barrett 2005).

Example

Box 4.1 shows an example of this process for the disease of melanoma. The only deviation from the series of studies described to date is the absence of studies attempting to distinguish between thecontribution of genes and environment, but for melanoma the relative rareness of the disease makes such studies infeasible. Box 4.2 indicates some of the software available for genetic analysis.

Box 4.1   **An example of the investigation of the contribution of genes and environment to disease showing the sequence of questions and approaches to addressing those questions for melanoma**

Question	Study design to address question	Conclusion
Does the disease aggregate in families?	Are close relatives of cases more likely to develop the same disease than the general population?	
	Within a Utah population database, disease risk among first-degree relatives of cases compared to disease risk among first-degree relatives of controls	First-degree relatives have 2.1 times the risk of melanoma compared with persons without a family history (Goldgar et at. 1994). Disease aggregates in families
What is the pattern of disease within families?	1. Are there particular extended families with many cases of melanoma? How is disease risk inherited within these families?	1. Yes, there are families worldwide (but these are rare overall) which show many cases of diseases, far beyond chance. The pedigrees show a pattern of inheritance consistent with autosomal dominant disease
	2. Outside these extended families, is there a pattern of disease risk definable from the families?	2. In most families with melanoma there are only two cases and there is no discernable pattern of risk. Not as likely to be due to a single gene but could involve a combination of risk factors
What genes are important in the dominant families?	Use linkage mapping to identify location of gene, using samples from these highly loaded extended families	Linkage mapping showed (primarily using one family with 27 cases of melanoma) that a gene was located on chromosome 9 (Cannon-Albright et al. 1994) and the gene was identified to be CDKN2A, involved in cell cycle regulation (Kamb et al. 1994)
	By examining many such families, we can see how many families have disease due to this gene	About 50% of the extended families have disease due to this gene. Another gene has been mapped to chromosome 1p22 but not identified (Gillanders et al. 2003)
What are the risk factors for melanoma in the general population?	Epidemiological case–control studies examining putative risk factors	Aetiological epidemiological studies show a number of risk factors, including large numbers of skin naevi, red hair, fair skin, having a pattern of intermittent sun exposure and perhaps sunburns (Gandini et al. 2005a, 2005b; 2005c)
Are there any genes which might have an effect on melanoma risk overall?	Many of the risk factors have a genetic component, in particular red hair and fair skin. One of these genes is MC1R (melanocortin-1 receptor). Examine this gene in a case–control study	Variants in this gene that are associated with red hair are found to be at increased frequency among cases as compared to controls (Sturm 2002)
How do the genetic risk factors and risk factors such as sun exposure relate to each other?	Further case–control studies looking at sun exposure and the genetic variants, e.g. in MC1R, to look at the joint effect of having one or more risk factor	Such studies are still under way

---

Box 4.2   Software for statistical analysis

Many analyses of data from genetic epidemiology studies can be carried out in standard statistical software, e.g. Stata (StataCorp. 2005), SAS (SAS Institute Inc. 2007) or the freely available R (R Development Core Team 2006). This applies particularly to population-based association studies, where methods such as (conditional) logistic regression are used. For more specialized genetic analyses, the field of statistical genetics is fortunate in having a tradition whereby many research groups make software freely available to other researchers. Many such software packages have been extremely widely used and can be used with as much confidence as commercial software. In addition, many specific programs are available that run in general software, such as Stata or R, including, for example, routines for testing HWE (Cleves 1999) or for estimation of haplotype frequencies (Mander 2001).

There are too many examples of good statistical genetic software to offer anything like a comprehensive list, but we include here a few examples:

Linkage analysis

Linkage and the faster version *Fastlink* (linkage analysis on pedigrees)
Genehunter and its extensions (includes non-parametric linkage analysis and many other features)
Merlin (Multipoint Engine for Rapid Likelihood Inference; especially appropriate for dense marker maps)
SOLAR (Sequential Oligogenic Linkage Analysis Routines; genetic variance components analysis, useful for linkage analysis of quantitative traits)

Association studies

Genepop (a population genetics software package; includes test for HWE)
SNPHAP (estimating frequencies of haplotypes from unphased genotype data)
Phase (reconstructs haplotypes from unphased genotype data)

Family-based association studies

FBAT and *PBAT* (Family-based Association Test: tests for association/linkage between disease phenotypes and haplotypes by utilizing family-based controls)
Transmit (tests for association between genetic marker and disease by examining the transmission of markers from parents to affected offspring)
A useful resource for genetic analysis software is the web-page http://linkage.rockefeller.edu/soft/, which (at December 2006) lists approximately 400 programs. All the programs listed above are documented on the list, with references to the papers describing the methods and links to the software.

4.3. GENETIC ASSOCIATION STUDIES

Genetic case–control studies

A common design for a genetic association study is the classical case–control design, where rates of exposure to potential risk factors in incident cases are compared with rates in controls (who do not have the disease) selected from the same population (Breslow and Day 1980). To investigate genetic association the potential risk factors are particular genotypes, rather than environmental factors such as meat consumption or sun exposure.

In the simplest case, to investigate association between a particular single nucleotide polymorphism (SNP) and disease, the frequencies of the three genotypes (AA, AT and TT in the case of an A-to-T SNP) can be compared between the cases and controls. This leads to a 3×2 contingency table, and association can be tested for using a chi-squared (χ^2) test for independence. Equivalently, logistic regression of case–control status on the three-level factor can be used, which has the advantages that odds ratios for disease are estimated for people with each alternate genotype (AT or TT) vs. the baseline group (AA), and that covariates can be included in the model. An alternative method of analysis sometimes used is to compare frequencies of the T allele in cases and controls. However, this assumes that the two alleles each individual contributes to the analysis are independent observations, i.e. assumes Hardy–Weinberg equilibrium (HWE) in both cases and controls, and so leads to incorrect test statistics when this assumption is not true (Sasieni 1997).

Power can potentially be increased by fitting a more restrictive genetic model to the data. For example, assuming a recessive mode of inheritance, the frequency of the TT genotype would be compared with the AA/AT frequency. Assuming an additive mode of inheritance, the Cochran–Armitage trend test can be applied, or equivalently logistic regression can be used, regressing case–control status on the number of T alleles (0, 1 or 2). These tests will also be valid under alternative modes of inheritance, but the power will be greater when the correct model is assumed.

To test for disease association with a more polymorphic genetic marker (or set of haplotypes, as discussed below) these methods are still generally applicable. However a marker with n alleles (or a region with n haplotypes) gives rise to $n(n + 1)/2$ distinct genotypes, so that the general test based on a contingency table of genotype by disease will usually have very low power. In addition, the data are likely to be sparse, in which case the significance of the resulting statistic must be evaluated by calculating the empirical distribution under the null hypothesis of no association, rather than relying on the (asymptotic) χ^2 distribution. For highly polymorphic loci, it is more common to compare allele or haplotype frequencies, despite the above caveat about HWE.

Differences in genotype frequencies between cases and controls can either be due to causation, i.e. the tested polymorphism itself increases disease risk (*direct association*), or to correlation between this and another causative polymorphism, i.e linkage disequilibrium (LD) (*indirect association*). Either of these is of great interest, since LD extends only over small regions, so that genuine association of this kind implies a very small region of the genome in disease susceptibility. In practice, there are other possible reasons for an apparent association between a SNP and disease arising from this type of study, including statistical type I error (especially from under-powered samples) and confounding.

Bias and confounding

Although the case–control design is simple in principle, there is much debate within epidemiology on the practical approaches to selecting cases to achieve as closely as possible a random sample from the population from which cases are drawn. Most methods of selecting controls (e.g. via family doctors, hospital outpatient clinics or even the electoral register) fail to perfectly capture the whole population in the sampling frame. A potentially greater problem is participation bias (for a fuller discussion of bias, refer to Chapter 2). Participation rates are sometimes low, and this can introduce differences between the case and control groups that are not due to disease. For example, differences in participation rates according to socio-economic status can lead to differences in frequency of the exposure of interest between cases and controls, leading to an apparent, but possibly spurious, association with disease (Mezei and Kheifets 2006).

One approach to control for confounders is to select a *matched* control (or set of controls) for each case. For example, age and sex are often potential confounding factors, being related to both disease risk and to the exposure under study. To control for sex and age, the matched control is selected to be of the same sex and within the same age-band as the case. One result of matching on confounders is that cases and controls in the study will be more similar in their exposure frequencies.

Thus 'over-matching' on too many factors can lead to cases and controls being virtually identical in their exposure frequencies and hence to an uninformative study. For the same reason, it is important that matching is taken into account in the analysis, generally by using conditional logistic regression, since otherwise effect estimates will tend to be biased towards the null.

These issues are of similar but not identical relevance in genetic association studies. Genetic association studies often of course include measurement of environmental as well as genetic factors, as interest may be in their joint effect (gene–environment interaction), and as such the same concerns apply. Focusing purely on genetic factors, some sources of bias (e.g. recall bias, where cases may differentially recall or report previous exposure compared with controls, influenced by their current disease) do not apply, and participation bias may be less likely to lead to confounding. However, in genetic studies there is concern about confounding that may arise through population stratification.

Genotype frequencies differ between human populations around the world. Even within one geographical region and broadly defined ethnic group, frequencies have been found to show variation. There have been various reports of trends in allele frequencies moving from Northern to Southern Europe (e.g. Panza *et al.* 2003) and even of differences within Finland, which is generally regarded as a highly homogeneous population (Ilonen *et al.* 2000). Superficially homogeneous populations may contain hidden strata, corresponding to groups with distinct ancestry. If disease rates also differ between these strata, the population stratification could lead to false-positive disease–gene associations if not properly controlled for.

Family-based study designs

Concern about population stratification has been one motivation for the development of family-based study designs. By choosing unaffected close relatives as controls in a matched design, it is possible to ensure that any observed association (if not a type I error) is due to direct or indirect association, as defined above.

A simple family-based design is to recruit for each case at least one sibling unaffected by the disease. Numerous methods of analysis have been proposed that take into account the matching between the affected (case) and unaffected (control) siblings, by conditioning on the genotypes of the sibship as a whole (Spielman and Ewens 1998; Siegmund *et al.* 2000). Disadvantages of the affected–unaffected sibling design include: the fact that a substantial proportion of cases may have no available unaffected sibling willing to participate; the difficulties of matching for sex and age; and the reduction in power to detect genetic association compared with using population-based controls (see Table 4.1).

For early-onset disease, an alternative approach is to use parents of cases as controls (irrespective of parental disease status). Here the two alleles not transmitted to the affected offspring from the parents are treated as the control genotype or 'pseudo-sib', and a matched analysis is carried out conditioning on the parental genotypes. Many variations on the method of analysis have been proposed, including McNemar's test for each parent–child transmission (Spielman *et al.* 1993) and extensions to cope with multi-allelic markers (Sham and Curtis 1995) or missing parental data (Clayton 1999).

Recently a more general framework has been suggested for the analysis of family-based genetic association studies. Whatever the family structure, it is possible to construct a valid test for association by conditioning on the genotypes of the founders in the family. Where these genotypes are not known, the analysis is instead conditioned on a sufficient statistic for the founder genotypes (Rabinowitz and Laird 2000). This method is valid (although not necessarily powerful) for general pedigree structures and patterns of missing data, and has been generalized to allow analysis of haplotypes (see below) and quantitative traits (reviewed in Laird and Lange 2006).

Haplotypes

Although genotyping can identify the alleles at several loci in a region, it does not reveal which alleles at successive loci lie on the same chromosome.

Table 4.1 Some advantages and disadvantages of population-based and family-based association studies

	Population-based	Family-based
Power to detect association	Good power compared with family designs	Power per subject genotyped generally lower than population-based methods
Population sub-structure	Potential for confounding by population stratification; methods to correct for this require large study and extensive genotyping	Robust to population stratification
Recruitment of controls	Large pool from which to select controls, but care must be taken to choose appropriately	Relative controls may not be available. Case–parent designs are particularly suitable for study of childhood diseases
Use of sample collections	In some instances a common control group can be used for comparison with different disease groups	May be able to make use of existing family collections from linkage studies
Genotype error	Limited information to detect genotype error. Non-differential genotype error leads to loss of power; differential genotype error can lead to inflation of false positives	Family relationships provide some additional checks on genotype error. Even non-differential genotype error can inflate false positive rate
Statistical analysis	Easily carried out using standard software	May require specialist software
Other points		In some designs can estimate parent-of-origin effects. Include both within- and between-family information

The alleles at a series of ordered loci on a chromosome are known as a haplotype. The two haplotypes across several loci in a genetic region therefore contain more information than the genotypes alone.

So far, when discussing the mapping of susceptibility loci using association, we have only considered studying a single locus at a time, dealing with both direct and indirect association. However, it may be more powerful to consider multiple loci simultaneously.

Other than the obvious statistical argument that multiple variables will contain more information than one, the logic behind such an approach can be seen more clearly when the history of a disease-predisposing allele is considered. Those disease-predisposing alleles that are descended from the same founder would initially have existed as a mutation on a single chromosome. As the mutation was passed from one generation to the next, recombination has broken up the chromosome on which the mutation lies, but the haplotype immediately surrounding the mutation is less likely to have undergone recombination and so may have been inherited intact from the founder. The size of this haplotype will depend on many factors and will vary from one person to the next, but immediately surrounding the disease-susceptibility allele there should exist a haplotype that is shared by all carriers. Thus, if we can characterize this haplotype we may be able to use this to identify the region in which the disease-susceptibility locus lies. Clearly this is a simplification, particularly if the disease-susceptibility allele has arisen multiple times, but even so such an approach may be useful.

It may be that one of the loci in the region is in such strong LD with the susceptibility locus that there is sufficient information to use just a single locus, but more usually there is more information in considering a haplotype consisting of several loci. While using haplotypes increases the amount of information, it also increases the complexity of any models fitted, which may reduce power (Chapman *et al.* 2003), so there will be a trade-off between the size of the haplotype and the degrees of freedom available.

Reconstructing haplotypes

Although it is possible in theory to observe individual haplotypes in the laboratory, this is not easy, and haplotypes are usually estimated from observed genotypes. If family members have been genotyped, this may provide enough information to reconstruct haplotypes with certainty, but such a situation is rare (see Figure 4.2).

The most common method for estimating haplotypes from genotype data is to use an estimation–maximization (E-M) algorithm on a sample of unrelated individuals from the population (Excoffier and Slatkin 1995). Here, an initial rough estimate is made of the population haplotype frequencies across the genotyped loci. These frequency estimates are used in a conditional likelihood to assign the most likely haplotype to each individual, assuming HWE.

(i) True allele configuration across three loci

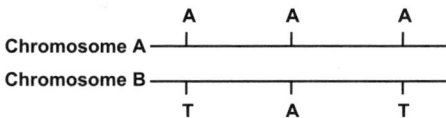

(ii) Genotyping results from the three loci

(iii) Possible haplotypes reconstructed from genotyping results

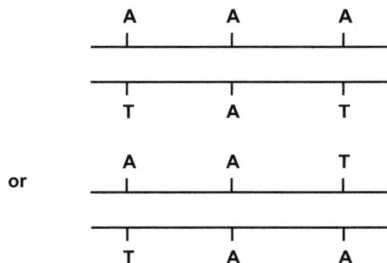

Figure 4.2 Every individual has two copies of each chromosome, one maternally and the other paternally inherited. These are labelled here as chromosomes A and B. Genotyping does not show which alleles lie on which chromosome, so multiple arrangements of the alleles may be possible

The assigned haplotypes are then used to estimate the haplotype frequencies in the population and the process is repeated iteratively until convergence. The method works well for small numbers of loci but difficulties arise when it is applied to many loci, as the number of possible haplotypes increases rapidly with increasing loci. Various methods have been proposed that avoid this problem, such as starting with shorter haplotypes and building them up slowly, discarding haplotypes whose frequencies are too low along the way (SNPHAP-1.3.1, http://www-gene.cimr.cam.ac.uk/clayton/software/), or possibly increasing accuracy by making use of population genetic assumptions in a coalescent model (Stephens *et al.* 2001).

Association studies with haplotypes

If the most likely individual haplotype assignments are treated as though they were observed, this can lead to bias. Differences in haplotype frequencies between cases and controls may instead be investigated by applying the E-M algorithm to cases and controls separately and comparing the final combined likelihood of these data with that of the data when the E-M algorithm is applied to the whole sample. Again, many methods exist for case–control haplotype analysis. Methods include those that are designed to avoid the problem of over-parameterization due to large numbers of haplotypes by haplotype grouping (Thomas *et al.* 2003; Durrant *et al.* 2004).

SNP selection

One aspect of genetic studies that has not yet been discussed is the selection of SNPs. Ideally, a comprehensive search for genetic influence on disease would be conducted by characterizing all genetic variation in a large study population. This may be possible in the future by means of sequencing, but at the current time this is not possible for reasons of both cost and technical limitations, other than in small candidate regions. Instead it is hoped that the genetic influences on disease (particularly common diseases) may be discovered by studying common genetic variants.

This is the common disease/common variant hypothesis.

Several databases now exist that catalogue large numbers of SNPs across the human genome. However, even if all the known SNPs in the region of interest are genotyped, it is not known how much of the variation in the region is captured (the coverage), and if it is not viable to genotype all the known variants in the region, usually for financial reasons, which SNPs should be chosen in order to best characterize the region?

One approach that aims to tackle the problems of sufficient coverage and limited genotyping availability is the use of tagSNPs (or tSNPs). Originally known as haplotype-tagging SNPs, these are a set of SNPs chosen to capture as much of the variation in a region as possible with the minimal number of SNPs (Chapman *et al.* 2003). As initially described, a small pilot sample is genotyped at as many polymorphic loci as possible in the region of interest, ideally identified exhaustively through sequencing. tSNPs are then selected one by one such that, in combination, they optimally predict the variation at the remaining non-tSNPs. tSNPs are chosen until some threshold of prediction at the remaining SNPs is achieved. Prediction was originally measured by the proportion of variation (r^2) at each of the non-tSNPs captured by the tSNPs in a linear regression on either tSNP genotypes or haplotypes. Only SNPs with a minor allele frequency above a given threshold (say, 0.03) were used, and the threshold for the variation explained was suggested as 0.8. Various other methods have been proposed, all based around a similar idea (Horne and Camp 2004; de Bakker *et al.* 2005). Tagging is sensitive to the initial data set used in terms of bias in marker frequency, sufficient density of markers and suitability of the sample in terms of size and population of origin (Weale *et al.* 2003; Iles, 2005; Ahmadi *et al.* 2005; Iles, 2006).

Whole-genome association studies

Until recently all genetic association studies investigated 'candidate' genes, i.e. genes selected on the basis of their supposed biological relevance or because they lay in a chromosomal region implicated in disease by linkage analysis. It has now become technically feasible to genotype markers at sufficient density to consider the whole genome, without any hypotheses about specific genes. Theoretically, there is no reason why standard association methods for candidate regions should not be extended in this way, although several new issues arise.

The main practical challenge has been to find markers at a sufficient density across the genome. This has given rise to the International HapMap Project (www.hapmap.org), an international consortium aiming to catalogue common variation across the human genome in different populations. At the time of writing, HapMap plans to include about 5 million SNPs (of an estimated 10 million in the human genome), giving one SNP every 600 bases. One of the outcomes of the HapMap Project is genome-wide dense marker data that can be used to select tagSNPs. For candidate regions it is clearly less desirable than sequencing to find SNPs in a pilot study, due to potential problems with frequency bias, differences between populations, etc. For use in genome-wide association studies, however, HapMap provides the only currently viable way to consistently select markers to give sufficient coverage across the entire genome. Indeed, it has been used to design the markers for genome-wide SNP chips. Such chips currently consist of several hundred thousand SNPs to test for association with disease, and this number can provide reasonable genome-wide coverage.

Second, handling and analysing such large amounts of data presents technical challenges, and numerous software systems are available or under development to tackle this. Moreover, the scale of these studies and their high cost means that serious attention is being given to designing studies that minimize the amount of genotyping while preserving power, by conducting the genotyping in two or more stages (Skol *et al.* 2006; Wang *et al.* 2006). Third, the huge amount of multiple testing means that interpretation of results is difficult. Two-stage designs and multiple testing are discussed in Chapter 11.

Finally, it is not clear what the best approaches to analysis might be. To date such studies have analysed each SNP singly, using one of the methods outlined above (see Box 4.3 for examples of studies

---

Box 4.3   Whole-genome association studies

Whole-genome association (WGA) studies were first proposed as a possible approach to locating low-penetrance disease genes by Risch and Merikangas (1996). It was almost 10 years until such an approach was first practically feasible and the viability of the approach could be properly tested. One of the earliest published papers (Klein *et al.* 2005) was a study of age-related macular degeneration and used a sample size (96 cases and 50 controls) and marker density (116 000 SNPs) much smaller than is now considered reasonable to identify a common disease susceptibility locus. Fortunately, a disease susceptibility locus was found with a high relative risk (RR; 7.4 in homozygotes) lying in a region with extensive LD. Following this success, several other WGA studies have been published, including a study of Parkinson's disease (Maraganore *et al.* 2005), using 200 000 SNPs in 443 sibling pairs and 332 case–controls, finding an odds ratio (OR) of 1.7; and a study of obesity (Herbert *et al.* 2006) as a quantitative trait, using 116 000 SNPs and 694 individuals, finding an OR of 1.22.

Sample sizes and marker numbers have increased over time and markers have been selected in a more informed way, using LD information to design SNP panels with improved coverage. With increasing numbers of SNPs giving greater opportunities for false positives, more emphasis has been placed on replication of results in an independent population as part of the initial study. Thus, a study on irritable bowel disease (Duerr *et al.* 2006) utilized over 500 cases and controls across 300 000 SNPs selected for their coverage, and replicated their findings in an independent sample of different ethnicity. Sladek (2007) studied type II diabetes, utilizing 400 000 SNPs chosen for coverage and to be gene-centric, in a two-stage case–control design of over 1300 cases and controls in the first stage, followed by a further 2600 to look at the most promising SNPs.

Although only a handful of studies have been published at the time of writing, many whole-genome association studies are now in progress, with an upcoming study from the Wellcome Trust looking at 500 000 SNPs across seven different disease panels, with 2000 individuals in each panel and 3000 controls (to be published in 2007). As more studies of this nature are published, we will acquire greater insight into the best approaches for the design of such whole-genome studies and what they might reveal about common human diseases.

---

to date). However, power may be increased by analysing haplotypes based on two or more neighbouring SNPs (Pe'er *et al.* 2006). It has also been suggested that, since common diseases are likely to be due to the joint effect of multiple SNPs, power may be increased by considering all possible two-way interactions, despite the necessary penalties for multiple testing (Marchini *et al.* 2005). A compromise may be to consider a more restricted set of hypothesis-driven interactions, e.g. interactions between each SNP and an established genetic or environmental risk factor.

The advent of whole-genome association studies has coincided with, and contributed to, renewed interest in population-based (rather than family) controls, since: (a) one population-based control group can be used for comparison with many diseases; and (b) there is enough genetic information

in such studies to investigate and correct for major population stratification. By examining the correlation pattern among unlinked genetic markers, inferences can be made about the population sub-structure, and this structure can be taken into account in the analysis of association (Pritchard *et al.* 2000; Hoggart *et al.* 2003).

Whole-genome association analysis is still in its infancy and many questions remain over its efficacy and the best approach to take to maximize the power to discover disease susceptibility genes.

4.4. DISCUSSION

Technological and scientific advances have produced the possibility of novel insights into the role of genes with respect to disease. Previously, the field had been dominated by family-based approaches, which sought to identify evidence consistent with an underlying genetic susceptibility. Linkage analysis then efficiently identified genes with major effects on risk, where such genes existed and contributed to a significant proportion of familial disease. Environmental exposures could then be evaluated among persons with a susceptibility genotype. Finding the rare, high-penetrance disease variants has not ubiquitously led to an understanding of the common diseases for which familial aggregation is a notable feature, but for which precise patterns of inheritance for an underlying susceptibility are not observed. For cancer, the identification of the dominantly inherited high-penetrance genes indicated that faults in DNA repair underlay a number of susceptibilities.

The natural and most straightforward hypothesis to explore, named the 'common disease/common variant' model, predicts that substantial proportions of the population carry common, deleterious variants. Such hypotheses will be tested through the developing SNP array systems available. The presence of common variants on the arrays, combined with the choice of tagging SNPs, should efficiently assess the hypothesis for each disease. As yet, it is not known which diseases are likely to be due to common variants (and observing families with affected members cannot meaningfully compare this hypothesis with competing hypotheses), and hence the likely outcomes of this approach cannot be predicted. Diseases for which a 'common disease/multiple rare variants' mechanism acts will not be efficiently tested, and investigation will have to wait until sequencing becomes so cheap that, instead of genotyping a defined set of genetic variants, it is more efficient to define the complete genomic sequence of each person. Efforts are being made to produce such technology but improvements have not been sufficient to date. There is also the potential that a considerable proportion of disease is due to the exposure of a susceptible person to a specific environmental exposure, especially if there are only small marginal genetic and environmental effects. Given the lack of definition of the contribution of genetic susceptibility and environmental exposure, and the joint effect of these factors, and with well-catalogued patient [and relative(s)] samples becoming increasingly more difficult to achieve, efforts at producing multi-purpose samples will become standard. One approach and study design that allows examination of genes, environmental exposures, joint effects of genes and environment and gene discovery is the case–control family study (Hopper *et al.* 2005), where probands and their relatives are recruited, interviewed and genotyped.

The next few years promise an interesting time as technology changes produce the capability to define more precisely the aetiology of disease. The question then remains how society will want to use this information, and in particular the extent to which it assists in reducing morbidity and mortality.

References

Ahmadi KR, Weale ME, Xue ZY *et al.* 2005. A single-nucleotide polymorphism tagging set for human drug metabolism and transport. *Nat Genet* **37**: 84–89.

Antoniou AC, Easton DF. 2006. Models of genetic susceptibility to breast cancer. *Oncogene* **25**: 5898–5905.

Breslow NE, Day NE. 1980. *Statistical Methods in Cancer Research, vol 1, The Analysis of Case–control Studies.* IARC Science: Lyon, France.

Cannon-Albright LA, Bishop DT, Goldgar C, Skolnick MH. 1991. Genetic predisposition to cancer. *Important Adv Oncol* **7**: 39–55.

Cannon-Albright LA, Goldgar DE, Neuhausen S *et al.* 1994. Localization of the 9p melanoma susceptibility locus (MLM) to a 2-cM region between D9S736 and D9S171. *Genomics* **23**: 265–268.

Chapman JM, Cooper JD, Todd JA *et al.* 2003. Detecting disease associations due to linkage disequilibrium using haplotype tags: a class of tests and the determinants of statistical power. *Hum Hered* **56**: 18–31.

Cleves M. 1999. Hardy–Weinberg equilibrium test and allele frequency estimation. *Stata Technical Bulletin Reprints*, vol 8. Stata Press, Texas; 280–284.

De Bakker PI, Yelensky R, Pe'er I *et al.* 2005. Efficiency and power in genetic association studies. *Nat Genet* **37**: 1217–1223.

Duerr RH, Taylor KD, Brant SR *et al.* 2006. A genome-wide association study identifies IL23R as an inflammatory bowel disease gene. *Science* **314**: 1461–1463.

Durrant C, Zondervan KT, Cardon LR *et al.* 2004. Linkage disequilibrium mapping via cladistic analysis of single-nucleotide polymorphism haplotypes. *Am J Hum Genet* **75**: 35–43.

Easton D, Peto J. 1990. The contribution of inherited predisposition to cancer incidence. *Cancer Surv* **9**: 395–416.

Elston RC, Anne Spence M. 2006. Advances in statistical human genetics over the last 25 years. *Stat Med* **25**: 3049–3080.

Excoffier L, Slatkin M. 1995. Maximum-likelihood estimation of molecular haplotype frequencies in a diploid population. *Mol Biol Evol* **12**: 921–927.

Gandini S, Sera F, Cattaruzza MS *et al.* 2005a. Meta-analysis of risk factors for cutaneous melanoma: I. Common and atypical naevi. *Eur J Cancer* **41**: 28–44.

Gandini S, Sera F, Cattaruzza MS *et al.* 2005b. Meta-analysis of risk factors for cutaneous melanoma: II. Sun exposure. *Eur J Cancer* **41**: 45–60.

Gandini S, Sera F, Cattaruzza MS *et al.* 2005c. Meta-analysis of risk factors for cutaneous melanoma: III. Family history, actinic damage and phenotypic factors. *Eur J Cancer* **41**: 2040–2059.

Gillanders E, Juo SH, Holland EA *et al.* 2003. Localization of a novel melanoma susceptibility locus to 1p22. *Am J Hum Genet* **73**: 301–313.

Goldgar DE, Easton DF, Cannon-Albright LA *et al.* 1994. Systematic population-based assessment of cancer risk in first-degree relatives of cancer probands. *J Natl Cancer Inst* **86**: 1600–1608.

Herbert A, Gerry NP, McQueen MB *et al.* 2006. A common genetic variant is associated with adult and childhood obesity. *Science* **312**: 279–283.

Hoggart CJ, Parra EJ, Shriver MD *et al.* 2003. Control of confounding of genetic associations in stratified populations. *Am J Hum Genet* **72**: 1492–1504.

Hopper JL, Bishop DT, Easton DF. 2005. Population-based family studies in genetic epidemiology. *Lancet* **366**: 1397–1406.

Hopper JL, Carlin JB. 1992. Familial aggregation of a disease consequent upon correlation between relatives in a risk factor measured on a continuous scale. *Am J Epidemiol* **136**: 1138–1147.

Horne BD, Camp NJ. 2004. Principal component analysis for selection of optimal SNP-sets that capture intragenic genetic variation. *Genet Epidemiol* **26**: 11–21.

Iles MM. 2005. The effect of SNP marker density on the efficacy of haplotype tagging SNPs – a warning. *Ann Hum Genet* **69**: 209–215.

Iles MM. 2006. Obtaining unbiased estimates of tagging SNP performance. *Ann Hum Genet* **70**: 254–261.

Ilonen J, Reijonen H, Green A *et al.* 2000. Geographical differences within Finland in the frequency of HLA-DQ genotypes associated with type 1 diabetes susceptibility. The Childhood Diabetes in Finland Study Group. *Eur J Immunogenet* **27**: 225–230.

Kamb A, Shattuck-Eidens D, Eeles R *et al.* 1994. Analysis of the p16 gene (*CDKN2*) as a candidate for the chromosome 9p melanoma susceptibility locus. *Nat Genet* **8**: 23–26.

Khoury MJ, Beaty TH, Cohen BH. 1993. *Fundamentals of Genetic Epidemiology*. Oxford University Press: New York.

King MC, Lee GM, Spinner NB, Thomson G, Wrensch MR. 1984. Genetic epidemiology. *Annu Rev Publ Health* **5**: 1–52.

Klein RJ, Zeiss C, Chew EY *et al.* 2005. Complement factor H polymorphism in age-related macular degeneration. *Science* **308**: 385–389.

Laird NM, Lange C. 2006. Family-based designs in the age of large-scale gene-association studies. *Nat Rev Genet* **7**: 385–394.

Mander AP. 2001. Haplotype analysis in population-based association studies. *Stata J* **1**: 58–75.

Marchini J, Donnelly P, Cardon LR. 2005. Genome-wide strategies for detecting multiple loci that influence complex diseases. *Nat Genet* **37**: 413–417.

Maranganore DM, De Andrade M, Lesnick TG *et al.* 2005. High-resolution whole-genome association study of Parkinson disease. *Am J Hum Genet* **77**: 685–693.

Mezei G, Kheifets L. 2006. Selection bias and its implications for case–control studies: a case study of magnetic field exposure and childhood leukaemia. *Int J Epidemiol* **35**: 397–406.

Morton NE. 1955. Sequential tests for the detection of linkage. *Am J Hum Genet* **7**: 277–318.

Morton NE. 1978. *Genetic Epidemiology*. Academic Press: New York.

Motulsky AG, Vogel F. 1997. *Human Genetics*. Springer-Verlag: New York.

Panza F, Solfrizzi V, D'Introno A *et al.* 2003. Angiotensin I converting enzyme ACE. gene polymorphism in centenarians: different allele frequencies between the North and South of Europe. *Exp Gerontol* **38**: 1015–1020.

Pe'er I, De Bakker PI, Maller J *et al.* 2006. Evaluating and improving power in whole-genome association studies using fixed marker sets. *Nat Genet* **38**: 663–667.

Pritchard JK, Stephens M, Rosenberg NA *et al.* 2000. Association mapping in structured populations. *Am J Hum Genet* **67**: 170–181.

R Development Core Team. 2006. R: A Language and Environment for Statistical Computing: http://www.R-project.org

Rabinowitz D, Laird N. 2000. A unified approach to adjusting association tests for population admixture with arbitrary pedigree structure and arbitrary missing marker information. *Hum Hered* **50**: 211–223.

Rice TK, Borecki IB. 2001. Familial resemblance and heritability. *Adv Genet* **42**: 35–44.

Risch N. 1990. Linkage strategies for genetically complex traits. I. Multilocus models. *Am J Hum Genet* **46**: 222–228.

Risch N. 2001. The genetic epidemiology of cancer: interpreting family and twin studies and their implications for molecular genetic approaches. *Cancer Epidemiol Biomarkers Prevent* **10**: 733–741.

Risch N, Merikangas K. 1996. The future of genetic studies of complex human diseases. *Science* **273**: 1516–1517.

SAS, version 9. 2007. SAS Institute: Cary, NC, USA.

Sasieni PD. 1997. From genotypes to genes: doubling the sample size. *Biometrics* **53**: 1253–1261.

Sham PC, Curtis D. 1995. An extended transmission/disequilibrium test (TDT) for multi-allele marker loci. *Ann Hum Genet* **59**: 323–336.

Siegmund KD, Langholz B, Kraft P *et al.* 2000. Testing linkage disequilibrium in sibships. *Am J Hum Genet* **67**: 244–248.

Skol AD, Scott LJ, Abecasis GR *et al.* 2006. Joint analysis is more efficient than replication-based analysis for two-stage genome-wide association studies. *Nat Genet* **38**: 209–213.

Sladek R, Rocheleau G, Rung J *et al.* 2007. A genome-wide association study identifies novel risk loci for type 2 diabetes. *Nature* **445**: 881–885.

Spielman RS, Ewens WJ. 1998. A sibship test for linkage in the presence of association: the sib transmission/disequilibrium test. *Am J Hum Genet* **62**: 450–458.

Spielman RS, McGinnis RE, Ewens WJ. 1993. Transmission test for linkage disequilibrium: the insulin gene region and insulin-dependent diabetes mellitus IDDM. *Am J Hum Genet* **52**: 506–516.

Stata statistical software: release 9. 2005. StataCorp LP: College Station, TX.

Stephens M, Smith NJ, Donnelly P. 2001. A new statistical method for haplotype reconstruction from population data. *Am J Hum Genet* **68**: 978–989.

Sturm RA. 2002. Skin colour and skin cancer – MC1R, the genetic link. *Melanoma Res* **12**: 405–416.

Susser M. 1985. Separating heredity and environment. *Am J Prevent Med* **1**: 5–23.

Teare MD, Barrett JH. 2005. Genetic linkage studies. *Lancet* **366**: 1036–1044.

Thomas DC. 2004. *Statistical Methods in Genetic Epidemiology*. Oxford University Press: Oxford.

Thomas DC, Stram DO, Conti D, Molitor J, Marjoram P. 2003. Bayesian spatial modeling of haplotype associations. *Hum Hered* **56**: 32–40.

Vogel F, Motulsky AG. 1979. *Human Genetics*. Springer-Verlag: Berlin.

Wang H, Thomas DC, Pe'er I, Stram DO. 2006. Optimal two-stage genotyping designs for genome-wide association scans. *Genet Epidemiol* **30**: 356–368.

Weale ME, Depondt C, Macdonald SJ *et al.* 2003. Selection and evaluation of tagging SNPs in the neuronal-sodium-channel gene SCN1A: implications for linkage-disequilibrium gene mapping. *Am J Hum Genet* **73**: 551–565.

5

Individual Susceptibility and Gene–Environment Interaction

Seymour Garte

University of Pittsburgh Cancer Institute, USA

5.1. INDIVIDUAL SUSCEPTIBILITY

It has always been clear that individual human beings differ widely in their susceptibility to the toxic effects of environmental agents (Bois 1995; Harris 1989). There are many factors that play a part in determining the particular degree of susceptibility for each individual to the effects of toxic agents. Susceptibility factors that originate from the person at risk are called host factors, and they include aspects of the individual's health, weight, habits and genetic make-up. Host factors include some characteristics that are under total or partial control by the individual, such as diet, weight, sexual habits, medication or drug use and general health status. Other host factors, such as hormonal levels and a great number of genetic factors, are beyond the control of the individual. Identifying risks associated with environmental and host factors that are subject to change is of obvious value, since preventive strategies can be identified. The identification of risks associated with host factors that are not subject to alteration by the will of the individual raises more complex issues (VanDamme *et al.* 1995; Schulte *et al.* 1999).

The study of host factor-related susceptibility has been mostly focused on the genetic variation that is known to be prevalent in human populations (Vineis *et al.* 1999; Fryer *et al.* 1999; Kelada *et al.* 2003). Research into the genetic basis of human variability in disease risk has the potential to produce a number of benefits, in addition to the identification of susceptible individuals. The identification of subpopulations who are more susceptible to chemically-induced disease increases our ability to recognize relevant risk factors (Garte 2001), and knowledge of genes involved in the metabolism of a putative aetiological agent strengthens the evidence for its role in toxicity. As discussed later, genetic polymorphisms may be particularly important for low exposure levels (Garte *et al.* 1997; Taioli *et al.* 1998; Vineis *et al.* 1994, 2004; Taioli and Garte 1999; Garte 2006), which could influence the whole process of risk assessment, a process that is now starting to take individual variability in susceptibility into account. This chapter focuses on the effects of genetic background on individual susceptibility and how genetic factors interact with toxic exposures to produce individual variability in the observed effects of such exposures.

5.2. GENETIC SUSCEPTIBILITY

Current knowledge of the human genome, which includes the identity and sequences of the genes that code for the enzymes involved in detoxification of toxins and carcinogens, as well as in drug activity and toxicity, is beginning to make it possible to investigate the basic sources of human variability using new technologies. This field of biomedical research is called 'pharmacogenetics' when dealing with the genes controlling drug

Molecular Epidemiology of Chronic Diseases, Edited by C. P. Wild, P. Vineis, and S. Garte

metabolism, or 'toxicogenetics' when the emphasis is on detoxification of chemical toxins. Often the same genes are involved in both processes (Idle *et al.* 1992). When the research is not related to a limited number of known genes, but encompasses the entire genome (e.g. when searching for new targets for a drug or a toxin), the area is defined as 'toxicogenomics' or 'pharmacogenomics'. There is a considerable degree of overlap between the terms 'pharmacogenetics' and 'pharmacogenomics' and they have often been used interchangeably.

Genetic susceptibility is a reality for the large majority of diseases. Because of the interests of the author, this chapter tends to highlight those genes associated with susceptibility to carcinogenesis and interaction of genetic factors with environmental carcinogens. However, it is critical to note that genetic susceptibility is a major and important area of research into many other human diseases, including heart disease (Kaab and Schulze-Bahr 2005; Watkins and Farrall 2006; Bianchi 2005; Ye 2006), infectious diseases (Clementi and Di Gianantonio 2006; Bellamy 2006; Kaslow *et al.* 2005; Fernando and Britton 2006), psychiatric disorders (Gogos and Gerber 2006), hepatic fibrosis (Day 2005), kidney disease and diabetes (Gohda *et al.* 2005; Freedman *et al.* 2006; Maier and Wicker 2005), osteoporosis (Zmuda *et al.* 2005) and many others (Khoury *et al.* 1995; Lidral and Murray 2004; Ban and Tomer 2005; Libby *et al.* 2005; Abel *et al.* 2006).

The genes responsible for differences in susceptibility may be divided into many groups, but each of these genes can be assigned to one of two broad categories. The category 1 genes act within the mechanistic cellular pathways, and do not generally interact with environmental agents in order to produce effects. For carcinogenesis, these genes, including tumour suppressor genes, such as *p53*, *Rb*, *BRCA1*, other genes involved in inherited cancer susceptibility syndromes, such as *ATM* or *RET*, and oncogenes such as *ras* or *myc*, may be called 'major cancer genes' (Friend *et al.* 1988; Frebourg and Friend 1992; Savitsky *et al.* 1995). The mechanisms of action of these genes do not depend on the presence of any environmental exposures, and they can exert their effects in the absence of such exposures. The second category of genes influence events in the mechanistic pathways of toxicity (such as metabolism or repair), but are not part of the pathway. Because of this, they differ in many ways from the first category of genes.

5.3. METABOLIC SUSCEPTIBILITY GENES

Genetic sequence polymorphisms have been found in these genes that directly affect the function of their gene products, which are usually enzymes that act within the classical two-phase biochemical detoxification pathway (Figure 5.1). In phase 1, a toxic agent is metabolized by a complex reaction, usually mediated by a cytochrome P450-containing enzyme, in which an oxygen-containing moiety such as a hydroxyl group is placed on the xenobiotic compound. This reaction has the effect of painting a target onto the foreign molecule and allowing it to be more easily attacked by the conjugating reactions of phase II enzymes. The oxidizing enzymes of phase I are now referred to as *CYP* genes (for cytochrome P450) and the number and variations of these genes is immense (Hukkanen *et al.* 2002; Ingelman-Sundberg *et al.* 2000). The phase II genes include those that code for enzymes that add glutathione, glucuronide or acetyl groups to the targeted (oxidized) xenobiotic agent, and thereby allow for their rapid elimination from the cell and ultimately the organism.

Reports during the past decade have suggested that certain alleles of these xenobiotic metabolizing

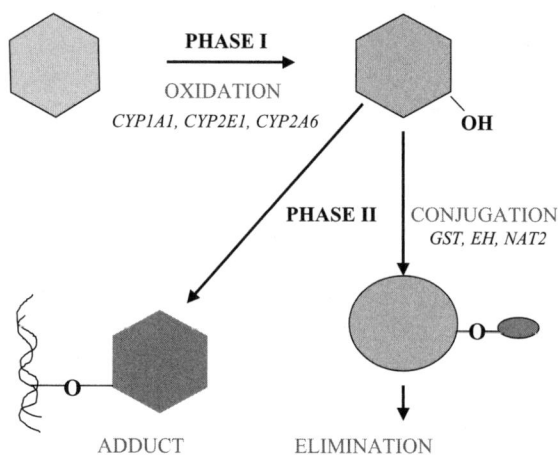

Figure 5.1 Schematic of two phase deionification, including the formation a reactive intramediate that can form a DNA ADDUCT

Table 5.1 Differences between categories of cancer susceptibility genes

	Category 1 Major genes	Category 2 Susceptibility genes
Act on mechanistic pathway	Yes	No
Somatic mutations	Yes	No
Germline polymorphisms	Rare	Common
Penetrance	High	Low
Phenotype requires exposure	No	Yes
Function (examples)	Biological processes of growth and regulation	Metabolism and detoxification of xenobiotics

genes, such as *CYP1A1*, *CYP1B1*, *CYP2A6*, *CYP2D6*, *CYP2E1*, glutathione-*S*-transferases (*GSTM1*, *GSTT1*, *GSTP1*), myeloperoxidase (*MPO*), nitroquinoline oxide (*NQO1*), *N*-acetyl transferase (*NAT1*, *NAT2*) and epoxide hydrolase (*EPHX*) may be associated with many types of common human epithelial and other cancers of several organs (Hein 2006; Garte *et al.* 2000; London *et al.* 2000; Ramachandran *et al.* 1999; Schnakenberg *et al.* 2000; Park *et al.* 2000; Boissy *et al.* 2000; Sweeney *et al.* 2000; Wadelius *et al.* 1999; Wu *et al.* 2004). There have also been reports of their association with a wide variety of other diseases, especially those in which environmental risk factors play a role in aetiology (Paracchini *et al.* 2006; Gao *et al.* 2006; Lea *et al.* 2005; Okubo *et al.* 2003; Masetti *et al.* 2003; Hukkanen *et al.* 2001; Takeyabu *et al.* 2000). Because of the relatively high frequency of these allelic variants in the population (1–50%; Garte *et al.* 2001), the attributable risk for these genes could be quite high, even if their penetrance is low. In some cases effects were only observed in the presence of two or more risk alleles.

Another category of genes that has elicited great interest as low-penetrance susceptibility genes are those that code for DNA repair enzymes (Berwick and Vineis 2000; Goode *et al.* 2002). These genes, like the metabolic genes, are generally highly polymorphic in humans, and a number of variants of several of these genes have been found to be associated with higher risk for cancer. Gene–environmental interaction between DNA repair gene polymorphisms and specific exposures have also been studied (Hung *et al.* 2005).

Table 5.1 illustrates many of the differences in the two categories of susceptibility genes. The effects of category II genes are dependent on the presence of toxic agents. Their actions affect the potency of such agents, but have no influence on toxicity by themselves if exposure is absent. Their effect is to increase the host's susceptibility to the effects of environmental toxicants. The category I genes tend to have high penetrance, meaning that the genotype at these loci is closely associated with an observable phenotype, such as the presence of a disease state. On the other hand, the metabolic and DNA repair genes that act outside the mechanistic pathways, and that require the presence of environmental exposures in order for their effects to be observed, generally exhibit a fairly low penetrance. Table 5.2 lists some examples of the two categories of susceptibility genes.

Table 5.2 Examples of cancer susceptibility genes

Category 1	Category 2
Oncogenes	Metabolic genes
c-myc	*CYP1A1*
H-ras	*CYP2D6*
c-abl	*CYP1B1*
c-fos	*GSTM1*
c-jun	*NQO1*
Tumour suppressor genes	DNA repair genes
p53	*XPG/ERCC5*
P16	*XRCC1*
Rb	*XPD/ERCC2*
VHL	*XPF/ ERCC4*

There are many differences between the two classes of susceptibility genes (Caporaso and Goldstein 1995), in addition to their differences in penetrance (Table 5.1). Changes in the category 1 or high-penetrance genes are usually caused by some mutation in one or a few cells, and these mutations then lead to the onset of a disease state, such as malignant transformation of the affected cell. On the other hand, for the low-penetrance genes, somatic mutations are rarely if ever seen, and instead, genetic variation is found throughout the population at these loci. These polymorphic genes may be involved in risk as either homozygotes or heterozyotes. By convention, the most common allele is considered to be the wild-type, although on occasion, the labelling of certain alleles as wild-type and others as variants appears to be arbitrary, especially when the population frequency of so-called variant alleles is very high in the population.

The low-penetrance susceptibility genes are important in understanding the basis of both individual differences in susceptibility and the mechanisms of gene–environment interactions in cancer and other chronic diseases. Most of these genes are polymorphic in the human population (Garte et al. 2001) and a large number of studies have investigated the role of germline variants in metabolic and DNA repair gene polymorphisms in gene–environmental interactions to cause disease (Hemminki and and Shields 2002; Brennan 2002; Thier et al. 2003; Geisler and Olshan 2001; Ishibe and Kelsey 1997; Smith et al. 1994; Cascorbi et al. 1999; Strange et al. 1998; Hirvonen 1999; Autrup 2000; D'Errico et al. 1996). Metabolic susceptibility genes follow the form of 'type 2' gene–environment interaction (Ottman 1996; Khoury 1998), whereby the polymorphic genetic risk factor functions only in the presence of an environmental exposure (see below).

Genes that confer differences in susceptibility to toxic or carcinogenic agents in the environment can code for proteins with a great variety of function. For the Category 2 genes, such functions can include the uptake, metabolism, excretion and binding of toxic chemicals or their metabolites. A project devoted to the interplay of genetic susceptibility factors in the context of epidemiological investigations was initiated several years ago under the auspices of the US Centers for Disease Control (CDC) (Khoury and Dorman 1998). This project, the Human Genome Epidemiology (HuGE) project, sponsors review articles that cover all aspects of genetics in the epidemiology of many human diseases, with a focus on both genetic susceptibility, by summarizing results of association studies, and on gene–environment interactions. A more recent outgrowth of the HuGE network is the development of the HuGE Published Literature database, recently described (Lin et al. 2006).

A promise of genetic susceptibility markers is the potential to identify high-risk groups for certain diseases and exposures. This is especially true for individuals who might already be at high risk because of unusual exposures, diet, smoking behaviour or other genetic backgrounds. This does not mean that it is necessarily a good idea to screen the whole population for these genetic markers, since there are good reasons to believe that such an exercise would not be fruitful. For single gene polymorphisms, the penetrance is usually so low and the population frequency is so high that it is generally difficult to make any use of large-scale screening for prevention or public health. For example, if the GSTM1 variant, a gene deletion which is present in roughly half the Caucasian population, confers a significant but minor risk of bladder cancer of < 10%, what message should be given to the 50% of the population who have inherited this variant? To suggest that they are all at higher risk of developing a rare tumour such as bladder cancer is clearly ridiculous and probably counterproductive. On the other hand, if the group being screened were limited to workers exposed to aniline dyes, and the risk of bladder cancer were elevated three- to five-fold in GSTM1 variants among these workers, then it would make sense to screen the subset of these people based on their specific exposure history. Those with the susceptible genotype could then be offered more frequent and regular surveillance to allow for earlier detection of the cancer, thus saving lives.

Combinations of several susceptibility genetic variants have been shown to increase the risk associated with the susceptible genotypes, sometimes to the point where relative risks can reach those found with the major cancer or other disease susceptibility genes (Hou *et al.* 2000; Bartsch *et al.* 2000; Abdel-Rahman *et al.* 1998). However, as the number of loci used in these composite genotypes increases, the population frequency decreases, also reaching that of the major cancer gene polymorphisms (Taioli and Garte 1999b). This means that whole-population screening is still not warranted, as is the case for such genes as *BRCA1*, whose rarity makes it non-feasible for routine screening. Once again, this situation changes if we are looking at people with higher than average expected risk. For *BRCA1* this would include women with a strong family history of breast or ovarian cancer, and women of Ashkenazi Jewish heritage, who have been shown to have a higher frequency of the risk allele. For the composite multi-locus genotypes of low-penetrance susceptibility genes, the same idea holds true. Individuals, such as heavy smokers, in some groups might be worth screening for these high-risk combinations of genotypes, if there is a useful strategy for intervention or early detection.

5.4. STUDY DESIGNS

Studies of susceptibility genes in human disease, particularly cancer and heart disease, generally have employed a case–control study design as the most common approach to investigate the role of these genes in disease aetiology, progression and prognosis (Caporaso *et al.* 1999; Gelatti *et al.* 2005). Some advantages of the case–control design are relatively low cost, lower investment of time than, for example, a cohort study, and the ability to use stored or archival DNA tissue banks, including tissue blocks, frozen blood or urine, or other materials such as saliva. This allows for the collection of material from deceased cases, and can allow for studying changes with time and exposure conditions, if the archival material is extensive enough.

A variation of the case–control design that is used in genetic susceptibility studies is the case–case design (Hamajima *et al.* 1999; Gatto *et al.* 2004). In this type of study, cases with a polymorphism are compared to cases without it; the outcome variable could be survival or time to relapse or response to therapy. Studies of cases only can also be used to investigate gene–environment interactions, e.g. to determine whether such interaction is present in cases that have better or worse clinical courses.

Just as the case-only design has been used in studies of genetic susceptibility, so has the control-only design. The object of such studies is not to understand the role of the genetic variant in the disease process directly. Instead, by studying healthy populations, it is possible to explore the mechanistic effects of putative susceptibility genes on non-disease end-points or intermediate markers, such as DNA adducts, levels of metabolites, repair processes, etc., that could be related to one or more diseases. This approach, which has been called biochemical susceptibility analysis (Garte *et al.* 2005), has the advantage of providing information on genotype–phenotype correlations, and giving insights into the mechanisms by which some genetic variants function to increase susceptibility. An example of such a study that would provide important information would be the relative levels of nicotine metabolism as a function of genotype in smokers. It is possible that people with some genotypes might be more or less addicted to tobacco, depending on genotype, and therefore at different levels of risk for continuing smoking and development of any of the numerous diseases associated with smoking exposure. This approach is also particularly useful in studies of intermediate markers that are likely to be involved in a disease process. For example, individuals with different genotypes could be compared as a function of exposure, with respect to the level of DNA adducts or other forms of DNA damage, such as strand breaks, toxic metabolite levels, tissue injury, antibody response and any number of biological effects (Teixeira *et al.* 2002; Butkiewicz *et al.* 2000; Georgiadis *et al.* 2004; Rojas *et al.* 2000; Neri *et al.* 2005; Paracchini *et al.* 2005; Schoket *et al.* 2001; Palli *et al.* 2004; Whyatt *et al.* 2000; Merlo *et al.* 1998; Fustinoni *et al.* 2005). This approach not only gives information on the probable identification of susceptibility genotypes, but also provides potentially very useful information regarding the mechanism of action of specific genes and their variants. When the phenotypes of the variants are known, it might be possible to determine mechanistic pathways and refine our knowledge of toxic mechanisms. Until recently such detailed studies of biochemical and molecular mechanisms of toxicity, involving specific gene products and genetic alterations, were only done in animal and cell-culture models. The use of genetic variability in the human population now make it possible to explore such mechanistic questions in human beings *in vivo*.

5.5. GENE–ENVIRONMENT INTERACTION

The interaction of environmental and genetic risk factors in the production of human disease is an important and well-studied area of molecular epidemiology. Many approaches have been taken in order to clarify these interactions and to try to understand the underlying basis of such interactions, as well as to further refine the identification of subpopulations who could be at higher risk. Small case–control studies are not usually useful for such investigations, since it is unlikely that sufficient power will be available to test hypotheses once stratification and division of a relatively small population (100–300 individuals) into subgroups is done. Larger case–control studies, as well as cohort and prospective studies involving populations in the thousands, can address issues of gene–environmental interactions, depending on the prevalence of the exposure as well as the frequency of the putative at-risk genotype (Matullo *et al.* 2006; Gormally *et al.* 2006). An approach that has proved extremely valuable in such studies is the use of very large networks of databases (Ioannidis *et al.* 2006) for pooled analyses, in which data from tens of thousands of cases and controls are collected from various laboratories, using the same or similar methodology for genotyping and exposure analysis. This approach is exemplified by the long-standing international project on Genetic Susceptibility to Environmental Carcinogenesis (GSEC) (Taioli 1999; Gaspari *et al.* 2001; Raimondi *et al.* 2005; Vineis *et al.* 2004b; Hung *et al.* 2003), which has resulted in a great many interesting results related to gene–environment interactions because of the unusually high power of the study. A more detailed discussion of this project is provided in Chapter 15.

Very few chronic diseases are caused solely by environmental or solely by genetic factors. Even in cases where one type of risk factor is clearly more important (such as smoking and lung cancer, or *BRCA1* mutation and breast cancer), there is usually some effect of the other type of risk factor in determining individual susceptibility to the disease. As we know, not all smokers get lung cancer, and not all women with an inherited *BRCA1* and *BRCA2* mutation develop breast cancer. In the former case it is quite likely that low-penetrance metabolic and DNA repair genes play a role in determining relative susceptibility to disease occurrence, and in the latter case there are a number of hypotheses, that some lifestyle or exposure factors, as well as other genes (Pharoah *et al.* 2002; de Jong *et al.* 2002; Conway *et al.* 1995), influence the likelihood of carriers coming down with the disease. Since for the great majority of human chronic illness, including cancer (Lichtenstein *et al.* 2000; Mucci *et al.* 2001; Shields and Harris 2000), heart disease (Imumorin *et al.* 2005; Hegele 1992), diabetes (Wareham *et al.* 2002), lung diseases (Kleeberger and Peden 2005; Sengler *et al.* 2002; Caramori and Adcock 2006) and others (Brennan and Silman 1994; Caspi and Moffitt 2006; Liangos *et al.* 2005; Whitcomb 2006; Cummings and Kavlock 2004), both genetics and environmental factors are presumed to play some role, it is prudent to be interested in the possible ways that such factors could interact with each other (Kraft and Hunter 2005; Daly *et al.* 1994). In addition to the field of gene–environment interaction, which has become a major field of research recently, there are studies on gene–gene, and gene–gene–environment interactions as well as gene–hormone–environment interaction, etc. (Manuguerra *et al.* in press; Szolnoki and Melegh 2006; Taylor *et al.* 1998). Not all publications or studies that examine more than one gene or single nucleotide polymorphism (SNP) at a time, necessarily find evidence for gene–gene interaction or synergistic effects between different loci. A number of examples of gene–gene interactions, where various genetic variants combine to give strong associations with a large variety of disease and other phenotypic end-points, are shown in Table 5.3.

Gene–environment interaction is not a new idea in medicine. The great geneticist J. B. S. Haldane published a classic paper entitled 'The interaction

Table 5.3 Examples of gene–gene interaction in disease susceptibility

Genes	Disease or end-point	Reference
CYP1A1 and *GSTM1* and *GSTT1*	Lung cancer	Taioli *et al.* 2003
CYP1A1 and *GSTM1* and *GSTT1*	Lung cancer	Vineis *et al.* 2004
GSTP1 and *GSTM1* and *GSTT1* and *CYP1A1*	DNA adducts	Teixeira *et al.* 2002
CYP1A1 and *GSTM1*	Lung cancer	Hung *et al.* 2003
GCR and *ESR2*	Osteoporosis	Xiong *et al.* 2006
XRCC1 and *APE1*	Cutaneous melanoma	Li *et al.* 2006
ALDH2 and *ADH2*	Colorectal cancer	Matsuo *et al.* 2006
MTHFR and *GSTM1*	Male infertility	Paracchini *et al.* 2006
FAS and *FASL*	Cervical carcinogenesis	Lai *et al.* 2005
GSTM1 and *GSTT1* and *GSTP1*	Bladder cancer	Srivastava *et al.* 2005
CARD15 and TNFα promoter	Crohn's disease	Linderson *et al.* 2005
Cyclin D1 and *XPD*	Aero-digestive tract cancer	Buch *et al.* 2005
PPARγ and *2 ADRβ3*	Obesity	Ochoa *et al.* 2004
CYP2D6 and *NAT2*	Multiple chemical sensitivity	McKeown *et al.* 2004
Tyrosine phosphatase 1B and *LEPR*	Body mass index	Santaniemi *et al.* 2004
GSTM1 and *matrix metalloproteinase 9*	COPD	Yanchina *et al.* 2004
Cyclin D1 and *thymidylate synthase*	Childhood ALL	Costea *et al.* 2003
Prothrombin and *factor XIII-A*	Myocardial infarction	Butt *et al.* 2003
CYP1A1 and *GSTM1*	DNA adducts	Firozi *et al.* 2002
NQO1 and *MPO* and *CYP2E1*	Childhood ALL	Krajinovic *et al.* 2002
NAT2 and *GSTM1*	Adducts, lung cancer	Hou *et al.* 2001
NAT2 and *GSTM1* and *CYP1A1*	Childhood ALL	Krajinovic *et al.* 2000
NAT2 and *GSTM1*	Bladder cancer	Schnakenberg *et al.* 2000
GSTM1 and *GSTT1*	Breast cancer	Park *et al.* 2000
GSTP1 and *CYP1A1*	DNA adducts	Whyatt *et al.* 2000
GSTM3 and *GSTM1*	Larynx cancer	Jourenkova-Mironova *et al.* 1999
NAT1 and *NAT2*	Bladder cancer	Taylor *et al.* 1998

[a]For details on gene and allele definitions, see the referenced publications.
[b]COPD, chronic obstructive pulmonary disease; ALL, acute lymphoblastic leukaemia.

of nature and nurture', published in 1946, on this topic (Haldane 1946). In later literature on the subject, Khoury and co-workers (Khoury and James 1993; Khoury and Wagener 1995) and Ottman (1995, 1990, 1996) described a number of types of gene–environment interactions, which differ from each other on the basis of the degree of independence of action of the two factors. For example, in some cases, disease causation is observed only when both risk factors are present; alternatively, one factor might act independently of the other, which can, however, exert a mediating influence. In one particular type of gene–environment interaction, termed 'Type 2', genetic risk factors function to exacerbate (or sometimes mollify) the toxic effects of an exposure (Garte *et al.* 1997; Taioli *et al.* 1998). The exposure by itself acts as a risk factor for the disease, even in the absence of the genetic factor. On the other hand, in the absence of exposure, simply being a carrier of the genetic variant that is considered to be the risk factor should have no effect on disease risk. This type of gene–environment interaction can be applied to the metabolic susceptibility genes, whose function is to modify the effective dose of an environmental exposure, increasing risk. Of course if the exposure is absent, then no increased risk should be seen. More details on the different types of gene–environment interaction is given in Box 5.1.

Box 5.1 Types of gene–environment interactions

Various forms of gene–environment interaction (GEI) have been applied to molecular epidemiological studies. Khoury and his co-workers have described six types of gene–environment interactions (Khoury *et al.* 1988, 1993, 1995; Khoury and James 1993; Khoury and Wagener 1995). In type 1, neither the environmental nor the genetic risk factor have any effect on disease (or other end-point) in the absence of the other. However, when both environmental and genetic risk factors are present, an interaction between them produces an effect. An example often given for type 1 GEI is that of phenylalanine exposure in the diet, and the phenylketonuria genotype. Type 2 GEI is defined as one in which the genetic variant has no effect on disease in the absence of the relevant environmental exposure, but the environmental risk factor by itself can effect risk, even in the absence of the genetic factor. This is the most important type of GEI for environmental carcinogenesis related to metabolic susceptibility genes. Type 3 GEI is the converse of the second, in that the genetic variant can produce an effect in the absence of environmental exposure, and the exposure by itself plays no role in disease aetiology. This type of interaction may be important in human disease if a specific toxicant act has an effect only on people with a particular genetic make-up. Type 4 GEI occurs when both the environmental exposure and the genetic factor carry some risk for disease, but the combination is interactive and/or synergistic. Cancers associated with DNA repair gene deficiencies, such as ataxia telangectasia or xeroderma pigmentosum, are examples. Most category 1 cancer genes (*c-myc*, *p53*, etc.) belong to this type of GEI, since the gene mutations themselves produce increased risk which is exacerbated by exposure to environmental carcinogens. Types 5 and 6 GEI refer to cases in which the genetic risk factor is protective.

Ottman (1990, 1994, 1995) has described five similar types of GEI. In the first, the disease may be caused by either the genetic or the environmental agent, but the genotype increases the expression of the agent. The second and third are the same as Khoury's. In the fourth type, both environmental and genetic risk factors must be present to cause the disease, equivalent to Khoury's type 1. In the final model, both factors influence risk by themselves, but with interaction between them.

Both groups make the point that the term 'interaction' covers a variety of biological phenomena. The type of gene–environment interaction is as important as the fact that an interaction occurs. Further understanding of mechanistic and proactical or clinical aspects of GEI depend on knowledge of which type of GEI is operative in any particular situation.

5.6. EXPOSURE DOSE EFFECTS IN GENE–ENVIRONMENT INTERACTIONS

There are a number of interesting theoretical and practical implications of the type 2 model of gene–environment interactions (Garte *et al.* 1997), one of which relates to the role of exposure dose. It has been found in many studies that some genetic risk variants seem to exert relatively more significant effects on disease or other end-point, when the interactive level of exposure is low. This phenomenon, which we have termed a low-exposure gene (LEG) effect (Taioli *et al.* 1998), is defined as the situation when the degree of interaction with a particular genetic variant decreases as a function of exposure. A number of examples of LEG effects

have been reported for a variety of diseases and for a number of gene and environmental exposure combinations (Belogubova *et al.* 2004; Kiyohara *et al.* 2003; Alexandrie *et al.* 2004; McNamara *et al.* 2004; Ordovas *et al.* 2002; Wang *et al.* 2001). The observation of a LEG effect is not equivalent to an observation of a protective effect, since the relative risk of the exposure associated with the genetic variant is diminished as a function of increasing exposure dose. Thus, individuals with the higher-risk genetic variant will be at a higher relative risk in persons with a low exposure than in persons with a high exposure.

It has been suggested that the phenomenon of the inverse dose or LEG effect might be simply

Box 5.2 Relationship of functional effects of polymorphisms to dose effects

The genetic variants (polymorphisms) of some genes code for enzymes with either a gain- or loss-of-function phenotype. Using classical Michaelis–Menten kinetics, it has been shown (Garte 2006) that the LEG effect can be predicted to always occur for any gain-of-function polymorphism that affects the K_m of the enzymatic gene product. On the other hand, loss-of-function polymorphisms, such as deletions or loss of enzymatic activity, should always produce a high dose effect. These conclusions are based on a theoretical treatment, and have yet to be confirmed or refuted by experimental data.

due to a saturation phenomenon, so that for people with very high exposure, having a genetic risk factor that increases the effective dose is irrelevant, since saturation of the exposure effect has already occurred. The problem with this explanation is that the LEG effect has also been seen when the exposure levels to environmental risk factors are not high enough to be saturating. An alternative hypothesis is based on the biochemistry of the enzymes coded for by the metabolic susceptibility genes, which make up the majority of genetic risk factors that follow type 2 gene–environment interaction (Vineis et al. 2004; Garte 2006). Box 2 provides some further thoughts on this topic.

5.7. MUTATIONAL EFFECTS OF GENE–ENVIRONMENT INTERACTIONS

In addition to the type 2 gene–environment interactions that are the dominant form of such interactions for the low-penetrance metabolic susceptibility genes, there are many other kinds of such interactions that are well known. The best example is that of somatic mutations caused by the direct chemical interaction of a toxic agent with a regulatory or active site of a critical gene, such as an oncogene or tumour suppressor gene in the case of cancer. It has been shown that certain environmental carcinogens, such as tobacco smoke, UV radiation, aflatoxin and (TCE), produce specific patterns of such mutations in specific genes. In this case, the genes

involved tend to be the category 1 high-penetrance pathway-related genes. A good example of such genes is *p53*, an important tumour suppressor gene, which has been found to be involved in a large number of cancer types and is frequently mutated in human cancer cases (Hernandez-Boussard et al. 1998; Olivier et al. 2004). A database of *p53* mutations in human tumours was originated and is maintained at the International Agency for Research in Cancer (IARC) in Lyon, France, and the website for this database (www-p53.iarc.fr) is very informative concerning the mutational spectra of mutations in this gene with respect to interactions with mutagenic chemicals and other environmental carcinogens. For this type of gene–environment interaction, unlike the type 2 described above, environmental exposure is not required for the gene to have a risk effect. Again, taking the example of *p53*, inherited mutations of this gene, which occur in the rare Li–Fraumeni syndrome, cause cancer in the absence of any environmental exposure. In this interaction, the effect of the exposure is to produce mutations, so that if mutations are present due to other causes, then exposure is no longer relevant.

5.8. CONCLUSIONS

In conclusion, individual susceptibility and gene–environment interaction are closely related areas of intense research into the origins of a number of chronic disease, with a good deal of emphasis placed on environmental carcinogenesis. The study of individual variability due to genetic differences is of critical importance in all aspects of human health care, from the avoidance of adverse drug effects, to protection from environmental exposure to toxic agents and regulations designed to protect human health. With the completion of the Human Genome Project and current progress in haplotype mapping, it is becoming possible to identify and study very large numbers of potential genetic risk factors and variants that could play a role in determining genetic susceptibility to disease. Successful efforts in this area will require careful planning and study designs that take advantage of, and are not confounded by, the vast amounts of data that can now be available using modern technologies related to whole-genome scanning. Of equal importance in this field are the myriad potential problems associated with ethical and social implications of genetic susceptibility testing, which are covered in other parts of this volume.

References

Abdel-Rahman SZ, Anwar WA, Abdel-Aal WE *et al.* 1998. *GSTM1* and *GSTT1* genes are potential risk modifiers for bladder cancer. *Cancer Detect Prevent* **22**: 129–138.

Abel K, Reneland R, Kammerer S *et al. 2006.* Genome-wide SNP association: identification of susceptibility alleles for osteoarthritis. *Autoimmun Rev* **5***: 258–263.*

Alexandrie AK, Nyberg F, Warholm M *et al.* 2004. Influence of *CYP1A1, GSTM1, GSTT1,* and *NQO1* genotypes and cumulative smoking dose on lung cancer risk in a Swedish population. *Cancer Epidemiol Biomarkers Prevent* **13**: 908–914.

Autrup H. 2000. Genetic polymorphisms in human xenobiotica metabolizing enzymes as susceptibility factors in toxic response. *Mutation Res* **464**: 65–76.

Ban Y, Tomer Y. *2005.* Genetic susceptibility in thyroid autoimmunity. *Pediatr Endocrinol Rev* **3***: 20–32.*

Bartsch H, Nair U, Risch A *et al.* 2000. Genetic polymorphism of *CYP* genes, alone or in combination, as a risk modifier of tobacco-related cancers. *Cancer Epidemiol Biomarkers Prevent* **9**: 3–28.

Bellamy R. *2006.* Genome-wide approaches to identifying genetic factors in host susceptibility to tuberculosis. *Microbes Infect* **8***: 1119–1123.*

Belogubova EV, Togo AV, Karpova MB *et al.* 2004. A novel approach for assessment of cancer predisposing roles of *GSTM1* and *GSTT1* genes: use of putatively cancer resistant elderly tumor-free smokers as the referents *Lung Cancer* **43**: 259–266.

Berwick M, Vineis P. 2000. Markers of DNA repair and susceptibility to cancer in humans: an epidemiologic review *J Natl Cancer Inst* **92**: 874–897.

Bianchi G. *2005.* Genetic variations of tubular sodium reabsorption leading to 'primary' hypertension: from gene polymorphism to clinical symptoms. *Am J Physiol Regulat Integr Comp Physiol* **289***: R1536–1549.*

Bois FY, Krowech G, Zeise L. 1995, Modeling human interindividual variability in metabolism and risk: the example of 4-aminobiphenyl. *Risk Anal* **15**: 205–213.

Boissy RJ, Watson MA, Umbach DM *et al.* 2000. A pilot study investigating the role of NAT1 and NAT2 polymorphisms in gastric adenocarcinoma *Int J Cancer* **87**: 507–511.

Brennan P, Silman AJ, 1994, An investigation of gene–environment interaction in the etiology of rheumatoid arthritis. *Am J Epidemiol* **140**: 453–460.

Brennan P. 2002. Gene–environment interactions and aetiology of cancer: what does it mean and how can we measure it? *Carcinogenesis* **23**: 381–387.

Buch S, Zhu B, Davis AG *et al.* 2005 Association of polymorphisms in the *cyclin D1* and *XPD* genes and susceptibility to cancers of the upper aero-digestive tract. *Mol Carcinogen* **42**: 222–228.

Butkiewicz D, Grzybowska E, Phillips DH *et al.* 2000. Polymorphisms of the *GSTP1* and *GSTM1* genes and PAH–DNA adducts in human mononuclear white blood cells. *Environ Mol Mutagen* **36**: 99–105.

Butt C, Zheng H, Randell E *et al.* 2003. Combined carrier status of *prothrombin 20210A* and *factor XIII-A Leu34* alleles as a strong risk factor for myocardial infarction: evidence of a gene–gene interaction. *Blood* **101**: 3037–3041.

Caporaso N, Goldstein A. 1995. Cancer genes: single and suseptibility: exposing the difference. *Pharmacogenetics* **5**: 59–63.

Caporaso N, Rothman N, Wacholder S. 1999. Case–control studies of common alleles and environmental factors. *J Natl Cancer Inst Monogr* **26**: 25–30.

Caramori G, Adcock I. 2006. Gene–environment interactions in the development of chronic obstructive pulmonary disease. *Curr Opin Allergy Clin Immunol* **6**: 323–328.

Cascorbi I, Brockmoller J, Mrozikiewicz PM *et al.* 1999. Arylamine *N*-acetyltransferase activity in man. *Drug Metab Rev* **31**: 489–502.

Caspi A, Moffitt TE. *2006.* Gene–environment interactions in psychiatry: joining forces with neuroscience. *Nat Rev Neurosci* **7***: 583–590.*

Clementi M, Di Gianantonio E. *2006.* Genetic susceptibility to infectious diseases. *Reprod Toxicol* **21***: 345–349.*

Conway K, Edmiston S, Fried DB *et al.* 1995. *Ha-ras* rare alleles in breast cancer susceptibility. *Breast Cancer Res Treat* **35**: 97–104.

Costea I, Moghrabi A, Krajinovic M. 2003. The influence of cyclin D1 (CCND1) 870A>G polymorphism and *CCND1*–thymidylate synthase (*TS*) gene–gene interaction on the outcome of childhood acute lymphoblastic leukaemia. *Pharmacogenetics* **13**: 577–580.

Cummings AM, Kavlock RJ. 2004. Gene–environment interactions: a review of effects on reproduction and development. *Crit Rev Toxicol* **34**: 461–485.

D'Errico A, Taioli E, Xhen X *et al.* 1996. Genetic metabolic polymorphisms and the risk of cancer: a review of the literature. *Biomarkers* **1**: 149–173.

Daly AK, Cholerton S, Armstrong M *et al.* 1994. Genotyping for polymorphisms in xenobiotic metabolism as a predictor of disease susceptibility. *Environ Health Perspect* **102** (suppl 9P): 55–56.

Day CP. 2005. Genetic studies to identify hepatic fibrosis genes and SNPs in human populations. *Methods Mol Med* **117**: 315–331.

de Jong MM, Nolte IM, te Meerman GJ *et al.* 2002. Genes other than *BRCA1* and *BRCA2* involved in breast cancer susceptibility. *J Med Genet* **39**: 225–242.

Fernando SL, Britton WJ. 2006. Genetic susceptibility to mycobacterial disease in humans. *Immunology and Cell Biology* **84**: 125–137.

Firozi PF, Bondy ML, Sahin AA *et al.* 2002. Aromatic DNA adducts and polymorphisms of *CYP1A1*, *NAT2*, and *GSTM1* in breast cancer. *Carcinogenesis* **23**: 301–306.

Frebourg T, Friend SH. 1992. Cancer risks from germline *p53* mutations. *J Clin Invest* **90**: 1637–1641.

Freedman BI, Bowden DW, Sale MM *et al.* 2006. Genetic susceptibility contributes to renal and cardiovascular complications of type 2 diabetes mellitus. *Hypertension* **48**: 8–13.

Friend SH, Dryja TP, Weinberg RA. 1988. Oncogenes and tumor-suppressing genes. *N Engl J Med* **318**: 618-622.

Fryer AA, Jones PW. 1999. Interactions between detoxifying enzyme polymorphisms and susceptibility to cancer. In *Metabolic Polymorphisms and Susceptibility to Cancer*, Vineis P, Malats N, Lang M *et al.* (eds). IARC Scientific Publications No. 148. Oxford University Press: Oxford; 303–322.

Fustinoni S, Buratti M, Campo L *et al.* 2005. Urinary *t,t*-muconic acid, *S*-phenylmercapturic acid and benzene as biomarkers of low benzene exposure. *Chem Biol Interact* **153–154**: 253–256.

Gao X, Yang H, ZhiPing T. 2006. Association studies of genetic polymorphism, environmental factors and their interaction in ischemic stroke. *Neurosci Lett* **398**: 172–177.

Garte S. 2001. Metabolic susceptibility genes as cancer risk factors: time for a reassessment? *Cancer Epidemiol Biomarkers Prevent* **10**: 1233–1237.

Garte S, Gaspari L, Alexandrie AK *et al.* 2001 Metabolic gene frequencies in control populations. *Cancer Epidemiol Biomarkers Prevent* **10**: 1239–1248.

Garte S, Taioli E, Raimondi S *et al.* 2005. Effects of metabolic genotypes on intermediary biomarkers in subjects exposed to PAHs. Results from the EXPAH study. *Mutation Res* **153–154**: 247–251.

Garte S. 2006. Dose effects in gene–environment interaction: an enzyme kinetics-based approach. *Med Hypotheses* **67**: 488–492.

Garte S, Taioli E, Crosti F *et al.* 2000. Deletion of parental *GST* genes as a possible susceptibility factor in the etiology of infant leukemia. *Leukemia Res* **24**: 971–974.

Garte SJ, Zocchetti C, Taioli E. 1997. Gene–environment interactions in the application of biomarkers of cancer susceptibility in epidemiology. In *Applications of Biomarkers in Cancer Epidemiology*, Toniolo P, Boffetta P, Shuker D *et al.* (eds). IARC Scientific Publication No. 142. IARC, Lyon; 251–264.

Gaspari L, Marinelli D, Taioli E. 2001. Collaborative Group on Genetic Susceptibility to Environmental Carcinogens; international collaborative study on genetic susceptibility to environmental carcinogens (GSEC): an update. *Int J Hygiene Environm Health* **204**: 39–42.

Gatto NM, Campbell UB, Rundle AG *et al.* 2004. Further development of the case-only design for assessing gene–environment interaction: evaluation of and adjustment for bias. *Int J Epidemiol* **33**: 1014–1024.

Geisler SA, Olshan AF, 2001. *GSTM1*, *GSTT1* and the risk of squamous cell carcinoma of the head and neck: a mini-HuGE review. *Am J Epidemiol* **154**: 95–105.

Gelatti U, Covolo L, Talamini R *et al.* 2005. *N*-Acetyltransferase-2, glutathione *S*-transferase M1 and T1 genetic polymorphisms, cigarette smoking and hepatocellular carcinoma: a case–control study. *Int J Cancer* **115**: 301–306.

Georgiadis P, Demopoulos NA, Topinka J *et al.* 2004. Impact of phase I or phase II enzyme polymorphisms on lymphocyte DNA adducts in subjects exposed to urban air pollution and environmental tobacco smoke. *Toxicol Lett* **149**: 269–280.

Gogos JA, Gerber DJ. 2006. Schizophrenia susceptibility genes: emergence of positional candidates and future directions. *Trends Pharmacol Sci* **27**: 226–233.

Gohda T, Tanimoto M, Watanabe-Yamada K *et al.* 2005. Genetic susceptibility to type 2 diabetic nephropathy in human and animal models. *Nephrology* **10**(suppl): S22.

Goode EL, Ulrich CM, Potter JD. 2002. Polymorphisms in DNA repair genes and associations with cancer risk. *Cancer Epidemiol Biomarkers Prevent* **11**: 1513–1530.

Gormally E, Vineis P, Matullo G *et al.* 2006. *TP53* and *KRAS2* mutations in plasma DNA of healthy subjects and subsequent cancer occurrence: a prospective study. *Cancer Res* **66**: 6871–6876.

Haldane JBS. 1946. The interaction of nature and nurture. *Ann Eugen* **13**: 197–205.

Hamajima N, Yuasa H, Matsuo K *et al.* 1999 Detection of gene–environment interaction by case-only studies. *Japan J Clin Oncol* **29**: 490–493.

Harris CC. 1989. Interindividual variation among humans in carcinogen metabolism, DNA adduct formation and DNA repair. *Carcinogenesis* **10**: 1563–1566.

Hegele RA. 1992. Gene–environment interactions in atherosclerosis. *Mol Cell Biochem* **113**: 177–186.

Hein DW. 2006. *N*-acetyltransferase 2 genetic polymorphism: effects of carcinogen and haplotype on urinary bladder cancer risk. *Oncogene* **25**: 1649–1658.

Hemminki K, Shields P. 2002. Skilled use of DNA polymorphisms as a tool for polygenic cancers. *Carcinogenesis* **23**: 379–380.

Hernandez-Boussard TM, Hainaut P. 1998. A specific spectrum of *p53* mutations in lung cancer from smokers: review of mutations compiled in the IARC *p53* database. *Environ Health Perspect* **106**: 385–391.

Hirvonen A. 1999. Polymorphisms of xenobiotic-metabolizing enzymes and susceptibility to cancer. *Environ Health Perspect* **107**(suppl 1): 37–47.

Hou SM, Ryberg D, Falt S *et al.* 2000. *GSTM1* and *NAT2* polymorphisms in operable and non-operable lung cancer patients. *Carcinogenesis* **21**: 49–54.

Hou SM, Falt S, Yang K *et al.* 2001. Differential interactions between *GSTM1* and *NAT2* genotypes on aromatic DNA adduct level and *HPRT* mutant frequency in lung cancer patients and population controls. *Cancer Epidemiol Biomarkers Prev* **10**: 133–140.

Hukkanen J, Pelkonen O, Raunio H. 2001. Expression of xenobiotic-metabolizing enzymes in human pulmonary tissue: possible role in susceptibility for ILD. *Eur Resp J Suppl* **32**: 122–126s.

Hukkanen J, Pelkonen O, Hakkola J *et al.* 2002. Expression and regulation of xenobiotic-metabolizing cytochrome P450 (CYP) enzymes in human lung. *Crit Rev Toxicol* **32**: 391–411.

Hung RJ, Brennan P, Boffetta P. 2005. Genetic polymorphisms in base excision repair pathways and cancer risk: a HuGE disease-association review. *Am J Epidemiol* **162**: 925–942.

Hung RJ, Boffetta P, Brockmöller J *et al.* 2003. *CYP1A1* and *GSTM1* genetic polymorphisms and lung cancer risk in Caucasian nonsmokers: a pooled analysis. *Carcinogenesis* **24**: 875–882.

Idle JR, Armstrong M, Boddy AV *et al.* 1992. The pharmacogenetics of chemical carcinogenesis. *Pharmacogenetics* **2**: 246–258.

Imumorin IG, Dong Y, Zhu H *et al.* 2005. Gene–environment interaction model of stress-induced hypertension. *Cardiovasc Toxicol* **5**: 109–132.

Ingelman-Sundberg M, Daly AK, Oscarson M *et al.* 2000 Human cytochrome P450 (*CYP*) genes: recommendations for the nomenclature of alleles. *Pharmacogenetics* **10**: 91–93.

Ioannidis JP, Gwinn M, Little J *et al.* 2006. Human Genome Epidemiology Network and the Network of Investigator Networks, a road map for efficient and reliable human genome epidemiology. *Nat Genet* **38**: 3–5.

Ishibe N, Kelsey KT. 1997. Genetic susceptibility to environmental and occupational cancers. *Cancer Causes Control* **8**: 504–513.

Jourenkova-Mironova N, Voho A, Bouchardy C *et al.* 1999. Glutathione *S*-transferase *GSTM3*, and *GSTP1* genotypes and larynx cancer risk. *Cancer Epidemiol Biomarkers Prevent* **8**: 185–188.

Kaab S, Schulze-Bahr E. 2005. Susceptibility genes and modifiers for cardiac arrhythmias. *Cardiovasc Res* **67**: 397–413.

Kaslow RA, Dorak T, Tang JJ. 2005. Influence of host genetic variation on susceptibility to HIV type 1 infection. *J Infect Dis* **191**(suppl 1): S68–77.

Kelada SN, Eaton DL, Wang SS *et al.* 2003. The role of genetic polymorphisms in environmental health. *Environ Health Perspect* **111**: 1055–1064.

Khoury MJ, Beaty TH, Hwang SJ. 1995. Detection of genotype–environment interaction in case–control studies of birth defects: how big a sample size? *Teratology* **51**: 336.

Khoury MJ, Dorman JS. 1998. The Human Genome Epidemiology Network. *Am J Epidemiol* **148**: 1–3.

Khoury MJ. 1998. Genetic and epidemiologic approaches to the search for gene–environment interaction: the case of osteoporosis. *Am J Epidemiol* **147**: 1–2.

Khoury MJ, James LM. 1993. Population and familial relative risks of disease associated with environmental factors in the presence of gene–environment interaction. *Am J Epidemiol* **137**: 1241–1250.

Khoury MJ, Wagener DK. 1995. Epidemiological evaluation of the use of genetics to improve the predictive value of disease risk factors. *Am J Hum Genet* **56**: 835–844.

Kiyohara C, Wakai K, Mikami H *et al.* 2003. Risk modification by *CYP1A1* and *GSTM1* polymorphisms in the association of environmental tobacco smoke and lung cancer: a case–control study in Japanese nonsmoking women. *Int J Cancer* **107**: 139–144.

Kleeberger SR, Peden D. 2005. Gene–environment interactions in asthma and other respiratory diseases. *Annu Rev Med* **56**: 383–400.

Kraft P, Hunter D. 2005. Integrating epidemiology and genetic association: the challenge of gene–environment interaction. *Phil Trans R Soc B Biol Sci* **360**: 1609–1616.

Krajinovic M, Sinnett H, Richer C *et al.* 2002. Role of *NQO1, MPO* and *CYP2E1* genetic polymorphisms in the susceptibility to childhood acute lymphoblastic leukemia. *Int J Cancer* **97**: 230–236.

Krajinovic M, Richer C, Sinnett H *et al.* 2000. Genetic polymorphisms of *N*-acetyltransferases 1 and 2 and gene–gene interaction in the susceptibility to childhood acute lymphoblastic leukemia. *Cancer Epidemiol Biomarkers Prevent* **9**: 557–562.

Lai HC, Lin WY, Lin YW *et al.* 2005. Genetic polymorphisms of *FAS* and *FASL* (*CD95/CD95L*) genes in cervical carcinogenesis: an analysis of haplotype and gene–gene interaction. *Gynecol Oncol* **99**: 113–118.

Lea RA, Ovcaric M, Sundholm J *et al.* 2005. Genetic variants of angiotensin converting enzyme and methylene tetrahydrofolate reductase may act in combination to increase migraine susceptibility. *Brain Res Mol Brain Res* **136**: 112–117.

Li C, Liu Z, Wang LE *et al.* 2006. Genetic variants of the *ADPRT, XRCC1* and *APE1* genes and risk of cutaneous melanoma. *Carcinogenesis* **27**: 1894–1901.

Liangos O, Balakrishnan VS, Jaber BL. 2005. DialGene Consortium Model for gene–environment interaction: the case for dialysis. *Semin Dialysis* **18**: 41–46.

Libby RT, Gould DB, Anderson MG *et al.* 2005. Complex genetics of glaucoma susceptibility. *Ann Rev Genom Hum* Genet **6**: 15–44.

Lichtenstein P, Holm N, Verkasalo P *et al.* 2000. Environmental and heritable factors in the causation of cancer. *N Engl J Med* **343**: 78–85.

Lidral AC, Murray JC. 2004. Genetic approaches to identify disease genes for birth defects with cleft lip/palate as a model. *Birth Defects Res* **70**: 893–901.

Lin BK, Clyne M, Walsh M *et al.* 2006. Tracking the epidemiology of human genes in the literature: the HuGE published literature database. *Am J Epidemiol* **164**: 1–4.

Linderson Y, Bresso F, Buentke E *et al.* 2005. Functional interaction of *CARD15/NOD2* and Crohn's disease-associated TNFα polymorphisms. *Int J Colorect Dis* **20**: 305–311.

London SJ, Smart J, Daly AK. 2000. Lung cancer risk in relation to genetic polymorphisms of microsomal epoxide hydrolase among African-Americans and Caucasians in Los Angeles County. *Lung Cancer* **28**: 147–155.

Maier LM, Wicker LS. 2005. Genetic susceptibility to type 1 diabetes. *Curr Opin Immunol* **17**: 601–608.

Manuguerra M, Matullo G, Veglia F *et al.* In press. Multi-factor dimensionality reduction applied to a large prospective investigation on gene–gene and gene–environment interactions. *Carcinogenesis*.

Masetti S, Botto N, Manfredi S *et al.* 2003. Interactive effect of the glutathione *S*-transferase genes and cigarette smoking on occurrence and severity of coronary artery risk. *J Mol Med* **81**: 488–494.

Matsuo K, Wakai K, Hirose K *et al.* 2006. Gene–gene interaction between *ALDH2 Glu487Lys* and *ADH2 His47Arg* polymorphisms regarding the risk of colorectal cancer in Japan. *Carcinogenesis* **27**: 1018–1023.

Matullo G, Dunning AM, Guarrera S *et al.* 2006. DNA repair polymorphisms and cancer risk in non-smokers in a cohort study. *Carcinogenesis* **27**: 997–1007.

McKeown-Eyssen G, Baines C, Cole DE *et al.* 2004. Case–control study of genotypes in multiple chemical sensitivity: *CYP2D6, NAT1, NAT2, PON1, PON2* and *MTHFR. Int J Epidemiol* **33**: 971–978.

McNamara DM, Holubkov R, Postava L *et al.* 2004. Pharmacogenetic interactions between angiotensin-converting enzyme inhibitor therapy and the angiotensin-converting enzyme deletion polymorphism in patients with congestive heart failure. *J Am Coll Cardiol* **44**: 2019–2026.

Merlo F, Andreassen A, Weston A *et al.* 1998. Urinary excretion of 1-hydroxypyrene as a marker for exposure to urban air levels of polycyclic aromatic hydrocarbons. *Cancer Epidemiol Biomarkers Prevent* **7**: 147–155.

Mucci LA, Wedren S, Tamimi RM *et al.* 2001. The role of gene–environment interaction in the aetiology of human cancer: examples from cancers of the large bowel, lung and breast. *J Intern Med* **249**: 477–493.

Neri M, Filiberti R, Taioli E *et al.* 2005. Pleural malignant mesothelioma, genetic susceptibility and asbestos exposure. *Mutat Res* **592**: 36–44.

Okubo T, Harada S, Higuchi S *et al.* 2003. Association analyses between polymorphisms of the phase II detoxification enzymes (*GSTM1, NQO1, NQO2*) and alcohol withdrawal symptoms. *Alcoholism Clin Exp Res* **27**(8 suppl): 68–71S.

Olivier M, Hussain SP, Caron de Fromentel C *et al.* 2004. *TP53* mutation spectra and load: a tool for generating hypotheses on the etiology of cancer. IARC Scientific Publication No. 157. IARC: Lyon; 247–270.

Ordovas JM, Corella D, Demissie S *et al.* 2002. Dietary fat intake determines the effect of a common polymorphism in the hepatic lipase gene promoter on high-density lipoprotein metabolism: evidence of a strong dose effect in this gene–nutrient interaction in the Framingham Study. *Circulation* **106**(18): 2315–2321.

Ochoa MC, Marti A, Azcona C *et al.* 2004. Gene–gene interaction between PPARγ2 and ADRβ3 increases obesity risk in children and adolescents. *Int J Obesity Rel Metab Disord* **28**(suppl 3): S37–41.

Ottman R. 1995. Gene–environment interaction and public health. *Am J Hum Genet* **56**: 821–823.

Ottman R. 1996. Gene–environment interaction: definitions and study designs. *Prevent Med* **25**: 764–770.

Ottman R. 1990. An epidemiologic approach to gene–environment interaction. *Genet Epidemiol* **7**: 177–185.

Palli D, Masala G, Peluso M *et al.* 2004. The effects of diet on DNA bulky adduct levels are strongly modified by *GSTM1* genotype: a study on 634 subjects. *Carcinogenesis* **25**: 577–584.

Paracchini V, Chang SS, Santella RM *et al.* 2005. *GSMT1* deletion modifies the levels of polycyclic aromatic hydrocarbon–DNA adducts in human sperm. *Mutat Res* **586**: 97–101.

Paracchini V, Garte S, Taioli E. 2006. *MTHFR C677T* polymorphism, *GSTM1* deletion and male infertility: a possible suggestion of a gene–gene interaction? *Biomarkers* **11**: 53–60.

Park SK, Yoo KY, Lee SJ *et al.* 2000. Alcohol consumption, glutathione *S*-transferase M1 and T1 genetic polymorphisms and breast cancer risk. *Pharmacogenetics* **10**(4): 301–309.

Pharoah PD, Antoniou A, Bobrow M *et al.* 2002. Polygenic susceptibility to breast cancer and implications for prevention. *Nat Genet* **31**: 33–36.

Raimondi S, Boffetta P, Anttila S *et al.* 2005. Metabolic gene polymorphism and lung cancer risk in non-smokers: an update of the GSEC study. *Mutat Res* **592**: 45–57.

Ramachandran S, Lear JT, Ramsey H *et al.* 1999. Presentation with multiple cutaneous basal cell carcinomas: association of glutathione *S*-transferase and cytochrome P450 genotypes with clinical phenotype. *Cancer Epidemiol Biomarkers Prevent* **8**: 61–67.

Rojas M, Cascorbi I, Alexandrov K *et al.* 2000. Modulation of benzo-α-pyrene diolepoxide–DNA adduct levels in human white blood cells by *CYP1A1*, *GSTM1* and *GSTT1* polymorphism. *Carcinogenesis* **21**: 35–41.

Santaniemi M, Ukkola O, Kesaniemi YA. 2004. Tyrosine phosphatase 1B and leptin receptor genes and their interaction in type 2 diabetes. *J Intern Med* **256**: 48–55.

Savitsky K, Bar-Shira A, Gilad S *et al.* 1995. A single ataxia telangiectasia gene with a product similar to PI-3 kinase. *Science* **268**: 1749–1753.

Schnakenberg E, Lustig M, Breuer R *et al.* 2000. Gender-specific effects of *NAT2* and *GSTM1* in bladder cancer. *Clin Genet* **57**: 270–277.

Schoket B, Papp G, Levay K *et al.* 2001. Impact of metabolic genotypes on levels of biomarkers of genotoxic exposure. *Mutat Res* **482**: 57–69.

Schulte PA, Lomax GP, Ward EM *et al.* 1999. Ethical issues in the use of genetic markers in occupational epidemiologic research. *J Occup Environ Med* **41**: 639–646.

Sengler C, Lau S, Wahn U, Nickel R. 2002. Interactions between genes and environmental factors in asthma and atopy: new developments. *Resp Res* **3**: 7.

Shields PG, Harris CC. 2000. Cancer risk and low-penetrance susceptibility genes in gene–environment interactions. *J Clin Oncol* **18**: 2309–2315.

Smith CM, Kelsey KT, Wiencke JK *et al.* 1994. Inherited glutathione-*S*-transferase defiency is a risk factor for pulmonary asbestosis. *Cancer Epidemiol Biomarkers Prevent* **3**: 471–477.

Srivastava DS, Mishra DK, Mandhani A *et al.* 2005. Association of genetic polymorphism of glutathione *S*-transferase M1, T1, P1 and susceptibility to bladder cancer. *Eur Urol* **48**: 339–344.

Strange RC, Lear JT, Fryer AA. 1998. Polymorphism in glutathione *S*-transferase loci as a risk factor for common cancers. *Arch Toxicol Suppl* **20**: 419–428.

Sweeney C, Farrow DC, Schwartz SM *et al.* 2000. Glutathione *S*-transferase *M1*, *T1*, and *P1* polymorphisms as risk factors for renal cell carcinoma: a case–control study. *Cancer Epidemiol Biomarkers Prevent* **9**: 449–454.

Szolnoki Z, Melegh B. 2006. Gene–gene and gene–environment interplay represent specific susceptibility for different types of ischaemic stroke and leukoaraiosis. *Curr Med Chem* **13**: 1627–1634.

Taioli E, Zocchetti C, Garte S. 1998. Models of interaction between metabolic genes and environmental exposure in cancer susceptibility. *Environ Health Perspect* **106**: 67–70.

Taioli E. 1999. International Collaborative Study on Genetic Susceptibility to Environmental Carcinogens. *Cancer Epidemiol Biomarkers Prevent* **8**: 727–728.

Taioli E, Garte S. 1999a. Low dose exposure to carcinogens and metabolic gene polymorphisms. *Adv Exp Med Biol* **472**: 223–230.

Taioli E, Garte S. 1999b. Genetic susceptibility to environmental carcinogens. *Recent Res Dev Cancer* **1**: 57–67.

Taioli E, Gaspari L, Benhamou S *et al.* 2003. Polymorphisms in *CYP1A1*, *GSTM1*, *GSTT1* and lung cancer below the age of 45 years. *Int J Epidemiol* **32**: 60–63.

Takeyabu K, Yamaguchi E, Suzuki I *et al.* 2000. Gene polymorphism for microsomal epoxide hydrolase and susceptibility to emphysema in a Japanese population. *Eur Resp J* **15**: 891–894.

Taylor JA, Umbach DM, Stephens E *et al.* 1998. The role of *N*-acetylation polymorphisms in smoking-associated bladder cancer: evidence of a gene–gene–exposure three-way interaction. *Cancer Res* **58**: 3603–3610.

Teixeira JP, Gaspar J, Martinho G *et al.* 2002. Aromatic DNA adduct levels in coke oven workers: correlation with polymorphisms in genes *GSTP1*, *GSTM1*, *GSTT1* and *CYP1A1*. *Mutat Res* **517**: 147–155.

Thier R, Bruning T, Roos PH *et al.* 2003. Markers of genetic susceptibility in human environmental hygiene and toxicology: the role of selected *CYP*, *NAT* and *GST* genes. *Int J Hyg Environ Health* **206**: 149–171.

VanDamme KL, Casteleyn E, Heseltine A *et al.* 1995. Individual susceptibility and prevention of occupational diseases: scientific and ethical issues. *J Occup Environ Med* **37**: 91–99.

Vineis P, Alavanja M, Garte S. 2004. Dose–response relationship in tobacco-related cancers of bladder and lung: a biochemical interpretation. *Int J Cancer* **108**: 2–7.

Vineis P, Malats N, Lang M *et al.* (eds). 1999. *Metabolic Polymorphisms and Susceptibility to Cancer*. IARC Scientific Publication No. 148. Oxford University Press: Oxford; 303–322.

Vineis P, Veglia F, Anttila S *et al.* 2004. *CYP1A1*, *GSTM1* and *GSTT1* polymorphisms and lung cancer: a pooled analysis of gene–gene interactions. *Biomarkers* **9**: 298–305.

Vineis P, Bartsch H, Caporaso N *et al.* 1994. Genetically based *N*-acetyltransferase metabolic polymorphism and low level environmental exposure to carcinogens. *Nature* **369**: 154–156.

Wadelius M, Autrup JL, Stubbins MJ *et al.* 1999. Polymorphisms in *NAT2*, *CYP2D6*, *CYP2C19* and *GSTP1* and their association with prostate cancer. *Pharmacogenetics* **9**: 333–340.

Wang Z, Chen C, Niu T *et al.* 2001 Association of asthma with β2-adrenergic receptor gene polymorphism and cigarette smoking. *Am J Resp Crit Care Med* **163**: 1404–1409.

Wareham NJ, Franks PW, Harding AH. 2002. Establishing the role of gene–environment interactions in the etiology of type 2 diabetes. *Endocrinol Metab Clin N Am* **31**: 553–566.

Watkins H, Farrall M. 2006. Genetic susceptibility to coronary artery disease: from promise to progress. *Nat Rev* Genet **7**: 163–173.

Whitcomb DC. 2006. Gene–environment factors that contribute to alcoholic pancreatitis in humans. *J Gastroenterol Hepatol* Suppl **3**: S52–55.

Whyatt RM, Perera FP, Jedrychowski W *et al.* 2000. Association between polyclic aromatic hydrocarbon–DNA adduct levels in maternal and newborn blood cells and glutathione *S*-transferase P1 and *CYP1A1* polymorphsisms. *Cancer Epidemiol Biomarkers Prevent* **9**: 207–212.

Wu X, Zhao H, Suk R *et al.* 2004. Genetic susceptibility to tobacco-related cancer. *Oncogene* **23**: 6500–6523.

Xiong DH, Shen H, Zhao LJ *et al.* 2006. Robust and comprehensive analysis of 20 osteoporosis candidate genes by very high-density single-nucleotide polymorphism (SNP) screen among 405 white nuclear families identified significant association and gene–gene interaction. *J Bone Min Res* **21**: 1678–1695.

Yanchina ED, Ivchik TV, Shvarts EI *et al.* 2004. Gene–gene interactions between glutathione *S*-transferase M1 and matrix metalloproteinase 9 in the formation of hereditary predisposition to chronic obstructive pulmonary disease. *Bull Exp Biol Med* **137**: 64–66.

Ye S. 2006. Influence of matrix metalloproteinase genotype on cardiovascular disease susceptibility and outcome. *Cardiovasc Res* **69**: 636–645.

Zmuda JM, Sheu YT, Moffett SP. 2005. Genetic epidemiology of osteoporosis: past, present, and future. *Curr Osteoporosis Rep* **3**: 111–115.

6

Biomarker Validation

Paolo Vineis[1] and Seymour Garte[2]

[1]*Imperial College London, UK, and* [2]*University of Pittsburgh Cancer Institute, USA*

6.1. VALIDITY AND RELIABILITY

To achieve accurate estimates of risk in molecular epidemiology, we need reliable and valid measurements (markers) of exposure, of covariates (potential confounders and effect modifiers) and of outcomes. Causal inference is impossible in the absence of such requirements. We will distinguish, in the following, between *marker* (any variable that can be measured in a biological medium and is informative for the purposes of the study), *assay* (a specific laboratory procedure that aims at measuring that marker) and *measurement* or *test* (the concrete act of measuring the value of a marker in an individual, by a specific assay). For example, polycyclic aromatic hydrocarbon (PAH)–DNA adducts are a type of marker, ^{32}P-postlabelling is a type of assay, and actual data are measurements.

Validity is defined as the (relative) lack of systematic measurement error when comparing the actual observation with a standard, i.e. a reference method that represents the 'truth'. While validity entails a 'standard', *reliability* concerns the extent to which an experiment or any measuring procedure yields the same results on repeated trials (Carmines and Zeller 1979). By 'the same results' we do not mean an absolute correspondence but a relative concept, i.e. 'a tendency towards consistency found in repeated measurements of the same phenomenon'. Reliability is relative to the type

and purpose of the measurement: for some purposes we may accept a level of unreliability that is unacceptable for other purposes. In addition to being reliable, the marker must be valid, i.e. provide an accurate representation of the underlying phenomenon.

There are several types of validity in the context of biomarkers. Analytical validity focuses on the ability of the assay method to measure accurately and reliably the marker of interest. The components of analytical validity are sensitivity, specificity and test reliability. Sensitivity evaluates how well the test measures the marker when it is present. Specificity, on the other hand, evaluates the test to determine how well the test identifies the *absence* of the marker. The positive predictive value (PPV), i.e. the proportion of individuals who develop the outcome given that they have the marker, is a useful measure of the ability of the marker to predict the outcome and can thus have clinical application. The reliability of a test measures how often the same results are obtained when a sample is retested. Such measures apply to all kinds of biomarkers, including markers of exposure, susceptibility or early response.

Other concepts refer only to certain types of markers. For example, clinical utility addresses the elements that need to be considered when evaluating risks and benefits associated with the introduction of the biomarker assay into routine practice. A test that has clinical utility, such as blood cholesterol, provides the individual with valuable information that can be used for prevention, treatment or life-planning, regardless of results.

Validity and reliability are independent: a measurement may be perfectly reliable (reproducible in different laboratories and repeatable at different times) but consistently wrong, i.e. far away from the true value. We are interested in both validity and reliability; however, since validity is often not measurable, reliability is sometimes used (incorrectly) as a surrogate.

6.2. BIOMARKER VARIABILITY

The most important issue in biomarker reliability is that of the variability of the results of assays done in human beings. Some of this variability, which is generally large for the majority of human biomarkers studied, is due to measurement error, but a large proportion of human biomarker assay variability may simply be due to variation among and within human beings.

The major components of biomarker variability that affect the design of epidemiological studies are therefore variability between subjects (inter-subject), within subjects (intra-subject) and due to measurement errors (see below, and definitions in Table 6.1). The impact of these three categories of variability on the biomarker response can be represented by a linear model of the following form (Taioli *et al.* 1994):

$$Y_{ijk} = u + a_i + b_j + e_{ijk}$$

where Y_{ijk} is the marker response for subject i at time j and replicate measurement k; u is the true population mean response; a_i is the offset in mean response for subject i (assumed to be normally distributed with mean = 0 and variance = s_i^2) (the variance represents the extent of inter-subject variability); b_j is the offset in response at time j (assumed to be normally distributed with mean = 0 and variance = sj^2) (this variance represents the extent of intra-subject variability) and e_{ijk} is the assay measurement error (Normally distributed, with mean = 0 and variance = s_{ijk}^2) (Taioli *et al.*

Table 6.1 Some definitions

Intra-individual variability: within-person variability of biomarkers over time; it can be expressed as standard deviation or -fold
Inter-individual variability: represents the between persons variability of biomarkers; it is generally expressed as standard deviation (SD)
Sample stability: the capacity of samples to keep the same characteristics during the time necessary for laboratory tests
Repeatability: ability to yield the same results (or closely similar results) each time the test is conducted in the same laboratory
Reproducibility: ability to yield the same results (or closely similar results) each time the test is conducted in different laboratories
Dose–response: association between external dose of agents and observed outcomes

1994). The Normality of distribution, assumed in the model, must be verified. In fact, many biomarkers have distributions that are far from being Normal; Normalization can be achieved through an appropriate transformation, e.g. log transformation.

The model is based on a linear (additive) assumption, which implies that measurement errors are independent of average measurements. Such assumption must be verified case-by-case, e.g. by checking whether errors are correlated with the mean.

Inter-subject variability in marker response may derive from factors such as ethnic group, gender, diet or other characteristics. Similarly, the marker response may vary within the same subject over time, due to diurnal variation, changes in diet, health status, variation in exposure to the compound of interest (for dietary items, season is an important variable) and variation in exposure to other compounds that influence the marker response.

Table 6.2 shows how methodological data can be organized according to biomarker type. Intra-individual and sampling variation are considered because of the extent of their influence on actual measurements for most markers.

Table 6.2 Examples of behaviour of biomarkers according to intra-individual or sampling variation

Biological biomarker category	Intra-individual variation	Sampling variation
Internal dose (blood)		
Hormones	Yes (diurnal variation)	No
Water-soluble nutrients	Yes (short half-life)	No
Organochlorine	No (long half-life)	No
Biologically effective dose		
Peripheral white blood cells	Yes (half-life weeks to months)	No
Exfoliated urothelial cells		
DNA adducts	Yes (half-life months)	Yes
Early biological effects		
Lymphocyte metaphase chromosome aberrations	More or less stable	?
Somatic cell mutations glycophorin A	Probably low	No (?)
Intermediate markers		
Cervical dysplasia	Yes	Yes
Colonic hyperproliferation	Yes	Yes
Genetic susceptibility		
Genotype assay	No	No
Non-inducible phenotype	No	No
Inducible phenotype	Yes	No
Tumour markers	Yes	Yes

6.3. MEASUREMENT OF VARIATION

The extent of variability in measurements can be measured itself in several ways. Let us distinguish between continuous measurements and categorical measurements. A general measure of the extent of variation for continuous measurements is the coefficient of variation (CV; = standard deviation/mean, expressed as a percentage). A more useful measure is the ratio between CVb and CVw: CVw measures the extent of laboratory variation within the same sample in the same assay, CVb measures the between-subject variation, and the CVb:CVw ratio indicates the extent of the between-subject variation relative to the laboratory error. Large degrees of laboratory error can be tolerated if between-person differences in the parameter to be measured are large.

A frequently used measure of reliability for continuous measurements is the intra-class correlation coefficient, i.e. the between-person variance divided by the total (between-subject + within-subject) variance. The intra-class coefficient is equal to 1.0 if there is exact agreement between the two measures on each subject (thus differing from the Pearson correlation coefficient, which takes the value 1.0 when one measure is a linear combination of the other, not only when the two exactly agree). A coefficient of 1.0 occurs when within-subject variation is null, i.e. laboratory measurements are totally reliable. The intra-class correlation coefficient can then be used to correct measures of association (e.g. relative risks) in order to allow for laboratory error (this is what is done, for example, in the calibration of dietary data – obtained through food frequency questionnaires – by using a more accurate method such as 24 h recall; see Bingham *et al.* 2003).

The intra-class correlation coefficient estimates the extent of between-subject variability in relation to total variability. The latter includes variation due to different sources (reproducibility, repeatability and sampling variation). To measure reproducibility, i.e. the ability of two laboratories to agree when measuring the same analyte in the same sample, the mean difference between observers, and the corresponding confidence interval (CI), has been proposed (Brennan and Silman 1992).

Another concept used in biomarker validation is inter-observer concordance in the classification of binary outcomes. It should be borne in mind that concordance between two observers can arise by pure chance. Therefore, agreement beyond chance must be measured. However, total potential agreement between two readers cannot be 100%, i.e. to be fair we must subtract chance agreement from 100% to have an estimate of total attainable agreement. The final measure is the difference between observed agreement and chance agreement, divided by the total possible agreement beyond chance; this measure is called the *kappa index*. There are other measures of agreement beyond chance, and the use of kappa has to be made cautiously, since there are some methodological pitfalls, e.g. the value of kappa strictly depends on the prevalence of the condition which is studied – with a high underlying prevalence we expect a higher level of agreement (Brennan and Silman 1992).

Until now we have considered reliability as a property of the assay in the hands of different readers (reproducibility, inter-observer agreement); we can also measure reliability as a property of the assay at repeat measurements (repeatability). Let us consider validity of assessment, i.e. correspondence with a standard. It is essential to bear in mind that two readers may show very high levels of agreement, e.g. as measured by Pearson correlation coefficient (i.e. $r = 0.9$), even if the first observer consistently records twice the value of the second observer. Alternatively (e.g. when using the intra-class correlation coefficient), two readers could show high levels of agreement (e.g. ICC = 0.9) but poor validity if the same errors repeat themselves for both raters. Now we are interested in the correspondence of the measurement with a conceptual entity, e.g. accumulation of the p53 protein as a consequence of gene mutation. Table 6.3 shows data on the correspondence between immunohistochemistry and *p53* mutations. Sensitivity of immunohistochemistry is estimated as 85%, i.e. false negatives are 15% of all samples containing mutations; specificity is estimated as 71%, i.e. 29% of samples not containing mutations are falsely positive at immunohistochemistry. A combined estimate of sensitivity and specificity is the area under the receiver operating curve (ROC), i.e. a curve that graphically represents the relationship between sensitivity and (1 – specificity). Each point in the curve represents the combination of a level of sensitivity and (1 – specificity), depending on the value of the variable that is measured. For extremely low levels of blood glucose, for example, the sensitivity in identifying diabetes is very high (few false negatives) but the specificity is very low (many false positives). The curve is a graphical representation of this concept.

It is usually believed (Fletcher *et al.* 1988) that sensitivity and specificity indicate properties of a test irrespective of the frequency of the condition to be detected; however, this is an assumption that requires verification. In the example of Table 6.3 (bladder cancer), the proportion of samples showing a mutation is high (32/73 = 44%); it would be

Table 6.3 Validity of p53 immunohistochemistry as compared to mutations in the *p53* gene (bladder cancer patients)

p53 mutations by SSCP	p53 nuclear reactivity (immunohistochemistry)			
	−	+	+ +	Total
No mutation	29	7	5	41
All mutations	5	8	19	32
Total	34	15	24	73

Sensitivity of immunohistochemistry (+ and + +) = 27/32 = 85%; specificity = 29/41 = 71%;
positive predictive value = (8 + 19)/(15 + 24) = 27/39 = 69%.
Data from Esrig *et al.* 1993).

much lower, for example, in patients with benign bladder conditions or in healthy subjects. A measure which is useful to predict how many subjects, among those testing positive, are really affected by the condition we aim to detect is the *positive predictive value*, i.e. how many subjects who test positive to immunohistochemistry actually have the condition of interest (in this case the mutation). In the example, 39 patients test positive at immunohistochemistry, and 27 actually have mutations, i.e. immunohistochemistry correctly predicts mutations in 69% of the positive cases. Let us suppose, however, that the prevalence of mutations is not 44%, but 4.4% (32/730). With the same sensitivity and specificity values (85% and 71%, respectively) we would have a positive predictive value of 11.8%, i.e. much lower. The predictive value is a very useful measure, because it indicates how many true positive cases we will obtain within a population of subjects who test positive with the assay we are applying. However, we must bear in mind that the predictive value is strongly influenced by the prevalence of the condition: a very low predictive value may simply indicate that we are studying a population in which very few subjects actually have the condition we want to identify.

6.4. OTHER ISSUES OF VALIDATION

Other issues need to be considered when measuring biomarkers. One is the use of surrogate markers instead of a marker that is strictly relevant to the biological phenomenon we wish to measure. Many biomarker studies are done on readily accessible biological samples, such as urine, peripheral blood or oral mucosa, since the more relevant target tissue (e.g. epithelial lining of the lung or bladder) is usually not available for study, especially in large population studies. These surrogate tissue studies may be useful for many purposes, but the validity of the use of such surrogate samples should be carefully considered. In some cases, e.g. when clinical specimens are being used, it might be possible to directly compare the results with target tissue obtained from clinical material with those from a selected surrogate source, such as blood or buccal cells. If such comparisons are not possible, then the issue of cross-tissue extrapolation can be

an important one for validation, and some justification of the choice of surrogate should be made. As an example, many enzymes have tissue-specific expression, so it would not be appropriate to use expression of a gene in blood lymphocytes as a surrogate for a tissue in which the gene is never expressed.

Another special issue is the shape of dose–reponse relationships. On one side, one expects a proportionality between exposure and effect, and this is a criterion for judging causality (in fact, it was part of Bradford–Hill's guidelines for causality assessment; Hill 1965). In practice, all sorts of dose–response relationships can be found in nature. Levelling-off of the curve associating, for example, an exposure with a marker such as DNA adducts can be related to enzyme saturation for enzymes involved in activation of carcinogens or, vice versa, an exponential rise in the effect can be due to saturation of detoxifying enzymes (Vineis *et al.* 2004).

6.5. MEASUREMENT ERROR

Errors of marker measurement may have different impacts depending on the error distribution. If the epidemiological study has been conducted blindly, i.e. laboratory analyses have been done with no knowledge of the exposed/unexposed or diseased/healthy status of the subjects, we expect that the measurement error will be evenly distributed across strata of exposure or disease. This is true only if the error is equally distributed across the scale of the exposure, i.e. a marker may be more difficult to characterize in smokers than in non-smokers.

Both underestimation and overestimation of the association of interest may occur when misclassification is not evenly distributed across the study variables. We may have a more general distortion of the aetiological relationship if classification of exposure depends on the outcome (diseased/healthy status). Distortion as a consequence of uneven distribution of misclassification can be in either direction, both toward and away from the null hypothesis.

An example of bias dependent on knowledge of the disease status by the researcher is related to degradation of analytes when biological samples are stored for a long time. If the samples from the cases affected by the disease of interest and, respectively, those from controls, are analysed at different times, bias can arise from differential degradation in the two series. For example, if the researcher analyses samples from the cases as soon as these arise in the cohort, while controls are analysed at the end of the study, serious bias may arise from differential timing of measurement in the two series, due to degradation of the marker being analysed. Any inference about the meaning of biomarker measures should be strictly time-specific, since time influences the results in several different ways. For this reason, biochemical analyses should be made after matching of cases and controls for time since sample collection.

Three examples of sources of error in laboratory measurements are shown in Boxes 6.1–6.3. Box 6.1 refers to one of the very few trials designed to measure inter-laboratory variability in the measurement of DNA adducts. Box 6.2 is one of the few examples of comparison among laboratories for genotyping. Box 6.3 shows that the levels of plasma DNA varied according to centre in a large multicentre study, and this was likely to be due to the methods of blood collection and storing in different centres.

Box 6.1   Variability in DNA adduct measurements

An interlaboratory trial (Phillips *et al.* 1999) was aimed at the standardization and validation of different ^{32}P-DNA postlabelling techniques and at the development of recommended protocols. The results are shown in the following table:

Results of interlaboratory trial on ^{32}P-postlabelling

Adduct analysed	Mean	SD	CV (%)
Low B(a)P	11.43	4.06	35.5
Low ABP	172.84	91.22	52.78
High B(a)P	22.15	9.43	42.57
High ABP	1227.20	578.30	47.12
O^6-MeG (b)	206.7	76.45	37

Adducts are per 10^8 nucleotides.
ABP, 4-aminobiphenyl; B(a)P, benzo(a)pyrene; MeG, methylguanine; SD, standard deviation; CV, coefficient of variation.

Examples of quality control for dietary markers are: (a) an international round-robin for folate involving 20 laboratories (Gunter *et al.* 1996) – CV was 27% for serum folate and 36% for whole-blood folate, with substantial inter-method variation; (b) an interlaboratory comparison of homocysteine in plasma samples (14 laboratories) – CV was 9% between laboratories and 6% within laboratories (Pfeiffer *et al.* 1999).

Sources of laboratory measurement error

When we organize and analyse an epidemiological study employing biomarkers, we want to minimize total intra-group variability in order to identify inter-group differences (e.g. between exposed and unexposed or between diseased and healthy subjects), if they exist. Total intra-group variation is the weighted sum of inter-subject, intra-subject, sampling and laboratory variation, with weights that are inversely correlated to the numbers of subjects, measurements per subject, and analytical replicates used in the study design, respectively. Obviously, if we do not have detailed information we cannot adjust for intra-group variation. This is the reason why in epidemiological studies employing biomarkers it is important to collect, whenever possible: (a) repeat samples (day-to-day, month-to-month or year-to-year variation may be relevant, depending on the marker); (b) potentially relevant information on subject characteristics that may influence inter-subject variation; (c) conditions under which samples have been collected and laboratory analyses have been conducted (batch, assay, specific procedures).

An example of a study that was designed to assess the different sources of laboratory variation is reported by Taioli *et al* (1994), using the model described above. In one experiment, they drew blood from five subjects three times in three different weeks, in order to measure DNA–protein crosslinks. The results indicated that variation between batch was quite important, and larger than variation between subjects.

Measurement variation may occur as a consequence of many different aspects that are related not only to the choice of the assay but also to

Box 6.2   Assessment of genotyping (courtesy of A Dunning)

An evaluation of genotyping methods has been performed for a known polymorphism, *BRCA2 N372H*. This polymorphism was initially typed by allele-specific oligonucleotide (ASO) hybridization, using radio-labelled oligos. Using 864 DNA samples of mixed quality, the authors determined the *N372H* genotype by ASO and also by Taqman. If the two results did not agree, they additionally used a forced *RsaI* digest. Thus, they developed a 'consensus genotype' of the same result by at least two methods for each sample (note that the 'consensus' could be called 'no-result' in the case of poor DNA templates; samples where all three methods gave a different genotype were excluded). Subsequently they typed a random subset of DNA samples with the *RsaI* forced digest and also typed 96 samples with the Invader Cleavase technology (Third Wave Inc.). These latter results were also related back to the original 'consensus genotype'. For each method they assessed sensitivity and specificity. The results obtained are given in the table below. Thus, although ASO and Taqman have essentially the same sensitivity on these DNA samples, the specificity of Taqman is the highest of all the methods evaluated.

Comparison of four genotyping methods at one laboratory. The standard is represented by a panel evaluation of all results

Method	Sensitivity	%	Specificity	%
ASO	836/864	97	753/836	90
Taqman	826/864	96	812/826	98
RsaI digest	125/173	72	103/125	82
Invader	62/92	67	45/62	73

Box 6.3   An example of variation by study centre

An analysis of the concentrations of plasma DNA from 1185 subjects is shown in the table below (Gormally et al. 2004). The amount of DNA was extremely variable across study centres. A high DNA concentration might be due to the type of population recruited, and/or the treatment of the samples. In Oxford, for example, samples were posted from general practices throughout the UK to one central laboratory. In the centres with highest plasma DNA concentrations, some blood samples were prepared and frozen many hours after blood drawing, probably allowing some lysis of white blood cells to occur and the release of their DNA into the plasma. In the centres where the concentrations of plasma DNA were lower, the blood was processed on the day of collection. No association was found with gender (Gormally et al. 2004).

Multivariate analysis: plasma DNA amount (logarithm transformation, dependent variable) by centre, age and gender*

Parameter	F Value	DF	*p* Value
Centre	11.23	22	< 0.0001
Age	5.21	1 (a)	0.023
Gender	0.52	1	0.47

*Controls only (*n* = 778).
Reproduced with permission from Gormally et al. (2004).

collection of the sample. Questions that may need to be addressed include: how and when a biological sample was drawn; the type of collection container (e g. test tube) utilized; the amount of biological material collected; whether or not the subject was fasting; and avoidance of exposure of the sample to light (as in the case of measurement of vitamin C). Methods of sample processing and storage are also important for understanding laboratory measurement variability.

The appropriate timing and method of collection depends on the measurement being made: for example, hormones have hourly, daily or monthly cycles. Prolonged venepuncture can induce release of prolactin or increase white cells. A very narrow needle causes haemolysis. Orthostatism decreases plasma volume, so that proteins and cholesterol levels can be lowered by 5–15% compared with the supine position.

The stability of compounds measured depends on the type of measurement and temperature of storage, e.g. fatty acids should be measured within 2 weeks at 4°C, a few months at –20°C, or after no more than 1 year at –80°C. A few studies have been conducted on the stability of different analytes, but the literature is far from being exhaustive. The effect of different storage conditions has been investigated for exposure markers (Riboli *et al.* 1995), hormones (Muti *et al.* 1996), cytokines (Thavasu *et al.* 1992), DNA (Steimberg *et al.* 2002), RNA (Ward *et al.* 2004), proteomics (Diamandis 2004) and a number of nutrients (Komaromy-Hiller *et al.* 1997). For some analytes, typically folate, there is a rapid diminution of stability over time at room temperature, even when preservatives are used.

Cell viability decreases rapidly after 48 h; afterwards cells must be cultured or cryopreserved in liquid nitrogen. In fact a conflict can arise, in the first hours after blood collection, between the goal of preserving cell viability and the goal of preserving labile analytes such as folic acid. The latter goal would imply freezing the samples as soon as possible, but this would hamper cell viability and isolation, while keeping the blood at room temperature for hours or days would threaten labile biomarkers. A good trade-off that has been proposed is keeping samples at 4°C until the cells have been isolated (Holland *et al.* 2003).

During storage, careful records of the location of all stored material, and of all important temperature fluctuations, should be kept. Samples located at the bottom of a freezer have lower temperature than at the top and are less exposed to fluctuations. In liquid nitrogen the temperature is –196°C, but at the liquid–vapour interface it is –120 to –150°C.

An important issue is safety, since biological samples collected from the general population can be infected and should always be treated as though they were. A manual of instructions should be made available to all personnel and carefully implemented (see e.g.: http://www.niehs.nih.gov/odhsb/manual/home.htm).

One type of laboratory error, called 'laboratory drift', is a consequence of changes in procedures and accuracy in the course of time, so that the first samples that are analysed tend to differ from subsequent samples. Avoidance of laboratory drift implies a monitoring programme, which consists of repeated quality controls. For example, measurements may be compared with a standard at different points in time. Another source of 'drift', which cannot be technically avoided, is degradation of analytes when they are stored for a long time. More about how the variability in laboratory measurements influences study design decisions can be found in Rundle *et al* (2005).

The practical reaction to measurement error is the establishment of standards, particularly when a biomarker is used for clinical purposes. For example, standards for the measurement of cholesterol have been in use for a long time (Box 6.4).

Box 6.4   Example of a quality control programme, the US National Cholesterol Education Program (NCEP)

Goals:

1. Attain analytical accuracy and precision ($< 3\%$ CV).
2. Identify individual determinants of cholesterol variation (lifestyle factors).
3. Identify clinical determinants of variation (metabolic states, illness).
4. Other sampling sources (fasting status, posture, serum vs. plasma).

The NCEP guidelines have proved adequate to ensure 90% correct classification. http://www.nhlbi.nih.gov/health/indexpro.htm

http://www.nhlbi.nih.gov/health/indexpro.htm

6.6. BLOOD COLLECTION FOR BIOMARKERS

The following is a minimum list of variables to be collected when blood is drawn for a biomarker study: time and date of draw (and whether the sample was from fasting individuals); volume and type of specimen (noting any additives); medical conditions and drug intake; reproductive/hormonal status (time since last menstrual period); time since last cigarette; other relevant information. Several additives can be added to blood: metaphosphoric acid for vitamin C; anticoagulants such as heparin, EDTA or citrate are needed for plasma collection (i.e. not needed if only serum is collected). There may be disadvantages: heparin binds to many proteins and influences T cell proliferation and EDTA interferes with cytogenetic analyses. Citrate-stabilized blood affords better quality of RNA and DNA than other anticoagulants (Holland *et al.* 2003). Other additives include protease inhibitors and RNAse inhibitors, to avoid degradation of proteins and RNA, respectively (Young *et al.* 1986; Pickard 1989).

From 10 ml of blood one can get: 4–6 ml of plasma or serum; $4.3–10.8 \times 10^6$ white blood cells(WBC)/ml (lymphocytes and mononuclear cells); $4.2–5.9 \times 10^9$ red blood cells (RBC)/ml; 100–150 mg/ml haemoglobin. Minimum (reasonable) requirements for analyses are: for haemoglobin adducts, 50 mg haemoglobin; for DNA adducts in WBC (^{32}P-postlabelling), 1–5 μg DNA; for genotyping, 1–2 μg DNA (10–20 genotypes by PCR–RFLP) – only 1–100 ng is required for high-throughput techniques. Centrifuging allows the separation of plasma, serum, RBC and buffy coat, i.e. the supernatant including white blood cells. It is better, instead of collecting just buffy coat, to use a gradient (e.g. Ficoll or Lymphoprep), which allows separation of granulocytes from lymphocytes. Separation of lymphocytes allows, for example, the extraction of high-quality RNA.

6.7. VALIDATION OF HIGH-THROUGHPUT TECHNIQUES

Validation of new biomarkers based on high-throughput technologies (often abbreviated as '-omics', which encompasses genomics, proteomics, metabolomics and others) represents a number of challenges and difficult issues that have only just begun to be addressed. The -omics biomarkers can be divided into two broad general categories. The first category includes single proteins, metabolites or genes which can serve as a biomarker. This category is quite similar in practice to other previously validated biomarkers, the chief difference being that the choice of the single component (such as the level of a protein or metabolite or mRNA) was made based on a high-throughput -omics technology that scanned a large number of potential markers. Presumably the choice of one of these to be further pursued and validated was based on its high sensitivity and specificity, and a reasonably high level of signal above background noise or control. The validation of any single marker identified in a scanning experiment requires testing the biomarker in a series of experiments, using positive and negative controls. For example, if a particular growth factor shows a five-fold increase in expression at the mRNA or protein level in a scanning experiment in subjects with high exposure to a particular agent, or with early-stage disease (or in some other experimental model), the next step would be to develop an assay for that particular factor, and apply it to a suitable large population (sample size depending on power calculation) of both exposed and unexposed individuals, or in people with and without the disease. It is standard practice to confirm any result showing changes in gene expression from a microarray experiment by a real-time PCR, or northern blot analysis, using probes specific for the gene in question. Based on current knowledge of the degree of variability and reproducibility of many -omics experiments, it is to be expected that some substantial proportion of such validation tests of single biomarkers initially identified from scanning will not confirm their utility.

The second category of -omics-related biomarkers includes those that depend on more than a single component. This could be a group of several components that all change in concert, or an even larger pattern of changes that include dozens or hundreds of components. Once again, multicomponent markers of exposure, effect or disease status must be rigorously validated, using control populations to determine the sensitivity and specificity of the assay. There have been examples of such gene expression or proteomic pattern changes that have been validated to provide diagnostic or prognostic indications of different pathologies or stages of disease (Allison *et al.* 2006; Baggerly *et al.* 2005). The advantage of the multi-component patterns is that there may be more biological plausibility attached to larger shifts in patterns of multi-gene (or multiple-metabolite) pathways than one would expect to see for single genes or proteins. On the other hand, the potential for artifactual results is especially high, since what is being evaluated is a fingerprint signature, without any depth of knowledge of the mechanistic contributions from most of the individual components or interactions between the components that constitute the pattern. In some fortuitous circumstances, one might find activation or repression of a series of inter-related components, such as a series of proteins all involved in a particular metabolic or cell regulation-related pathway, or a group of metabolites that all stem from a common biochemical pathway (see Chapter 13). In this case, the task of providing the key ingredient for biomarker validation of biological plausibility is considerably simpler (D'Haeseeler 2005).

When particular patterns of global changes are seen repeatedly to be associated with a particular biological or epidemiological state, it may be feasible to attempt to demonstrate the validity of such a pattern as a useful biomarker, even in the absence of detailed knowledge of biological plausibility or mechanistic understanding. Such biomarkers are still rare, although some have been proposed based on repeated experiments (Ransohoff 2005a, 2005b).

Acknowledgements

This chapter was made possible by a grant from the European Community (6th Framework Programme, Grant No. 513943) for the Network of Excellence – Environmental Cancer Risk, Nutrition and Individual Susceptibility (ECNIS) (WP4 and WP8).

References

Allison DB, Cui X, Page GP. 2006. Sabripour M. Microarray data analysis: from disarray to consolidation and consensus. *Nat Rev Genet* **7**: 55–65.

Baggerly KA, Morris JS, Edmonson SR *et al.* 2005. Signal in noise: evaluating reported reproducibility of serum proteomic tests for ovarian cancer. *J Natl Cancer Inst* **97**: 307–309.

Beach AC, Gupta RC. 1992. Human biomonitoring and the ^{32}P-postlabeling assay. *Carcinogenesis* **13**: 1053–1074.

Bingham SA, Day NE, Luben R *et al.* 2003. European Prospective Investigation into Cancer and Nutrition. Dietary fibre in food and protection against colorectal cancer in the European Prospective Investigation into Cancer and Nutrition (EPIC): an observational study. *Lancet* **3**: 1496–1501.

Brennan P, Silman A. 1992. Statistical methods for assessing observer variability in clinical measures. *Br Med J* **304**: 1491–1494.

Carmines EG, Zeller RA. 1979. *Reliability and Validity Assessment*. Sage: London.

D'Haeseleer P. How does gene expression clustering work? 2005. *Nat Biotechnol* **23**: 1499–1501.

Diamandis EP. 2004. Analysis of serum proteomic patterns for early cancer diagnosis: drawing attention to potential problems. *J Natl Cancer Inst* **96**: 353–356.

Esrig D, Spruck CH III, Nichols PW. 1993. p53 nuclear protein accumulation correlates with mutations in the *p53* gene, tumor grade and stage in bladder cancer. *Am J Pathol* **143**: 1389–1397.

Fletcher RH, Fletcher SW, Wagner EH. 1988. *Clinical Epidemiology – The Essentials*, 2nd edn. Williams and Wilkins: Baltimore, MD.

Gormally E, Hainaut P, Caboux E *et al.* 2004. Amount of DNA in plasma and cancer risk: a prospective study. *Int J Cancer* **111**: 746–649.

Gunter EW, Bowman BA, Caudill SP *et al.* 1996. Results of an international round robin for serum and whole-blood folate. *Clin Chem* **42**: 1689–1694.

Hill AB. The environment and disease: association or causation? 1965. *Proc R Soc Med* **58**: 295–300.

Holland NT, Smith MT, Eskenazi B *et al.* 2003. Biological sample collection and processing for molecular epidemilogal studies. *Mutat Res* **543**: 217–234.

Komaromy-Hiller G, Nuttall KL, Ashwood ER. 1997. Effect of storage on serum vitamin B_{12} and folate stability. *Ann Clin Lab Sci* **27**: 249–253.

Muti P, Trevisan M, Micheli A *et al.* 1996. Reliability of serum hormones in premenopausal and postmenopausal women over a one-year period. *Cancer Epidemiol Biomarkers Prevent* **5**: 917–922.

Pfeiffer CM, Huff DL, Smith SJ *et al.* 1999. Comparison of plasma total homocysteine measurements in 14 laboratories: an international study. *Clin Chem* **45**: 1261–1268.

Phillips DH, Castegnaro M, Autrup H *et al.* 1999. Standardization and validation of DNA adduct postlabelling methods: report of interlaboratory trials and production of recommended protocols. *Mutagenesis* **14**: 301–315.

Pickard NA. 1989. Collection and handling of patients specimens. In *Clinical Chemistry: Theory, Analysis and Correlation*, 2nd edn, Kaplan LA, Pesce AJ (eds). C. V. Moody: St. Louis, MO.

Ransohoff DF. 2005a. Lessons from controversy: ovarian cancer screening and serum proteomics. *J Natl Cancer Inst* **16**: 315–319.

Ransohoff DF. 2005b. Bias as a threat to the validity of cancer molecular-marker research. *Nat Rev Cancer* 2005; **5**: 142–149.

Riboli E, Haley NJ, De Waard F *et al*. 1995. Validity of urinary biomarkers of exposure to tobacco smoke following prolonged storage. *Int J Epidemiol* **24**: 354–358.

Rundle AG, Vineis P, Ahsan H. 2005. Design options for molecular epidemiology research within cohort studies. *Cancer Epidemiol Biomarkers Prevent* **14**: 1899–1907.

Steinberg K, Beck J, Nickerson D *et al*. 2002. DNA banking for epidemiologic studies: a review of current practices. *Epidemiology* **13**: 246–254.

Taioli E, Kinney P, Zhitkovich A *et al*. 1994. Application of reliability models to studies of biomarker validation. *Environ. Health Perspect* **102**: 306–309.

Thavasu PW, Longhurst S, Joel SP *et al*. 1992. Measuring cytokine levels in blood. Importance of anticoagulants, processing, and storage conditions. *J Immunol Methods* **153**: 115–124.

Vineis P, Alavanja M, Garte S. 2004. Dose–response relationship in tobacco-related cancers of bladder and lung: a biochemical interpretation. *Int J Cancer* **1**: 2–7.

Young DS, Bermes EW. 1986. Specimen collection and processing: sources of biological variation. In *Textbook of Clinical Chemistry*, Tiertz NW (ed.). W. B. Saunders: Philadelphia, PA.

Yanagisawa K, Shyr Y, Xu BJ *et al*. 2003. Proteomic patterns of tumour subsets in non-small-cell lung cancer. *Lancet* **362**: 433–439.

Ward CL, Dempsey MH, Ring CJ *et al*. 2004. Design and performance testing of quantitative real time PCR assays for influenza A and B viral load measurement. *J Clin Virol* **29**: 179–188.

7

Exposure Assessment

Mark J. Nieuwenhuijsen

Centre for Research in Environmental Epidemiology (CREAL), Barcelona, Spain

7.1. INTRODUCTION

'Exposure' is a substance or factor affecting human health, either adversely or beneficially. Exposure variables used in practice in epidemiology usually have to be regarded as *approximations* to the 'true' exposure of the subjects who are being studied. The accuracy and precision with which 'true' exposure is being approximated may vary widely from one 'surrogate' exposure variable to the next. Exposure misclassification and/or measurement error can lead to attenuation in health risk estimates and/or a loss of power. A simple example illustrates the potential effects of exposure misclassification and shows how important a good exposure assessment is. Let's assume a hypothetical example of a study with 2000 subjects, of whom 1000 are exposed to an environmental pollutant and the other 1000 are unexposed. In the exposed group, 200 have developed the disease, while only 100 have done so in the unexposed group (Table 7.1). The relative risk in this case is 2. However, the risk estimate would be only 1.5 if 20% of the subjects were misclassified (20% of the exposed become unexposed and vice versa; see numbers in brackets in Table 7.1). This example shows that even if the exposure misclassification is relatively minor (20%), the effect on the relative risk is considerable. Of course, this is a fairly simple example, and the effect of exposure misclassification/measurement error depends on other factors, including the type of error model (Classical or Berkson error; Nieuwenhuijsen, 2003).

7.2. INITIAL CONSIDERATIONS OF AN EXPOSURE ASSESSMENT STRATEGY

A major aim of epidemiological studies is to determine whether or not there is an association between a particular substance of interest, the exposure and morbidity and/or mortality. If there is an association, it is desirable to be able to show an exposure–response (or dose–response) relationship, i.e. a relationship in which the rate of disease increases when the level of exposure (or dose) increases. This will aid in the interpretation of such studies.

Over recent years there has been increasing interest in the field of exposure assessment causing it to develop rapidly. We know more than ever to what, where and how people are exposed and improvements have been made to methods for assessing the levels of exposure, its variability and the determinants. New methods have been developed or newly applied throughout this field, including analytical, measurement, modelling and statistical methods. This has led to a considerable improvement in exposure assessment in epidemiological studies, and therefore improvement in the epidemiological studies themselves.

In epidemiology, there are different study designs (e.g. cohort, case–control) to assess the association between exposure and disease (see Chapter 2). All the study designs require exposure estimates or exposure indices to be able to estimate the risk associated with the substance of interest, but they may differ depending on the study design. The design and interpretation of epidemiological

Table 7.1 Hypothetical example of the effect of exposure misclassification. The numbers in brackets are after 20% exposure misclassification

	Diseased	Non-diseased	Total
Exposed	200 (180)	800 (820)	1000
Non exposed	100 (120)	900 (880)	1000
Total	300	1700	2000

studies is often dependent on the exposure assessment and therefore needs careful consideration. Quantification of the relation between exposure and adverse human health effects requires the use of exposure estimates that are accurate, precise, biologically relevant, apply to the critical exposure period and show a range of exposure levels in the population under study. Furthermore, there is generally also a need for the assessment of confounders, i.e. substances that are associated with both exposure and disease and may bias the study results. Assessment of confounders should be in similar detail to the assessment of exposure indices, since measurement error in confounders may also affect the health risk estimates (Nieuwenhuijsen 2003).

In epidemiology we often deal with large population sizes, with the population spread over large distances. This may make estimating exposure for the subjects in the study more difficult, for example in environmental epidemiological studies, since we cannot go out and visit each of the subjects. Therefore, one often relies on some form of modelling or surrogate of exposure (e.g. distance to a source). Small sample sizes, on the other hand, may allow some more refined exposure assessment, such as personal monitoring. The size of the population determines how refined the exposure assessment could be. Increasing the population size could allow for cruder exposure estimates, while smaller population sizes will require more refined exposure estimates to have similar statistical power. Armstrong (1996) has provided a general framework that can help individual researchers to decide which measures of exposure to include in their study in order to obtain maximum statistical power, and which validity and reliability substudies to include to assess the quality of the exposure assessment methods used in the full study.

The basic premise of Armstrong's considerations is that it is better, but more expensive, to measure 'true' exposure than to measure 'approximate' exposure. When the correlation between the 'approximate' and the 'true' exposure variable is high, the loss of power by using the 'approximate' rather than the 'true' exposure variable is small. Therefore, if the cost per study subject of measuring 'approximate' exposure is clearly lower than the cost of measuring 'true' exposure, a study using the 'approximate' measure of exposure will be more efficient. Alternative study designs can be compared by calculating the so-called 'asymptotic relative efficiency' (ARE), defined as the ratio of the sample size necessary to achieve equal power to detect an association (Armstrong 1996). When r is the correlation between the 'approximate' and the 'true' exposure, C_i is the basic cost of including a subject in the study (e.g. related to assessment of disease status) and C_a and C_t are the cost of measuring 'approximate' and 'true' exposure, respectively, per study subject, then the ARE can be expressed as:

$$ARE_{a/t} = r^2 [(C_i + C_t)/(C_i + C_a)]$$

So, when the correlation between the 'approximate' and the 'true' exposure variable is equal to 0.5 ($r^2 = 0.25$), then the use of the 'approximate' variable is more efficient than the use of the 'true' variable when the total cost of including subjects by measuring the 'approximate' variable is more than four times lower than the cost of including subjects in the study by measuring the 'true' exposure variable. Differences in cost of this or even a larger magnitude can easily occur when the choice is between doing personal exposure measurements on all subjects compared to collecting information on sources, habits and/or occupations by questionnaire, and when the 'fixed' cost of including study subjects (C_i) is not too high compared to the cost of making exposure measurements *per se*.

A decision to use 'approximate' exposure variables needs to be based on knowledge of the correlation between these variables and the 'true' exposure, and often, this information needs to be obtained in a pilot study preceding the full study. This adds to the cost of the total study, and Armstrong (1996) also provides guidance for the relative allocation of budget for pilot and the main study.

Besides sample size and costs, other considerations to be taken into account in designing a specific exposure assessment strategy for an epidemiological study are, for example, accessibility to the subjects and availability of tools and measurement methods, particularly for historical assessments. For example, many environmental epidemiological studies use routinely collected health outcome data from subjects for whom only postcode (or zip code) location of residence is available and no contact with the subject can be made. Questionnaires, personal or biomonitoring cannot be used or conducted in this case and one is often restricted to modelling of environmental concentrations at the location of residence. A further disadvantage of this is that no or little information is available where the subject spends his/her time outside the residence, which may be important for obtaining information on total exposure. Estimating past exposure of subjects is difficult and the tools available are sparce. Subjects often tend to recall only limited information and personal monitoring is not possible. Biomonitoring is often not informative because the biological half-life for many substances is too short to provide helpful information on past exposures, and therefore one has to rely on some form of environmental modelling.

A further consideration is whether to obtain individual or group estimates. In the individual approach, exposure estimates are obtained at the individual level, e.g. every member of the population is monitored either once or repeatedly. In the group approach, the group is first split into smaller subpopulations, more often referred to as 'exposure groups', based on specific determinants of exposure, and group or ecological exposure estimates are obtained for each exposure group. In environmental epidemiological studies, exposure groups may be defined, e.g. on the basis of presence or absence of an exposure source (such as gas cooker or smoker in the house), distance from an exposure source (such as roads or factories) or activity (such as playing sport or not); in occupational studies exposure groups are often defined by job title. The underlying assumption is that subjects within each exposure group experience similar exposure characteristics, including exposure levels and vari-

ation. A representative sample of members from each exposure group can be personally monitored, either once or repeatedly. If the aim is to estimate mean exposure, the average of the exposure measurements is then assigned to all the members in that particular exposure group. Alternatively, other exposure estimates can be assigned to the groups, e.g. data from ambient air pollution monitors in the area where the subjects live. Ecological and individual estimates can be combined, e.g. in the case of chlorination by-products, where routinely collected trihalomethane measurements providing ecological estimates are sometimes combined with individual estimates on actual ingestion, showering and bathing (Nieuwenhuijsen *et al.* 2000).

Intuitively, it is expected that the individual estimates provide the best exposure estimates for an epidemiological study. This is often not true, however, because of within-subject variability in exposure and the limited number of samples on each individual. In general, in epidemiological studies, individual estimates lead to attenuated, although more precise, health risk estimates than ecological estimates. The ecological estimates, in contrast, result in less attenuation of the risk estimates, albeit with some loss of precision (Kromhout *et al.* 1996; Seixas and Sheppard 1996; Heederik *et al.* 1996; Nieuwenhuijsen 1997). These differences can be explained by Classical and Berkson-type error models. The between group, between subjects and within subject variance, can be estimated using analysis of variance (ANOVA) models and this information can be used to optimize the ability to detect an exposure–response relationship, e.g. by changing the distribution of exposure groups (Kromhout and Heederik 1995; Nieuwenhuijsen 1997; van Tongeren *et al.* 1997). In this case the aim is to increase the contrast of exposure between exposure groups, expressed as the ratio between the between group variance and the sum of the between- and within-group variance, while maintaining a reasonable precision of the exposure estimates of the exposure groups.

7.3. EXPOSURE PATHWAYS AND ROUTES

Generally there are a number of different pathways of exposure to a given contaminant, e.g. food, indoor and outdoor air, water, soil, workplace, and these may all need to be considered to obtain estimates of total exposure for subjects. Furthermore, there are three possible exposure routes for substances: inhalation through the respiratory system; ingestion through the gastrointestinal system; and absorption through the skin. The exposure route(s) of a substance depends on e.g. the biological, chemical and physical characteristics of the substances, the location and the activities of the person. Inhalation and deposition of particles through the respiratory system depends on the particle diameter and the breathing characteristics of the person. Smaller particles are more often inhaled and penetrate deeper into the lungs. Furthermore, inhalation depends on the breathing rate of the subject; those carrying out heavy work may inhale much more air and breathe more deeply (20 l/min for light work vs. 60 l/min for heavy work).

Skin absorption can play an important role for uptake of substances such as solvents, pesticides and trihalomethanes. Trihalomethanes are volatile compounds that are formed when water is chlorinated and the chlorine reacts with organic matter in the water. In this context there are a number of possible exposure pathways and routes (Figure 7.1). The main pathway of ingestion is generally drinking tap water or tap water-based drinks (e.g. tea, coffee and squash). Swimming, showering, bathing, and dish washing may all result in considerable uptake through inhalation and skin absorption and, for the former three, ingestion to minor extent. Water standing or flushing in the toilet may lead to uptake by inhalation, through volatilization of the chloroform. The total uptake of trihalomethanes may be assessed using the concentration measured in exhaled breath or serum.

In the human body, the uptake, distribution, transformation and excretion of substances such as trihalomethanes can be modelled using physiologically-based pharmacokinetic (PBPK) models (Nieuwenhuijsen 2003). These models are becoming more sophisticated, although they are still rarely used in environmental epidemiology. They can be used to estimate the contribution of various exposure pathways and routes to the total uptake and model the dose of a specific target organ. For example, where ingested trihalomethanes may

mostly be metabolized rapidly in the liver and not appear in blood, uptake through inhalation and skin increases the blood levels substantially. Furthermore, metabolic polymorphisms may lead to different dose estimates under similar exposure conditions.

7.4. EXPOSURE DIMENSIONS

The exposure to a substance can depend upon the following factors:

Duration (e.g. in hours or days) or amount (kg/day ingested)
Concentration (e.g. in mg/m^3 in air or mg/l in water)
Frequency (e.g. times/week)

It is important to recognize that there may be considerable variability in these factors, both temporally and geographically, which can be exploited by epidemiological studies. Any of these factors can be used as an exposure index in an epidemiological study, but they can also be combined to obtain a new exposure index, e.g. by multiplying duration and concentration to obtain an index of cumulative exposure. The choice of index depends on the health effect of interest. For substances that cause acute effects, such as ammonia (irritation), the short-term concentration is generally the most relevant exposure index, while for substances that cause chronic effects, such as asbestos (cancer), long-term exposure indices, such as cumulative exposure, may be a more appropriate exposure index.

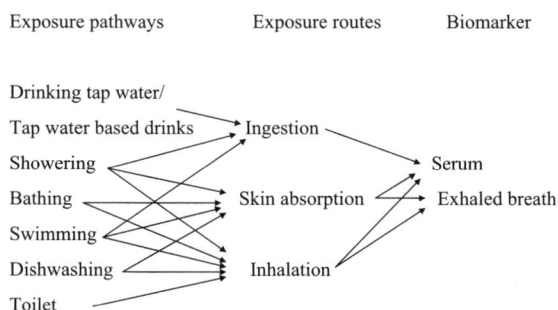

Figure 7.1   Examples of exposure pathways, routes and biomarkers for trihalomethanes

7.5. EXPOSURE CLASSIFICATION, MEASUREMENT OR MODELLING

Different tools, such as questionnaires, monitors and statistical techniques, are available for classifying exposures. The methods are often divided into direct and indirect methods (Figure 7.2).

The main aim of an exposure assessment is generally to obtain accurate, precise and biologically relevant exposure estimates in the most efficient and cost-effective way. As discussed before, the cost of the exposure assessment generally increases with increasing accuracy and precision, and therefore the assessment is often a balancing act between cost on one side and accuracy and precision on the other (Armstrong 1996). The choice of a particular method depends on the aim of the study and, more often, on the financial resources available.

Subjects in an epidemiological study can be classified to a particular substance on an ordinal scale, for example as exposed:

Yes/no
No, low, medium, high

This can be achieved by:

- *Expert assessment.* A member of the research team decides, based on prior knowledge, whether the subject in the study is exposed or unexposed, e.g. lives in an area with highly contaminated soil or not. Some environmental epidemiological studies have used simple proxies, e.g. distance from a (point) source, such as a factory (Dolk *et al.* 1999), radio and TV transmitters (Dolk *et al.* 1997), incinerators (Elliott *et al.* 1996), emissions from roads (Livingstone *et al.* 1996; English *et al.* 1999; Hoek *et al.* 2002) or landfill (Elliott *et al.* 2001), while others have categorized exposure by industrial sources, land use or urban zone (Barbone *et al.* 1995).

- *Self-assessment by questionnaire.* The subject in the study is asked to fill out a questionnaire in which he/she is asked about a particular substance, e.g. pesticides. Questionnaires are often used to ask a subject whether he/she is exposed to a particular substance and also for an estimation of the duration of exposure. They are also often used in nutritional epidemiology, in the form of food frequency questionnaires or diaries.

Questionnaires can be used not only to ask the subject to estimate his/her exposure but also to obtain information related to the exposure, such as where people spend their time (time microenvironment diaries), work history, including the jobs and tasks they carry out, what they eat and drink, and where they live. These variables could be used as exposure indices in epidemiological studies or translated into a new exposure index, e.g. by multiplying the amount of tap water people drink and the contaminant level in the tap water to obtain the total ingested amount of the substance.

Expert and self-assessment methods are generally the easiest and cheapest, but can suffer of a lack of objectivity and knowledge and may therefore bias the exposure assessment. Both experts and subjects may not know exactly what the subjects are exposed to or at what level, and therefore misclassify the exposure, while diseased subjects may recall certain substances better than subjects without disease (recall bias) and cause differential misclassification, leading to biased health risk estimates.

A more objective way to assess the exposure is by measuring the level of the contaminant in air, water or food. Here are some examples of such measurements:

- *Levels of outdoor air pollution* can be measured by ambient air monitors (Dockery *et al.* 1993; Katsouyanni *et al.* 1995; Dockery and Pope 1997).

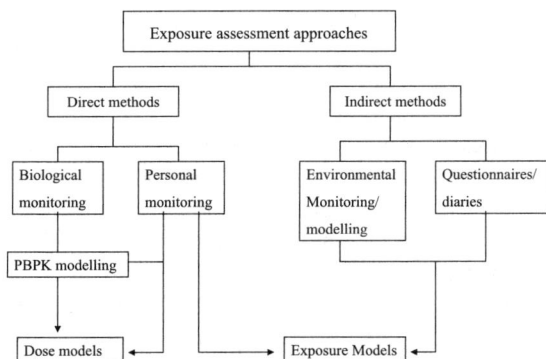

Figure 7.2 Different approaches to human exposure assessment. PBPK, physiological pharmacokinetic

These monitors are placed in an area and measure the particular substance of interest in this area. Subjects living within this area are considered to be exposed to the concentrations measured by the monitoring station. This may or may not be true, depending on, for example, where the subject in the study lives, works or travels. The advantage of this method is that it could provide a range of exposure estimates for a large population. Nowadays there are many monitors that are routinely monitoring air pollutants for regulatory purposes, particularly in cities in the developed world but also elsewhere.

- *Levels of air pollution* can be measured by personal exposure monitors (Magnus *et al.* 1998; Krämer *et al.* 2000; Magari *et al.* 2002; Brant *et al.* 2005). These monitors are lightweight devices that are worn by the subject in the study. They are often used in occupational studies and are becoming more frequently used in environmental studies. The advantage of this method is that it is likely to estimate the subject's exposure better than, for example, ambient air monitoring. The disadvantage is that this method is often labour intensive and expensive and can often only be used for relatively small populations. However, personal monitoring is ideal for validation studies, e.g. of modelled exposure estimates.

- *Levels of water pollutants and soil contaminants* can be estimated by taking water samples and soil samples, respectively, and analysing these for a substance of interest in the laboratory. Often these need to be combined with behavioural factors, such as water intake, contaminated food intake or hand-to-month contact, to obtain a level of exposure (Nieuwenhuijsen *et al.* 2000).

- *Levels of uptake of the substance into the body* can be estimated by biomonitoring, e.g. for lead (Bellinger and Schwartz 1997; Nieuwenhuijsen 2003). Biomonitoring consists of taking biological samples, such as urine, exhaled breath, hair, adipose tissue or nails, e.g. the measurement of lead in serum. The samples are subsequently analysed in a laboratory for the substance of interest or a metabolite. Biomonitoring is expected to estimate the actual uptake (dose) of the

substance of interest, rather than the exposure. Biomonitoring can be very informative, particularly for substances that have multiple pathways and routes. A major drawback is often the fairly short biological half-life of many substances, which makes this method only useful for estimating current exposures/doses. However, infectious disease epidemiology has extensively employed exposure biomarkers. In the case of infections and cancer, developmental work to establish validated laboratory assays for antibodies to viral or bacterial antigens, e.g. hepatitis viruses, human papilloma virus and *Helicobacter pylori*, has been central to understanding the aetiological role of these agents in epidemiological studies. Furthermore, the more recent emphasis on DNA and protein adducts has undoubtedly contributed substantially to establishing the biological plausibility of exposure–disease associations. Examples include investigations of the association between genotypes for carcinogen-metabolizing enzymes and adduct levels; or the use of biomarkers as modifiable end points in short-term intervention studies (Groopman *et al.* 1999). In contrast, the application of exposure biomarkers of this nature to aetiological studies is far more limited. Aflatoxin is perhaps the prime example in which the exposure biomarker permitted categorization of this environmental agent as a human carcinogen (Qian *et al.* 1994; Wang *et al.* 1996), others include polycyclic aromatic hydrocarbon–DNA adducts in lung cancer (Tang *et al.* 2001) and arylamine– haemoglobin adducts in bladder cancer (Gan *et al.* 2004). Further developments are envisaged (Wild *et al.* 2005).

The measurement of exposure is generally expensive, particularly for large populations, and, as mentioned above, can be restricted due to inaccessibility to subjects or the need for a historical assessment of exposure, rather than assessment of current exposure. It may be very useful for validation purposes.

Modelling of exposure can be carried out, preferably in conjunction with exposure measurements, either to help to build a model and/or to validate a model. It is in particularly important that the model estimates are validated.

Modelling can be divided into:

- *Deterministic modelling* (i.e. physical), in which the models describe the relationship between variables mathematically on the basis of knowledge of the physical, chemical and/or biological mechanisms governing these relationships (Brunekreef 1999; see Box 7.1).
- *Stochastic modelling* (i.e. statistical), in which the statistical relationships are modelled between variables. These models do not necessarily require fundamental knowledge of the underlying physical, chemical and/or biological relationships between the vari-

ables. Examples are regression and Bayesian modelling.

For regression modelling, a statistical regression model can be constructed, expressed in the form:

$$\mathrm{Ln}(C_{ij}) = \beta_0 + \beta_1 var_i + \beta_2 var_j + E$$

in which $\ln(C_{ij})$ denotes the log-transformed exposure concentration, β_0 the background level, var_x the potential determinant of exposure, β_x the regression coefficient of var_x providing the magnitude of the effect, and E a random variable with mean 0, often called the error term (for examples, see Boxes 7.2 and 7.3).

Box 7.1 Deterministic modelling

Hodgson *et al.* (2006) used an atmospheric dispersion modelling system (ADMS) to assess mercury dispersion in Runcorn in north-western England. ADMS uses algorithms that take account of stack height and diameter, volume flow rate, temperature and emission rates of pollutants, as well as meteorology, local geography, atmospheric boundary layer and deposition parameters, to calculate concentrations of pollutants at ground level. Three authorized processes were included in the model, a chloralkali plant, an associated multi-fuel power station, and a coal-fired power station. Compared to using distance as a proxy for exposure, the model identified a much smaller exposed population (Figure 7.3).

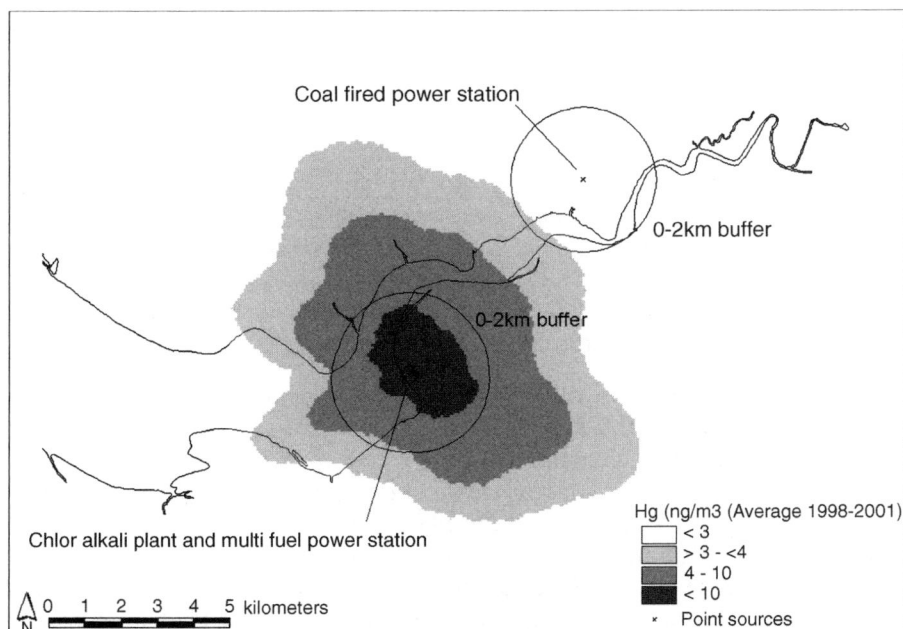

Figure 7.3 Comparison of modelled exposure output (average 1998–2001) to exposure analysis based on distance as a proxy for exposure for a study of mercury in the north-west of England

Box 7.2   Regression modelling

Harris *et al* (2002) measured 2,4-D (2,4-dichlorophenoxyacetic acid), mecopop [2-(4-chloro-2 methylphenoxy)propionic acid, MCPP] and dicamba (3,6-dichloro-*o*-anisic acid) in urine (two consecutive 24 h periods) collected from a group of 98 professional turf applicators from 20 companies across south-western Ontario. The group also filled out questionnaires to acquire information on all known variables that could potentially increase or decrease pesticide exposure to the amount handled, to build models for epidemiological studies. They used linear regression to assess the relationship between the concentrations of the substances in urine and the questionnaire data. They found that the volume of pesticide (active ingredient) applied was only weakly related to the total dose of 2,4-D absorbed ($R^2 = 0.21$). Two additional factors explained a large proportion of the variation in measured pesticide exposure, the type of spray nozzle used and the use of gloves while spraying. Individuals who used a fan-type nozzle had significantly higher doses than those who used a gun-type nozzle. Glove use was associated with significantly lower doses. Job satisfaction and current smoking influenced the dose but were not highly predictive. In the final multiple regression model, it was concluded that approximately 64% of the variation in doses could be explained by the small number of variables identified (Table 7.2). Biological monitoring in this case was important to be able to determine the true effect of wearing protective equipment, such as gloves. This study provided extremely useful information for epidemiological and health risk assessment studies, which could focus on obtaining information on these particular variables in a larger population.

Table 7.2   Regression models predicting the log of total dose of 2, 4 d in 94 volunteers ($r^2 = 0.64$)

Variable	Estimate	SE	p Value	Partial r^2
Intercept	−1.09	0.01	0.29	
Log spray	0.96	0.12	0.001	0.44
Nozzle	1.37	0.23	0.001	0.29
Glove wear	−1.50	0.25	0.001	0.29
Satisfaction	−0.39	0.17	0.021	0.06
Smoke	0.51	0.22	0.02	0.06

Reprinted by permission from Macmillan Publishers Ltd: Harris *et al.* (2002) *Journal of Exposure Science and Environmental Epidemiology.*

Box 7.3   Modelling using sparse data

Trihalomethane (THM) concentrations were used as the marker for chlorination by-products in a study of chlorination by-products and birth outcomes. In the UK, where the study was conducted, water samples are routinely collected and analysed from each water zone (population up to 50 000 people), using random samples at the tap (an average of four measurements per zone). Because of the small number of THM measurements in some water zones, the need for quarterly (3-monthly) estimates (to allow for trimester-weighted exposure estimates) and the problem of measurements below the limit of detection, it was necessary to model the raw THM data to obtain more robust estimates of the mean THM concentration in each zone. This was done using a hierarchical mixture model in the software WinBUGS (Bayesian inference using Gibbs sampling) (Spiegelhalter *et al.* 1996), as described in detail elsewhere (Whitaker *et al.* 2005). A three-component mixture model was fitted, in which zones were assumed to belong to one or some mixture of three components, which were labelled 'ground', 'lowland surface' and 'upland surface' waters (the components may not strictly correspond to these three water source types, and simply aimed to group waters with similar THM profiles, which are more likely to be shared among water of the same source type; Figure 7.4). The hierarchical model was assigned over the zone-specific mean individual THM concentrations, enabling zones to 'borrow' information from other zones with the same water source type. This resulted in more stable estimates for zones where few samples were taken. Seasonal variation was taken into account by estimating a quarterly effect common to all zones supplied by the same source type. The modelling provided estimates of exposures for various seasons (Figure 7.5). The modelled exposure estimates provided a better exposure–response relationship than when using estimates based on the mean of the raw THM concentrations for each zone.

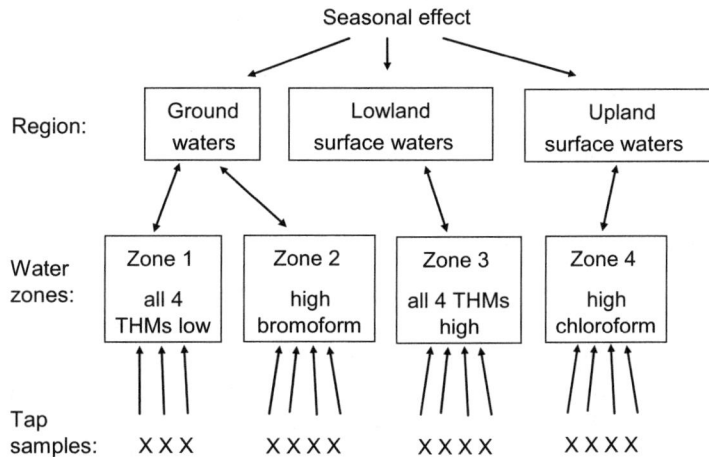

Figure 7.4 Hierarchical mixture model to estimate the water zone means of THMs by water source, using tap water samples and applying a common seasonal effect

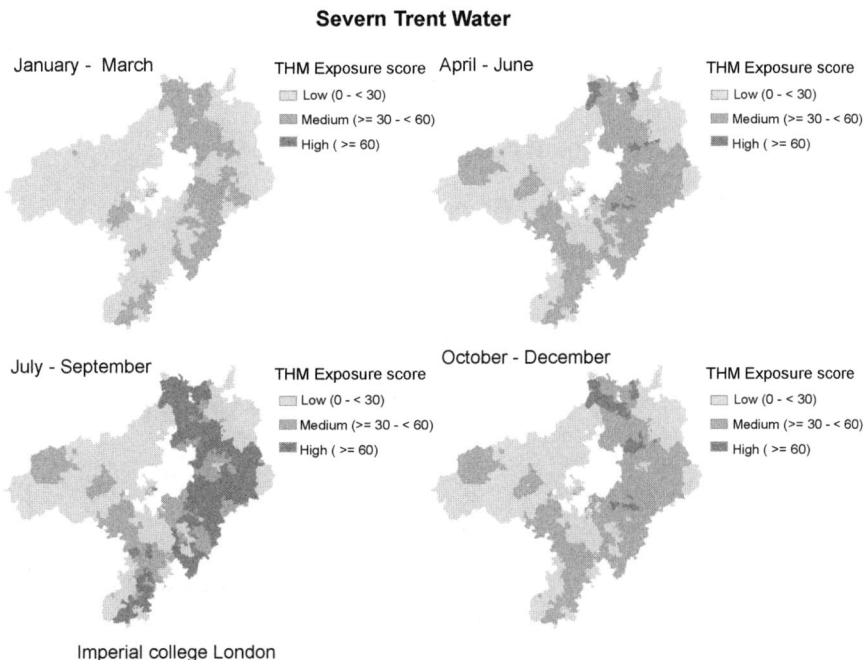

Figure 7.5 Modelled THM levels by water zone, using Bayesian mixture modelling

A problem in exposure assessment is that often few routinely collected measurements are available to model exposure estimates and therefore more sophisticated statistical techniques need to be used, as was demonstrated in a study of chlorination by-products (see Box 7.3).

In recent years, in environmental epidemiology many of the modelling methods have been greatly strengthened by the use of geographical information system (GIS) techniques (Nuckols *et al.* 2004; Briggs 2005). Looked at simply, GISs are computerized mapping systems. GISs, however, can do more than simply map data. They also provide the capability to integrate the data into a common spatial form and to analyse the data geographically. It is these capabilities that give GISs their special power in relation to exposure assessment. However, there are often some problems in acquiring the data needed to carry out geographic methods of exposure assessment. Also, a GIS requires that all data be *georeferenced*. Various GIS and geostatistical techniques have been used to model local pollution patterns, on the basis of the monitored data, e.g. using inverse distance weighting, kriging or focal sum methods. These essentially fit a surface through the available monitored data, in order to predict pollutant concentrations at unmeasured sites. It is an approach that has been

---

Box 7.4. GIS-based regression modelling

Regression techniques were used as part of the (SAVIAH) study, for example, to model exposures to NO_2 (as a marker for traffic-related air pollution) in four study cities (Briggs *et al.* 1997). Data from 80 monitoring sites were used to construct a regression equation, using information on road traffic (e.g. road network, road type, traffic volume), land cover/use, altitude and monitored NO_2 data. The results showed that the maps produced extremely good predictions of monitored pollution levels, both for individual and for the mean annual concentrations, with $r^2 \approx 0.79$–0.87 across 8–10 reference points, although the accuracy of the predictions for individual periods was more variable (Figure 7.6). Subsequently it was shown that regression models developed in one location could be applied successfully, with local calibration using only a small number of sites, to other study areas or periods (Briggs *et al.* 2000). More recently, the same approach has been further developed to assess exposures to particles in a number of different cities as part of the TRAPCA study (Brauer *et al.* 2003), and to model traffic-related air pollution in Munich (Carr *et al.* 2002).

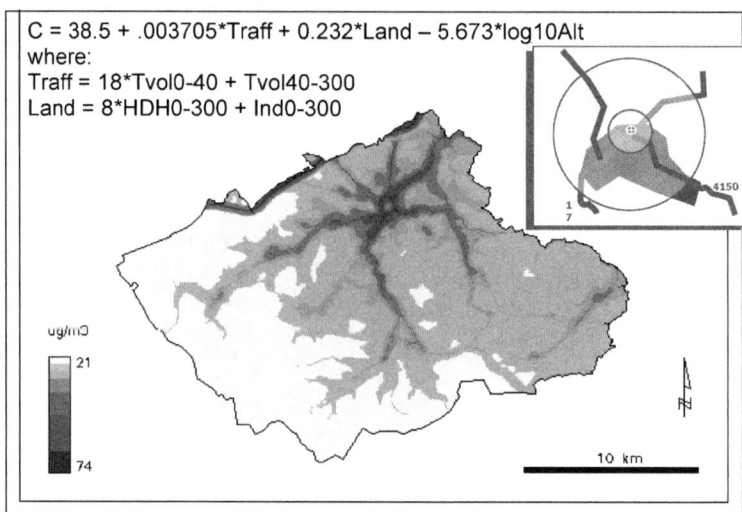

Figure 7.6   Modelled NO_2 levels in Huddersfield, using the SAVIAH approach

greatly facilitated through the development of GIS (Bayer-Oglesby 2004).

The approach appears to work well in areas where there is relatively gentle variation in air pollution and/or where the density of the monitoring network is high; conditions that are often not fulfilled. In other situations, it is helpful to supplement the available monitoring data through the use of covariates, i.e. variables that correlate with monitored concentrations and can be more readily obtained than quantitative measurements (Bayer-Oglesby 2004). Both cokriging and regression methods enable this (see Box 7.4).

People typically move about during the day and this mobility can greatly affect exposure. The importance of this was clearly shown by a study in Helsinki (Kousa *et al.* 2002), which used dispersion modelling to predict nitrogen dioxide concentrations across the city at different times of the day. These were then overlaid onto data showing where people were at different times, in order to build up a picture of exposure variations throughout the day. An interesting development in this area is that it is possible to track people through their environment using global positioning systems (GPS) with enough resolution to be useful for this type of study (Phillips *et al.* 2001; Elgethun *et al.* 2003). There are some restrictions as a result of the limitations in the technology, e.g. reception of the satellite signals can be adversely impacted by shielding from buildings of certain materials (concrete, steel), electrical power stations, and to some extent vehicle body panels. However, combining pollution maps with information on where people spend their time may greatly improve the exposure estimates if further improvements to the technology can be made.

All these different approaches are not exclusive and often are combined to obtain the best exposure index. It is often difficult or impossible to measure the exposure to the actual substance of interest and therefore exposure to an 'exposure surrogate' is estimated. The National Research Council (NRC) in the USA came up with a ranking of exposure data and surrogate measures around point sources, such as landfill sites (Table 7.2). The data at the top of the hierarchy shown in this table provide some fairly good information on the exposure of the subjects, while those at the bottom may not be helpful in the interpretation of exposure levels in an epidemiological study. Of course, there are still other issues that are important in the ranking, e.g. many area measurements may still be better than a few personal exposure measurements if exposure arises from a general source in the area.

7.6. RETROSPECTIVE EXPOSURE ASSESSMENT

A special challenge in epidemiological studies is studying disease with a long latency time, e.g. cancer; in such cases it is not the current exposures that are of most interest, but those that occurred in the past. A reconstruction of historical exposure, often referred to as retrospective exposure assessment, is therefore needed, and often involves some extensive modelling and specific expertise. Retrospective exposure assessment is difficult because there are often many changes occurring over time.

A good recent example of retrospective exposure assessment is provided by the study of air pollution and lung cancer in Stockholm, where the investigators used emission data, dispersion models and GIS to assess historical exposures to air pollutants and compared these estimates with actual measurements (Bellander *et al.* 2001). For NO_2 they used a detailed regional database, which included information on approximately 4300 line sources related to traffic and 500 point sources, including major industries and energy plants as well as small industry and ferries in ports. Limited diffuse emission sources, e.g. air traffic and merchant vessels in commercial routes, were treated as area sources, and several population density-related sources, such as local heating, were mapped as grid sources, as were work machine emissions. They collected information on the growth of urban areas, the development of district heating, and the growth and distribution of the road traffic over time. They used the Airviro model, together with population data, to derive population-weighted average exposures during 1955–1990. In the case of traffic-related NO_2, exposures were seen to increase over this period from about 15 $\mu g/m^3$ in 1955 to about 24 $\mu g/m^3$ in 1990, showing the effect of increasing traffic

Table 7.3. Hierarchy of exposure data and surrogates for fixed source contaminants

Type of data	Approximation to actual exposure
1. Quantified personal measurement	Best
2. Quantified area measurements in the vicinity of the residence or sites of activity	
3. Quantified surrogates of exposure (e.g. estimates of drinking water use)	
4. Distance from the site and duration of exposure	
5. Distance or duration of residence	
6. Residence or employment in the geographical area in reasonable proximity to the site where exposure can be assumed	
7. Residence or employment in a defined geographical area (e.g. a county) of the site	Worst

Adapted from NRC (1991).

volumes. In contrast, modelled SO_2 exposures fell from > 90 $\mu g/m^3$ to < 20 $\mu g/m^3$ as a result of improvements in fuel technology, emission controls and a shift to district heating.

7.7. VALIDATION STUDIES

In epidemiological studies it is often not possible to obtain detailed exposure information on each subject in the study. For example, in a large cohort study it is generally not feasible to take measurements on each subject and administer a detailed exposure questionnaire. In this case it is desirable to carry out a small validation study on a subset of the population that is representative of the larger population. Ideally this will be carried out before the main study starts and can make use of information from the literature. Questions in the questionnaire could be validated with measurements and exposure models could be constructed. The exposure assessment in the whole population could focus on key questions that have a large influence on the exposure estimates and thereby reduce the length of the questionnaire. Information on key determinants will also provide a better understanding of the exposure and how it may affect exposure–response relationships in epidemiological studies. Besides assessing the validity of the exposure surrogates, the reproducibility of the surrogates can also be evaluated in the subsample.

7.8. QUALITY CONTROL ISSUES

A well-designed and well-thought-out exposure assessment strategy carried out by well-trained personnel is essential for a successful exposure assessment. Issues such as cost, feasibility, accuracy, precision, validity, sample size, power, sensitivity, specificity, robustness and reproducibility always need to be addressed (e.g. during sampling, storage and analysis), while feasibility and pilot studies always need to take place before the actual study. Any form of bias (e.g. bias in sampling, selection, participation, monitoring, information, measurement error and exposure misclassification) should be avoided where possible, or if it takes places it should be clearly described.

Clear protocols for sampling, storage and analysis, including quality control should be written and be available at any time and researchers in the study should be properly trained. Potential sources of bias should be addressed at every stage.

Control measurement, e.g. air pollution filters that are not exposed but are otherwise treated as exposed filters, should be included (5–10% of total samples) particularly where measurements are close to the detection limit.

Samplers can measure with different accuracy, e.g. over- or under-sampling the true level, and this should be addressed when different samplers are used in order to reduce or avoid bias. This can be easily done by comparative sampling and adjusting for any difference observed.

References

Armstrong, B. 1996. Optimizing power in allocating resources to exposure assessment in an epidemiologic study. *Am J Epidemiol* **144**: 192–197.

Barbone F, Boveni M, Cavallieri F *et al.* 1995. Air pollution and lung cancer in Trieste, Italy. *Am J Epidemiol* **141**: 1161–1169.

Bayer-Oglesby L, Briggs D, Hoek G *et al.* 2004. Airnet Exposure Assessment Workgroup Report: http://airnet.iras.uu.nl/products/pdf/airnet_wg1_exposure_report.pdf

Bellinger D, Schwartz J. Effects of lead in children and adults. In *Topics in Environmental Epidemiology*, Steenland K, Savitz D (eds). Oxford University Press: New York; 314–349.

Bellander T, Berglind N, Gustavsson P *et al.* 2001. Using geographic information systems to assess individual historical exposure to air pollution from traffic and house heating in Stockholm. *Environ Health Perspect* **109**: 633–639.

Brant A, Berriman J, Sharp C *et al.* 2005. The changing distribution of occupational asthma: work-related respiratory symptoms and specific sensitisation in supermarket bakery workers. *Eur Resp J* **25**: 303–308.

Brauer M, Hoek G, van Vliet P *et al.* 2003. Estimating long-term average particulate air pollution concentrations: application of traffic indicators and geographic information systems. *Epidemiology* **14**: 228–239.

Briggs DJ, Collins S, Elliott P *et al.* 1997. Mapping urban air pollution using GIS: a regression-based approach. *Int J Geogr Inform Sci* **11**: 699–718.

Briggs DJ, de Hoogh C, Gulliver J *et al.* 2000. A regression-based method for mapping traffic-related air pollution: application and testing in four contrasting urban environments. *Sci Total Environ* **253**: 151–167.

Briggs D. 2005. The role of GIS: coping with space and time in air pollution exposure assessment. *J Toxicol Environ Health A* **68**: 1243–1261.

Brunekreef B. 1999. Exposure assessment. In *Environmental Epidemiology: A Textbook on Study Methods and Public Health Applications*. WHO: Geneva.

Carr D, von Ehrenstein O, Weiand S *et al.* 2002. Modelling annual benzene, toluene, NO_2, and soot concentrations on the basis of road traffic characteristics. *Environ Res* **A90**: 111–118.

Dolk H, Elliott P, Shaddick G *et al.* 1997. Cancer incidence near high power radio and TV transmitters in Great Britain: II. All transmitter sites. *Am J Epidemiol* **145**: 10–17.

Dolk H, Thakrar B, Walls P *et al.* 1999. Mortality among residents near cokeworks in Great Britain. *Occup Environ Med* **56**: 34–40.

Dockery DW, Pope A III, Xu X *et al.* 1993. An association between air pollution and mortality in six US cities. *N Engl J Med* **329**: 1753–1759.

Dockery DW, Pope CA. 1997. Outdoor air I: particulates. In *Topics in Environmental Epidemiology*, Steenland K, Savitz D (eds). Oxford University Press: New York; 119–166.

Elliott P, Arnold R, Cockings S *et al.* 2000 Risk of mortality, cancer incidence, and stroke in a population potentially exposed to cadmium. *Occup Environ Med* **57**: 647–648.

Elliott P, Shaddick G, Kleinschmidt I *et al.* 1996. Cancer incidence near municipal solid waste incinerators in Great Britain. *Br J Cancer* **73**: 702–710.

Elliott P, Briggs DJ, Morris S *et al.* 2001. Risk of adverse birth outcomes in populations living near landfill sites. *Br Med J* **323**: 363–368.

English P, Neutra R, Scalf R *et al.* 1999. Examining associations between childhood asthma and traffic flow using a geographic information system. *Environ Health Perspect* **107**: 761–767.

Gan JP, Skipper PL, Gago-Dominguez M *et al.* 2004. Alkylaniline–hemoglobin adducts and risk of non-smoking-related bladder cancer. *J Natl Cancer Inst* **96**: 1425–1431.

Groopman JD, Kensler TW. 1999. The light at the end of the tunnel for chemical-specific biomarker: daylight or headlight? *Carcinogenesis* **20**: 1–11.

Harris SA, Sass-Kortsak AM, Corey PN *et al.* 2002. Development of models to predict dose of pesticides in professional turf applicators. *J Expos Anal Environ Epidemiol* **12**: 130–144.

Heederik D, Kromhout H, Braun W. 1996. The influence of random exposure estimation error on the exposure–response relationship when grouping into homogeneous exposure categories. *Occup Hygiene* **3**: 229–241.

Hoek G, Brunekreef B, Goldbohm S *et al.* 2002. Associations between mortality and indicators of traffic-related air pollution in The Netherlands: a cohort study. *Lancet* **360**: 1203–1209.

Hodgson S, Nieuwenhuijsen MJ, Colvile R *et al.* In press. Identifying populations at risk of mercury exposure from industrial emissions. *Occup Environ Med*.

Katsouyanni K, Zmirou D, Spix C *et al.* 1995. Short-term effects of air pollution on health – a European approach using epidemiologic time-series data – the Aphea project – background, objectives, design. *Eur Resp J* **8**: 1030–1038.

Kousa A, Monn C, Rotko T *et al.* 2001. Personal exposures to NO_2 in the EXPOLIS study: relation to residential indoor, outdoor and workplace concentrations in Basel, Helsinki and Prague. *Atmos Environ* **35**: 3405–3412.

Krämer U, Koch T, Ranft U *et al.* 2000. Traffic-related air pollution is associated with atopy in children living in urban areas. *Epidemiology* **11**: 64–70.

Kromhout H, Heederik D. 1995. Occupational epidemiology in the rubber industry; implications of exposure variability. *Am J Indust Med* **27**: 171–185.

Kromhout H, Tielemans E, Preller L *et al.* 1996. Estimates of individual dose from current measurements of exposure. *Occup Hyg* **3**: 23–29.

Livingstone AE, Shaddick G, Grundy C *et al.* 1996. Do people living near inner city main roads have more asthma needing treatment? Case control study. *Br Med J* **312**: 676–677.

Magari SR, Schwartz J, Williams PL *et al.* 2002. The association between personal measurements of environmental exposure to particulates and heart rate variability. *Epidemiology* **13**: 305–310.

Magnus P, Nafstad P, Øie L *et al.* 1998. Exposure to nitrogen dioxide and the occurrence of bronchial obstruction in children below 2 years. *Int J Epidemiol* **27**: 995–999.

National Research Committee (NRC). 1991. *Environmental Epidemiology I. Public Health and Hazardous Waste.* National Academy Press: Washington, DC.

Nieuwenhuijsen MJ (ed.). 2003. *Exposure Assessment in Occupational and Environmental Epidemiology.* Oxford University Press: New York.

Nieuwenhuijsen MJ, Toledano MB, Elliott P. 2000. Uptake of chlorination disinfection by-products; a review and a discussion of its implications for epidemiological studies. *J Expos Anal Environ Epidemiol* **10**: 586–599.

Nuckols JR, Ward M, Jarup L. 2004. Using geographic information systems for exposure assessment in environmental epidemiological studies. *Environ Health Perspect* **112**: 1007–1015.

Qian GS, Ross RK, Yu MC *et al.* 1994. A follow-up study of urinary markers of aflatoxin exposure and liver cancer risk in Shanghai, People's Republic of China. *Cancer Epidemiol Biomarkers Prevent* **3**: 3–10.

Ranft U, Miskovic P, Pesch B *et al.* 2003. Association between arsenic exposure from a coal-burning power plant and urinary arsenic concentrations in Priedvidza District, Slovakia. *Environ Health Perspect* **111**: 889–894.

Seixas NS, Sheppard L. 1996. Maximazing accuracy and precision using individual and grouped exposure assessments. *Scand J Environ Work Health* **22**: 94–101.

Spiegelhalter D, Thomas A, Best N *et al.* 1996. BUGS 0.5; Bayesian inference using Gibbs sampling. Manual, Version ii. Available at: http://www.mrc-bsu.cam.ac.uk/bugs

Tang DL, Phillips DH, Stampfer M *et al.* 2001. Association between carcinogen–DNA adducts in white blood cells and lung cancer risk in the Physicians Health Study. *Cancer Res* **61**: 6708–6712.

van Tongeren M, Gardiner K, Calvert I *et al.* 1997. Efficiency of different grouping schemes for dust exposure in the European carbon black respiratory morbidity study. *Occup Environ Med* **54**: 714–719.

Wang LY, Hatch M, Chen CJ *et al.* 1996. Aflatoxin exposure and risk of hepatocellular carcinoma in Taiwan. *Int J Cancer* **67**: 620–625.

Ward MH, Nuckols JR, Weigel SJ *et al.* 2000. Identifying populations potentially exposed to agricultural pesticides using remote sensing and a Geographical Information System. *Environ Health Perspect* **108**: 5–12.

Whitaker H, Best N, Nieuwenhuijsen MJ *et al.* 2005. Hierarchical modelling of trihalomethane levels in drinking water. *J Expos Anal Environ Epidemiol* **15**: 138–146.

Wild CP. 2005. Complementing the genome with an 'exposome': the outstanding challenge of environmental exposure measurement in molecular epidemiology. *Cancer Epidemiol Biomarkers Prevent* **14**: 1847–1850.

8

Carcinogen Metabolites as Biomarkers

Stephen S. Hecht

The Cancer Center, University of Minnesota, Minneapolis, MN, USA

8.1. INTRODUCTION

Why would one want to use carcinogen metabolites as biomarkers in epidemiological studies of cancer? To investigate this question, let us first define some terms. A carcinogen is any agent, chemical, physical or viral, that causes cancer or increases the incidence of cancer (Schwab 2001). This chapter will consider only chemical agents. A metabolite is a substance produced from the carcinogen when the carcinogen is introduced into a living system, in this case a human. A biomarker is a distinctive biological or biologically derived indicator (as a metabolite) of a process, event or condition (*Merriam-Webster's Collegiate Dictionary* 2003).

Carcinogen metabolite biomarkers have been most widely used as monitors of exposure. The major advantage of using these biomarkers is that many of the uncertainties associated with exposure assessment can be bypassed. For example, measurements of carcinogens in the workplace or the general environment do not tell us much about individual exposure, which will depend on many other factors. Dietary questionnaires depend on recall and may not provide accurate carcinogen exposure assessment, even when levels of carcinogens in specific types of foods are known, and exposure to carcinogens from cigarette smoke will depend on how an individual smokes each cigarette, which may be quite different from measurements made by smoking machines. Assessment of exposure of non-smokers to secondhand smoke would be even more uncertain.

Carcinogen metabolite biomarkers can be used to determine individual uptake of specific carcinogens, and thus have great potential for exposure assessment in molecular epidemiological studies. A disadvantage of carcinogen metabolite biomarkers is that metabolism differs among individuals and can be affected in a given individual by genetic and environmental factors. This is one reason why there is seldom a perfectly linear relationship between carcinogen exposure and carcinogen metabolite biomarker measurements. On the other hand, individual differences in metabolism, as assessed by carcinogen metabolite biomarkers, also have the potential to be useful in identifying high-risk individuals via a carcinogen metabolite phenotyping approach.

This chapter first presents an overview of carcinogen metabolism. A clear understanding of carcinogen metabolism is critical for intelligent use of carcinogen metabolite biomarkers. Then it discusses in some detail two types of carcinogen metabolite biomarkers – one that is essentially an exposure marker that has been used in studies on carcinogen exposure from tobacco products and a second that has the potential to be used for identifying individuals at high risk for cancer upon exposure to carcinogenic combustion products. Finally, some examples of currently used carcinogen metabolite biomarkers are presented.

Molecular Epidemiology of Chronic Diseases, Edited by C. P. Wild, P. Vineis, and S. Garte
© 2008 John Wiley & Sons, Ltd

8.2. OVERVIEW OF CARCINOGEN METABOLISM

Carcinogens, drugs, pollutants and other foreign chemicals are generally classified as 'xenobiotics', a word derived from the Greek for 'foreign to the body' (Guengerich 1997). The metabolism of xenobiotics has been classified into phases I, II and III. All of these enzyme-mediated processes have one purpose: to facilitate removal of the xenobiotic from the system, usually by making it more water-soluble. The study of carcinogen metabolism has focused almost exclusively on phase I and II reactions (see Chapter 5), which are summarized in Table 8.1. Phase I reactions

Table 8.1 Classification of pathways of drug metabolism

Phase I reactions (functionalization reactions)
Oxidation via the hepatic microsomal P450 system

Aliphatic oxidation
Aromatic hydroxylation
N-Dealkylation
O-Dealkylation
S-Dealkylation
Epoxidation
Oxidative deamination
Sulphoxide formation
Desulphuration
N-Oxidation and *N*-hydroxylation
Dehalogenation

Oxidation via non-microsomal mechanisms
Alcohol and aldehyde oxidation
Purine oxidation
Oxidative deamination (monoamine oxidase and
    diamine oxidase)

Reduction
Azo- and nitro-reduction

Hydrolysis
Ester and amide hydrolysis
Peptide bond hydrolysis
Epoxide hydration

Phase II reactions (conjugation reactions)
Glucuronidation
Acetylation
Mercapturic acid formation
Sulphate conjugation
N-, *O*- and *S*-methylation
trans-Sulphuration

Reproduced from Pratt and Taylor (1990) with permission from Elsevier.

are often simple oxidation reactions, frequently catalysed by cytochrome P450 enzymes, found in the microsomal fractions (fragments of the endoplasmic reticulum) of liver and many other tissues. A number of other reactions are also observed, some of them catalysed by non-microsomal or cytosolic enzymes (Table 8.1). Phase II reactions add polar groups to phase I metabolites, increasing their water-solubility and therefore facilitating excretion. Phase II reactions include glucuronidation, acetylation and mercapturic acid formation (the end-product of glutathione-*S*-transferase conjugation). Phase III reactions are mediated by transporters such as P-glycoprotein and mutltidrug resistance-associated proteins.

Carcinogen metabolism is usually described in terms of metabolic activation and detoxification, based on the classic concepts of the Millers (Miller and Miller 1981; Miller 1994) (Figure 8.1). Metabolic activation reactions are most commonly phase I reactions, although there are exceptions. In metabolic activation, the carcinogen is converted to a more electrophilic form. Electrophiles are 'electron-seeking' and thus will readily react with electron-rich nucleophilic sites in DNA (and other cellular macromolecules). The reactions with DNA result in the formation of a covalent bond between the carcinogen metabolite and DNA. This covalently bound carcinogen metabolite is commonly known as a 'DNA adduct'. In some cases involving electrophilic carcinogens, DNA adducts form directly by reaction of the carcinogen with DNA. DNA adducts are absolutely critical in the carcinogenic process. Cellular repair enzymes can remove DNA adducts and return DNA to its normal state, but if adduct levels overwhelm the cellular repair mechanisms or certain tissues are deficient in the appropriate DNA repair enzyme, then the DNA adducts can persist. When cell replication occurs in the presence of DNA adducts, miscoding can result. Thus, DNA polymerases can misread DNA containing an adduct resulting in insertion of an incorrect base (e.g. adenine instead of cytosine opposite a guanine adduct). This produces a permanent mutation. If that mutation occurs in a critical region of a critical gene, such as *K-ras* or *p53*, the result can be loss of normal cellular growth control mechanisms, genomic instability and the development of cancer (Hecht 2003a).

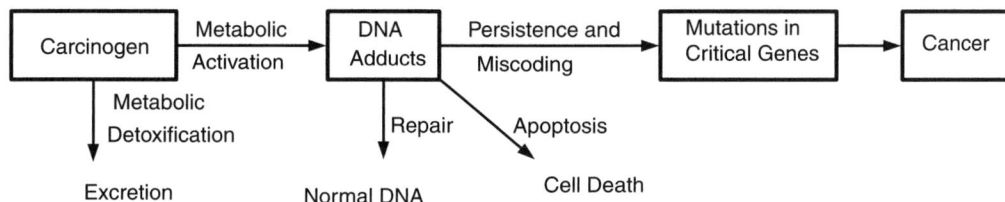

Figure 8.1 Overview of carcinogen metabolic activation to DNA adducts, resulting in mutations in critical genes and ultimately in cancer. Non-horizontal pathways are protective

In metabolic detoxification, the carcinogen is converted to a non-reactive, water-soluble form which is excreted. Some carcinogens are directly detoxified by phase I or II reactions, while in other cases phase II reactions follow phase I reactions.

Some examples of carcinogen metabolic activation and detoxification by phase I and II reactions are illustrated in Figure 8.2. In each case where metabolism is required, only some of the metabolites are shown as examples. More detailed descriptions of the metabolism of these carcinogens have been published (Hecht 1998; Cooper *et al.* 1983; Turesky 2002).

The first example is the tobacco-specific carcinogen 4-(methylnitrosamino)-1-(3-pyridyl)-1-butanone (known commonly as NNK, for nicotine-derived nitrosamino ketone). When NNK is introduced into almost all living systems, it is converted in part to NNAL (nicotine-derived nitrosamino alcohol) by carbonyl reductase and related enzymes (phase I). NNAL is then conjugated as glucuronides by uridine diphosphate glucuronosyltransferases (UGTs) (a phase II reaction) to produce NNAL-*O*-Gluc and other glucuronides, which are detoxification products. Hydroxylation of NNK at its methyl carbon, catalysed by cytochrome P450s (phase I), produces an unstable intermediate which spontaneously is converted to an electrophile that reacts with DNA to produce adducts.

The second example is the prototypic polycyclic aromatic hydrocarbon (PAH) benzo(*a*)pyrene (BaP), formed in the incomplete combustion of organic matter and a common environmental pollutant. It undergoes a three-step metabolic activation process involving phase I enzymes. First, it is converted to BaP-7,8-epoxide by cytochrome P450s. Then the epoxide is hydrated with catalysis

by epoxide hydrolase (EH). The resulting BaP-7,8-diol is then further oxidized by cytochrome P450s and other enzymes to give BaP-7,8-diol-9,10-epoxide (BPDE), an electrophile that reacts with DNA to produce adducts. Competing with this are detoxification reactions that proceed by way of 3-HOBaP and BaP-4,5-epoxide, as examples.

The third example is ethylene oxide, used in hospital and industrial sterilization. Ethylene oxide is a carcinogen which does not require metabolic activation. The epoxide ring is electrophilic and reacts with DNA directly.

The fourth example is 2-amino-1-methyl-6-phenylimidazo(4,5-*b*)-pyridine (PhIP), a carcinogen found in cooked meats. Cytochrome P450s catalyse its metabolic activation to an electrophile, *N*-hydroxy-PhIP, which can react with DNA to form adducts. This intermediate can be conjugated as a glucuronide, representing metabolic detoxification.

8.3. EXAMPLES OF CARCINOGEN METABOLITE BIOMARKERS

Total NNAL (NNAL plus its glucuronides): an established biomarker of exposure to the tobacco-specific lung carcinogen NNK

NNAL in its unconjugated form, together with its glucuronides (Figure 8.2), are quantitatively significant metabolites of the tobacco-specific carcinogen NNK. NNK is considered to be one of the most important carcinogens in tobacco products and, together with the related compound *N*'-nitrosonornicotine, has been evaluated as a human carcinogen by the International Agency for Research on Cancer (IARC 2004, 2008). The lung is the major target organ of NNK in laboratory animals, independent of the route of administration (Hecht 1998). Low

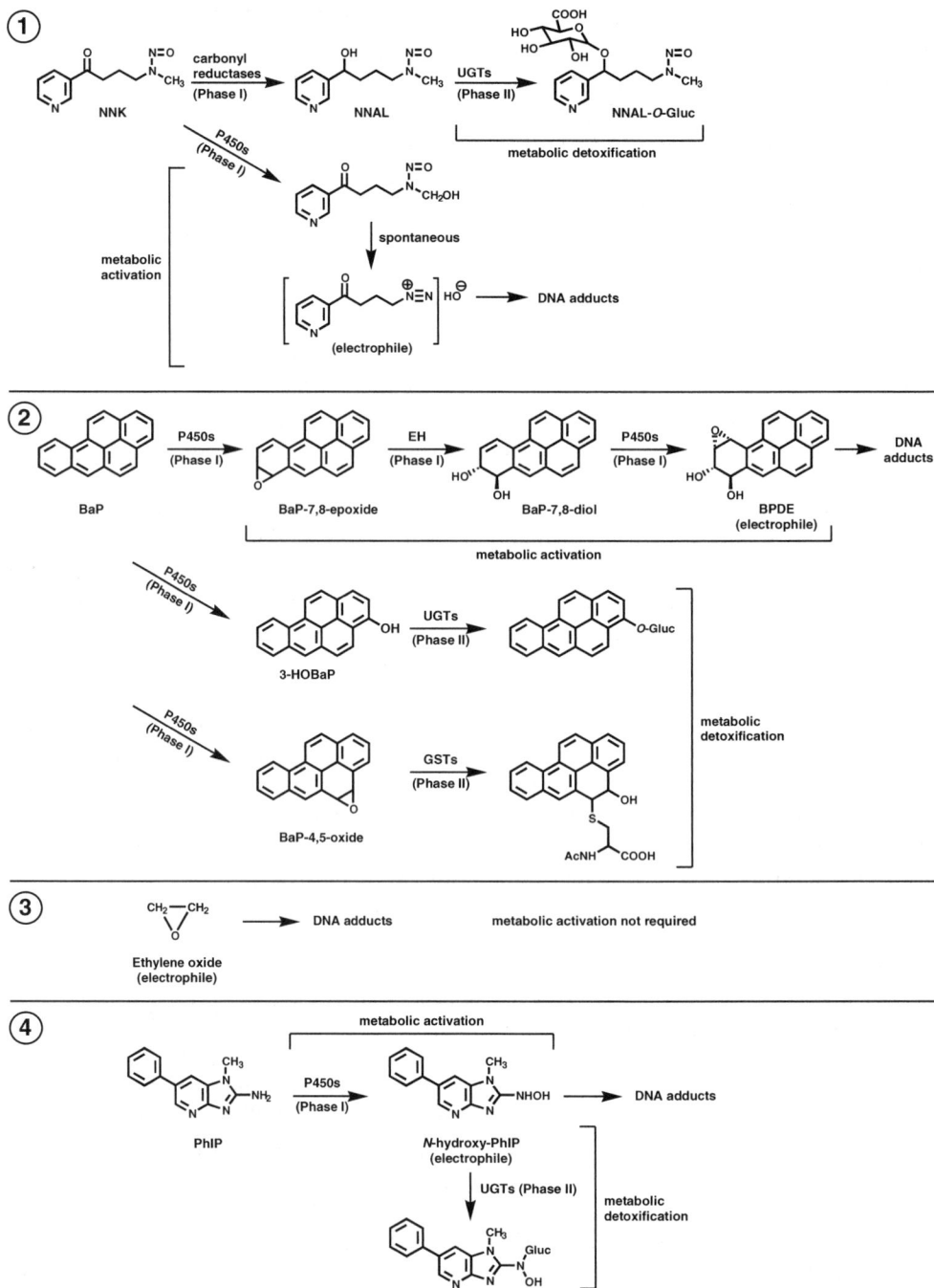

Figure 8.2 Examples of metabolic activation and detoxification of representative carcinogens. The schemes are oversimplified; more detailed accounts can be found (Hecht 1998; Cooper *et al.* 1983; Turesky 2002). P450s, cytochrome P450s; NNK, 4-(methylnitrosamino)-1-(3-pyridyl)-1-butanone; NNAL, 4-(methylnitrosamino)-1-(3-pyridyl)-1-butanol; UGTs, uridine diphosphate glucuronosyltransferases; Gluc, glucuronide; BaP, benzo(*a*)pyrene; BPDE, BaP-7,8-diol-9,10-epoxide; EH, epoxide hydrolase; GSTs, glutathione-*S*-transferases; PhIP, 2-amino-1-methyl-6-phenylimidazo(4,5-*b*)pyridine

doses of NNK reproducibly produce lung adenoma and adenocarcinoma in rodents, with rats being particularly sensitive. Because NNK is a chemical relative of nicotine, it is specific to tobacco products. While NNK itself is undetectable in human urine, total NNAL can be reliably detected in the urine of smokers, smokeless tobacco users and non-smokers exposed to secondhand cigarette smoke (Hecht 2002a). Since NNAL cannot arise from exposures other than those that are tobacco-related, it is a powerful and specific biomarker for estimating carcinogen uptake from tobacco products.

The first requirement for a useful carcinogen metabolite biomarker is analytical specificity and sensitivity. One must be certain of analyte identity and quantitation. Published methods for total NNAL are highly specific and have adequate sensitivity (Carmella *et al.* 2003, 2005). Urine is treated with β-glucuronidase to release NNAL from its glucuronide conjugates and total NNAL (the sum of the released NNAL and that which was already present as free NNAL) is determined, after a purification step, by either gas chromatography with nitrosamine-selective detection (GC–TEA) or liquid chromatography–electrospray ionization–tandem mass spectrometry. Both methods are highly specific and produce clean peaks which can be assigned to NNAL with complete certainty. Both methods also have adequate sensitivity for quantitation of total NNAL in relatively small urine samples, or in plasma, and are accurate and precise.

Total NNAL in urine has all the hallmarks of a carcinogen metabolite biomarker useful for assessing carcinogen exposure in people who use tobacco products or are exposed to secondhand cigarette smoke. There are several important questions to answer in evaluating such a biomarker. First, is there a difference between tobacco users and non-users? Second, does the biomarker decrease on cessation of tobacco use? Third, is there a dose–response with use? And finally, does the biomarker change with reduced use (Hatsukami *et al.* 2006)?

With respect to total NNAL, the answer to each of these questions is yes. Levels of total NNAL are typically about 2–4 pmol/ml urine in smokers and smokeless tobacco users and about 0.05 pmol/ml urine in non-smokers exposed to secondhand smoke (Hecht 2002b). Total NNAL is

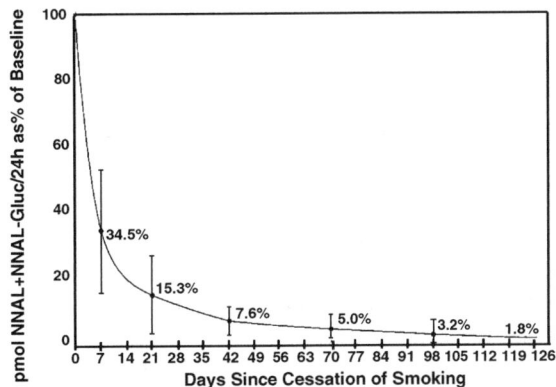

Figure 8.3 Mean levels of total NNAL in the urine of 27 subjects at various intervals after smoking cessation

not detected in non-exposed non-smokers. Total NNAL gradually disappears from urine upon cessation of smoking, as shown in Figure 8.3 (Hecht *et al.* 1999). Interestingly, the decrease is relatively gradual, surprising for a low molecular weight water-soluble compound. This suggests that NNAL binds to a receptor in the body, but the nature of this proposed receptor is currently unknown. Levels of total NNAL correlate significantly ($r = 0.5$, $p < 0.0001$) with cigarettes smoked per day, but there is considerable scatter (Joseph *et al.* 2005). This is likely due to metabolic differences, inaccurate reporting of cigarettes per day, and differences in smoking intensity at different levels of smoking. Total NNAL decreases with reduced number of cigarettes per day, but not to the extent expected based purely on the number of cigarettes (Hecht *et al.* 2004). This is illustrated in Figure 8.4. In this study, smokers reduced their cigarettes by 25% during the first 2 weeks, 50% in the next 2 weeks and 75% in the next 2 weeks of the study, and were then encouraged to remain at this level. As shown in Figure 8.4, there was significant reduction of total NNAL at each time point, but it was not as great as the reduction in cigarettes per day. For example, at the 6 week time point, there was a 73% reduction in cigarettes per day, but only a 30% reduction in total NNAL. This was subsequently shown to be due to the well known phenomenon of compensation, in which smokers who smoke fewer cigarettes per day will smoke those cigarettes differently, presumably to compensate for the lower dose of nicotine.

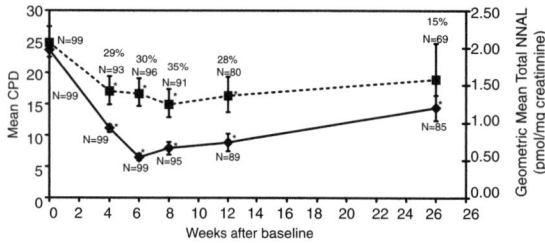

Figure 8.4 Cigarettes/day (CPD; solid line) and total NNAL/mg creatinine (dashed line) in the urine of smokers who reduced their smoking. Percentage figures are the mean extent of reduction of total NNAL from baseline. Values with asterisks are significantly lower than baseline values

Consistently, total NNAL correlates with total cotinine (cotinine plus its glucuronides, metabolites of nicotine). In one of the larger studies, the correlation coefficient was 0.62 ($p < 0.0001$; Joseph *et al.* 2005). Total cotinine is an accepted biomarker of exposure to tobacco products (Lee 1999). Its levels in urine are about 10 000 times as great as those of total NNAL. As a metabolite of nicotine, it has many features in common with NNAL, a metabolite of NNK. Prominent among these is its tobacco-specificity (although it is also observed in people using nicotine replacement products). The correlation of total NNAL and total cotinine is consistent with the data described above characterizing total NNAL as an exposure marker. The main difference between total NNAL and total cotinine is that total NNAL directly measures exposure to a lung carcinogen, NNK, while total cotinine is a metabolite of nicotine, which is addictive but non-carcinogenic (Hecht 2003b).

While total NNAL is an excellent biomarker of exposure, there may be potential in examining the ratio of NNAL-Glucs to free NNAL. There is a wide range of NNAL-Glucs:NNAL ratios in smokers (Hecht 2002b). Since NNAL-Glucs is a detoxification product of NNAL and NNK, but free NNAL is carcinogenic, with activity similar to that of NNK, a higher NNAL-Gluc:NNAL ratio may be indicative of protection against cancer. This hypothesis requires further investigation.

Total NNAL has been particularly useful to document carcinogen uptake in non-smokers exposed to secondhand cigarette smoke (Hecht 2003b). The tobacco-specificity of total NNAL is particularly important in this respect. Finding NNAL and its glucuronides in the urine of non-smokers unambiguously demonstrates uptake of NNK from secondhand smoke, because total NNAL cannot have dietary or other environmental sources. The high analytical sensitivity of the total NNAL assay is also important because levels of this biomarker in the urine of non-smokers exposed to secondhand smoke are generally quite low.

As shown in Figure 8.5, there is NNK uptake throughout life in non-smokers exposed to secondhand smoke in various circumstances. While a range of exposures has been observed, the amount of total NNAL found in the urine of individuals exposed to secondhand smoke is typically about 0.05 pmol/ml urine, based on analyses of nearly 500 exposed non-smokers. This translates to excretion of about 75 pmol total NNAL per day. Based on 20% metabolic conversion of NNK to NNAL, daily uptake of NNK by non-smokers exposed to secondhand smoke would typically be about 375 pmol (78 ng). Integrated over 30 years of exposure, this amounts

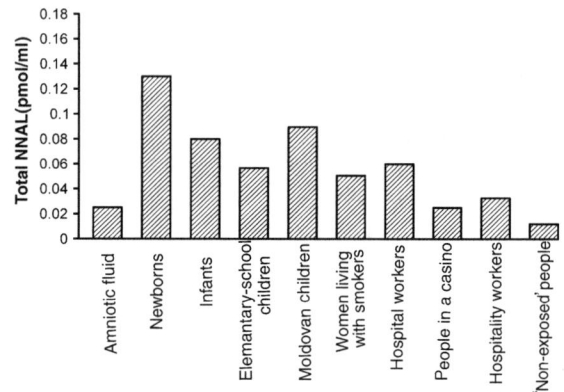

Figure 8.5 Mean levels of total NNAL in the urine of non-smokers exposed to secondhand smoke at various stages of life and in different environments (Anderson *et al.* 2001; Milunsky *et al.* 2000; Parsons *et al.* 1998; Lackmann *et al.* 1999; Anderson *et al.* 2003; Stepanov *et al.* 2006; Hecht SS *et al.* 2001; Tulunay *et al.* 2005)

to about 0.9 mg NNK uptake, or 0.01 mg/kg. The lowest total dose of NNK shown to induce lung tumours in rats, as part of a dose–response, was 1.8 mg/kg, about 200 times higher than the estimated dose to secondhand smoke-exposed humans (Belinsky 1990). It is plausible that there may be some individuals, with high susceptibility to the effects of NNK and other lung carcinogens in secondhand smoke, who would ultimately develop lung cancer after exposure to even these relatively low carcinogen levels. Looked at in another way, the amount of total NNAL in the urine of people exposed to secondhand smoke is about 2–3% as great (0.05/2 × 100) as that in the urine of smokers. The risk of a non-smoker exposed to secondhand smoke to develop lung cancer is about 1–2% as great as that of a smoker, remarkably consistent with the total NNAL biomarker data (Anderson *et al.* 2001).

Total NNAL allows a rough estimation of carcinogen exposure from secondhand smoke which would not be possible in the absence of biomarkers. Total NNAL has emerged as the leading biomarker for determining carcinogen exposure in non-smokers exposed to secondhand smoke. The only other routinely used secondhand smoke biomarker that has comparable tobacco-specificity is total cotinine.

Phenanthrene metabolites: developing biomarkers of PAH exposure and metabolism

The purpose of this section is to illustrate how carcinogen metabolite biomarkers could provide essential information pertinent to inter-individual differences in carcinogen metabolism. As illustrated in Figure 8.2, the metabolism of the prototypic PAH carcinogen, BaP, is quite complex. The major metabolic activation pathway to DNA adducts proceeds through BPDE as illustrated (Conney 1982). Other metabolic activation pathways have also been proposed, but the one illustrated is best supported by currently available data from laboratory animals and humans (Yu *et al.* 2002; Cavalieri and Rogan 1995). Figure 8.2 also illustrates some detoxification pathways of BaP metabolism. There are other detoxification pathways as well, not

shown for simplicity. Multiple forms of several enzymes are involved in both the metabolic activation and detoxification of BaP and other PAH. For example, cytochrome P450s catalyse the initial and final steps of the metabolic activation pathway but these same enzymes also catalyse the detoxification of BaP through formation of 3-HOBaP and other HOBaP isomers.

The metabolic activation of PAH to DNA adducts varies widely among individuals (Harris *et al.* 1976; Nebert 2000; Alexandrov *et al.* 2002; Sabadie *et al.* 1981; Nowak *et al.* 1988; McLemore *et al.* 1990; Kiyohara *et al.* 1998). This observation leads to the major hypothesis that people who efficiently metabolically activate PAH will be at a higher risk for cancer. This hypothesis has been extensively tested in molecular epidemiology studies that have investigated polymorphisms in genes involved in the metabolic activation and detoxification of PAH. The results of these studies have been mixed but there are indications that certain polymorphisms in *CYP1A1* and *GST* genes may, for example, lead to enhanced lung cancer risk (Houlston 2000; Bartsch *et al.* 2000; Benhamou *et al.* 2002; Hung *et al.* 2003; Le Marchand *et al.* 2003; Vineis *et al.* 2003; Lee *et al.* 2002; Taioli and Pedotti 2005; Liu *et al.* 2005; Vineis *et al.* 2004.

Due to the complexity of PAH metabolism, it is unlikely that inter-individual differences in metabolism can be assessed simply by measurement of variants in one or a few genes involved in the process. This leads to the proposal that carcinogen metabolite phenotyping through the measurement of carcinogen metabolite biomarkers resulting from PAH activation and detoxification may be a more reliable way to identify those individuals who efficiently metabolically activate PAH and therefore may be at higher risk for cancer. This line of research has been pursued by quantifying phenanthrene (Phe) metabolites in human urine (Hecht *et al.* 2003; Carmella *et al.* 2004; Hecht *et al.* 2005. Phe is the simplest PAH molecule with a 'bay region', a feature closely associated with carcinogenicity, and illustrated in Figure 8.6 for Phe and BaP (Conney 1982; Thakker *et al.* 1985). Phe metabolism is similar in many ways to that shown for BaP (Figure 8.2), but the products are far more abundant in human urine, and therefore

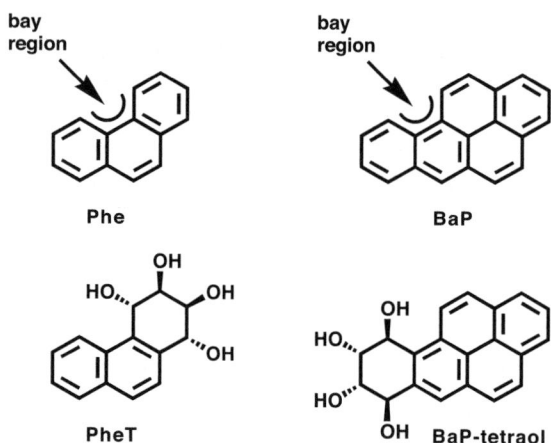

Figure 8.6   Structures of Phe, PheT, BaP, and BaP-tetraol

Phe metabolites are practical biomarkers (Hecht *et al.* 2003). As shown in Figure 8.2, a major metabolic activation pathway of BaP proceeds through formation BPDE, which forms adducts with DNA (Cooper *et al.* 1983; Baird and Ralston 1997). However, most of the BPDE which is produced reacts with H_2O, yielding BaP-tetraol (structure in Figure 8.6). The analogous pathway of Phe metabolism yields *r*-1,*t*-2,3,*c*-4-tetrahydroxy-1,2,3,4-tetrahydrophenanthrene (PheT, Figure 8.6), which can be considered as a measure of PAH metabolic activation by the diol epoxide pathway (Shou *et al.* 1994). Phe and BaP are both metabolized to phenols (HOBaP and HOPhe), which are considered to be detoxification products (Cooper *et al.* 1983; Shou *et al.* 1994). The ratio PheT:HOPhe should be characteristic of an individual, and it is proposed that individuals with higher PheT:HOPhe ratios would be at higher risk for cancer upon exposure to PAH.

As in the case of total NNAL, the first requirement for a useful carcinogen metabolite biomarker is analytical specificity and sensitivity. PheT and HOPhe can be measured with absolute specificity and high sensitivity by gas chromatography–negative ion chemical ionization–mass spectrometry (GC–NICI–MS) and positive-ion chemical ionization (GC–PICI–MS), respectively (Hecht *et al.* 2003; Carmella *et al.* 2004). Urine is treated with β-glucuronidase and aryl sulfatase to release the free metabolites and, after appropriate purification

steps, PheT and HOPhe are quantified. A method for the quantitation of PheT in plasma has also been described (Carmella *et al.* 2006). These methods are accurate and precise, and provide reliable data on levels of PheT and HOPhe in human samples. Another advantage of measuring Phe metabolites is that they are mainly excreted in urine, whereas metabolites of higher molecular weight PAH are excreted predominantly in faeces (Cooper *et al.* 1983).

Initial studies on levels of PheT in the urine of people exposed to PAH from different sources indicated that PheT is a good carcinogen metabolite biomarker of exposure (Hecht *et al.* 2003). Thus, the highest levels of PheT were found in psoriasis patients who were receiving Goekerman coal tar therapy, a rich source of PAH exposure. The next highest levels were found in coke oven workers exposed to PAH, followed by smokers and then non-smokers. Levels of PheT in these samples were consistent with levels of PAH exposure and with those of 1-hydroxypyrene (1-HOP), a well established biomarker of PAH exposure, discussed further in the next section. Similar conclusions were reached in a study of workers exposed to residual fly ash (Kim *et al.* 2005). Thus, PheT appears to be a good biomarker of exposure but, since its levels reflect the extent of the diol epoxide metabolic activation pathway, it also has the potential to be a biomarker of exposure *plus* metabolic activation, representing the total flux of activated metabolites through this pathway in a given individual.

The ratio PheT:HOPhe can be used to achieve a purer measure of metabolic differences. One major goal would be to determine the longitudinal consistency of this ratio in a person. This would provide essential information with respect to the design of epidemiological studies using this biomarker. Unfortunately, such studies are seldom carried out for many of the currently used biomarkers (Groopman and Kensler 1999). Levels of PheT, HOPhe and PheT:HOPhe ratios were relatively constant in most individuals, with mean coefficients of variation in the range 29.3–45.7% (Hecht *et al.* 2005). There were no significant changes over time in levels of the metabolites or in ratios. These results indicated that a single urine sample should be sufficient when comparing

Phe metabolites in different groups. PheT:HOPhe ratios were significantly higher in smokers than in non-smokers demonstrating that smoking induces the diol epoxide metabolic activation pathway of Phe. This finding is consistent with previous studies indicating that inducibility of PAH metabolism contributes to cancer risk in smokers (Nebert *et al.* 2004).

In order to determine whether there was a relationship between urinary PheT:HOPhe ratios and genetic polymorphisms, these parameters were measured in 346 smokers who were also genotyped for 11 polymorphisms in genes involved in PAH metabolism: *CYP1A1Msp*I, *CYP1A1*I462V, *CYP1B1*R48G, *CYP1B1*A119S, *CYP1B1*L432V, *CYP1B1*N453S, *EPHX1*Y113H, *EPHX1*H139R, *GSTP1*I105V, *GSTP1*A114V and *GSTM1* null. *CYP1A1* and *CYP1B1* code for cytochromes P450 1A1 and 1B1, involved in *both* the metabolic activation of BaP and Phe by the diol epoxide pathway, and their detoxification by phenol formation (Figure 8.2) (Shou *et al.* 1994; Bauer *et al.* 1995; Shou *et al.* 1994; Kim *et al.* 1999; Shimada *et al.* 1999; Shimada *et al.* 1996). *EPHX1* codes for microsomal epoxide hydrolase (EH) which is involved in the metabolic activation of BaP and Phe through the diol epoxide pathway (Cooper *et al.* 1983; Guenther and Oesch 1981). *GSTM1* and *GSTP1* code for glutathione *S*-transferases (GSTs), which catalyse the detoxification of BPDE and other PAH diol epoxides (Sundberg *et al.* 1997, 1998, 2002; Robertson *et al.* 1986). The results showed that about 10% of the smokers had relatively high PheT:3-HOPhe ratios. There was a significant association between the presence of the *CYP1A1*I462V polymorphism and high PheT:3-HOPhe ratios and the effect was particularly strong in females and in combination with the *GSTM1* null polymorphism. In contrast, the *CYP1B1*R48G and *CYP1B1*A119S polymorphisms were associated with significantly lower PheT:3-HOPhe ratios, particularly in blacks. There were no consistent significant effects of any of the other polymorphisms on PheT:3-HOPhe ratios. The highest 10% of PheT:3-HOPhe ratios could not be predicted by the presence of any of the 11 polymorphisms individually or by certain combinations. The effects of the *CYP1A1*I462 polymorphism, particularly in combination with

GSTM1 null, are quite consistent with reports in the literature. However, the overall results of the study indicate that genotyping is not an effective way to predict PAH metabolism, at least as represented by PheT:HOPhe ratios. Carcinogen metabolite phenotyping by the measurement of carcinogen metabolite biomarkers should be a more accurate approach, in spite of its greater technical difficulty at present. However, high-throughput methods are being developed and it is plausible that carcinogen metabolite phenotyping could be applied in the future in large molecular epidemiology studies.

Other examples of carcinogen metabolite biomarkers

Pyrene is a non-carcinogenic component of all PAH mixtures. 1-Hydroxypyrene (1-HOP), excreted as its glucuronide in human urine, is a well established biomarker of pyrene exposure, which has been measured in hundreds of studies of occupational and environmental exposures (Jongeneelen 1994, 2001). One advantage of the 1-HOP biomarker is its relative ease of measurement, by HPLC with fluorescence detection, equipment that is generally considerably less expensive than mass spectrometers.

Naphthalene is a component of all PAH mixtures and is carcinogenic in rats when administered by inhalation (National Toxicology Program 2000). Naphthols are another useful biomarker of PAH exposure (Nan *et al.* 2001; Yang *et al.* 1999). There is some indication that urinary naphthols may be particularly appropriate as biomarkers of inhalation exposure to PAH, because of the volatility of naphthalene.

S-Phenylmercapturic acid (S-PMA) is a metabolite of the human leukaemogen benzene. It is formed by normal degradation of the glutathione conjugate of benzene oxide, which is a phase I metabolite of benzene. S-PMA appears to have greater specificity for benzene exposure than *trans,trans*-muconic acid, another benzene metabolite that has been widely measured, but has sources other than benzene. Other benzene metabolites investigated as biomarkers include phenol, catechol and hydroquinone (Hecht 2002*a*; Scherer *et al.* 1998; Kim *et al.* 2006). Another mercapturic

---

Box 8.1   4-(Methylnitrosamino)-1-(3-pyridyl)-1-butanol (NNAL) and its glucuronides (total NNAL), a biomarker of exposure to a tobacco-specific lung carcinogen

The tobacco-specific lung carcinogen 4-(methylnitrosamino)-1-(3-pyridyl)-1-butanone (NNK) is considered to be one of the causes of lung cancer in smokers and non-smokers exposed to secondhand tobacco smoke. When humans are exposed to NNK, NNAL and its glucuronides are formed as metabolites and are excreted in the urine. Highly sensitive and specific methods are available to quantify total NNAL in urine. Total NNAL is a biomarker of exposure to NNK. It has been widely applied to investigate changes in carcinogen exposure when smokers reduce the number of cigarettes they smoke per day or switch to different types of tobacco products. It has also been particularly useful in quantifying carcinogen exposure in non-smokers. Detection of total NNAL in the urine confirms tobacco exposure because NNK is a tobacco-specific compound.

Box 8.2.   Phenanthrene metabolites, biomarkers of exposure plus metabolism of carcinogenic polycyclic aromatic hydrocarbons (PAHs)

PAHs are well-established carcinogens which are products of incomplete combustion. They are believed to play a significant role as causes of lung cancer in smokers and in certain occupational settings. There are major inter-individual differences in PAH metabolism among humans, and it is theorized that those who metabolically activate to a greater extent are at higher risk for cancer. This theory has been widely tested by genotyping approaches but the results are not too clear. Carcinogen metabolite phenotyping – the actual measurement of PAH metabolites – is perhaps a better way to test this hypothesis. Measurement of phenanthrene metabolites in human urine is a straightforward approach to phenotype PAH metabolism. All humans have these metabolites in their urine. Sensitive and specific methods are available to quantify products of phenanthrene metabolic activation and detoxification in human urine, and it is proposed that a ratio of such metabolites could be predictive of cancer risk upon exposure to PAHs.

---

acid which has been suggested as a carcinogen metabolite biomarker is *N*-acetyl-*S*-(2-hydroxy-3-butenyl)-L-cysteine, formed by degradation of the glutathione conjugate of the mono-epoxide metabolite of the carcinogen 1,3-butadiene (Urban 2003). However, this biomarker has found limited application to date.

Heterocyclic aromatic amines, such as PhIP, 2-amino-9*H*-pyrido(2,3-*b*)-indole, and 2-amino-3,8-dimethylimidazo(4,5-*f*)-quinoxaline, have been quantified in human urine treated with acid or base to release these compounds from their conjugates (Holland *et al*. 2004). These compounds are good exposure biomarkers for carcinogens formed in cooked meat.

AFM_1 is a primary oxidative metabolite of the powerful carcinogen aflatoxin B_1. AFM_1 can be readily quantified in urine and has been detected in urine samples from individuals exposed to aflatoxin B_1 in the diet. AFM_1 has been used as a biomarker in studies designed to assess the effects of the chemopreventive agent oltipraz on aflatoxin B_1 metabolism in exposed humans (Groopman and Kensler 1999). Another potentially useful biomarker for

such studies is the mercapturic acid of aflatoxin B_1 epoxide (Groopman and Kensler 1999).

8.4. SUMMARY

Carcinogen metabolite biomarkers offer great promise for integration into molecular epidemiological studies that investigate individual risk for cancer. These biomarkers are the best way to estimate individual exposure to particular carcinogens. Their use bypasses many of the uncertainties inherent in determining the doses of carcinogens to which people are exposed. Rapid advances in analytical chemistry, particularly mass spectrometry, have made possible the routine measurement of relatively low levels of carcinogens and their metabolites in urine and plasma. An example given here involves the tobacco-carcinogen metabolite NNAL and its glucuronides. Levels of these metabolites in the urine of non-smokers exposed to secondhand smoke provide accurate dose information that could not reasonably be obtained in any other way. Similar considerations apply to carcinogen uptake from environmental, occupational and dietary exposures.

There is bedrock evidence that most carcinogens require a metabolic activation process leading to DNA adducts, and that this process is a requirement for cancer development. The flux of carcinogen plus activated metabolites through this pathway can be reliably determined with the appropriate carcinogen metabolite biomarkers. Ratios of metabolites resulting from metabolic activation and detoxification could also be constructed in a carcinogen metabolite phenotyping approach. This approach would undoubtedly provide a robust method to test the hypothesis that individual differences in carcinogen metabolism are related to cancer risk.

While carcinogen metabolite biomarkers have been used to estimate exposure and to investigate carcinogen metabolism in humans, they have yet to be widely applied in molecular epidemiology studies. Since available technology now makes this feasible, carcinogen metabolite biomarkers should be part of the next wave of molecular epidemiology study designs.

Acknowledgement

The author's research on carcinogen biomarkers is supported by grants CA-81301, CA-85702, CA-92025 and ES-11297 from the US National Institutes of Health and RP-00-138 from the American Cancer Society.

References

Alexandrov K, Cascorbi I, Rojas M *et al.* 2002. *CYP1A1* and *GSTM1* genotypes affect benzo(*a*)pyrene DNA adducts in smokers' lung: comparison with aromatic/hydrophobic adduct formation. *Carcinogenesis* **23**: 1969–1977.

Anderson KE, Carmella SG, Ye M *et al.* 2001. Metabolites of a tobacco-specific lung carcinogen in the urine of nonsmoking women exposed to environmental tobacco smoke in their homes. *J Natl Cancer Inst* **93**: 378–381.

Anderson KE, Kliris J, Murphy L *et al.* 2003. Metabolites of a tobacco-specific lung carcinogen in nonsmoking casino patrons. *Cancer Epidemiol Biomarkers Prev* **12**: 1544–1546.

Baird WM, Ralston SL. 1997. Carcinogenic polycyclic aromatic hydrocarbons. In *Comprehensive Toxicology: Chemical Carcinogens and Anticarcinogens*, vol 12, Bowden GT, Fischer SM (eds). Elsevier Science: New York; 171–200.

Bartsch H, Nair U, Risch A *et al.* 2000. Genetic polymorphism of *CYP* genes, alone or in combination, as a risk modifier of tobacco-related cancers. *Cancer Epidemiol Biomarkers Prevent* **9**: 3–28.

Bauer E, Guo Z, Ueng YF *et al.* 1995. Oxidation of benzo(*a*)pyrene by recombinant human cytochrome P450 enzymes. *Chem Res Toxicol* **8**: 136–142.

Belinsky SA, Foley JF, White CM *et al.* 1990. Dose–response relationship between O^6-methylguanine formation in Clara cells and induction of pulmonary neoplasia in the rat by 4-(methylnitrosamino)-1-(3-pyridyl)-1-butanone. *Cancer Res* **50**: 3772–3780.

Benhamou S, Lee WJ, Alexandrie AK *et al.* 2002. Meta- and pooled analyses of the effects of glutathione *S*-transferase *M1* polymorphisms and smoking on lung cancer risk. *Carcinogenesis* **23**: 1343–1350.

Carmella SG, Chen M, Yagi H *et al.* 2004. Analysis of phenanthrols in human urine by gas chromatography–mass spectrometry: potential use in carcinogen metabolite phenotyping. *Cancer Epidemiol Biomarkers Prevent* **13**: 2167–2174.

Carmella SG, Han S, Fristad A *et al.* 2003. Analysis of total 4-(methylnitrosamino)-1-(3-pyridyl)-1-butanol (NNAL) in human urine. *Cancer Epidemiol Biomarkers Prevent* **12**: 1257–1261.

Carmella SG, Han S, Villalta PW *et al.* 2005. Analysis of total 4-(methylnitrosamino)-1-(3-pyridyl)-1-butanol (NNAL) in smokers' blood. *Cancer Epidemiol Biomarkers Prevent* **14**: 2669–2672.

Carmella SG, Yoder A, Hecht SS. 2006. Combined analysis of *r*-1,*t*-2,3,*c*-4-tetrahydroxy-1,2,3,4-tetrahydrophenanthrene and 4-(methylnitrosamino)-1-(3-pyridyl)-1-butanol in smokers' plasma. *Cancer Epidemiol Biomarkers Prevent* **15**: 1490–1494.

Cavalieri EL, Rogan EG. 1995. Central role of radical cations in metabolic activation of polycyclic aromatic hydrocarbons. *Xenobiotica* **25**: 677–688.

Conney AH. 1982. Induction of microsomal enzymes by foreign chemicals and carcinogenesis by polycyclic aromatic hydrocarbons: G. H. A. Clowes Memorial Lecture. *Cancer Res* **42**: 4875–4917.

Cooper CS, Grover PL, Sims P. 1983. The metabolism and activation of benzo(*a*)pyrene. *Prog Drug Metab* **7**: 295–396.

Groopman JD, Kensler TW. 1999. The light at the end of the tunnel for chemical-specific biomarkers: daylight or headlight? *Carcinogenesis* **20**: 1–11.

Guengerich FP. 1997. Introduction and Historical Perspective. In *Comprehensive Toxicology: Biotransformation*, vol 3, Guengerich FP (ed.). Elsevier Science: New York; 1–6.

Guenthner TM, Oesch F. 1981. Microsomal epoxide hydrolase and its role in polycyclic aromatic hydrocarbon biotransformation. In *Polycyclic Hydrocarbons and Cancer*, vol 3, Gelboin HV, Ts'o POP (eds). Academic Press: New York; 183–212.

Harris CC, Autrup H, Connor R *et al*. 1976. Interindividual variation in binding of benzo(*a*)pyrene to DNA in cultured human bronchi. *Science* **194**: 1067–1069.

Hatsukami DK, Benowitz NL, Rennard SI *et al*. 2006. Biomarkers to assess the utility of potential reduced exposure tobacco products. *Nicotine Tobacco Res* **8**: 169–191.

Hecht SS, Carmella SG, Chen M *et al*. 1999. Quantitation of urinary metabolites of a tobacco-specific lung carcinogen after smoking cessation. *Cancer Res* **59**: 590–596.

Hecht SS, Chen M, Yagi H *et al*. 2003. r-1,t-2,3,c-4-Tetrahydroxy-1,2,3,4-tetrahydrophenanthrene in human urine: a potential biomarker for assessing polycyclic aromatic hydrocarbon metabolic activation. *Cancer Epidemiol Biomarkers Prevent* **12**: 1501–1508.

Hecht SS, Chen M, Yoder A *et al*. 2005. Longitudinal study of urinary phenanthrene metabolite ratios: effect of smoking on the diol epoxide pathway. *Cancer Epidemiol Biomarkers Prevent* **14**: 2969–2974.

Hecht SS, Murphy SE, Carmella SG *et al*. 2004. Effects of reduced cigarette smoking on uptake of a tobacco-specific lung carcinogen. *J Natl Cancer Inst* **96**: 107–115.

Hecht SS, Ye M, Carmella SG *et al*. 2001. Metabolites of a tobacco-specific lung carcinogen in the urine of elementary school-aged children. *Cancer Epidemiol Biomarkers Prevent* **10**: 1109–1116.

Hecht SS. 1998. Biochemistry, biology, and carcinogenicity of tobacco-specific *N*-nitrosamines. *Chem Res Toxicol* **11**: 559–603.

Hecht SS. 2002a. Human urinary carcinogen metabolites: biomarkers for investigating tobacco and cancer. *Carcinogenesis* **23**: 907–922.

Hecht SS. 2002b. Cigarette smoking and lung cancer: chemical mechanisms and approaches to prevention. *Lancet Oncol* **3**: 461–469.

Hecht SS. 2003a. Tobacco carcinogens, their biomarkers, and tobacco-induced cancer. *Nat Rev Cancer* **3**: 733–744.

Hecht SS. 2003b. Carcinogen derived biomarkers: applications in studies of human exposure to secondhand tobacco smoke. *Tobacco Control* **13**(suppl 1): i48–56.

Holland RD, Taylor J, Schoenbachler L *et al*. 2004. Rapid biomonitoring of heterocyclic aromatic amines in human urine by tandem solvent solid phase extraction liquid chromatography electrospray ionization mass spectrometry. *Chem Res Toxicol* **17**: 1121–1136.

Houlston RS. 2000. *CYP1A1* polymorphisms and lung cancer risk: a meta-analysis. *Pharmacogenetics* **10**: 105–114.

Hung RJ, Boffetta P, Brockmoller J *et al*. 2003. *CYP1A1* and *GSTM1* genetic polymorphisms and lung cancer risk in Caucasian non-smokers: a pooled analysis. *Carcinogenesis* **24**: 875–882.

International Agency for Research on Cancer. 2004. *Tobacco Smoke and Involuntary Smoking*. IARC Monographs on the Evaluation of Carcinogenic Risks to Humans No. 83. IARC: Lyon; 53–119.

International Agency for Research on Cancer. 2008. *Smokeless Tobacco and Tobacco-specific Nitrosamines*. IARC Monographs on the Evaluation of Carcinogenic Risks to Humans No. 89. IARC: Lyon.

Jongeneelen FJ. 1994. Biological monitoring of environmental exposure to polycyclic aromatic hydrocarbons; 1-hydroxypyrene in urine of people. *Toxicol Lett* **72**: 205–211.

Jongeneelen FJ. 2001. Benchmark guideline for urinary 1-hydroxypyrene as biomarker of occupational exposure to polycyclic aromatic hydrocarbons. *Ann Occup Hyg* **45**: 3–13.

Joseph AM, Hecht SS, Murphy SE *et al*. 2005. Relationships between cigarette consumption and biomarkers of tobacco toxin exposure. *Cancer Epidemiol Biomarkers Prevent* **14**: 2963–2968.

Kim JH, Stansbury KH, Walker NJ *et al*. 1999. Metabolism of benzo(*a*)pyrene and benzo(*a*)pyrene-7,8-diol by human cytochrome P450 1B1. *Carcinogenesis* **19**: 1847–1853.

Kim JY, Hecht SS, Mukherjee S *et al*. 2005. A urinary metabolite of phenanthrene as a biomarker of polycyclic aromatic hydrocarbon metabolic activation in workers exposed to residual oil fly ash. *Cancer Epidemiol Biomarkers Prevent* **14**: 687–692.

Kim S, Vermeulen R, Waidyanatha S *et al*. 2006. Using urinary biomarkers to elucidate dose-related patterns of human benzene metabolism. *Carcinogenesis* **27**: 772–781.

Kiyohara C, Nakanishi Y, Inutsuka S *et al*. 1998. The relationship between CYP1A1 aryl hydrocarbon hydroxylase activity and lung cancer in a Japanese population. *Pharmacogenetics* **8**: 315–323.

Lackmann GM, Salzberger U, Tollner U *et al*. 1999. Metabolites of a tobacco-specific carcinogen in the urine of newborns. *J Natl Cancer Inst* **91**: 459–465.

Le Marchand L, Guo C, Benhamou S *et al*. 2003. Pooled analysis of the *CYP1A1* exon 7 polymorphism and lung cancer (United States). *Cancer Causes Control* **14**: 339–346.

Lee PN. 1999. Uses and abuses of cotinine as a marker of tobacco smoke exposure. In *Analytical Determination of Nicotine and Related Compounds and Their Metabolites*, Gorrod JW, JacobP III (eds). Elsevier: Amsterdam; 669–719.

Lee WJ, Brennan P, Boffetta P *et al*. 2002. Microsomal epoxide hydrolase polymorphisms and lung cancer risk: a quantitative review. *Biomarkers* **7**: 230–241.

Liu G, Zhou W, Christiani DC. 2005. Molecular epidemiology of non-small cell lung cancer. *Semin Resp Crit Care Med* **26**: 265–272.

McLemore TL, Adelberg S, Liu MC *et al*. 1990. Expression of *CYP1A1* gene in patients with lung cancer: evidence for cigarette smoke-induced gene expression in normal lung tissue and for altered gene regulation in primary pulmonary carcinomas. *J Natl Cancer Inst* **82**: 1333–1339.

Merriam-Webster's Collegiate Dictionary. 2003. Merriam-Webster: Springfield, MA.

Miller EC, Miller JA. 1981. Searching for the ultimate chemical carcinogens and their reactions with cellular macromolecules. *Cancer* **47**: 2327–2345.

Miller JA. 1994. Research in chemical carcinogenesis with Elizabeth Miller – a trail of discovery with our associates. *Drug Metab Dispos* **26**: 1–36.

Milunsky A, Carmella SG, Ye M *et al*. 2000. A tobacco-specific carcinogen in the fetus. *Prenat Diagn* **20**: 307–310.

Nan HM, Kim H, Lim HS *et al*. 2001. Effects of occupation, lifestyle and genetic polymorphisms of *CYP1A1*, *CYP2E1*, *GSTM1* and *GSTT1* on urinary 1-hydroxypyrene and 2-naphthol concentrations. *Carcinogenesis* **22**: 787–793.

National Toxicology Program. *Toxicology and Carcinogenesis Studies of Naphthalene in F344/ N Rats*. National Toxicology Program Technical Report Series No. 500. US Department of Health and Human Services: Research Triangle Park, NC, 2000; 5–7.

Nebert DW, Dalton TP, Okey AB *et al*. 2004. Role of aryl hydrocarbon receptor-mediated induction of the CYP1 enzymes in environmental toxicity and cancer. *J Biol Chem* **279**: 23847–23850.

Nebert DW. 2000. Drug-metabolizing enzymes, polymorphisms and interindividual response to environmental toxicants. *Clin Chem Lab Med* **38**: 857–861.

Nowak D, Schmidt-Preuss U, Jorres R *et al*. 1988. Formation of DNA adducts and water-soluble metabo-lites of benzo(*a*)pyrene in human monocytes is genetically controlled. *Int.J Cancer* **41**: 169–173.

Parsons WD, Carmella SG, Akerkar S *et al*. 1998. A metabolite of the tobacco-specific lung carcinogen 4-(methylnitrosamino)-1-(3-pyridyl)-1-butanone (NNK) in the urine of hospital workers exposed to environmental tobacco smoke. *Cancer Epidemiol Biomarkers Prevent* **7**: 257–260.

Pratt WB, Taylor P. 1990. *Principles of Drug Action: The Basis of Pharmacology*, 3rd edn. Churchill Livingstone: New York; 369.

Robertson IGC, Guthenberg C, Mannervik B *et al*. 1986. Differences in stereoselectivity and catalytic efficiency of three human glutathione transferases in the conjugation of glutathione with 7β,8α-dihyroxy-9α,10α-oxy-7,8,9,10-tetrahydrobenzo(*a*)pyrene. *Cancer Res* **46**: 2220–2224.

Sabadie N, Richter-Reichhelm HB, Saracci R *et al*. 1981. Inter-individual differences in oxidative benzo(*a*)pyrene metabolism by normal and tumorous surgical lung specimens from 105 lung cancer patients. *Int J Cancer* **27**: 417–425.

Scherer G, Renner T, Meger M. 1998. Analysis and evaluation of *trans,trans*-muconic acid as a biomarker for benzene exposure. *J Chromatogr B Biomed Sci Appl* **717**: 179–199.

Schwab M (ed.). 2001. *Encyclopedic Reference of Cancer*. Springer-Verlag: Berlin; 157.

Shimada T, Gillam EMJ, OdaY *et al*. 1999. Metabolism of benzo(*a*)pyrene to *trans*-7,8-dihydroxy-7,8-dihydrobenzo(*a*)pyrene by recombinant human cytochrome P450 1B1 and purified liver epoxide hydrolase. *Chem Res Toxicol* **12**: 623–629.

Shimada T, Hayse CL, Yamazaki H *et al*. 1996. Activation of chemically diverse procarcinogens by human cytochrome P450 1B1. *Cancer Res* **56**: 2979–2984.

Shou M, Korzekwa KR, Crespi CL *et al*. 1994. The role of 12 cDNA-expressed human, rodent, and rabbit cytochromes P450 in the metabolism of benzo(*a*)pyrene and benzo(*a*)pyrene *trans*-7,8-dihydrodiol. *Mol Carcinogen* **10**: 159–168.

Shou M, Korzekwa KR, Krausz KW *et al*. 1994. Regio- and stereo-selective metabolism of phenanthrene by twelve cDNA-expressed human, rodent, and rabbit cytochromes P-450. *Cancer Lett* **83**: 305–313.

Stepanov I, Hecht SS, Duca G *et al*. 2006. Uptake of the tobacco-specific lung carcinogen 4-(methylnitrosamino)-1-(3-pyridyl)-1-butanone (NNK) by Moldovan children. *Cancer Epidemiol Biomarkers Prevent* **15**: 7–11.

Sundberg K, Dreij K, Seidel A *et al*. 2002. Glutathione conjugation and DNA adduct formation of dibenzo[*a*,l]pyrene and benzo(*a*)pyrene diol epoxides in V79 cells stably expressing different human glutathione transferases. *Chem Res Toxicol* **15**: 170–179.

Sundberg K, Johansson AS, Stenberg G *et al.* 1998. Differences in the catalytic efficiencies of allelic variants of glutathione transferase P1–1 towards carcinogenic diol epoxides of polycyclic aromatic hydrocarbons. *Carcinogenesis* **19**: 433–436.

Sundberg K, Widersten M, Seidel A *et al.* 1997. Glutathione conjugation of bay- and fjord-region diol epoxides of polycyclic aromatic hydrocarbons by glutathione transferase M1–1 and P1–1. *Chem Res Toxicol* **10**: 1221–1227.

Taioli E, Pedotti P. 2005. Pooled analysis on metabolic gene polymorphisms and lung cancer. *Exp Lung Res* **31**: 217–222.

Thakker DR, Yagi H, Levin W *et al.* 1985. Polycyclic aromatic hydrocarbons: metabolic activation to ultimate carcinogens. In *Bioactivation of Foreign Compounds*, Anders MW (ed.). Academic Press: New York; 177–242.

Tulunay O, Hecht SS, Carmella SG *et al.* 2005. Urinary metabolites of a tobacco-specific lung carcinogen in nonsmoking hospitality workers. *Cancer Epidemiol Biomarkers Prevent* **14**: 1283–1286.

Turesky RJ. 2002. Heterocyclic aromatic amine metabolism, DNA adduct formation, mutagenesis, and carcinogenesis. *Drug Metab Rev* **34**: 625–650.

Urban M, Gilch G, Schepers G *et al.* 2003. Determination of the major mercapturic acids of 1,3-butadiene in human and rat urine using liquid chromatography with tandem mass spectrometry. *J Chromatogr B Analyt Technol Biomed Life Sci* **796**: 131–140.

Vineis P, Veglia F, Anttila S *et al.* 2004. *CYP1A1*, *GSTM1* and *GSTT1* polymorphisms and lung cancer: a pooled analysis of gene–gene interactions. *Biomarkers* **9**: 298–305.

Vineis P, Veglia F, Benhamou S *et al.* 2003. *CYP1A1 T3801* C polymorphism and lung cancer: a pooled analysis of 2451 cases and 3358 controls. *Int J Cancer* **104**: 650–657.

Yang M, Koga M, Katoh T *et al.* 1999. A study for the proper application of urinary naphthols, new biomarkers for airborne polycyclic aromatic hydrocarbons. *Arch Environ Contam Toxicol* **36**: 99–108.

Yu D, Berlin JA, Penning TM *et al.* 2002. Reactive oxygen species generated by PAH *o*-quinones cause change-in-function mutations in *p53*. *Chem Res Toxicol* **15**: 832–842.

9

Biomarkers of Exposure: Adducts

David H. Phillips

Institute of Cancer Research, Sutton, UK

9.1. INTRODUCTION

A biomarker (sometimes called a biological marker) is any substance (or its products), structure or process that can be measured in the human body that may influence or predict the incidence or outcome of disease. In the field of molecular epidemiology, the general objective of using biomarkers is to increase the accuracy of measurements of susceptibility, exposure, modifying influences or prognostic factors.

Genotoxic carcinogens undergo metabolic activation to reactive species that bind covalently to cellular macromolecules, including DNA and protein. The possible consequences of DNA adduct formation are that DNA repair processes may excise the damage and restore the normal DNA sequence, that the damage may induce processes that lead to cell death (by necrosis or apoptosis), or that the DNA will replicate erroneously, with potentially mutagenic and carcinogenic consequences. Thus, as binding to DNA can be a premutagenic event that can initiate carcinogenesis if erroneous replication of the DNA ensues, the measurement of DNA adducts provides a biologically relevant indicator of carcinogen exposure. A molecular epidemiological view of the stages of cancer is shown in Figure 9.1, in which the process from exposure to clinical disease is divided into several intermediate stages, defined by the potential for intermediate biomarkers to monitor the process. These collectively fall into the three categories of biomarkers of exposure, effect and susceptibility, although there is some overlap (e.g. micronucleus formation can

be considered a biomarker of exposure *and* of risk). In general, biomarkers of exposure impinge on earlier stages of the carcinogenic process, prior to biomarkers of effect, while biomarkers of genetic susceptibility may influence all stages of the process.

Measurement of DNA adducts provides an indication of the 'biologically effective dose' as an integrated measure of human intake, uptake by target tissue, metabolic activation by phase I and/or phase II enzymes, delivery to critical cellular target (DNA), and influence of protective cellular mechanisms, such as detoxification pathways (i.e. other phase I/II enzymes) and DNA repair (Phillips 2005; Santella *et al.* 2005). Although carcinogens can modify all four bases in DNA as well as the phosphodiester backbone, guanine is the most commonly modified base, followed by adenine (Phillips 2007). The structures of some of the DNA adducts mentioned in this chapter are shown in Figure 9.2. Polycyclic aromatic hydrocarbons (PAHs) react predominantly with the exocyclic amino groups at N^2-guanine, but also at N^6-adenine, while aromatic amines react predominantly at the C-8 of guanine. Not all adducts are chemically stable, e.g. N-7-guanine adducts, hence the excretion of aflatoxin–N-7-guanine adducts in the urine. It is also important to note that, although DNA adduct formation is essential to the induction of cancer by genotoxic carcinogens, it is not in itself sufficient and DNA adducts may be formed in both target and non-target tissues (a useful property for the purposes of human biomonitoring).

Molecular Epidemiology of Chronic Diseases, Edited by C. P. Wild, P. Vineis, and S. Garte
© 2008 John Wiley & Sons, Ltd

Biomarkers of genetic susceptibility

Metabolism genes　　DNA repair/stability genes　　Immunocompetence genes

Exposure → Internal dose → Biologically effective dose → Early biological effects → Altered structure and function → Clinical disease

Chemicals, metabolites in blood, urine, tissues

Protein adducts DNA adducts

Somatic mutations, chromosomal aberrations, micronuclei, aneuploidy, sister chromatid exchanges

Mutation spectra in tumours

Biomarkers of exposure

Biomarkers of effect

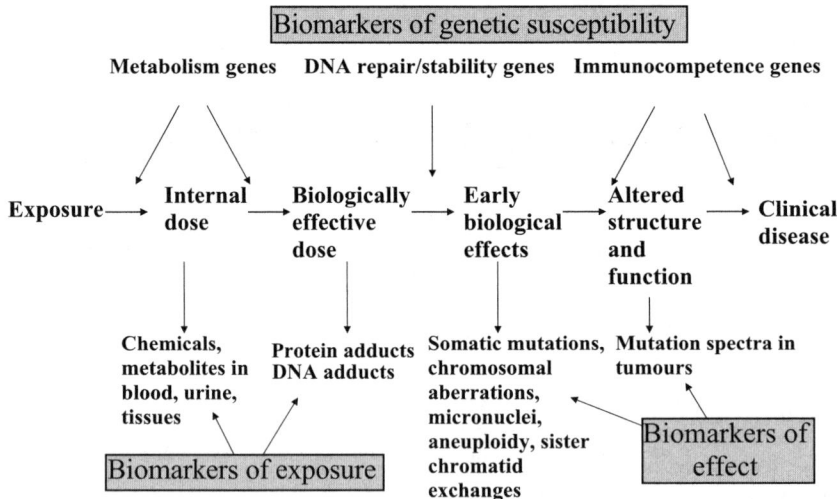

Figure 9.1 The carcinogenic process seen from a molecular epidemiological point of view, illustrating the influence of environmental and host factors in the process, and the opportunities for biomonitoring, including DNA and protein adduct analysis

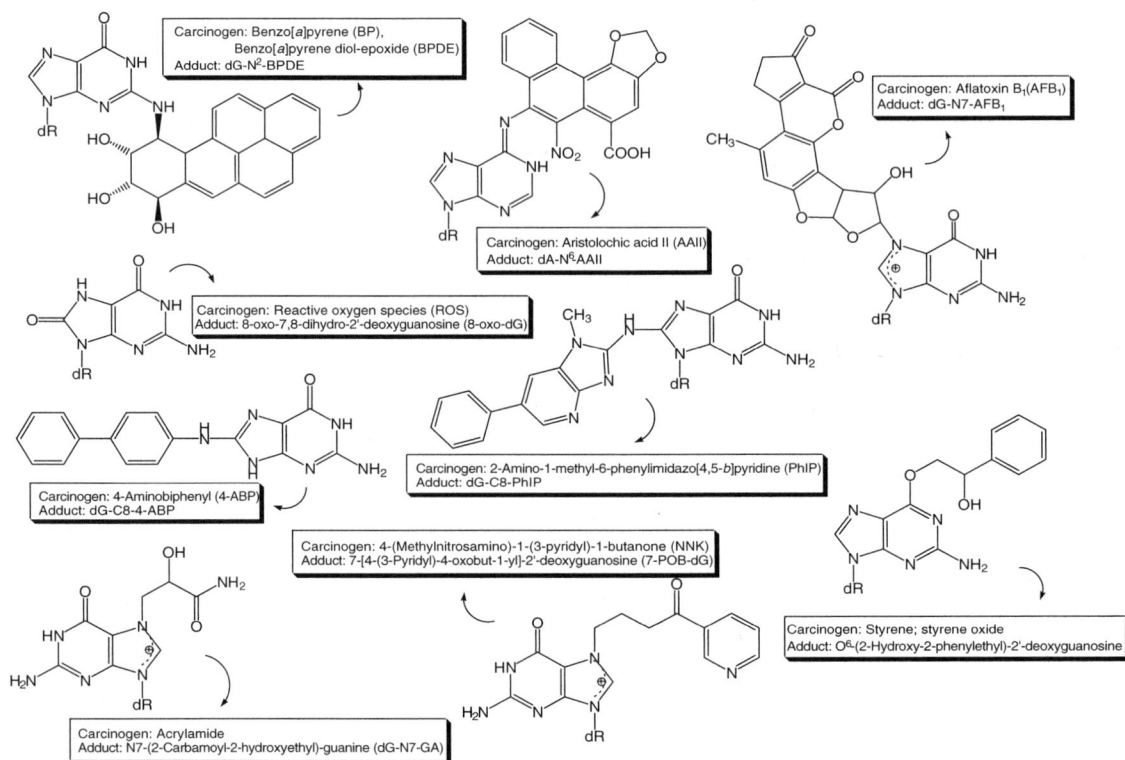

Carcinogen: Benzo[a]pyrene (BP), Benzo[a]pyrene diol-epoxide (BPDE) Adduct: dG-N²-BPDE

Carcinogen: Aristolochic acid II (AAII) Adduct: dA-N⁶-AAII

Carcinogen: Aflatoxin B₁(AFB₁) Adduct: dG-N7-AFB₁

Carcinogen: Reactive oxygen species (ROS) Adduct: 8-oxo-7,8-dihydro-2'-deoxyguanosine (8-oxo-dG)

Carcinogen: 2-Amino-1-methyl-6-phenylimidazo[4,5-b]pyridine (PhIP) Adduct: dG-C8-PhIP

Carcinogen: 4-Aminobiphenyl (4-ABP) Adduct: dG-C8-4-ABP

Carcinogen: 4-(Methylnitrosamino)-1-(3-pyridyl)-1-butanone (NNK) Adduct: 7-[4-(3-Pyridyl)-4-oxobut-1-yl]-2'-deoxyguanosine (7-POB-dG)

Carcinogen: Styrene; styrene oxide Adduct: O⁶-(2-Hydroxy-2-phenylethyl)-2'-deoxyguanosine

Carcinogen: Acrylamide Adduct: N7-(2-Carbamoyl-2-hydroxyethyl)-guanine (dG-N7-GA)

Figure 9.2 Examples of the DNA adducts formed by the environmental carcinogens discussed in this chapter. In each case one representative nucleoside adduct is shown. dR, deoxyribose

The formation of protein adducts, although not directly involved in the carcinogenic process, can also be used as a measure of biologically effective dose and can be useful in some circumstances because of the greater cellular abundance of protein and the potentially longer persistence of these adducts compared with DNA adducts (Farmer and Emeny 2006). The life time of haemoglobin in human blood is about 4 months, while albumin has a half-life of about 3 weeks (Farmer and Emeny 2006), and their use as biomarkers of human exposure to carcinogens has been extensively reviewed (Strickland et al. 1993; Skipper et al. 1994; Tornqvist and Landin 1995; Wild and Pisani 1998; Boogaard 2002). Potentially longer-lived proteins are collagen (Jonsson et al. 1995) and histones (SooHoo et al. 1994), but neither has been widely used as yet as sources for human biomonitoring.

Box 9.1 Aristolochic acid and renal disease

Aristolochic acid is a constituent of plants of the genus *Aristolochia*, which have been used since antiquity in herbal remedies. Yet aristolochic acid is genotoxic, and induces renal tumours in experimental animals. In the 1990s a cluster of cases of renal failure occurred among clients at a Belgian clinic where *Aristolochia fangchi* was accidentally included in a herbal slimming concoction instead of another plant (*Stephania tetrandra*), apparently because they have very similar names in traditional Chinese medicine (*han fang-ji* and *guang fang-ji*, respectively). Within months of the resulting ingestion of high doses of aristolochic acid caused, there occurred renal failure in about 5% of individuals and this was followed by rapid onset of urothelial cancer in about half of these affected individuals. ^{32}P-postlabelling analysis of urothelial DNA from such patients has shown clearly the presence of aristolochic acid–DNA adducts, thereby strongly implicating aristolochic acid as the aetiological agent (Arlt et al. 2002). The condition, initially termed 'Chinese herbs nephropathy' (CNN), is now more commonly referred to as 'aristolochic acid nephropathy' (AAN) as a consequence of this evidence. Although there has been a call for the worldwide banning of aristolochic acid from all herbal products, recent surveys have indicated that this has not yet been achieved.

The similarity of the pathology of AAN to another renal condition, Balkan endemic nephropathy (BEN), has led to aristolochic acid being implicated in the aetiology of that disease also. *Aristolochia clematitis* has been found growing wild in wheat fields in the Balkan region, and its seeds have been found to contaminate wheat grain used to make bread. It has recently been found, again by ^{32}P-postlabelling analysis, that patients with BEN have AA–DNA adducts in their urogenital tissues, which provides strong evidence that the compound may be at least partly responsible for this condition (Arlt 2006). Another compound that has been implicated in the aetiology of BEN is the fungal metabolite ochratoxin A (OTA). There has been some debate over whether OTA is primarily a genotoxic carcinogen reacting directly with DNA to form adducts, or whether it induces oxidative damage in urothelial tissues (Pfohl-Leszkowicz and Castegnaro 2005; Schilter et al. 2005). Further studies incorporating adduct biomonitoring analyses and other analytical and epidemiological approaches will be required to elucidate the roles of aristolochic acid and OTA in the aetiology of BEN.

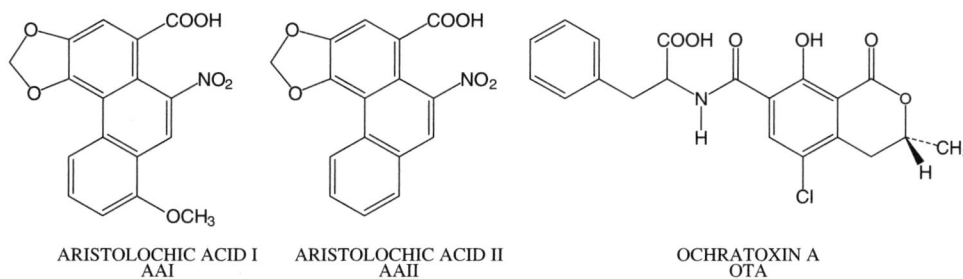

ARISTOLOCHIC ACID I
AAI

ARISTOLOCHIC ACID II
AAII

OCHRATOXIN A
OTA

Aristolochia species

Aspergillus ochraceus

Box 9.2   Human exposure to acrylamide

ACRYLAMIDE   GLYCIDAMIDE

An incident of occupational exposure to acrylamide in Sweden led to the detection of acrylamide-derived protein adducts in the exposed workers, but also to the detection of lower levels of the adducts in apparently 'unexposed' controls. This in turn led to the suspicion that there might be a dietary source of acrylamide, which was confirmed first by monitoring adduct levels in animals fed different diets of cooked food and then by the discovery that cooked food high in carbohydrate and asparagine content could be a source of high levels of acrylamide (Dybing *et al.* 2005). This has resulted in worldwide concern about human exposure to acrylamide from manufactured foods (e.g. cakes, bread, potato crisps). Further human studies are called for, in which biomonitoring by means of protein adducts, viz. the adduct of acrylamide with the N-terminal valine of haemoglobin [N-(2-carbamoylethyl)valine] and the corresponding adduct of the metabolite glycidamide [N-(2-carbamoyl-2-hydroxyethyl)valine], and possibly also DNA adducts [glycidamide reacts with ring nitrogens in guanine and adenine to produce N-7-(carbamoyl-2-hydroxyethyl)guanine (see Figure 9.2), N-1-(carbamoyl-2-hydroxyethyl)adenine and N-3-(carbamoyl-2-hydroxyethyl)adenine, while acrylamide is several orders less reactive than glycidamide and binds primarily to the N-1 position of adenine and the N-3 position of cytosine (Farmer and Emeny 2006)] will undoubtedly support estimates of exposure and risk.

This chapter considers methods for adduct detection, identified adducts in human tissue and sources of DNA for human biomonitoring. Adducts as biomarkers of human exposure to occupational and environmental carcinogens, and to tobacco smoke, are discussed. Examples of DNA adduct measurements in prospective epidemiological studies and in relation to DNA repair studies are described. Two specific examples of chemical agents where there is currently concern over human exposure, aristolochic acid (Box 9.1) and acrylamide (Box 9.2), are also presented.

9.2. METHODS FOR ADDUCT DETECTION

Methods for detecting DNA adducts that can be applied to the analysis of human DNA in molecular epidemiology studies have been reviewed recently (Poirier *et al.* 2000; Farmer and Emeny 2006). While these methods adopt very different approaches to detecting DNA adducts, and differ in their limits of sensitivity and requirements for quantities of DNA for analysis, sensitivities can often be achieved in the range of 1 adduct/10^8–10^9 nucleotides, using µg quantities of DNA for analysis (Table 9.1). Each method has its strengths and weaknesses, which have been widely discussed (Wild and Pisani 1998; Poirier *et al.* 2000; Farmer 2004), so the choice of method has to be on a case-by-case basis, depending on the type of exposure being investigated and the quantities of human material available.

^{32}P-Postlabelling analysis has been applied to the detection of many different carcinogen–DNA adducts, its chief attractions being its ultrasensitivity and the fact that little knowledge of the nature of an adduct or adducts is needed in order to detect it/them (Phillips 1997), although this can lead to uncertainties as to what is actually being detected (as discussed below). Immunochemical methods have to date been applied to the detection of about 20 different carcinogen–DNA adducts, limited by the need to raise antibodies against synthetically

Table 9.1 DNA adduct detection methods applicable to human biomonitoring and their limits of detection

Method	Variations	Amount of DNA required	Approximate detection limits
^{32}P-postlabelling	Nuclease P_1 digestion, butanol extraction, HPLC	1–10 μg	1 adduct/10^9–10^{10} nucleotides
Immunoassay	ELISA, DELFIA, CIA, immunohistochemistry	20 μg	1.5 adducts/10^9 nucleotides
Fluorescence	HPLC fluorescence, SFS	100–1000 μg	1 adduct/10^9 nucleotides
Mass spectrometry		Up to 100 μg	1 adduct/10^8 nucleotides
AMS[a]		Up to 100 μg	1 adduct/10^{11}–10^{12} nucleotides

[a]Accelerator mass spectrometry; requires use of radiolabelled compounds.

prepared and characterized adducts in order to detect them in human tissues. Nevertheless, cross-reactivity within classes of carcinogens, e.g. in the case of antibodies raised against benzo(*a*)pyrene diol-epoxide (BPDE)-modified adducts recognizing PAH–DNA adducts other than benzo(*a*)pyrene (BP)–DNA adducts, can be advantageous for human biomonitoring of exposure to environmental PAHs (Poirier 2004). Most of the immunological studies have been based on enzyme-linked immunosorbent assay (ELISA) but more recently modified assays, including dissociation-enhanced lanthanide fluoroimmunoassay (DELFIA; Schoket *et al.* 1993) and chemiluminescence immunoassay (CIA; Divi *et al.* 2002) have been applied with sensitivities approaching those of ^{32}P-postlabelling. Immunohistochemistry can be used to detect adducts in cells and tissue sections visually, thereby providing valuable information on their localization within a tissue, although the method is less sensitive than the other immunological methods and should be regarded as semi-quantitative in its assessment of adduct levels. Mass spectrometry has been less frequently applied to human biomonitoring, although advances in its methods indicate its wider use in the future (Singh and Farmer 2006). Physicochemical methods such as mass spectrometry have, until recently, been best suited to the detection of specific well-characterized lesions, although a recent approach has heralded the simultaneous detection of multiple adducts by mass spectrometry to define the so-called 'adductome' approach to detecting DNA adducts in humans (Kanaly *et al.* 2006).

Adducts with fluorescent chromophores, e.g. etheno adducts and those formed by PAHs and aflatoxin B_1, can be detected by means of their fluorescent emissions (Weston 1993). For biomonitoring studies, the application of fluorescence spectroscopy has, in some cases, necessitated the analysis of fairly large quantities of DNA in order to achieve the required sensitivity; nevertheless, it has been successfully applied in a number of exposure scenarios, utilizing HPLC separation of adducts or their hydrolysis products prior to spectrometric analysis (Farmer and Emeny 2006). With synchronous fluorescence spectrophotometry (SFS), where samples are scanned with a fixed difference between the excitation and emission wavelengths (e.g. 34 nm for BPDE–DNA adducts), it is theoretically possible to resolve multiple components in a complex mixture without separation. Another application of fluorescence spectroscopy is the fluorescent labelling of non-fluorescent DNA adducts, combined with capillary electrophoresis and laser-induced fluorescence (CE–LIF), although improvements in sensitivity will be required for most human biomonitoring scenarios (Schmitz *et al.* 2002).

Accelerator mass spectrometry (AMS) is currently the most sensitive method of DNA adduct detection, with a limit of detection as low as 1 adduct/10^{12} nucleotides, but it requires the use of radioisotopes (e.g. ^{14}C) and its use in biomonitoring is therefore limited to circumstances in which it is ethically acceptable to administer very low doses of isotopically-labelled carcinogens to human volunteers (Turteltaub and Dingley 1998).

Substances that can be oxidized or reduced can be detected by electrochemical detection. When combined with HPLC, adducts such as 7,8-dihydo-8-oxo-2′-deoxyguanosine (8-oxo-dGuo) can be detected by this means, although the sample is destroyed in the process (ESCODD 2002). DNA damage, including adducts, can be detected indirectly in individual cells using the single-cell gel electrophoresis assay (comet assay), whereby overt DNA strand breaks, repair-induced strand breaks or lesions that are hydrolysed by alkali result in DNA fragmentation that increases the mobility of nuclear DNA in an electric current. The fragmented DNA thus has the appearance of the tail of a comet streaming from the nucleus (comet head) when viewed under a microscope, and the length or density of the comet tail can be used to calculate the level of damage in the cells (Fairbairn *et al.* 1995; Farmer and Emeny 2006).

Methods for the detection of protein adducts are mostly based on immunochemical, HPLC-fluorescence or mass spectrometric methods, following the release of the carcinogen moiety from protein by mild acid or base hydrolysis, or release of the modified N-terminal valine of haemoglobin (Phillips and Farmer 1995; Tornqvist *et al.* 2002; Farmer and Emeny 2006).

For many carcinogen–DNA adducts, there are several different analytical methods available. For example, PAH–DNA adducts can be detected by ^{32}P-postlabelling, by fluorescence spectroscopy, by immunochemical methods (using antibodies raised against BPDE-modified DNA, or by mass spectrometry; Phillips 2005). Furthermore, PAH–protein adducts can be detected by using antibodies, mass spectrometry or HPLC fluorescence to detect the bound (to albumin or haemoglobin) or hydrolysed products.

9.3. ADDUCTS IDENTIFIED IN HUMAN TISSUE

It is clear from numerous studies that DNA from 'unexposed' or 'control' subjects contains 'background' levels of adducts or damage. Typical levels

that might be expected are shown in Table 9.2. These can be considered to be approximate background levels of DNA damage in human tissues due to normal environmental levels of DNA-damaging agents or to endogenous DNA-damaging agents. Because of the potential for multiple origins of background DNA damage (e.g. diet, air pollution and, of course, smoking), it is important that careful selection of control groups of individuals for comparison with a putative 'exposed' population is made. The two groups need to be matched for as many lifestyle and exposure parameters as possible, apart from the exposure of interest, to maximize the possibility of detecting differences in adduct levels due to this exposure and not due to other effects, and to minimize the possibility of failing to detect such differences because the control group has higher levels of adducts deriving from other sources.

Measurements of oxidative damage to DNA often use 8-oxo-dGuo as a surrogate for all types of oxidative DNA damage, because it is the most abundant and easily detected. Etheno adducts can result from lipid peroxidation, which represents an endogenous source, but they are also formed by some exogenous agents, such as vinyl halides.

Human biomonitoring has also revealed the presence of 'background' levels of a variety of adducts in human protein. These include methyl, ethyl, 2-hydroxyethyl, 2-carbamoylethyl (from acrylamide) and cyanoethyl (from acrylonitrile) adducts, as well as those of benzene and styrene metabolites (Farmer and Emeny 2006). In some cases these appear to originate from exogenous sources (e.g. acrylamide in the diet; see below) or they may arise from both exogenous and endogenous sources (e.g. 2-hydroxyethyl adducts from endogenous ethylene or exposure to ethylene oxide).

As already mentioned, the greater abundance in cells of protein and the longer persistence of protein

Table 9.2 Background levels of damage in human DNA

Type of lesion	Possible origins	Typical levels
8-Hydroxyguanine	Oxidative processes, hydroxyl radical	1 in 10^6–10^7 nucleotides
Etheno-dG, etheno-dC	Lipid peroxidation, vinyl chloride	1 in 10^7–10^8 nucleotides
O^6-Methylguanine	Alkylating agents	1 in 10^6–10^7 nucleotides
Bulky/hydrophobic adducts	Smoking, pollution, diet	1 in 10^7–10^8 nucleotides

adducts, relative to DNA and DNA adducts, respectively, has made measuring protein adducts resulting from occupational exposures a viable and attractive proposition. Among the exposures monitored in this way are those involving ethylene oxide, styrene oxide, acrylamide, acrylonitrile, BP (and PAHs), aniline and 4,4'-methyldianiline (MDA) (Phillips and Farmer 1995).

Sources of DNA for biomonitoring

It is not always practicable to measure DNA adducts in the target tissue and the use of a surrogate is often either preferable or essential. As evident from the occupational and environmental biomonitoring studies described below, blood is a common source of DNA for analysis. However, a wide range of different sources of DNA either have been or can be used for biomonitoring purposes, ranging from easily obtainable sources, such as blood, saliva, cervical mucosa, sperm, exfoliated bladder cells (in urine), hair roots and buccal scrapes; somewhat less accessible tissue, such as skin biopsies, bladder biopsies, bone marrow cells and bronchoalveolar lavage cells; through to tissue removed during major surgery or autopsy (Phillips 1996).

Despite the widespread use of surrogate sources of DNA, there have been relatively few studies comparing adducts in target and surrogate tissues. Among these, it has been shown that there is a correlation between adduct levels in the lymphocytes and induced sputum cells of smokers (Besaratinia et al. 2000), and while some studies show a correlation between adducts in lung tissue and blood of smokers, others do not (Phillips 2002). More studies validating the use of surrogate DNA sources are needed. It is probably preferable that biomonitoring studies should be carried out on uniform subfractions of blood cells, e.g. lymphocytes, rather than whole white blood or buffy coat cells (Godschalk et al. 2003), yet most biobanks currently being established are storing buffy coat cells (or the DNA isolated from them) because of cost constraints. Unfortunately, this may limit the power and quality of future adduct biomarker studies of these cohorts.

Adducts may arise in blood cells through routes different from those that occur in the target tissue. Thus, while inhalation may be the major route of exposure of the lung, adducts in blood can arise from inhalation, adsorption or ingestion. Indeed, dietary exposure to PAHs appears to have been a confounding factor in some studies monitoring occupational exposure to the carcinogens: in one study of woodland fire-fighters, adduct levels in white blood cells were correlated with recent ingestion of barbecued hamburgers, rather than activity in combating forest fires (Rothman et al. 1993), and in another study of military personnel extinguishing oilfield fires in Kuwait, adduct levels in blood were significantly lower during this activity than before and after, possibly due to the ingestion of a more restricted diet, lower in PAHs, while the military personnel were in the desert (Poirier et al. 1998). Thus, it is important when designing molecular epidemiology studies that due consideration be given to all potential sources and routes of exposure to carcinogens in both 'cases' and 'controls'.

DNA adducts that can be lost from DNA through depurination may be detectable in urine. Lesions detectable by this means include the N-7-guanine adduct of aflatoxin B_1, and the oxidative lesions 8-oxodG and 8-oxodGuo (Shuker and Farmer 1992). Such analyses can be combined with biomonitoring of metabolites of carcinogens (e.g. aflatoxin B_1 and PAHs) in urine. However, it should be borne in mind that base-substituted adducts in urine could be derived from RNA as well as, or instead of, from DNA. Alternatively, urine may be a source of significant numbers of exfoliated bladder epithelial cells, from which DNA can be isolated and analysed for adducts.

9.4. ADDUCTS AS BIOMARKERS OF OCCUPATIONAL AND ENVIRONMENTAL EXPOSURE TO CARCINOGENS

The development of sensitive methods for the detection of DNA adducts has revealed that human DNA contains many adducts of both endogenous and exogenous origin. In some cases the methods have detected increased levels of ill-defined and uncharacterized adducts (e.g. as a consequence of tobacco smoking, see below) and in other cases the lesion is well characterized (e.g. methylated DNA bases, or oxidized bases) but the reactive agent or species responsible for forming the adduct cannot be unequivocally identified. In relatively few instances has a DNA adduct been identified and its origin also elucidated. Examples of where this has been achieved, usually using mass spectrometry but also involving fluorescence analysis, include the detection of: BP adducts in lung (Beland *et al.* 2005); 2-amino-1-methyl-6-phenylimidazo(4,5-β)pyridine (PhIP) adducts in colon (Friesen *et al.* 1994); 4-aminobiphenyl (ABP) adducts in bladder (Lin *et al.* 1994); 4(*N*-methyl-*N*-nitrosoamino)-1-(3-pyridyl)-1-butanone (NNK) adducts in lung (Foiles *et al.* 1991); and aflatoxin B_1 adducts in urine (Groopman *et al.* 1992).

The many studies that have measured human exposure to PAHs can be considered to illustrate the principles of human biomonitoring by DNA adduct determination. Not all of these studies have detected higher levels of adducts in the 'exposed' populations relative to the 'unexposed' subjects, although most of them have (Phillips 2005). A summary of exposures and the methods that have been used is given in Table 9.3. In many of the situations where occupational exposure to PAHs has occurred, there is epidemiological evidence for increased cancer risk. Likewise, in the environmental exposures studied there has been evidence, or at least suspicion, of adverse health effects including cancer.

Table 9.3 Occupational and environmental exposures to PAHs monitored by DNA adduct determination (reviewed in Phillips 2005)

Type of exposure	Exposed population	Analytical methods used	Selected references
Occupational	Iron and steel production workers	^{32}P-postlabelling, immunoassay	(Phillips *et al.* 1988; Hemminki *et al.* 1990)
	Aluminium production workers	^{32}P-postlabelling, immunoassay, SFS	(Vahakangas *et al.* 1985; Schoket *et al.* 1993)
	Coke oven workers	^{32}P-postlabelling, immunoassay, SFS, HPLC/fluorescence	(Haugen *et al.* 1986; van Schooten *et al.* 1990; Binkova *et al.* 1998; Rojas *et al.* 2000)
	Graphite electrode manufacturing workers	^{32}P-postlabelling, immunoassay	(Arnould *et al.* 1999)
	Roofers	^{32}P-postlabelling	(Herbert *et al.* 1990)
	Chimney sweeps	^{32}P-postlabelling, HPLC/fluorescence	(Ichiba *et al.* 1994)
	Petrol refinery workers	^{32}P-postlabelling	(Abdel-Rahman *et al.* 2001)
	Miners[a]	^{32}P-postlabelling	(Scheepers *et al.* 2002)
	Policemen[a]	^{32}P-postlabelling, immunoassay	(Galati *et al.* 2001; Ruchirawat *et al.* 2002)
	Bus maintenance workers and drivers[a]	^{32}P-postlabelling	(Hemminki *et al.* 1994)
Environmental	Inhabitants of Upper Silesia, Poland	^{32}P-postlabelling, immunoassay	(Perera *et al.* 1992)
	Inhabitants of Northern Bohemia, Czech Republic	^{32}P-postlabelling	(Binkova *et al.* 1995)
	Inhabitants of Xuan Wei province, China	^{32}P-postlabelling	(Mumford *et al.* 1993)
	Denmark (urban vs. rural dwellers)	^{32}P-postlabelling	(Nielsen *et al.* 1996)
	Greece (urban vs. rural dwellers)	^{32}P-postlabelling	(Nielsen *et al.* 1996)
	Thailand (urban vs. rural dwellers)	^{32}P-postlabelling	(Ruchirawat *et al.* 2007)
Medicinal	Psoriasis patients[b]	^{32}P-postlabelling, immunoassay	(Santella *et al.* 1995; Godschalk *et al.* 1998)

[a]Exposed to vehicle exhaust emissions.
[b]Treated with coal tar therapy.

A number of studies have also monitored workers occupationally exposed to PAHs for PAH–protein adducts (Phillips 2005). As mentioned above, occupational exposure to other chemicals, including ethylene oxide, styrene oxide, acrylamide, acrylonitrile, aniline and MDA, has been monitored by means of protein adduct detection (Phillips and Farmer 1995).

9.5. SMOKING-RELATED ADDUCTS

The causal association between tobacco smoking and around 15 different cancers is evident from many epidemiological analyses (IARC 2004). Numerous biomonitoring studies have demonstrated elevated levels of DNA adducts in many smokers' tissues and also protein adducts in blood (Phillips 2002), reflecting the systemic nature of the exposure of the human body to carcinogens inhaled in tobacco smoke. Analysis by ^{32}P-postlabelling has identified a complex mixture of DNA adducts, with chromatographic characteristics of bulky and/or hydrophobic adducts, similar to those formed by complex mixtures of PAHs. Immunochemical methods of analysis have also detected the presence of PAH–DNA adducts, although in both cases the precise adduct-forming compounds remain to be elucidated. Indeed, although fluorescence detection methods have shown the presence of the pyrene chromophore (suggesting that BP–DNA adducts are present; Phillips 2002), recent mass spectrometry studies have suggested that there are fewer BP–DNA adducts in the lungs of smokers than the fluorescence evidence, or immunoassays using BPDE–DNA antibodies (notwithstanding their cross-reactivity with other PAH–DNA adducts), would suggest (Beland *et al.* 2005). While it is clear that tobacco smoke induces elevated levels of adducts in many human tissues, results obtained with different methods do not necessarily correlate well with each other, e.g. between values obtained by ^{32}P-postlabelling and by BPDE–DNA immunoassay (Gyorffy *et al.* 2004). Overall there are still uncertainties as to the nature of 'smoking-related adducts' detected in human tissues by the various methods.

In most of the studies comparing DNA adducts in smokers and non-smokers, there is considerable inter-individual variation, with an overlap in values between the two groups. However comparisons of 4-aminobiphenyl–haemoglobin adducts between smokers and non-smokers show no such overlap, making this biomarker capable of distinguishing smokers from non-smokers. Protein adduct measurements can also distinguish between non-smokers exposed to second-hand tobacco smoke from those unexposed, whereas DNA adduct measurement appears to be a less sensitive biomarker of passive smoking (Phillips 2002).

An example of molecular epidemiology studies providing clues (albeit not proof) to a possible aetiological association between exposure and disease is the case of tobacco smoking and cervical cancer. That smoking was a risk factor for cervical cancer has long been noted, but the consensus was that this was a confounder for some other risk factor. Only recently has it been concluded from epidemiological evidence that smoking is in fact an independent risk factor, and thus a causative agent (IARC 2004). However, biomarker evidence that lends mechanistic support to the hypothesis that smoking is a cause of cervical cancer predates this new epidemiological conclusion by many years: in smokers, the cervical mucosa has detectable levels of cotinine, a major metabolite of nicotine, and cervical epithelial cells have been found by three different methods to contain higher levels of smoking-related DNA adducts (Phillips 2005).

9.6. DNA ADDUCTS IN PROSPECTIVE STUDIES

In order to determine whether DNA adducts have predictive value in cancer risk, it is necessary to conduct prospective studies in which DNA samples are collected and stored from a large cohort of individuals, who are then followed up to determine who does and who does not develop cancer in the future. It is then possible to perform a nested case–control study within the cohort to determine whether DNA adduct analysis of the stored samples reveals whether differences between the two groups were evident prior to the onset of disease.

In a study conducted in Shanghai, China, a region with high incidence of liver cancer associated with dietary exposure to the mycotoxin aflatoxin B_1,

18 244 men each provided a single urine sample and completed a detailed dietary questionnaire data, in addition to which food analyses were carried out (Qian *et al.* 1994). Subsequently, when 55 cases of liver cancer had arisen in the cohort, these were matched to 267 disease-free controls and their urine samples analysed by HPLC–fluorescence to detect the presence of aflatoxin derivatives. A significant association was found between the presence of aflatoxin metabolites, including the aflatoxin–N-7-guanine adduct, and liver cancer (adjusted relative risk (RR), 9.1; 95% confidence interval (CI), 2.9–29.2). Interestingly, when data obtained from questionnaires and food analyses were considered without the biomarker data, no association between estimates of aflatoxin exposure and liver cancer was evident. Thus, in this case the power of biomarkers of exposure showed a clear advantage over more traditional means of exposure assessment to show a causal association.

The ability of DNA adducts to predict lung cancer risk was investigated in the Physicians Health Study (Tang *et al.* 2001). From a follow-up of a cohort of 15 700 males who had provided blood samples at the outset of the study, 93 cases of lung cancer were identified and matched to 173 controls. Analysis of white blood cell DNA by ^{32}P-postlabelling revealed that smokers who got lung cancer had two-fold higher levels of bulky/hydrophobic DNA adducts than smokers who did not. The smokers who had elevated levels of adducts were approximately three times more likely to be diagnosed with lung cancer 1–13 years later than the smokers with lower adduct concentrations (odds ratio (OR), 2.98; 95% CI, 1.05–8.42; $p = 0.04$).

Two recent studies that investigated bulky DNA adducts in leukocytes by ^{32}P-postlabelling have also shown the predictive power of DNA adducts to distinguish groups of individuals who developed cancer from those who did not. In the first, 115 cases of lung cancer were matched with twice the number of controls from European cohorts totalling more than 500 000 people (Peluso *et al.* 2005). Detectable DNA adducts were significantly more common in non-smokers and ex-smokers of more than 10 years abstinence who developed lung cancer than in those who did not (OR ratio, 4.04; 95% CI, 1.06–15.42). The second study investigated 245 individuals with lung cancer and 255 without

from a population-based cohort of 53 689 men and women (Bak *et al.* 2006). There was a higher median level of DNA adducts among smokers who developed lung cancer than among those that did not (RR, 1.61; 95% CI, 1.04–2.49). However, it should be noted that, although adduct levels were statistically significantly higher in the cases in both these studies, the numerical differences from the controls were somewhat small. Taking into account the extent of inter-individual variation in adduct levels in all these studies, the ability to predict cancer risk on an individual basis would be considered very limited, despite the collective differences between the cases and the controls. Nevertheless, DNA adduct analysis should have applications in investigating the efficacy of chemoprevention strategies. For example, it should be possible in interventions in an occupationally-exposed cohort to document a reduction in adduct levels concomitant with a reduction in cancer risk.

In addition to these prospective studies, a meta-analysis of (mostly) case–control studies measuring bulky DNA adducts found that current smokers with high levels of adducts have an increased risk of lung and bladder cancers (Veglia *et al.* 2003); no such correlation was evident for ex- or non-smokers. This study noted a wide variation between studies in the 'background' levels of adducts in controls, due possibly to population variation but also to the lack of standardized assay protocols.

DNA adducts in human DNA repair studies

Since DNA adducts are premutagenic and their formation is an early step in carcinogenesis, it is a plausible hypothesis that the DNA repair capacity is a host factor that may modulate individual susceptibility to cancer. While many studies have attempted to identify an influence of genetic polymorphisms in DNA repair genes on cancer susceptibility, results have been fairly disappointing and there has been a poor correlation between genotype and phenotype. Given the complexity of DNA repair processes, it is perhaps not surprising that the reductionist approach of looking at selected repair gene single nucleotide polymorphisms (SNPs) has failed to reveal strong associations with cancer risk. Nevertheless, there are still compelling reasons for believing that inter-individual differences in DNA repair capacity modulate human cancer risk; e.g. the involvement of *BRCA1* and *BRCA2* in double-strand break repair. Also, in the case of breast cancer, several studies investigating repair *phenotype* have suggested that breast cancer patients have poorer ability to repair DNA than controls. In a recent study, lymphoblastoid cells derived from breast cancer patients and treated with BPDE were compared for their ability to excise the DNA damage with cells derived from their sisters (Kennedy *et al.* 2005) and found to be significantly less effective at DNA repair. Similarly, BPDE-induced chromatid breaks in lymphocytes were significantly more frequent in breast cancer cases than in controls (Xiong *et al.* 2001) and, in another study (Shi *et al.* 2004), DNA repair capacity measured in the host-cell reactivation assay was significantly lower in lymphocytes of untreated breast cancer patients than in cells of matched controls. These studies all show a significantly lower repair capacity of breast cancer cases compared to controls and the pathway implicated is nucleotide excision repair (NER). It can also be hypothesized that there is an as-yet-unidentified environmental DNA damaging agent, or agents, whose lesions are processed by NER, involved in the aetiology of breast cancer.

However, these studies are all case–control studies and could have been confounded by disease status and/or consequences of therapy. It would be better in future to carry out nested case–control studies within prospective cohorts, comparing the repair capacity of cancer cases and controls, using samples taken well before disease onset and diagnosis.

Correlations between DNA and protein adducts

The many different assays available for biomonitoring adducts raises the question concerning validation and correlations between different measured endpoints. These are still matters of ongoing research, although a number of studies have sought to compare different biomarkers in the same subjects and/or tissues (Farmer and Emeny 2006). For example, no correlation was found between PAH–DNA adduct levels and PAH–albumin adduct levels in the blood of psoriasis patients treated with coal tar (Santella *et al.* 1995), although in a study of lamina-

tion workers exposed to styrene there was a correlation between the levels of the styrene-O^6–guanine adduct in lymphocytes and of the N-terminal valine adduct of styrene in haemoglobin (Vodicka *et al.* 1999). As already mentioned, there can be differences between smoking-related DNA adduct levels when assayed by different methods (Gyorffy *et al.* 2004). A wider comparison of adducts with other biomarkers of exposure, such as urinary mutagenicity, shows highly variable levels of correlation (Farmer and Emeny 2006); such observations are not unexpected, given the widely differing half-lives of the biomarkers being compared.

9.7. CONCLUSIONS

It is well established that DNA damage is the initiating event in chemical carcinogenesis. From many studies to date it is evident that measuring DNA adducts is useful for monitoring exposure to carcinogens. Many such studies rely on a surrogate source of DNA, normally from blood cells, and from a limited number of studies there are indications that there is a reasonable correlation between adduct levels in such tissues and in the target tissue, e.g. lung. There can be wide inter-individual variations in adduct levels, in part due to genetic polymorphisms, but other non-genetic environmental factors may be equally influential. At the same time there is little information on intra-individual variations in adduct levels, and yet studies to date have for the most part relied on a single determination.

In suitably designed studies, DNA adducts can be valid biomarkers of cancer risk, although the ability to predict risk on an individual basis would appear to be limited on the basis of current information. Nevertheless, refinement of predictability may be expected from future studies that combine exposure measurements with determination of host susceptibility factors, such as functional genetic polymorphisms. The detection and characterization of DNA adducts has already given important insight into the environmental aetiology of cancer in a number of important instances. The establishment of large biobanks of human samples will undoubtedly provide new and important opportunities for using biomarkers such as DNA and protein adducts for investigating the aetiology of cancer in future molecular epidemiology studies. Nevertheless, for such a rich source of material to be fully exploited, it will be necessary to develop higher-throughput analytical procedures than are currently available.

References

Abdel-Rahman AG, Allam MF, Mansour MT *et al.* 2001. PAH–DNA adducts in a petrol refinery in Egypt. *Eur J Cancer Prevent* **10**: 469–472.

Arlt VM. 2006. Traditional herbal medicines containing plant species of the genus *Aristolochia* are carcinogenic to humans. *Eur J Gen Mol Toxicol*: http://www.swan.ac.uk/cget/ejgt/issues.htm

Arlt VM, Stiborova M, Schmeiser HH. 2002. Aristolochic acid as a probable human cancer hazard in herbal remedies: a review. *Mutagenesis* **17**: 265–277.

Arnould JP, Pfohl-Leszkowicz A, Bach V *et al.* 1999. Biological monitoring exposure of workers from plant producing carbon electrodes: quantification of benzo(α)pyrene–DNA adducts in leukocytes, by a ^{32}P-postlabelling method and an immunoassay. *Hum Exp Toxicol* **18**: 314–321.

Bak H, Autrup H, Thomsen BL *et al.* 2006. Bulky DNA adducts as risk indicator of lung cancer in a Danish case–cohort study. *Int J Cancer* **118**: 1618–1622.

Beland FA, Churchwell MI, Von Tungeln LS *et al.* 2005. High-performance liquid chromatography electrospray ionization tandem mass spectrometry for the detection and quantitation of benzo(*a*)pyrene–DNA adducts. *Chem Res Toxicol* **18**: 1306–1315.

Besaratinia A, Maas LM, Brouwer EM *et al.* 2000. Comparison between smoking-related DNA adduct analysis in induced sputum and peripheral blood lymphocytes. *Carcinogenesis* **21**: 1335–1340.

Binkova B, Lewtas J, Miskova I *et al.* 1995. DNA adducts and personal air monitoring of carcinogenic polycyclic aromatic hydrocarbons in an environmentally exposed population. *Carcinogenesis* **16**: 1037–1046.

Binkova B, Topinka J, Mrackova G *et al.* 1998. Coke oven workers study: the effect of exposure and *GSTM1* and *NAT2* genotypes on DNA adduct levels in white blood cells and lymphocytes as determined by ^{32}P-postlabelling. *Mutat Res* **416**: 67–84.

Boogaard PJ. 2002. Use of haemoglobin adducts in exposure monitoring and risk assessment. *J Chromatogr B Analyt Technol Biomed Life Sci* **778**: 309–322.

Divi RL, Beland FA, Fu PP *et al.* 2002. Highly sensitive chemiluminescence immunoassay for benzo(*a*)pyrene–

DNA adducts: validation by comparison with other methods, and use in human biomonitoring. *Carcinogenesis* **23**: 2043–2049.

Dybing E, Farmer PB, Andersen M *et al.* 2005. Human exposure and internal dose assessments of acrylamide in food. *Food Chem Toxicol* **43**: 365–410.

ESCODD. 2002. Comparative analysis of baseline 8-oxo-7,8-dihydroguanine in mammalian cell DNA, by different methods in different laboratories: an approach to consensus. *Carcinogenesis* **23**: 2129–2133.

Fairbairn DW, Olive PL, O'Neill KL. 1995. The comet assay: a comprehensive review. *Mutation Res* **339**: 37–59.

Farmer PB. 2004. Exposure biomarkers for the study of toxicological impact on carcinogenic processes. *IARC Sci Publ* **157**: 71–90.

Farmer PB, Emeny JM (eds). 2006. *Biomarkers of Carcinogen Exposure and Early Effects*. ECNIS: Lodz.

Foiles PG, Akerkar SA, Carmella SG *et al.* 1991. Mass spectrometric analysis of tobacco-specific nitrosamine–DNA adducts in smokers and nonsmokers. *Chem Res Toxicol* **4**: 364–368.

Friesen MD, Kaderlik K, Lin D *et al.* 1994. Analysis of DNA adducts of 2-amino-1-methyl-6-phenylimidazo(4,5-*b*)pyridine in rat and human tissues by alkaline hydrolysis and gas chromatography/electron capture mass spectrometry: validation by comparison with ^{32}P-postlabeling. *Chem Res Toxicol* **7**: 733–739.

Galati R, Zijno A, Crebelli R *et al.* 2001. Detection of antibodies to the benzo(*a*)pyrene diol epoxide–DNA adducts in sera from individuals exposed to low doses of polycyclic aromatic hydrocarbons. *J Exp Clin Cancer Res* **20**: 359–364.

Godschalk RW, Ostertag JU, Moonen EJ *et al.* 1998. Aromatic DNA adducts in human white blood cells and skin after dermal application of coal tar. *Cancer Epidemiol Biomarkers Prevent* **7**: 767–773.

Godschalk RW, Van Schooten FJ, Bartsch H. 2003. A critical evaluation of DNA adducts as biological markers for human exposure to polycyclic aromatic compounds. *J Biochem Mol Biol* **36**: 1–11.

Groopman JD, Jiaqi Z, Donahue PR *et al.* 1992. Molecular dosimetry of urinary aflatoxin–DNA adducts in people living in Guangxi Autonomous Region, People's Republic of China. *Cancer Res* **52**: 45–52.

Gyorffy E, Anna L, Gyori Z *et al.* 2004. DNA adducts in tumour, normal peripheral lung and bronchus, and peripheral blood lymphocytes from smoking and non-smoking lung cancer patients: correlations between tissues and detection by ^{32}P-postlabelling and immunoassay. *Carcinogenesis* **25**: 1201–1209.

Haugen A, Becher G, Benestad C *et al.* 1986. Determination of polycyclic aromatic hydrocarbons in the urine, benzo(*a*)pyrene diol epoxide–DNA adducts in lymphocyte DNA, and antibodies to the adducts in sera from coke oven workers exposed to measured amounts of polycyclic aromatic hydrocarbons in the work atmosphere. *Cancer Res* **46**: 4178–4183.

Hemminki K, Randerath K, Reddy MV *et al.* 1990. Postlabeling and immunoassay analysis of polycyclic aromatic hydrocarbons – adducts of deoxyribonucleic acid in white blood cells of foundry workers. *Scand J Work Environ Health* **16**: 158–162.

Hemminki K, Soderling J, Ericson P *et al.* 1994. DNA adducts among personnel servicing and loading diesel vehicles. *Carcinogenesis* **15**: 767–769.

Herbert R, Marcus M, Wolff MS *et al.* 1990. Detection of adducts of deoxyribonucleic acid in white blood cells of roofers by ^{32}P-postlabeling. Relationship of adduct levels to measures of exposure to polycyclic aromatic hydrocarbons. *Scand J Work Environ Health* **16**: 135–143.

IARC. 2004. *Tobacco Smoke and Involuntary Smoking.* IARC Monographs on the Evaluation of Carcinogenic Risks to Humans No. 83. IARC: Lyon.

Ichiba M, Hagmar L, Rannug A *et al.* 1994. Aromatic DNA adducts, micronuclei and genetic polymorphism for *CYP1A1* and *GST1* in chimney sweeps. *Carcinogenesis* **15**: 1347–1352.

Jonsson BA, Wishnok JS, Skipper PL *et al.* 1995. Lysine adducts between methyltetrahydrophthalic anhydride and collagen in guinea pig lung. *Toxicol Appl Pharmacol* **135**: 156–162.

Kanaly RA, Hanaoka T, Sugimura H *et al.* 2006. Development of the adductome approach to detect DNA damage in humans. *Antioxid Redox Signal* **8**: 993–1001.

Kennedy DO, Agrawal M, Shen J *et al.* 2005. DNA repair capacity of lymphoblastoid cell lines from sisters discordant for breast cancer. *J Natl Cancer Inst* **97**: 127–132.

Lin D, Lay JO Jr, Bryant MS *et al.* 1994. Analysis of 4-aminobiphenyl–DNA adducts in human urinary bladder and lung by alkaline hydrolysis and negative ion gas chromatography–mass spectrometry. *Environ Health Perspect* **102**(suppl 6): 11–16.

Mumford JL, Lee X, Lewtas J *et al.* 1993. DNA adducts as biomarkers for assessing exposure to polycyclic aromatic hydrocarbons in tissues from Xuan Wei women with high exposure to coal combustion emissions and high lung cancer mortality. *Environ Health Perspect* **99**: 83–87.

Nielsen PS, Andreassen A, Farmer PB *et al.* 1996. Biomonitoring of diesel exhaust-exposed workers. DNA and hemoglobin adducts and urinary 1-hydroxypyrene as markers of exposure. *Toxicol Lett* **86**: 27–37.

Peluso M, Munnia A, Hoek G *et al.* 2005. DNA adducts and lung cancer risk: a prospective study. *Cancer Res* **65**: 8042–8048.

Perera FP, Hemminki K, Gryzbowska E *et al.* 1992. Molecular and genetic damage in humans from environmental pollution in Poland. *Nature* **360**: 256–258.

Pfohl-Leszkowicz A, Castegnaro M. 2005. Further arguments in favour of direct covalent binding of ochratoxin A (OTA) after metabolic biotransformation. *Food Addit Contam* **22**(suppl 1): 75–87.

Phillips DH. 1996. DNA adducts in human tissues: biomarkers of exposure to carcinogens in tobacco smoke. *Environ Health Perspect* **104**(suppl 3): 453–458.

Phillips DH. 1997. Detection of DNA modifications by the ^{32}P-postlabelling assay. *Mutation Res* **378**: 1–12.

Phillips DH. 2002. Smoking-related DNA and protein adducts in human tissues. *Carcinogenesis* **23**: 1979–2004.

Phillips DH. 2005. DNA adducts as markers of exposure and risk. *Mutat Res* **577**: 284–292.

Phillips DH. 2005. Macromolecular adducts as biomarkers of human exposure to polycyclic aromatic hydrocarbons. In *The Carcinogenic Effects of Polycyclic Aromatic Hydrocarbons*, Luch A (ed.). Imperial College Press: London; 137–169.

Phillips DH. 2007. The formation of DNA adducts. In *The Cancer Handbook*, Allison MR (ed.). Wiley: Chichester.

Phillips DH, Farmer PB. 1995. Protein and DNA adducts as biomarkers of exposure to environmental mutagens. In *Environmental Mutagenesis*, Phillips DH, Venitt S (eds). Bios: Oxford; 367–395.

Phillips DH, Hemminki K, Alhonen A *et al.* 1988. Monitoring occupational exposure to carcinogens: detection by ^{32}P-postlabelling of aromatic DNA adducts in white blood cells from iron foundry workers. *Mutat Res* **204**: 531–541.

Poirier MC. 2004. Chemical-induced DNA damage and human cancer risk. *Nat Rev Cancer* **4**: 630–637.

Poirier MC, Santella RM, Weston A. 2000. Carcinogen macromolecular adducts and their measurement. *Carcinogenesis* **21**: 353–359.

Poirier MC, Weston A, Schoket B *et al.* 1998. Biomonitoring of United States Army soldiers serving in Kuwait in 1991. *Cancer Epidemiol Biomarkers Prevent* **7**: 545–551.

Qian GS, Ross RK, Yu MC *et al.* 1994. A follow-up study of urinary markers of aflatoxin exposure and liver cancer risk in Shanghai, People's Republic of China. *Cancer Epidemiol Biomarkers Prevent* **3**: 3–10.

Rojas M, Cascorbi I, Alexandrov K *et al.* 2000. Modulation of benzo(a)pyrene diolepoxide–DNA adduct levels in human white blood cells by *CYP1A1*, *GSTM1* and *GSTT1* polymorphism. *Carcinogenesis* **21**: 35–41.

Rothman N, Correa-Villaseñor A, Ford DP *et al.* 1993. Contribution of occupation and diet to white blood cell polycyclic aromatic hydrocarbon–DNA adducts in wildland firefighters. *Cancer Epidemiol Biomarkers Prevent* **2**: 341–347.

Ruchirawat M, Mahidol C, Tangjarukij C *et al.* 2002. Exposure to genotoxins present in ambient air in Bangkok, Thailand – particle-associated polycyclic aromatic hydrocarbons and biomarkers. *Sci Total Environ* **287**: 121–132.

Ruchirawat M, Settachan D, Navasumrit P *et al.* 2007. Assessment of potential cancer risk in children exposed to urban air pollution in Bangkok, Thailand. *Toxicol Lett* **168**: 200–209.

Santella RM, Gammon M, Terry M *et al.* 2005. DNA adducts, DNA repair genotype/phenotype and cancer risk. *Mutat Res* **592**: 29–35.

Santella RM, Perera FP, Young TL *et al.* 1995. Polycyclic aromatic hydrocarbon–DNA and protein adducts in coal tar treated patients and controls and their relationship to glutathione *S*-transferase genotype. *Mutat Res* **334**: 117–124.

Scheepers PT, Coggon D, Knudsen LE *et al.* 2002. BIOMarkers for occupational diesel exhaust exposure monitoring (BIOMODEM) – a study in underground mining. *Toxicol Lett* **134**: 305–317.

Schilter B, Marin-Kuan M, Delatour T *et al.* 2005. Ochratoxin A: potential epigenetic mechanisms of toxicity and carcinogenicity. *Food Addit Contam* **22**(suppl 1): 88–93.

Schmitz OJ, Worth CC, Stach D *et al.* 2002. Capillary electrophoresis analysis of DNA adducts as biomarkers for carcinogenesis. *Angew Chem Int Ed Engl* **41**: 445–448.

Schoket B, Doty WA, Vincze I *et al.* 1993. Increased sensitivity for determination of polycyclic aromatic–DNA adducts in human DNA samples by dissociation-enhanced lanthanide fluoroimmunoassay (DELFIA). *Cancer Epidemiol Biomarkers Prevent* **2**: 349–353.

Schoket B, Phillips DH, Poirier MC *et al.* 1993. DNA adducts in peripheral blood lymphocytes from aluminum production plant workers determined by ^{32}P-postlabeling and enzyme-linked immunosorbent assay. *Environ Health Perspect* **99**: 307–309.

Shi Q, Wang LE, Bondy ML *et al.* 2004. Reduced DNA repair of benzo(a)pyrene diol epoxide-induced adducts and common XPD polymorphisms in breast cancer patients. *Carcinogenesis* **25**: 1695–700.

Shuker DEG, Farmer PB. 1992. Relevance of urinary DNA adducts as markers of carcinogen exposure. *Chem Res Toxicol* **5**: 450–460.

Singh R, Farmer PB. 2006. Liquid chromatography–electrospray ionization–mass spectrometry: the future of DNA adduct detection. *Carcinogenesis* **27**: 178–196.

Skipper PL, Peng X, Soohoo CK *et al.* 1994. Protein adducts as biomarkers of human carcinogen exposure. *Drug Metab Rev* **26**: 111–124.

SooHoo CK, Singh K, Skipper PL *et al.* 1994. Characterization of benzo(a)pyrene anti-diol epoxide adducts to human histones. *Chem Res Toxicol* **7**: 134–138.

Strickland PT, Routledge MN, Dipple A. 1993. Methodologies for measuring carcinogen adducts in humans. *Cancer Epidemiol Biomarkers Prev* **2**: 607–619.

Tang D, Phillips DH, Stampfer M *et al.* 2001. Association between carcinogen–DNA adducts in white blood cells and lung cancer risk in the Physicians Health Study. *Cancer Res* **61**: 6708–6712.

Tornqvist M, Fred C, Haglund J *et al.* 2002. Protein adducts: quantitative and qualitative aspects of their formation, analysis and applications. *J Chromatogr B Analyt Technol Biomed Life Sci* **778**: 279–308.

Tornqvist M, Landin HH. 1995. Hemoglobin adducts for *in vivo* dose monitoring and cancer risk estimation. *J Occup Environ Med* **37**: 1077–1085.

Turteltaub KW, Dingley KH. 1998. Application of accelerated mass spectrometry (AMS) in DNA adduct quantification and identification. *Toxicol Lett* **102–103**: 435–439.

Vahakangas K, Trivers G, Rowe M *et al.* 1985. Benzo(a)pyrene diolepoxide–DNA adducts detected by synchronous fluorescence spectrophotometry. *Environ Health Perspect* **62**: 101–104.

van Schooten FJ, van Leeuwen FE, Hillebrand MJ *et al.* 1990. Determination of benzo(a)pyrene diol epoxide–DNA adducts in white blood cell DNA from coke-oven workers: the impact of smoking. *J Natl Cancer Inst* **82**: 927–933.

Veglia F, Matullo G, Vineis P. 2003. Bulky DNA adducts and risk of cancer: a meta-analysis. *Cancer Epidemiol Biomarkers Prevent* **12**: 157–160.

Vodicka P, Tvrdik T, Osterman-Golkar S *et al.* 1999. An evaluation of styrene genotoxicity using several biomarkers in a 3-year follow-up study of hand-lamination workers. *Mutat Res* **445**: 205–224.

Weston A. 1993. Physical methods for the detection of carcinogen–DNA adducts in humans. *Mutation Res* **288**: 19–29.

Wild CP, Pisani P. 1998. Carcinogen DNA and protein adducts as biomarkers of human exposure in environmental cancer epidemiology. *Cancer Detect Prevent* **22**: 273–283.

Xiong P, Bondy ML, Li D *et al.* 2001. Sensitivity to benzo(*a*)pyrene diol-epoxide associated with risk of breast cancer in young women and modulation by glutathione *S*-transferase polymorphisms: a case–control study. *Cancer Res* **61**: 8465–8469.

10

Biomarkers of Mutation and DNA Repair Capacity

Marianne Berwick[1] and Richard J. Albertini[2]

[1]University of New Mexico, Albuquerque, NM, and [2]University of Vermont, Burlington, VT, USA

10.1. INTRODUCTION

Orderly growth and cell division in all living organisms requires that the genetic information encoded in DNA be precisely replicated. While the process of DNA replication is characterized by high fidelity, errors occasionally occur, altering the primary structure of this critical molecule. In addition, environmental insults can induce DNA damage, either directly or indirectly. DNA repair can reverse some of these changes but, if they persist, then a change in DNA structure will translate to a change in information content, i.e. a mutation. Once a mutation occurs, assuming it is not lethal, the information change is permanent and persists in the progeny of the original mutant, unless again changed by mutation.

Mutations, although rare, are always occurring (Van Houten and Albertini 1995). Those arising in the absence of an identifiable external cause are termed 'spontaneous', although this is often a designation of ignorance. For this reason, the term 'background mutations' seems more appropriate. Some background mutations result from simple mistakes in a cell's replication machinery, such as polymerase slippage. Background mutations also arise when molecules in the DNA, such as the bases, deoxyribose moieties or phosphate backbone, are damaged by endogenously produced chemicals arising from normal metabolism or inflammation, or by other cellular processes. Examples of such damage to bases include hydrolytic deaminations of cytosine or 5-methylcytosine, or covalent adductions from reactive oxygen species.

Superimposed on this background are the mutations caused by exogenous agents. There are several classes of genotoxic agents, including the physical, such as ionizing or ultraviolet radiations, the biological, such as viruses, and the chemical. Genotoxic chemicals constitute a broad class that includes large and small molecules that covalently bond to elements in the DNA, that intercalate into the DNA, that misincorporate into the DNA or that otherwise change DNA's primary structure. Chemicals that induce mutations are considered to be directly genotoxic if they or their metabolites directly react with the DNA; or to be indirectly genotoxic if neither they or their metabolites react directly with the DNA. Indirect mutagenic effects may be mediated by the generation of reactive oxygen species or the lowering of endogenous defences against oxidative damage, by modifications of endogenous deoxynucleotide precursor pools that might tip the balance of individual base incorporations during DNA replication, or by altering one of the many other cellular processes (e.g. damage to repair enzymes) that protect from or enhance mutations. The sum of mutations in a cell, therefore, includes both the background and those induced by exogenous agents (Strauss 1992).

Molecular Epidemiology of Chronic Diseases, Edited by C. P. Wild, P. Vineis, and S. Garte
© 2008 John Wiley & Sons, Ltd

Regardless of cause, DNA damage has a variety of potential outcomes. It may be repaired before an irreversible change in primary structure occurs, with no change in information content, i.e. with no mutation. If not repaired prior to DNA replication, damage to specific bases in the DNA, i.e. guanine (G), adenine (A), cytosine (C) or thymine (T), may alter their base-pairing characteristics, resulting in a mismatch during DNA replication (e.g. A pairing with G, or C rather than T, etc.) and a change in the sequence of bases newly synthesized DNA. More complex outcomes, such as DNA strand breaks, rearrangements of large portions of DNA, deletions, insertions, etc., may also occur. In all cases, the progeny inheriting the changed DNA have an alteration in their received information content, and are themselves changed. The first part of this chapter covers the topic of DNA damage and mutation. We then go on to consider how inter-individual differences in the DNA repair processes that can counteract these effects may be assessed in relation to cancer risk.

10.2. CLASSIFICATION OF MUTATIONS

There are many ways to classify mutations (Loeb and Christians 1996). Mutations, when restricted to a definition of submicroscopic changes as opposed to larger chromosomal aberrations, can be classified according to type as large or small alterations. The former may involve hundreds to thousands of base changes and encompass more than a single gene. At the extreme, such changes merge with the chromosome aberrations. At the other end of the size spectrum, small mutations usually involve changes of single or only a few bases in the DNA. These are usually termed 'point mutations', although only the single base changes precisely fit this designation. Point mutations that involve single base changes are usually base-pair substitutions. These changes are further subdivided into a subclass termed transitions, where a purine is substituted for a purine (i.e. A for G or vice versa) or a pyrimidine is substituted for a pyrimidine (i.e. C for T or vice versa) and a subclass termed transversions, where a purine is substituted for a pyrimidine or vice versa. There are six possible transversions, all of which reverse the orientation of purines and pyrmidines in the DNA strand.

Classifying mutations according to type has utility for determining causes.

For multicellular organisms, mutations are also classified according to cell lineage, i.e. their occurrence in germline or somatic cells. The relevance of making this distinction has to do with the consequences of the mutation. For the most part, germinal mutations do not affect the individuals in whom they occur. Rather, these mutations are passed to the next generation, where they affect 100% of the cells and the entire individual. Somatic mutations, on the other hand, affect only the individual in whom they occur, having no consequences for subsequent generations. Since, at the DNA level, mechanisms that produce DNA changes are likely to be operative in both cell types, germinal and somatic mutations are often considered together.

Many of the environmental agents that cause cancer also induce mutations. It follows, therefore, that exposures to these genotoxic carcinogens will increase the frequencies of somatic mutations over background levels, and that the magnitude of the increases will be proportional to the levels of the adverse exposures. In theory, monitoring DNA damage and mutagenicity in human populations could afford an early warning of deleterious, cancer causing exposures (Albertini et al. 1993; Albertini 1998). It might even be possible to estimate risk for the subsequent development of cancer, given that certain validations are made. An ability to characterize mutations at the molecular level might allow identification of the causes of the mutations if some agents damage the DNA in identifiable ways.

10.3. MUTATIONS IN MOLECULAR EPIDEMIOLOGY

Somatic mutations may be detected in disease relevant (cancer) genes or in reporter genes. Potentially informative cancer gene mutations for monitoring include those in oncogenes, e.g. ras, and in tumour suppressor genes, e.g. p53, which may be detected in normal, non-cancer cells or in actual tumours (Greenblatt et al. 1994; Hussain and Harris 1998). Studies in tumours that include characterizations of the kinds of mutations present (determinations of molecular mutational spectra) have been used to infer causation. Measuring mutations in cancer genes in normal individuals has the advantage of directly monitoring carcinogens. However, there is ambiguity with such mutations, because it is not always clear if the mutations are truly arising in non-cancer cells or if one is engaged in early diagnosis rather then environmental monitoring. A database of *p53* mutations has been developed by IARC (Oliver *et al.* 2002) and a number of studies have linked specific exposures, e.g. UV light, aflatoxin (Hussain and Harris 1998; Hussain *et al.*, 2007; Pfeiffer *et al.*, 2002) to specific mutations in this gene. The example of aflatoxin is developed in more detail in Chapter 25. A recent advance has been the ability to measure some of these mutations in plasma DNA in a non-invasive manner (Kirk *et al.*, 2005).

In contrast to mutations in cancer genes, those in reporter genes, which also serve to quantify and characterize *in vivo* mutational events, have no role in disease processes. These events can be monitored without the concern that what is being detected is, in reality, an early malignancy. Four reporter gene/cell systems are in some degree of use at present for human *in vivo* studies (Albertini *et al.* 1998; 2001; Relton *et al.* 2004). These include the glycophorin-A (*GPA*) gene, studied in red blood cells, the *HPRT* gene, the *HLA* genes and the T cell receptor (*TCR*) genes, all measured in T lymphocytes. Although different cell types can be studied, for human *in vivo* studies, blood provides the only practically accessible source of cells. The relative advantages and disadvantages of the different blood cell bypes for human muta-genicity monitoring are given in Box 10.1. Some examples of the application of mutation analysis using reporter genes in human population studies are given in Box 10.2.

Box 10.1   Reporter genes for human mutagenicity monitoring

Cells	Gene	Location	Size	Assay type
RBC	Glycophorin A (*GPA*)	Autosome 4q	>44 kb	Cytometry
T cell	Hypoxanthine-X guanine transferase (*HPRT*)	Chromosome Xq	>44 kb	Cloning Auto-phosphoribosyl radiography
T cell	*HLA*	Autosome 6p	>5 kb	Cloning
T cell	T cell receptor (TCR)	Multi-gene 7q, 14q		Cytometry

RBC, red blood cells; T cell, peripheral blood T lymphocytes; *HPRT* assays, cloning employs 6-thioguanine selection for mutants; autoradiography is a slide assay employing tritiated thymidine to label cycling cells. *HLA* assay, cloning employs immunoselection with *HLA*-specific antibodies.

Box 10.2   Somatic reporter mutations in humans

Cells	Advantages	Disadvantages
Red blood cells	Abundance: small samples Examines bone marrow Simple, inexpensive automated assays Simple *in vivo* kinetics Potentially long 'memory'	No DNA for genetic analyses Mutation compartment limited to bone marrow
T lymphocytes	DNA for genetic analyses Molecular targets well characterized Mutations may occur in all body sites	Relatively small numbers in blood Events limited to T cells Large samples required Complex *in vivo* kinetics Frequently require tissue culture

Gene mutations arising *in vivo* demonstrate not only that a potential mutagen (carcinogen) has penetrated to critical targets, but that it, or its biologically active metabolite(s), has escaped host protective mechanisms to produce the kinds of irreversible genetic effects that underlie cancer. *In vivo* mutational effects are expected for a carcinogenic agent that induces cancer by a mutagenic Mode of Action (MOA). One of the major protective agencies that serve to protect against these mutations is effective DNA repair.

10.4. DNA REPAIR

DNA repair is a system of defences designed to protect the integrity of the genome. As discussed above, DNA damage is constantly occurring in a cell as a result of normal endogenous metabolic processes, errors in replication and exposure to exogenous insults; one would typically expect 1000–1 000 000 DNA damage events/day in a cell, depending on the exposure that a person has and on that individual's endogenously produced DNA damage (Setlow D, personal communication). For example, in Texas a person in the noonday sun in the summertime might anticipate having 10 000 damages/hour/skin cell. Unfortunately, a person with compromised DNA repair, such as a person with the autosomal recessive disease xeroderma pigmentosum, does not repair damage as quickly as it occurs and thus has a very high rate of skin cancer, including a 1000-fold increase in cutaneous malignant melanoma.

The difficulty in studying DNA repair capacity in relation to disease aetiology stems from two issues: (a) the costs inherent in conducting whole cell assays, which represent an integrated response to DNA damage; and (2) the lack of extensive genotype–phenotype correlation data, which generally focus on the role of single nucleotide polymorphisms (SNPs). One reason for this is that DNA repair genotypes are merely surrogates to measure DNA repair capacity, and it is unlikely that a single SNP would alter the DNA repair phenotype significantly. Complicating the study of DNA repair is the existence of multiple proteins with similar activities (Lindahl and Wood, 1999). When considering using aberrant DNA repair capacity as a 'marker'

for disease, it is important to investigate the overall genetic and cellular facets of DNA repair.

The remainder of this chapter will focus on nuclear and cellular DNA repair and not discuss mitochondrial repair, which is also important to normal cellular function. Studies of DNA repair capacity and disease aetiology require careful consideration of design, potential bias and confounding and, importantly, assessment of assay variability, both intra- and inter-individual, as well as biological plausibility (Berwick and Vineis 2000). A helpful overview (Machado and Menck 1997) of diseases clearly associated with DNA repair deficiencies is given online at: http://www.scielo.br/scielo.php?script=sci_arttext&pid=S0100-4551997000400032&lng=en&nrm=iso

10.5. CLASSES OF DNA REPAIR

There are more than 150 genes in a number of pathways of DNA repair, with some in overlapping pathways (http://www.cgal.icnet.uk/DNA_Repair_Genes.html) (Wood *et al.* 2001, 2005).

The pathways operate on four broad classes of DNA damage which can all result in the types of mutation described above. These include:

1. *Base damage*, the result of reactive oxygen species (ROS) produced by normal metabolic processes, radiation or base modification through other mechanisms induced by environmental chemical exposures (e.g., O6-methyl-dG via exposure to methylating agents).

2. *Helix-distorting bulky adducts*, including UV-induced cyclobutane pyrimidine dimers (CPDs) and 6,4-pyrimidone pyrimidine dimers (6-4PDs), chemical adducts and intra- and inter-strand DNA cross-links.

3. *Strand breaks*, which can be either single- or double-strand and also result from endogenous as well as exogenous insult. Strand breaks develop during normal DNA metabolism, at blocked replication forks, by nucleases, and by radiation (Paques and Haber 1999; Jung *et al.* 2006; Ward 1998).

4. *Mismatches*, created during DNA replication, usually consisting of inaccurate pairing of bases.

Although some lesions can be repaired by direct reversal, most are repaired by one or a combination of pathways with multiple steps, including:

1. *Base excision repair (BER) proteins*, which remove and replace DNA bases damaged by endogenous oxidative and hydrolytic decay of DNA (Lindahl and Wood 1999). The first step involves recognition and removal of damage by specific DNA glycosylases. This is followed by cleavage of the sugar–phosphate chain, excision of the basic residue and finally repair of the AP site by local DNA synthesis and ligation. One of the most common lesions repaired through BER is the oxidized base, 8-oxoguanine, repaired with the enzyme OGG1. In addition, BER removes uracil formed by deamination of cytosine or by misincorporation of dUMP opposite adenine residues.

2. *Nucleotide excision repair (NER)*, which is complex and involves a single group of proteins – as many as 40 (e.g. xeroderma pigmentosum group D protein) – that identify and repair bulky lesions caused by environmental agents, such as ultraviolet light (UV) and cigarette smoke, by excising nucleotides with photoproducts and bulky adducts attached and then repairing through synthesis and ligation steps. In general, NER repairs damage from exogenous sources that results in helix distortion, i.e., bulky adducts.

3. *Mismatch repair (MMR)*, which corrects errors of DNA replication and recombination, and can correct small deletions and insertions formed due to polymerase infidelity or slippage. In this way, the MMR pathway improves the fidelity of DNA synthesis 100- to 1000-fold (Mohrenweiser *et al.* 2003), and prevents the occurrence of heterologous (or non-conservative) DNA exchanges (i.e. recombination). NER and BER detect endogenous damage to DNA by normal cellular processes and exogenous damage from external agents, whereas MMR generally corrects errors due to replication and recombination.

4. *Double-strand break repair (DSBR)*, including homologous recombination (HR) and non-homologous end joining (NHEJ) repair pathways. HR (e.g. *RAD51* gene) repairs DSBs with a template, such as a sister chromatid or homologous chromosome, found elsewhere in the genome, whereas NHEJ (e.g. *Ku80* gene) has no repair template at all. Multiple subpathways act on different DSBs. NHEJ can be mutagenic; HR is more accurate and increases in S–G_2 cell cycle phases. HR also can result in deletions and rearrangements.

5. *Direct reversal of DNA damage* can be performed by a few enzymes, such as O6-methylguanine methyltransferase (MGMT), which removes methyl groups and other small alkyl groups from the O^6 position of guanine.

6. *Chromatin remodelling factors*, such as histone H2AX, are not DNA repair genes, but many gene products that alter chromatin structure could conceivably modify repair pathways and this represents an interesting new area for research.

Recent reviews (Osley *et al.* 2006; Jagannathan *et al.* 2006) have evaluated the role of chromatin in BER, NER and DSB repair (HR and NHEJ). Interestingly, as DNA repair systems are needed at all times and all places, wherever DNA damage occurs in the cell nucleus, there are questions as to the spatial and temporal organization for repair (Essers *et al.* 2006). Key to the repair pathways is the role of ATP-dependent chromatin remodelling factors. Chromatin is present on nucleosomes which are arrayed into folded structures that inhibit the interaction of protein factors with DNA. Factor access is allowed by two classes of chromatin remodelling factors. Two sets of enzymes may: (a) catalyse a wide variety of post-translational modifications (acetylation, methylation, phosphorylation, ubiquitylation) that may signal downstream regulatory factors or 'loosen' the structure of chromatin (Jenuwein and Allis 2001); or (b) use energy from ATP hydrolysis to disrupt histone–DNA contacts to modify the nucleosomes. This process is not well understood but appears to allow access for factors involved in transcription, replication and repair (Geng and Laurent 2004; Kasten *et al.* 2004; Narlikar *et al.* 2002; Hassan *et al.* 2001).

One proposed model for the repair of DNA damage in a chromatin context is 'access–repair–restore'

(Smerdon and Conconi 1999; Green and Almouzni 2002), which functions as suggested in a stepwise fashion. The repair factors must overcome the barriers to DNA in chromatin, particularly the nucleosome cores. Lesion recognition is the first step and new data suggest that chromatin factors as well as specific lesions are critical for lesion recognition. Lesions repaired by NER may stimulate changes to only a few nucleosomes while DSBs lead to changes over Mb domains.

The multiple DNA repair pathways mentioned above are not truly distinct from each other; there is substantial functional overlap, cooperation and competition among the pathways, e.g. MMR interfaces with both BER and NER. These pathways are well integrated with other response systems and signalling networks. One rationale for multiple factor involvement in DNA repair may be due to differential targeting to specific DNA damage sites; another suggestion is that different chromatin structures occur at various points in repair processes and these structures can only be acted on by factors with different remodelling activities. At this point, a great deal more needs to be investigated in terms of the steps in NER, DSB and BER that are regulated by chromatin. However, as epidemiologists, it behooves one to follow this literature and to adjust planned experiments and analyses accordingly. Live cell imaging has provided new avenues of research into the way in which enzymes work together. Nuclear factors are seen to move in small complexes to quickly identify targets for repair (Houtsmuller *et al.* 1999; Phair and Misteli 2000).

10.6. COMMON ASSAYS TO MEASURE DNA REPAIR CAPACITY

In epidemiological studies, there have been numerous approaches to markers for DNA repair. These generally fall into the category of either 'genetic markers', using single nucleotide polymorphisms (SNPs) as surrogates to measure DNA repair capacity, or 'cellular markers', using time-consuming assays for which there has been some, but not much, inter-observer testing done. The major functional activities for DNA repair that have been used in epidemiological studies are summarized in the following sections.

Cellular biomarkers for DNA repair

Berwick and Vineis (2000) summarized the epidemiological studies up to 1998 that evaluated the putative association between DNA repair capacity and cancer. Such studies were limited by the fact that most individuals who were 'cancer cases' already had cancer when their blood was drawn for the assay, making it somewhat difficult to tell whether the results are due to the cancer or to a predisposition to develop cancer. Nevertheless, there was sufficient evidence overall to suggest that a lower DNA repair capacity was likely to be associated with an increased risk to develop cancer. A number of assays were discussed (see below) and unfortunately few new cellular assays have been recently developed. There remains, therefore, a clear need for high-throughput functional assays to assess DNA repair with sufficient power and reliability that appropriate conclusions may be drawn. None of the assays are truly 'high-throughput' and this is a major problem for epidemiology.

Berwick and Vineis (2000) placed the DNA repair assays into five categories and these are listed in Table 10.1. Below we comment on the published data on DNA repair analysis in relation to variability within individuals, between individuals and, in some cases, between laboratories.

Assays based on induced DNA damage:

Mutagen sensitivity assay

The bleomycin-induced mutagen sensitivity assay is an *in vitro* measure of DNA repair capacity developed by Hsu (1985). This assay is an indirect measure of both DNA damage and DNA repair, expressed as 'breaks per cell' (b/c) in short-term cultured lymphocytes. It is a relatively simple test in which a higher number of bleomycin-induced chromatid breaks reflects a higher 'mutagen sensitivity' and lower DNA repair. For example, if the number of 'breaks per cell' is > 0.8, the examined subject is considered 'mutagen-sensitive'. Erdei *et al.* (2006) compared DNA repair using this assay in individuals over time, rates at different times, and differences between two laboratories in performing the assay. They found excellent intra-individual, inter-individual and inter-laboratory reliability. Lee *et al.* (1996) had previously reported very good

Table 10.1 Examples of functional assays to test DNA repair capacity

Type of assay	Examples
DNA damage induced with chemicals or physical agents	Mutagen sensitivity assay, the G2-radiation assay, induced micronuclei and the Comet assay
Indirect tests of DNA repair	Unscheduled DNA synthesis
Direct measures of repair kinetics	Host cell reactivation assay
Measures of genetic variation associated with DNA repair	SNP association studies
Combination of more than one category of assay	Chromosome aberrations and SNPs in the metabolic polymorphism pathway

correlations between the first 50 metaphases read and the second, with attenuation in the odds ratio of only 15%, indicating that reading only 50 metaphases was an acceptable approach.

The G2-radiation assay is similar to the mutagen sensitivity assay but, as its name implies, it is measuring cellular sensitivity to ionizing radiation. Vral *et al.* (2004) reported that the intra-individual variability is very similar to the inter-individual variation, limiting the utility of the assay. He concludes that it is critical to sample an individual multiple times prior to drawing conclusions about individual DNA repair capacity. Unfortunately, this is seldom possible and thus seldom accomplished.

Comet assay

The Comet assay, also known as single cell gel electrophoresis (SCGE), has been used to measure DNA repair capacity in a number of studies by inducing damage and then measuring the time for repair and the amount of repair. Methods to automate this assay include not only a 96-well plate format, but also the development of methods to measure buccal cells as well as lymphocytes (Szeto *et al.* 2005).

Although it is essentially a method for measuring DNA breaks, the introduction of lesion-specific endonucleases allows detection of, for example, ultraviolet (UV)-induced pyrimidine dimers, oxidized bases and alkylation damage. This assay is currently the most popular assay for measurement of DNA damage and has been modified to assess DNA repair capacity. The most common form of this assay now used is an alkaline Comet assay. Lymphocytes (generally) are lysed and electrophoresed at a high pH. DNA loops containing

breaks escape form the comet head to form a 'tail'. According to Collins (2006), the use of alkali makes comet tails more pronounced and improves the resolution of the assay without affecting the sensitivity. Although many use it and it seems simple, sensitive, versatile and economical, it has often been misused. Care must be taken that the placement in the agarose gel does not confound for the measurement of the comet, as the results are sensitive to placement and can vary by site of placement.

When using the Comet assay as a DNA repair assay, the repair of damage induced by ionizing radiation or hydrogen peroxide can be measured using an appropriate endonuclease on the gel. Unfortunately, the treatment of the lymphocytes may interfere with this assay by exposing the lymphocytes to atmospheric oxygen, which causes substantial additional damage.

When using the Comet assay as a molecular epidemiology biomarker, it is critical to have appropriate standards. These could consist of frozen aliquots of lymphocytes that are used each time the assay is run.

Unscheduled DNA synthesis (UDS) assay

The UDS assay measures the ability of an agent to induce DNA lesions by measuring the increase in DNA repair (Pero *et al.* 1983). This assay measures DNA repair synthesis after excision and removal of a portion of DNA containing damage. When using lymphocytes from subjects, one is attempting to ascertain the 'native' ability of the individual to repair different types of damage, such as induced by tobacco-like exposures or ultraviolet radiation. Basically, tritium-labelled thymidine is incorporated into DNA which is not in S phase of the cell cycle. The uptake of the labelled thymidine is measured by either autoradiography (considered the 'gold standard') or scintillation counting (less reliable, but much easier) in the cultured cells.

Host cell reactivation assay

This assay, developed by Athas *et al.* (1991), is a somewhat more direct measure of repair kinetics. Separate sets of fresh or cryopreserved lymphocytes are transfected with both a damaged plasmid and an undamaged plasmid. Either a chloramphenicol acetyltransferase (CAT) or a luciferase gene (LUC) is incorporated into the plasmid. Repair is then measured as a rate, revealed by the amount of radiation or fluorescence at specific time points.

Other assays

Various laboratories have been developing new assays for DNA repair in order to develop specificity and reliability. Examples are described below.

OGG activity

Oxidative damage is a common suspect in multiple diseases, including cancer and heart disease. The DNA repair enzyme 8-oxyguanine DNA *N*-glycosylase (OGG) repairs the most common, pro-mutagenic, lesion, 8-oxyguanine. An assay was developed by Paz-Elizur *et al.* (2003) in which enzyme activity is measured in protein extracts from individuals' lymphocytes or lung tissue by evaluating the cleavage products from a radio-labelled synthetic DNA oligonucleotide containing an 8-oxyguanine residue. To date, no other laboratory has replicated these potentially important studies.

Combination studies

A number of combination studies (Santella *et al.* 2005) have measured several types of damage and repair to assess individual DNA repair capacity. These will not be discussed here. Basically, investigators might evaluate a functional assay for DNA repair capacity at the same time that they evaluate intervening variables, such as DNA repair genotypes (Shen *et al.* 2006), or metabolizing genotypes which may have an effect on the damage. For example,

Knudsen *et al.* (1999) evaluated chromosomal aberrations as indicative of the effects of air pollution, at the same time controlling for SNPs glutathione *S*-transferase and *N*-acetyltransferase 2.

10.7. INTEGRATION OF DNA REPAIR ASSAYS INTO EPIDEMIOLOGICAL STUDIES

A number of issues need to be taken into consideration when measuring 'functional' DNA repair capacity: (1) study design, including time from measurement to disease state, and intervening changes after baseline measures and their potential effect on the DNA repair capacity, as well as potential biases and confounding effects of multiple exposures; (2) assay variability – intra-individual, inter-individual, inter-laboratory; and (3) biological plausibility of the relationship being measured.

1. *Study design.* Issues of study design centre on the difficulty of selecting appropriate subjects for study. A major problem with almost all studies of DNA repair capacity and its role in disease aetiology is that susceptible individuals should be identified before the development of cancer, rather than after cancer has been diagnosed and treated. This is because the cancer itself, and probably any resulting treatment, may result in observations slanted toward lower DNA repair capacity among case patients, who may have been treated with radiation or chemicals, which may modify DNA repair capacity. Potential biases and confounding in studies of DNA repair capacity can result from several factors: an inability to identify all relevant exposures; the inducibility of some DNA repair genes (Berwick and Vineis 2000), which can occur as a result of exposure to many different agents; and those that represent biological cross-reactivity, age, dietary habits and exposure to pro-oxidants (Pero *et al.* 1990).

2. *Assay variability.* Measurement of assay variability is critical for assessing the true, underlying risk estimates. Because DNA repair capacity may be modified by environmental exposures, which could be useful as we consider

the potential of DNA repair capacity to be a target for intervention (Moller and Loft 2004) and a potential therapeutic target (Chow *et al.* 2004), it is critical to understand the inter- and intra-individual variability in any particular set of assays. Until recently, few investigators have paid attention to this critical set of values, probably because it is somewhat difficult to find funding for such time-consuming work. However, as stated above, if there is similar inter-individual and intra- individual variation, then 'signal' or the actual DNA repair capacity is difficult to assess and reproduce.

3. *Biological plausibility*. A major limitation of many tests is that DNA repair capacity, as currently measured, is only indirectly inferred from cellular DNA damage remaining after exposure to mutagens for a specific time period. In many of these studies, the mutagen used to induce damage is not known to initiate tumours in people, and, methodologically, it would be extremely useful to extend DNA repair assays to carcinogens specific to particular tumour types. Many of the DNA repair studies are necessarily small, due to the expense in terms of reagents and time, and so generally have limited power. Concomitant studies of gene expression and/or function may be beneficial.

To draw firm conclusions about cause–effect relationships, more evidence about the biological underpinning of the current assays is necessary. For example, assays that measure 'unrepaired DNA' are in fact measuring DNA repair processes.

10.8. GENETIC MARKERS FOR DNA REPAIR CAPACITY

There have been a great many studies and summaries of DNA repair gene variants and their relationship to a variety of diseases. Due to the same issues that plague all genetic association studies, it is difficult to evaluate genetic associations of DNA repair capacity and disease outcome. A recent review by Hirschhorn *et al.* (2002) assessed SNP association studies reporting only one DNA repair SNP associated with disease that was reproducible,

XRCC1. This issue of reproducibility of SNPs is being intensively studied but is still difficult to evaluate.

Criteria to evaluate whether genetic associations with disease outcome are real have to date included low *p* values, replication in multiple samples and avoidance of population stratification. Hirschhorn *et al* (2002) suggest that any publication of an association study should be accompanied by a meta-analysis from existing studies. Publication bias is, however, a major concern in this respect – omission of small negative studies can bias the results toward positive associations.

One such meta-analysis (Manuquerra *et al.* 2006) focused on the nucleotide excision repair gene *XPD* and the double-strand break repair gene *XRCC3*; this meta-analysis will be used to illustrate these issues. *XRCC3* participates in DNA soluble-strand break/recombinational repair. It is a member of a family of Rad-51-related proteins that likely participate in homologous recombination to maintain chromosome stability and repair DNA damage (Tebbs *et al.* 1995). Four coding SNPs (one synonymous and three amino acid substitutions) and 109 intronic SNPs have been described. Manuquerra *et al.* (2006) conducted a meta-analysis for the *XRCC3* variant Thr241Met. This meta-analysis (http://www.aje.oxfordjournals.org) illustrates the complexity of understanding the early SNP association studies.

In the first place, the studies were generally relatively small, ranging in size from 85 subjects to 1245. In the second place, the allele frequencies differ among ethnic groups, so that for African-Americans the minor allele frequency (MAF) is 4.6%, for Asians it is 0.2%, and for Caucasians 12.4%. These differences are statistically significant ($p < 0.001$). Finally, there is no functional evidence that this SNP leads to a diminution of repair, or results in greater sensitivity to DNA damage. Araujo *et al.* (2002) evaluated the impact of this SNP variant on repair phenotype as measured by a quantitative fluorescence assay. They found that cells containing this variant were still functionally active for homology-directed repair. In another assay, the authors examined cells expressing the variant for sensitivity to mitomycin C (MMC), an inter-strand cross-linking agent, and found that such cells were not more sensitive than those that were wild-type. Thus, relevant functional information shows no altered activity in cells containing this variant.

XPD is involved in the nucleotide excision repair pathway. This pathway recognizes and repairs many structurally unrelated lesions, such as bulky adducts and thymidine dimers (Flejter *et al.* 1992). The protein formed by this gene has a role in the initiation of RNA transcription by RNA polymerase II (Coin *et al.* 1998; Keriel *et al.* 2002). Eight coding SNPs (four synonymous and four amino acid substitutions) and 138 intronic SNPs have so far been reported. Manuquerra *et al.* conducted meta-analyses for the *XPD* variants Arg156Arg (C/A), Asp312Asn (G/A) and Lys751Gln (A/C).

More research has involved the *XPD 312* and *751* variants, with generally similar observations. The studies are generally relatively small, ranging from 13 subjects to 1240. The MAF range in *XPD 312* was from 0% in Asians to 11.1% in Caucasians ($p < 0.001$). The MAF range in *XPD 751* was from 1.1% in Asians, 6.9% in African-Americans and 13.4% in Caucasians ($p < 0.001$). A number of epidemiological studies have shown that there is unlikely to be a functional effect on DNA repair capacity from the *XPD* variants (Lunn *et al.* 2000; Au *et al.* 2003; Matullo *et al.* 2001; Qiao *et al.* 2002). Clarkson and Wood (2005), after considering such studies as well as evolutionary analyses that predict that both polymorphisms are benign, find that there is no 'credible' evidence for functional effects of such polymorphisms.

In order to evaluate the functional effects of genetic polymorphisms, Clarkson and Wood (2005) suggest that direct genetic mapping using haplotype analysis is much more likely to be fruitful, citing three independent studies of age-related macular degeneration (Klein *et al.* 2005), which all led to the same candidate SNP, that can now be tested for function. They suggest that a two-step method would enhance the work of those interested in DNA repair and its effect on disease. First, biochemists should identify polymorphisms that alter the function of a gene product. Then, epidemiological studies can evaluate the extent and importance for human disease of such changes. The problem, they note, is that *in vitro* assays still need to be developed to detect small changes in activity. They

also suggest the development of isogenic cell lines where the only difference is the SNP in the test gene, and then evaluating characteristics such as survival after DNA damage, NER capacity or the propensity to mutagenesis.

More robust methods for genetic association studies of DNA repair genes have been developed. These are better discussed elsewhere but, briefly, they include haplotype analyses, use of Bayes' theorem (Wacholder *et al.* 2004) and the standard pooling of data and meta-analyses (see Chapter 15).

Xi *et al.* (2004) suggest that approximately 62% of the 520 amino acid substitution variants identified in screening DNA repair genes for sequence variation are likely to impact protein activity. They used bioinformatics tools, such as sorting intolerant from tolerant (SIFT) genes and polymorphism phenotyping (PolyPhen). They used the example of *APE1* polymorphisms, where both algorithms accurately predicted the role of 26 functionally characterized amino acid substitutions in the APE1 protein activity, with one exception.

Additionally, creative approaches by Wu *et al.* (2006) may serve as a stimulus to other investigators to use new analytic methods to understand the way in which SNPs may work together. They used a 'multigenic' analytic approach, measuring 44 different SNPs, to DNA repair genes, as well as cell cycle genes to evaluate the risk to develop bladder cancer. The analytic approach was to use a classification and regression tree approach, CART, to evaluate individual SNPs and interactions among genes. They found 'gene dosage' effects that increased with increasing numbers of high-risk alleles. Contrary to other reports (e.g. Zhou *et al.* 2005), genetic effects were found only in ever smokers. Such a combined approach for SNP analyses makes good biological sense in relationship to the way in which DNA repair pathways work. It also *may* overcome the problems posed by multiple small effects from SNPs.

In summary, DNA repair capacity is widely understood to be an intrinsic function of individuals that may predispose them to disease. However, the measurement of DNA repair capacity is still under-developed and needs far more robust and creative work in order to bring epidemiological studies of DNA repair capacity in its relationship to disease to fruition. Similarly, the detection of mutations as a consequences of a lack of effective DNA repair can further contribute to the characterization of cancer risk and the aetiology of the disease.

References

Albertini RJ. 2001. HPRT mutations in humans: biomarkers for mechanistic studies. Mutat Res Rev Mutat 489: 1–16.

Albertini RJ. 1998. The use and interpretation of biomarkers of environmental genotoxicity in humans. *Biotherapy* **11**: 155–167.

Albertini RJ, Nicklas JA, Fuscoe JC *et al*. 1993. *In vivo* mutations in human blood cells: biomarkers for molecular epidemiology. *Environ Health Perspect* **99**: 135–141.

Araujo FD, Pierce AJ, Stark JM *et al*. 2002. Variant XRCC3 implicated in cancer is functional in homology-directed repair of double-strand breaks. *Oncogene* **21**: 4176–4180.

Athas WF, Hedayati MA, Matanoski GM *et al*. 1991. Development and field-test validation of an assay for DNA repair in circulating human lymphocytes. *Cancer Res* 1991 **51**: 5786–5793.

Au WW, Salama SA, Sierra-Torres CH. 2003. Functional characterization of polymorphisms in DNA repair gnees using cytogenetic challenge assays. *Environ Health Persp* **1**: 1843–1850.

Berwick M, Vineis P. 2000. Markers of DNA repair and susceptibility to cancer in humans: an epidemiological review. *J Natl Cancer Inst* **92**: 874–897.

Chow TY, Alaouli-Jamali MA, Yeh C *et al*. 2004. The DNA double-stranded break repair protein endo–exonuclease as a therapeutic target for cancer. *Mol Cancer Ther* **3**: 911–919.

Clarkson SG, Wood RD. 2005. Polymorphisms in the human *XPD* (*ERCC2*) gene, DNA repair capacity and cancer susceptibility: an appraisal. *DNA Repair* **4**: 1068–1074.

Coin F, Marinoni JC, Rodolfo C *et al*. 1998. Mutations in the *XPD* helicase gene result in XP and TTD phenotypes, preventing interaction between XPD and the p44 subunit of TFIIH. *Nat Genet* **20**: 184–188.

Collins AR. 2006. The Comet assay: principles, applications and limitations. In In *In Situ Detection of DNA Damage: Methods and Protocols*, Didenko VV (ed.). Methods in Molecular Biology, vol. 203. Humana: Totowa, NJ; 163–177.

Erdei E, Lee S-J, Wei Q *et al*. 2006. Reliability of mutagen sensitivity assay: an inter-laboratory comparison. *Mutagenesis* **21**: 261–264.

Essers J, Vermeulen W, Houtsmuller AB. 2006. DNA damage repair: anytime, anywhere? *Curr Opin Cell Biol* 2006; **18**: 240–246.

Flejter WL, McDaniel LD, Johns DR. 1992. Correction of xeroderma pigmentosum complementation group D mutant cell phenotypes by chromosome and gene transfer: involvement of the human *ERCC2* DNA repair gene. *Proc Natl Acad Sci USA* **89**: 261–265.

Geng F, Laurent BC. 2004. Roles of *SWI/SNF* and *HAT*s throughout the dynamic transcription of a yeast glucose-repressible gene. *EMBO J* **23**: 127–137.

Green CM, Almouzni G. 2002. When repair meets chromatin. First in series on chromatin dynamics. *EMBO Rep* **3**: 8–33.

Greenblatt MS, Bennett WP, Hollstein M *et al*. 1994. Mutations in the *p53* tumor suppressor gene: clues to cancer etiology and molecular pathogenesis. *Cancer Res* **54**: 4855–4878.

Hassan AH, Neely KE, Workman JL. 2001. Histone acetyltransferase complexes stabilize swi/snf binding to promoter nucleosomes. *Cell* **104**: 817–827.

Hirschhorn JN, Lohmueller K, Byrne E *et al*. 2002. A comprehensive review of genetic association studies. *Genet Med* 2002; **4**: 45–81.

Houtsmuller AB, Rademakers S, Nigg AL *et al*. 1999. Action of DNA repair endonuclease ERCC1/XPF in living cells. *Science* 1999; **284**: 958–961.

Hsu TC, Cherry LM, Samaan NA. 1985. Differential mutagen susceptibility in cultured lymphocytes of normal individuals and cancer patients. *Cancer Genet Cytogenet* **17**: 305–313.

Hussain SP, Harris CC. 1998. Molecular epidemiology of human cancer: contribution of mutation spectra studies of tumor suppressor genes. *Cancer Res* **58**: 4023–4037.

Hussain SP, Schwank J, Staib F *et al*. TP53 mutations and hepatocellular carcinoma: insights into the etiology and pathogenesis of liver cancer. *Oncogene* **26**: 2166–2176.

Jagannathan I, Cole HA, Hayes JJ. 2006. Base excision repair in nucleosome substrates. *Chrom Res* **14**: 27–37.

Jenuwein T, Allis CD. 2001. Translating the histone code. *Science* **293**: 1074–1080.

Jung D, Giallourakis C, Mostoslavsky R *et al*. 2006. Mechanism and control of V(D)J recombination at the immunoglobulin heavy chain locus. *Annu Rev Immunol* **24**: 541–570.

Kasten M, Szerlong H, Erdjument-Bromage H *et al*. 2004. Tandem bromodomains in the chromatin remodeler RSC recognize acetylated histone H3Lys14. *EMBO J* **23**: 1348–1359.

Keriel A, Stary A, Sarasin A *et al.* 2002. *XPD* mutations prevent TFIIH-dependent transactivation by nuclear receptors and phosphorylation of RARα. *Cell* **109**: 125–135.

Kirk GD, Lesi OA, Mendy M *et al.* 2005. 249(ser) *TP53* mutation in plasma DNA, hepatitis B viral infection, and risk of hepatocellular carcinoma. *Oncogene* **24**: 5858–5867.

Klein RJ, Zeiss C, Chew EY *et al.* 2005. Complementation factor H polymorphism in age-related macular degeneration. *Science* **308**: 385–389.

Knudsen LE, Norppa H, Gamborg MO *et al.* 1999. Chromosomal aberration in humans induced by urban air pollution's influence of DNA repair and polymorphisms of glutathione *S*-transferase 1 and *N*-acetyltransferase 2. *Cancer Epidemiol Biomarkers Prevent* 1999; **8**: 303–310.

Lee JJ, Trizna Z, Hus TC *et al.* 1996. A statistical analysis of the reliability and classification error in application of the mutagen sensitivity assay. *Cancer Epidemiol Biomarkers Prevent* 1996; **5**: 191–197.

Lindahl T, Wood RD. 1999. Quality control by DNA repair. *Science* 1999; **286**: 1897–1905.

Loeb LA, Christians FC. 1996. Multiple mutations in human cancer. *Mutat Res* **350**: 279–286.

Lunn R, Helzlsouer KJ, Parshad R *et al.* 2000. *XPD* polymorphisms: effects on DNA repair proficiency. *Carcinogenesis* 2000; **21**: 551–555.

Machado CR, Menck Carlos FM. 1997. Human DNA repair diseases: from genome instability to cancer. *Braz J Genet* **20**(4): http://www.scielo.br/scielo.php?script=sci_arttext&pid=S0100–84551997000400032&lng=en&nrm=iso-

Manuquerra M, Saletta F, Karagas MR *et al.* 2006. *XRCC3* and *XPD/ERCC2* single nucleotide polymorphisms and the risk of cancer: a HuGE review. *Am J Epidemiol* **164**: 297–302.

Matullo G, Palli D, Peluso M *et al.* 2001. *XRCC1*, *XRCC3*, *XPD* gene polymorphisms, smoking and ^{32}P–DNA adducts in a sample of healthy subjects. *Carcinogenesis* **22**: 1437–1445.

Mohrenweiser HW, Wilson DM III, Jones IM. 2003. Challenges and complexities in estimating both the functional impact and the disease risk associated with the extensive genetic variation in human DNA repair genes. *Mutat Res* **526**: 93–125.

Moller P, Loft S. 2004. Interventions with antioxidants and nutrients in relation to oxidative DNA damage and repair. *Mutat Res* **551**: 79–89.

Narlikar GJ, Fan Y, Kingston RE. 2002. Cooperation between complexes that regulate chromatin structure and transcription. *Cell* **108**: 475–487.

Olivier M, Eeles R, Hollstein M *et al.* 2002. The IARC TP53 database: new online mutation analysis and recommendations to users. *Hum Mutat* **19**: 607–614.

Osley MA, Tsukuda T, Nickoloff J. Submitted. ATP-dependent chromatin remodelling factors and DNA damage repair. *Mutat Res.*

Paques F, Haber JE. 1999. Multiple pathways of recombination induced by double-strand breaks in *Saccharomyces cerevisiae*. *Microbiol Mol Biol Rev* **63**: 349–404.

Paz-Elizur T, Krupsky M, Blumenstein M *et al.* 2003. DNA repair activity for oxidative damage and lung cancer. *J Natl Cancer Inst* **95**: 1312–1319.

Pero RW, Miller DG, Lipkin M *et al.* 1983. Reduced capacity for DNA repair synthesis in patients with or genetically predisposed to colorectal cancer. *J Natl Cancer Inst* **70**: 867–875.

Pero RW, Anderson MW, Doyle GA *et al.* 1990. Oxidative stress induces DNA damage and inhibits the repair of DNA lesions induced by *N*-acetoxy-2-acetylaminofluorene in human peripheral mononuclear leukocytes. *Cancer Res* **50**: 4619–4650.

Pfeifer GP, Denissenko MF, Olivier M *et al.* 2002. Tobacco smoke carcinogens, DNA damage and *p53* mutations in smoking-associated cancers. *Oncogene* 2002; **21**: 7435–7451.

Phair RD, Misteli T. 2000. High mobility of proteins in the mammalian cell nucleus. *Nature* **404**: 604–609.

Qiao Y, Spitz MR, Shen H *et al.* 2003. Modulation of repair of ultraviolet damage in the host cell reactivation assay by polymorphic *XPC* and *XPD/ERCC2* genotypes. *Carcinogenesis* **23**: 295–299.

Relton CL, Daniel CP, Hammal DM et al. 2004. DNA repair gene polymorphisms, pre-natal factors and the frequency of somatic mutations in the *glycophorin-A* gene among healthy newborns. *Mutat Res* **545**: 49–57.

Santella RM, Gammon M, Terry MB *et al.* 2005. DNA adducts, DNA repair genotype/phenotype and cancer risk. *Mutat Res* 2005; **592**: 29–35.

Shen J, Desai M, Agrawal M *et al.* 2006. Polymorphisms in nucleotide excision repair genes and DNA repair capacity phenotype in sisters discordant for breast cancer. *Cancer Epidemiol Biomarkers Prevent* **15**: 1614–1619.

Smerdon MJ, Conconi A. 1999. Modulation of DNA damage and DNA repair in chromatin. *Prog Nucleic Acid Res Mol Biol* **62**: 227–255.

Strauss BS. 1992. The origin of point mutations in human tumor cells. *Cancer Res* **52:** 249–253.

Szeto YT, Benzie IF, Collins AR *et al.* 2005. A buccal cell model comet assay: development and evaluation for human biomonitoring and nutritional studies. *Mutat Res* **278**: 371–381.

Tebbs RS, Zhao Y, Tucker JD *et al.* 1995. Correction of chromosomal instability and sensitivity to diverse mutagens by a cloned cDNA of the *XRCC3* DNA repair gene. *Proc Natl Acad Sci USA* **92**: 6354–6358.

Van Houten B, Albertini RJ. 1995. DNA damage and repair. In *Pathology of Environmental and Occupational Disease*, Craighead JE (ed.). Mosby: London; 311–327.

Vral A, Thierens H, Baeyens A *et al.* 2004. Chromosomal aberrations and *in vitro* radiosensitivity: intra-individual vs. inter-individual variability. *Toxicol Letters* **149**: 345–352.

Wacholder S, Chanock S, Garcia-Closas M *et al.* 2004. Assessing the probability that a positive report is false: an approach for molecular epidemiology studies. *J Natl Cancer Inst* **96**: 434–442.

Ward J. 1998. The nature of lesions formed by ionizing radiation. In *DNA Damage and Repair: DNA Repair in Higher Eukaryotes*, Nickoloff JA, Hoekstra MF (eds). Humana: Totowa, NJ; 65–84.

Wood RD, Mitchell M, Lindahl T. 2005. Human DNA repair genes, 2005: review. *Mutat Res* **577**: 275–283.

Wood RD, Mitchell M, Sgouros J *et al.* 2001. Human DNA repair genes. *Science* **291**: 1284–1289.

Wu X, Gu J, Grossman HB *et al.* 2006. Bladder cancer predisposition: a multigenic approach to DNA-repair and cell-cycle-control genes. *Am J Hum Gen* **78**: 464–479.

Xi T, Jones IM, Mohrenweiser HW. 2004. Many amino acid substitution variants identified in DNA repair genes during human population screenings are predicted to impact protein function. *Genomics* 2004; **83**: 970–979.

Zhou W, Liu G, Park S *et al.* 2005. Gene–smoking interaction associations for the *ERCC1* polymorphisms in the risk of lung cancer. *Cancer Epidemiol Biomarkers Prevent* **14**: 49149–49146.

11

High-Throughput Techniques – Genotyping and Genomics

Alison M. Dunning and Craig Luccarini

Strangeways Research Laboratories, University of Cambridge, UK

11.1. INTRODUCTION

The new field of genomics has emerged from conventional human genetics and is, at present, moving extremely rapidly. As a result, this chapter will be a snapshot in time, capturing a moment in the state of development. For this reason it will not dwell in detail on individual technologies or machines, as many of these will be obsolete in a very short time, but rather on the principles of high-throughput working, which should remain valid for longer. Technologies for high-throughput genotyping have existed for some years and there have been many studies on individual genes or specific loci. It has been possible to add the '-omics' suffix, since the completion of the sequencing of the human genome. We now have the ability to consider the genome as a unified whole rather than as many tiny, and largely unconnected, parts of chromosomes. The first genome-wide scans, using single nucleotide polymorphisms (SNPs) to find multiple genes contributing to single common human diseases are, in early 2007, on the verge of reporting their findings. The first has potentially found four loci for type 2 diabetes (Sladek *et al.* 2007). Now they have been clearly demonstrated to work, the same techniques will no doubt be applied to many other common diseases and complex genetic traits within a very short time.

The tools required for genome-wide studies; bioinformatics resources, high-throughput technologies and a detailed knowledge of population genetics and evolution, come from previously unrelated scientific fields and yet are all moving forward together and stimulating each other. Effective scientists will need an understanding of all three tools to make proper use of any one. The combined use of them to best effect is still under debate and will only be refined once the outcomes of the first scans have been considered.

11.2. BACKGROUND

Present knowledge of human genetics has grown out of the study of monogenetic disorders, which were dissected by examination of the disease's segregation in families using pedigree linkage studies. Genetics departments in hospitals and universities frequently developed from branches of paediatrics departments where children with rare, highly penetrant diseases arrived for treatment. The metabolic errors, generated by genetic mutations in their families, were severe and affected children from birth or in early life. In addition, the mode of inheritance of these diseases was often recessive, so that the pattern was only spotted by people who examined multiple children from big families. Usually the nature of the disease was so debilitating that it was rarely propagated through many generations and the causative mutations did not become common in the population. The high penetrance and rarity of these monogenic disorders meant that linkage studies on a relatively small number of family members (no more than a few hundred individuals and often fewer) were successful. Key to the success of linkage studies is the need to differentiate which family members have inherited which segment of any given chromosome and whether this segment is co-inherited with disease. The first genetic markers to be used were identified as changes in protein-folding. Later SNPs were recognized (although it was some years later before the term SNP was coined) as changes in the DNA-binding sites of particular enzymes, called restriction fragment length polymorphisms (RFLPs). However, the most informative polymorphisms for family studies were the microsatellite markers (see Figure 4.1, Chapter 4). Studies could take years and frequently stalled for want of a new marker in a poorly covered chromosome or a new family with a recombination site nearer the disease gene, enabling a large segment of chromosome to be excluded from harbouring the mutation. Most recent, successful linkage studies have been carried out using panels of microsatellite markers that were detected by electrophoresis, a technique that is slow and difficult to automate. Despite this problem, the small sample sizes meant that studies could be managed using manual skills in small research laboratories.

The genetic contribution to complex diseases that arise from a combination of environmental and multiple different genetic causes have not been resolved by pedigree linkage studies, although these have been attempted. The genetic association study, which compares the frequency of any allele between cases (with the disease) and controls (without) in a given population is more appropriate. Association studies were first carried out in the early 1980s – at much the same time as the first pedigree linkage studies – but have taken much longer to bear fruit. Some quite early ones were successful, e.g. the association between the *ApoE4* allele and increased susceptibility to late-onset Alzheimer's disease was first reported in 1994 (Tsai *et al.* 1994). This has served as a proof of principle, demonstrating that under the correct circumstances the association approach will work, although the *APOE* gene had first been identified by linkage (Corder *et al.* 1993). However, the field of genetic association has been dogged

by small, underpowered studies and consequently with reports of (false) positive associations that could not be confirmed. There are multiple reasons why, until recently, most association studies have been much too small, but one of the main driving forces has undoubtedly been the inability of most laboratories to handle sufficiently large sample sizes in an efficient and cost-effective manner.

The association study is only anticipated to work when the trait being examined complies with the 'common disease–common variant' hypothesis – the idea that if a disease is common within a population, its causes (whether genetic or environmental) must be similarly common. The lack of success from so many association studies has led to debate about whether a 'multiple rare variant' hypothesis may more usually be appropriate. This postulates that the genetic contribution to a common disease may be composed of very many different rare mutations, each likely to be found only in a single pedigree and each with a different penetrance. If this second hypothesis holds true for many diseases, it will not be easy to find their causative mutations by either pedigree linkage studies or case–control association. However, as yet, too few association studies have been carried out sufficiently well to make a conclusion about which hypothesis is likely to hold more often.

Properly undertaken association studies require much bigger resources than small research groups have usually been able to commit to them. Having an adequate sample size of cases and controls is the key to their success. The sample size must be large enough not only to detect a significant difference between cases and controls when the causative mutation is tested directly, but also to cope with the problems of multiple testing (hence false-positive associations) when many thousands of potentially causative alleles are tested in a single study and, furthermore, to cope with the dilution of statistical power when neutral markers (tagSNPs) are used in place of directly testing the causative mutations.

Figure 11.1 is a graph showing the sample size (total number of cases plus controls) required to detect alleles of different frequencies that confer double the risk of disease [odds ratio(OR) = 2] at different levels of statistical significance. It also

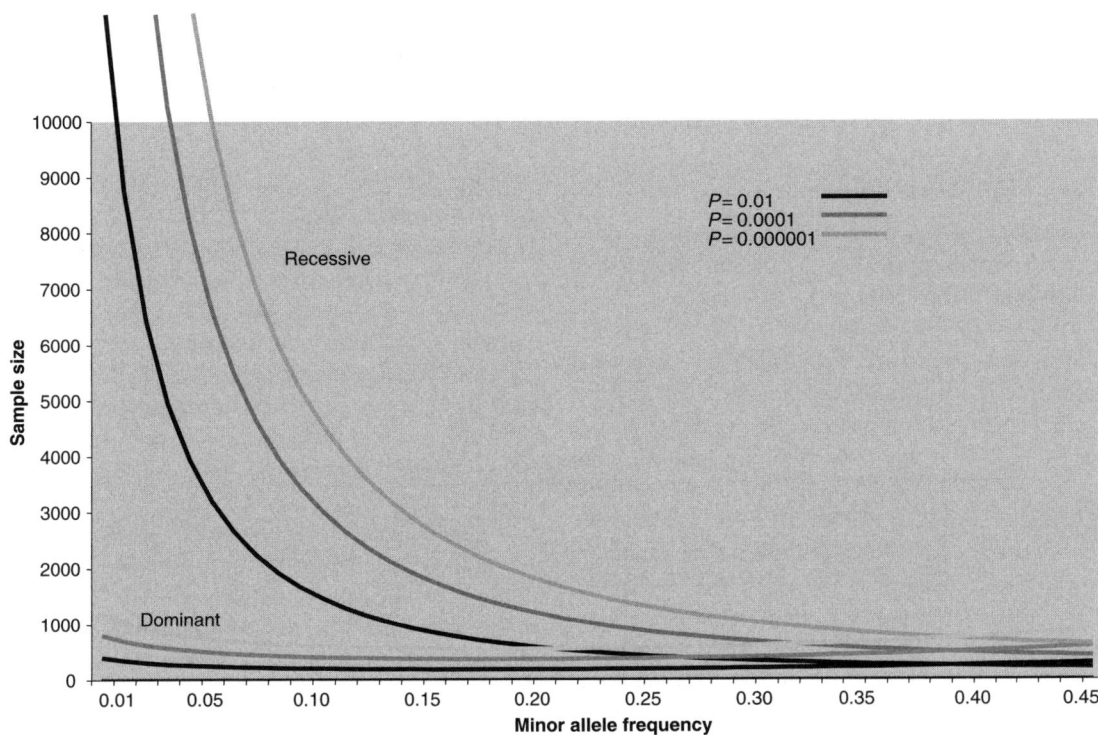

Figure 11.1 Sample size and power

shows the different numbers needed, depending on whether the disease-causing allele has a dominant or recessive mode of action. The situation does not appear bad if one expects to find an allele with, for instance, a frequency of 20% in the population with only a single test – a sample size of under 200 would generate a conventionally significant p value (< 0.05) if its effect was dominant and, even if it proved to be recessive, it could be found with < 1000 subjects (in association studies, unlike in pedigree linkage, it is rarely possible to know in advance whether the mode of action of the causative allele will be dominant, co-dominant or recessive). However, in most situations it is going to be necessary to test lots of potentially causative variants in order to find the right one(s). A genome-wide scan will typically test 300 000–500 000 SNPs in one go and, when this is done, the level of statistical proof needs to rise to $p < 10^{-7}$ to exclude the large number of conventionally significant p values that will have arisen by chance alone (by definition, $p < 0.05$ will be seen by chance once in every

20 tests). To find the same causative mutation in these circumstances will require 1500–3000 subjects, depending again on whether the effect is dominant or recessive. Furthermore, a conventional genome-wide scan relies on a tagSNP approach (described in more detail later), so that every known SNP in the genome is represented by another that is correlated with it, but that correlation may not be perfect. If the correlation between the truly causative SNP and the marker SNP representing it is, for instance, only 0.7, then the sample size will have to increase again to cope with the loss of power due to the imperfect correlation.

It is only possible to handle large sample sizes with adequate technology, systems and funding. This includes the ability to collect sufficiently large numbers of cases with the disease (or appropriate phenotype) and properly matched controls. It is likely that for rarer diseases the biggest problem will be waiting long enough for sufficient people to fall prey to the disease, particularly in small populations. For this as well as other reasons, the successful studies in the future are likely to come from large, international collaborations rather than small, competing and therefore highly secretive research groups.

11.3. SNP DATABASES

In addition to DNA from study subjects, the other raw material necessary for association studies is informative SNPs. Although SNP databases, such as dbSNP (http://www.ncbi.nlm.nih.gov/SNP/), have existed for several years and are improving all the time, they are still far from perfect. They are currently neither comprehensive (we have found that up to 25% of common SNPs in loci we have studied are still not documented in them) nor fully validated (very many of the 'SNPs' listed in them are merely artefacts from sequencing or mapping errors, rather than true polymorphic variants). These glitches are understandable, given the very rapid pace of recent developments, but now we have the possibility of carrying out genome-wide SNP scans it becomes all too clear that what we now lack is comprehensive and accurate SNP databases where all the documented SNPs have genotype data in a defined set of individuals of known ethnicity.

Two other types of database currently exist, which are useful, but each has its own drawbacks. The HAPMAP project (www.HAPMAP.org) provides the ideal raw material to find a genetic association anywhere in the genome, but it is not comprehensive enough to allow the identification of the causative variant in the same study. HAPMAP has identified common SNPs, which are reasonably evenly spaced across the entire genome, and provides the genotype data for each SNP in three different ethnic groups: 90 subjects from each of Yoruba Africans, Europeans, and Chinese and Japanese. This enables the correlations between groups of SNPs to be calculated and the recombination hotspots (defining blocks of linkage disequilibrium) to be mapped so that an efficient minimal tagging set of SNPs can be designed using the integral Tagger program (http://www.broad.mit.edu/mpg/tagger/). HAPMAP coverage is now detailed enough to reliably identify disease- or trait-associated loci from their tag SNPs. However,

once the locus has been identified, it is likely that the entire linkage-disequilibrium block, surrounding the associated tag SNP, will still have to be resequenced in sufficient human subjects to find all the SNPs in the region.

Other databases, e.g. Seattle SNPs (http://pga.mbt.washington.edu/) and NIEHS (http://egp.gs.washington.edu/), have set out to provide full resequencing data across key candidate genes in sets of up to 90 human individuals. Theoretically, these databases have the advantage of being comprehensive (all variants that can be detected by Sanger sequencing in the 90 study subjects should be identified) and the comparative raw data on the individuals still enables efficient SNP tagging. The disadvantage of these sites is that they are candidate gene-based, and therefore useless if they do not include the obvious genes for the disease being studied. Additionally, even within a candidate gene under study, the larger introns have not been resequenced (presumably because resources are limited). However, as we find and learn more about the true causative variants in common disease, it becomes clear that assuming they are more likely to be in exons or to affect splice sites is probably fallacious.

Now the human genome has been fully sequenced in the equivalent of a handful of individuals, it becomes clear that the rate-limiting step for future development of association studies is the resequencing of the entire genome in 50–100 subjects in each of several ethnic groups, so that a comprehensive catalogue of all common genetic variation can be made.

11.4. STUDY TYPES

Until recently, almost all human gene searches, whether by linkage or association, have had the aim of finding disease genes. The cost of the studies and the large scientific effort required has been simply too high to encourage the investigation of many other human genetic traits, which, although interesting, might be considered more frivolous. For instance, remarkably little is known about the number and types of loci that contribute to hair and eye colour, although most lay people would be able to name these as obvious genetic traits.

Two main study types are commonly undertaken: the binary outcome, where people are classified as diseased (cases) or normal/unaffected (controls); and the quantitative trait, which is often statistically more powerful. In a binary outcome study, the genotype distribution of a given SNP is compared between cases and controls, while with a quantitative trait, the mean level of the trait (e.g. a circulating hormone) is compared between the three genotype classes. In some situations a quantitative trait can be an intermediate risk factor for a disease (e.g. raised serum cholesterol levels are known to increase the risk of having a heart attack). In these situations it can be more revealing to tackle the genetics of the risk factor first, by studying the association of a SNP with mean circulating levels of cholesterol (which would have to have been measured in all the study subjects) and then ask how much contribution that SNP would make to the risk of heart attack. Examining the triangular relationship between a SNP, a quantitative risk factor and a binary disease outcome is sometimes referred to as a Mendelian randomization study, and this can reveal whether a quantitative risk factor is truly a cause of the disease or simply a covariate that is commonly associated with having the disease. A real example of a SNP effect on circulating oestradiol levels and risk of breast cancer is illustrated in Box 11.1.

Box 11.1   Triangulation – the effects of a SNP on both circulating oestradiol levels and risk of breast cancer

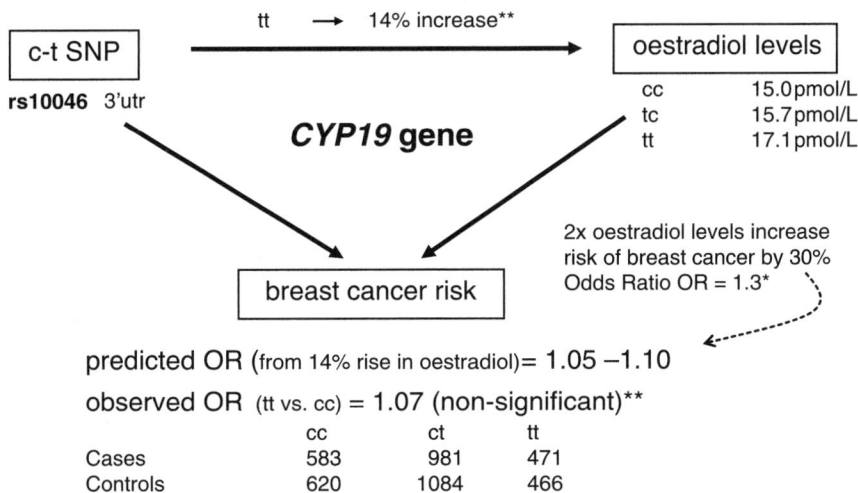

predicted OR (from 14% rise in oestradiol) = 1.05 –1.10

observed OR (tt vs. cc) = 1.07 (non-significant)**

	cc	ct	tt
Cases	583	981	471
Controls	620	1084	466

The *CYP19* gene is responsible for the conversion of testosterone to the most active female hormone, oestradiol. A large meta-analysis (Key *et al.* 2002) has shown that doubling circulating oestradiol levels increases the risk of breast cancer by approximately 30%. A *t-c* SNP in the *CYP19* gene, which helps regulate the stability of the RNA and subsequently the amount of Cyp19 protein that is made, affects levels of circulating oestradiol – homozygotes for the *t* allele have 14% higher levels than homozygotes for the *c* allele (Dunning *et al.* 2004). By increasing oestradiol levels in this way, the *t* allele would also be expected to increase risk of breast cancer by 5–10%. However, in practice, it would only be possible to detect such a small increase with a significant *p* value by using an enormous case–control set (in the order of 34 000 cases and a similar number of controls) and as yet a study of this size has not been attempted.

Increasingly, as clinical trials reach sizes that are comparable with those needed for genetic studies, it is possible to look at the genetics of treatment–response – why are some people cured by a particular dose of drug and others not, or why do certain individuals develop specific, possibly severe, side-effects, while others remain unaffected? Within the most severe diseases, such as cancer, it is also possible to study the genetics of length of survival after diagnosis.

Once the aim of the study has been decided, there are next a number of styles of SNP study that can be used, which are listed in Box 11.2. Until recently, the only practicable one in an association study format was the candidate-gene approach. This chooses genes based on prior knowledge of the biological basis of the disease. Such knowledge is often very limited and the method was rarely successful when it had previously been attempted in linkage studies. It did work for linkage in haemophilia, where it was reasonable to suppose from the nature of the disease that the defect might be in the clotting cascade. It was not successful for the more complex disorders, such as cystic fibrosis,

---

Box 11.2 Genetic study designs

FAMILIAL LINKAGE STUDIES
Look for co-segregation of microsatellite alleles (length differences in DNA repeats) with a disease in families with a Mendelian genetic trait.

ASSOCIATION STUDIES
Look for SNP alleles that occur more frequently in cases (people with trait under study) than in normal controls from the same population.
Candidate gene – only study SNPs in genes guessed from what is known about the biology of the trait.
Genome-wide non-synonymous SNP – study all SNPs known to generate amino acid substitutions in any gene; currently 6500+ documented.
Genome-wide gene-based – SNPs in and around the 20 000+ known genes.
Genome-wide SNP tagging – set of 300–500K SNPs designed to tag all known common variations in the genome regardless of position or function.

---

where multiple biological systems are affected and the common protein link between all of them was not clear.

An adaptation of the candidate-gene approach is the study of all functional SNPs. There exists a fairly comprehensive catalogue of all non-synonymous SNPs in the genome – those that generate amino acid substitutions in proteins. Since these might be expected to have an effect on biological systems, it is not unreasonable to test all the known ones in a single panel for association with a particular disease. One of these studies has already been completed for type 1 diabetes. Smyth *et al.* (2006) compared the genotype frequencies of > 6.5K non-synonymous SNPs in 2000 cases (with type 1 diabetes) and 1750 controls and ranked the SNPs in order of the level of statistical significance of the association. The top 10 SNPs of all those tested had p values between 10^{-4} and 10^{-7}. Four loci causing type 1 diabetes were already known when the study began, and so it helped confirm the value of this approach to find that the top two SNPs from the genome-wide study fell into two of these already-known loci and the third was in a biologically sensible candidate gene.

However, the non-synonymous SNP approach excludes the possibility that many of the variants in the genome that affect biology may alter the timing or levels of a particular protein product, rather than its composition. The natural successor to this is the true genome-wide scan, but even these presently come in at least three different styles:

1. The least comprehensive is the gene-based study – choosing just the SNPs that fall in and around each of the approximately 20 000 known genes (Stein 2004) but purposely excluding the often very large intergenic tracts of DNA in the genome, on the basis that we cannot yet imagine what purpose these stretches serve. This method has been used in a small myocardial infarction association study, in which 93 000 gene-based SNPs were examined in 94 cases and 658 controls (Ozaki and Tanaka 2005). However, data is already accumulating that SNPs in so called 'gene deserts' can be associated with disease; SNP rs1447295, which appears to be at least

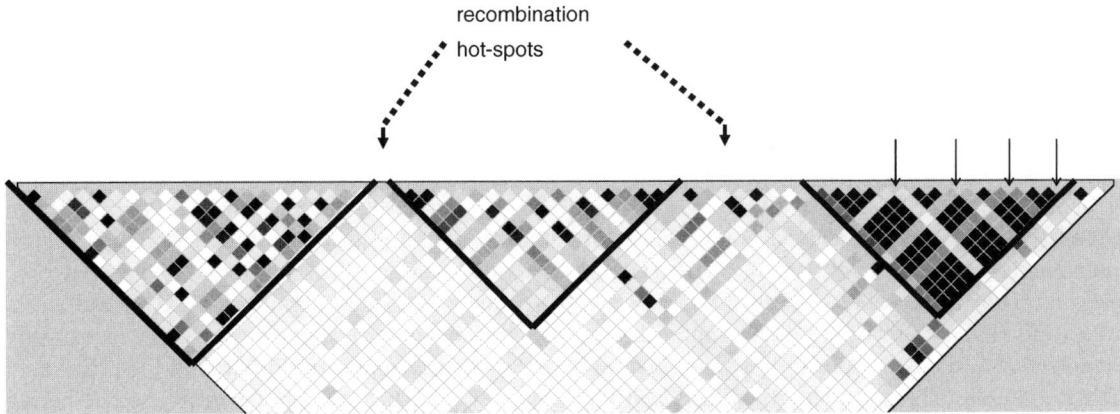

Figure 11.2 Typical HAPMAP matrix, indicating r^2 values between all pairs of SNPs genotyped across a locus. The positions of recombination hot-spots can clearly be seen. Between the hot-spots the SNPs tend to be highly correlated (dark squares) and can all be represented by a just few tag SNPs (marked with arrows on the right-hand LD block).

500 kb from the nearest known gene (*MYC*), has now been independently confirmed by several studies to be associated with an increased risk of developing prostate cancer (Amundadottir *et al.* 2006; Freedman *et al.* 2006).

2. More comprehensive studies aim to cover the *intergenic stretches as well as the known genes*. Some work by attempting to choose a common SNP (one with a minor allele frequency > 10%) at regular spaced intervals across all the chromosomes. In practice, the spacing needs to be in the order of one SNP every 3–5kb, in the expectation that this will be representative of all the other SNPs around it. For this purpose, several companies have produced arrays of about 500 000 SNPs which aim to cover the whole genome. They work on the principle that new mutation and recombination events occur evenly across the genome (although we already know that this is not always the case) and so common SNPs spaced every few kb apart are likely to be representative of their neighbours.

3. A more sophisticated approach still is the *true SNP-tagging approach*. This approach takes, as its starting point, the knowledge that between recombination hotspots, groups of SNP alleles tend to be highly correlated (see Figure 11.2). The correlation between the SNPs within a given region can be examined by comparing

the genotypes that they generate across a panel of individuals. From this, the minimum, most informative tagging set of SNPs can be chosen to best represent all the other known SNPs (see also Chapter 4). This is theoretically a better approach, since it takes into account that the positioning of recombination and new mutation hotspots are not even in different parts of the genome, so there are some very large areas of chromosomes that show very little variation and can be represented by just a few tag SNPs and others, where large numbers of tags are needed over a small area. This sophisticated tag SNP approach has been enabled by the collaborative HAPMAP project and also attempted independently by some companies (e.g. Perlegen Inc.). However, it remains to be seen whether in practice this is any better than the evenly spaced, random SNP approach or even the gene-based SNP approach. The big difference is that all the genome-wide methods are empirical – they aim to find all known variants associated with a trait or disease without prior knowledge of the biological basis of that trait.

11.5. STUDY DESIGN

Even in genome-wide studies that examine up to 500 000 SNPs at a time, the ideal study design would be to study all of the SNPs in all of the study samples. However, as yet this has not been attempted, for good economic and practical reasons. Not only is the cost prohibitive but the time required to study one person per chip would be very long and the amount of data generated and requiring management and analysis would be enormous. A phased approach, suggested by Satagopan and Elston (2003), is instead a good compromise for maintaining reasonable statistical power whilst keeping down the cost of running many subjects across individual chips. In a phased approach, the entire set of SNPs is examined in a small but sufficiently representative sample of subjects and the best (most significantly associated) SNPs are identified. These SNPs, which may consist of the best 5–10% from phase 1, will be further examined in a larger set of study subjects (phase 2) and a smaller set of even more strongly associated SNPs will be picked again. Further rounds of smaller SNP sets in bigger sample sizes may continue until the chance of false-positive associations is minimal. However, some false negatives will occur – SNPs dropped

in the early phases that would have progressed to significance – and so phased studies are unlikely to be able to claim that they have captured *all the possible* variants in the genome that are associated with a particular trait. From a practical point of view, the phased approach will require numerous different sample sets to be collected and arrayed (although luckily not all at the same time). It may also have a requirement for several different SNP-genotyping technologies, which might not all be present in a single laboratory. The first two phases will probably be most efficiently carried out with the chip-based technologies, but the subsequent phases, requiring progressively fewer SNPs to be genotyped on larger sample sizes, may be more efficiently carried out by the technologies to the right of the graph in Figure 11.3 (see next section). If the sample sets are dispersed across the world, it may also be practical to carry out the final phases of the genotyping on more than one site, a situation that puts extra emphasis on the employment of good quality control (QC) measures.

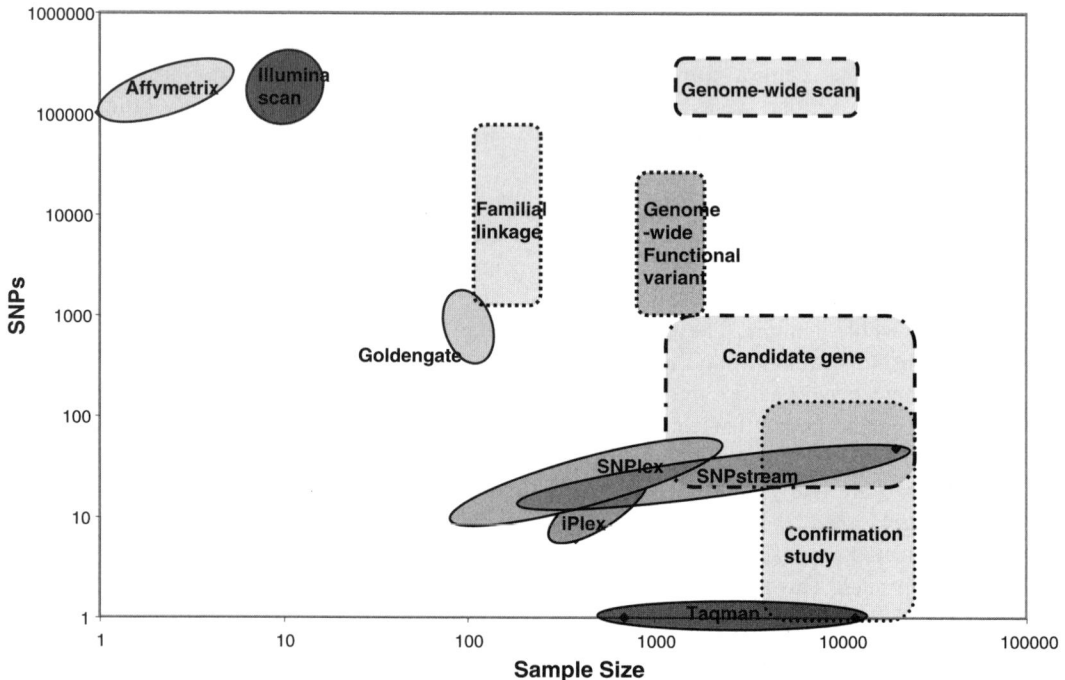

Figure 11.3. Dimensions of studies and technologies

11.6. GENOTYPING TECHNOLOGIES

There are now several different technologies available for high-throughput genotyping. Among the most well known are: Applied Biosystems Taqman; Sequenom iPlex; Beckman SNPstream, Applied Biosystems SNPlex, Illumina Goldengate; and Infinnium and Affymetrix chips. Unfortunately, for the smaller research laboratory with a limited budget, no single technology will be suitable for all studies. In fact, few are ideal for any study design and it is necessary to take their different limitations into account when designing a study that will be practicable over a reasonable timescale.

Figure 11.3 shows the dimensions of different study types: The number of SNPs needed vs. sample size for adequate coverage and statistical power of different designs are shown as grey boxes. In the ovals it also shows the number of SNPs and samples each technology can handle in a single run. The different systems generally divide into those that can handle many SNPs on few samples (top left of the graph) vs. those good for a few SNPs on much larger sample sizes (bottom right).

The first stages of genome-wide scans require the technologies that can handle many thousands of SNPs at one time on a few thousand individuals (i.e. in the top right of the graph). In practice, the only technologies currently available for handling thousands of SNPs can only deal with a handful of subjects in a day (top left of the graph). It is therefore necessary to run the same experiment many times to get the desired sample numbers. With sufficient end-point chip-readers, it is possible to run several chips at once (speeding up the sample throughput), but most laboratories are limited in the numbers of detectors they can afford and the temptation to reduce sample numbers to below those needed for adequate statistical power is great.

The later, confirmatory, stages of the scans require the genotyping of far fewer SNPs (most often in the range of the best 50–200 SNPs emerging from the early stages), but to be truly confirmatory these need to be tested in several thousand case and control subjects – particularly if the estimates of the ORs obtained from the

earlier stages are small. Technologies to handle these study dimensions are already available (bottom right of Figure 11.2) but the expertise needed to run these systems is often quite different from that needed to handle chips and may not always be available in the same laboratory.

11.7. SAMPLE AND STUDY MANAGEMENT AND QC

Crucial to the efficient running of all high-throughput genotyping is a pre-prepared supply of good quality, normalized, arrayed DNA from all the subjects to be tested. The aim is to treat all the samples in a single study as uniformly as possible, so as to minimize introducing differences in genotype calling that are simply the result of differences in the way the DNA was treated. Using robotics, it is often possible to take entire study sets of tens of thousands of DNA samples through large parts of the process in a single batch, and this is the most desirable way of working for both uniformity of results and time-efficiency. However, such huge experiments often consume very expensive reagents in a single run and so it is necessary to minimize the chances of things going wrong. As an added safeguard, proper QC steps have to be incorporated to make sure that, if there is a failure, it can be quickly diagnosed and corrected. In the past, with much smaller-scale linkage projects, it was not unusual to keep repeating a particular assay on one or more 'difficult' DNA samples until they yielded a satisfactory result. This kind of approach is neither desirable nor usually possible in a high-throughput setting. For each gain in throughput (i.e. by the introduction of arrayed DNA or robotics) there is, unfortunately, a parallel loss of flexibility in the way the samples can be handled, and this has to be factored into the study design and planning.

DNA extraction and normalization

As alluded to earlier, the principle philosophy behind high-throughput studies is the aim to carry out any assay only once and achieve the optimum result on all the samples without needing to pick out individual failed samples to repeat them. To achieve this requires careful planning and preparation as well as properly incorporated QC steps, so that if things do go wrong they can be quickly diagnosed and corrected.

Critical to the success of any study is to have ready, in advance, adequate quantities of high-quality DNA from all the study subjects. This can be problematical, especially in large collaborations where DNA (often quite old and poorly stored) has been donated from many different sources. The best-quality DNA for these studies is extracted from anti-coagulated whole blood by a process that leaves it free from contamination by protein (including nucleases) and by-products of the process, such as salt or chloroform. There are on the market several kits and processes that can extract good quality DNA as well as service companies that can extract it to a standard quality. For large study sets, use of a DNA extraction service can be the most economical approach. Scientists are often tempted to try to extract DNA from archived material, such as paraffin-embedded tissue on microscope slides or Guthrie spots, and whilst it is true that DNA can be extracted from these materials (forensic laboratories can extract minute amounts of DNA from almost anywhere) in practice this DNA is no use in a high-throughput laboratory, where the quality of the results is reduced to the lowest common denominator, and that is most usually DNA quality.

Once the DNA has been extracted, each sample needs to be quantified by a method that reliably measures intact strands of DNA rather than degraded nucleotides. Presently the methods that use intra-collated dye are the most accurate.

In the final step before use, the DNA samples from all the subjects in a study need to be arrayed together in 96- or 384-well plates, so that every well contains the same volume of dissolved DNA at the same concentration (e.g. 100µl at 4ng/µl) – a process commonly referred to as 'normalization'. This stage can be particularly laborious and time consuming, although there are now robotic systems available that can take the output readings from a spectrophotometer and dilute the samples appropriately. However, in practice this stage is the most crucial

to the future reliability of the genotyping, as illustrated in Figure 11.4.

Figure 11.4 shows the output of a Taqman genotyping experiment carried out on three 384-well plates of DNA genotyped in a single batch. The source of DNA on each plate was from a different laboratory; each plate contained two negative control samples. In the panel on the far right, the DNA had been properly normalized and the three clusters, corresponding to the three genotypes, are very distinct – there is no ambiguity in calling them the three genotype 'classes'. In the middle panel, the DNA had not been quite so well normalized and a few degraded DNA samples (trailing towards the origin of the graph) can clearly be seen. However in the panel on the left, the DNA had been so poorly normalized that the clusters merge and up to a third of the DNA samples cannot be reliably called. Compounding the problem, if the reader decided not to call the points that do not fall into a cluster, they would fail to call more of the samples that must be carrying an allele 2 than those that must carry allele 1. This results in the call rate being non-random with respect to a

---

Box 11.3 High-throughput genotyping quality control

Key points for success:

DNA – high yield, high quality, uniform quality and uniform concentration. Extractions from Guthrie spots, paraffin sections, mouthwash, etc., will not work in a high-throughput setting.

Whole-genome amplification of DNA– can increase yield without detriment to quality, so long as starting material is satisfactory.

Array – arrange normalized DNA from cases and controls alternately or randomly in 96- or 384-well plates.

Controls – leave empty wells in each plate – negative controls and duplicate some DNA samples (reproducibility check).

Robotics – use automated liquid-handlers and plate-sealers to improve uniformity and give even volume delivery, less evaporation and greater accuracy (robots don't tire ot get distracted).

Barcoding – and consider using LIM system to avoid mislabelling and to track samples.

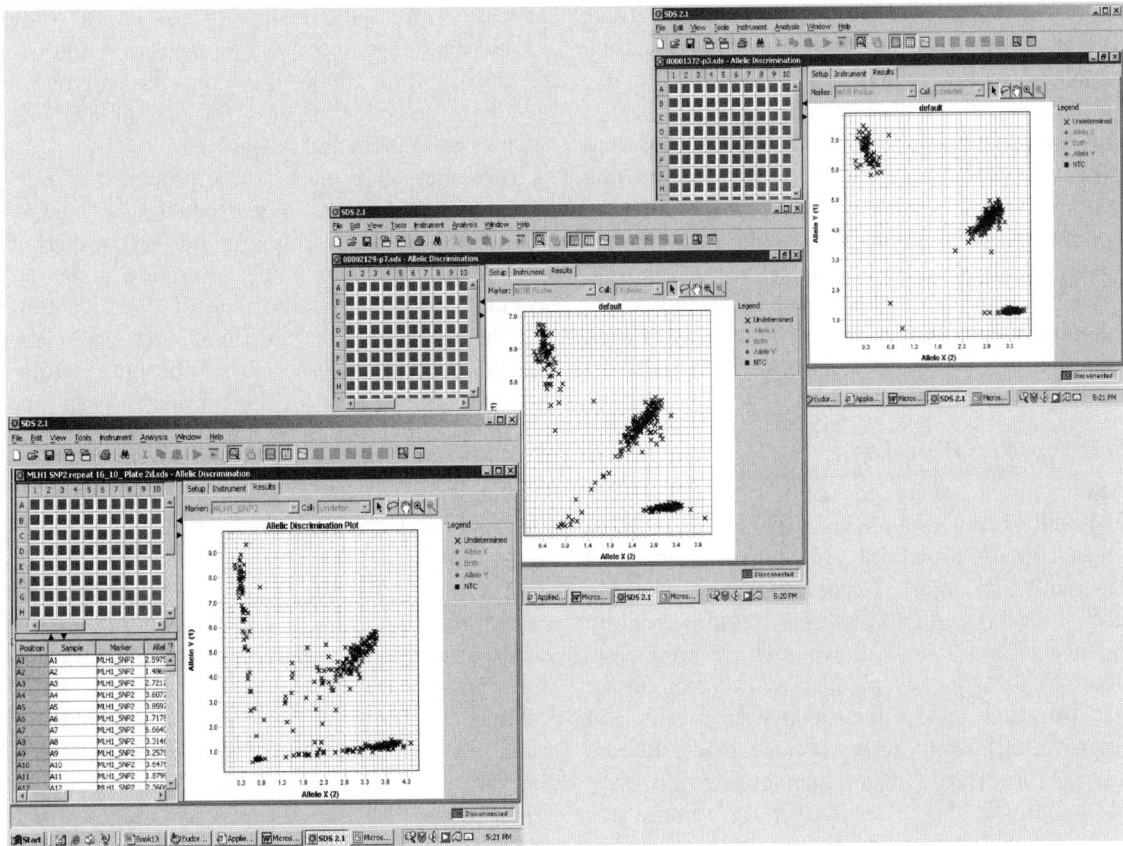

Figure 11.4 The effect of DNA normalization on genotype clusters

samples genotype, which is almost certain to bias the final result.

The problem was remedied by remeasuring the concentration of all the samples and rediluting them in a single batch of buffer. This is often the best way of treating pre-extracted DNA that comes from multiple sources before combining it into a single study set.

DNA arraying

Presently, microtitre plates containing 384 wells are the best for maximizing throughput and minimizing reagent volumes and hence cost, although with more accurate robotics, a 1536-well plate may become the standard format.

At the point of arraying, thought must be put into the format of randomizing or distributing DNA samples across the plates to avoid subtle bias in the way the samples are handled. This is particularly important in case–control study designs to minimize the chance of position in the plate from affecting the cases and controls in different ways. Although traditional epidemiologists find randomization of cases and controls very appealing, we have found that using a chequerboard pattern of alternate cases and controls ensures very even treatment and is highly practical, since a 384-well array is constructed from four 96-well arrays in exactly that pattern. There is a worry that a chequerboard will allow the laboratory scientist to work out which samples are the cases and which the controls and therefore not be truly 'blind' to status. This worry is unfounded in such large studies since, as illustrated in Figure 11.3, most genotyping systems rely upon the data-point from any sample falling into a tight cluster of others with the same genotype. The genotype call is then given to the entire cluster, rather than to one sample at a time. This not only vastly increases that speed of calling genotypes on many thousands of samples but also ensures that, at the time of calling, the scientist is looking at data-points from mixed cases and controls in a single cluster and has no way of guessing from whom any one point might be derived.

For later QC purposes, some wells on each plate should be left blank (to form negative controls) and a set proportion of samples should be duplicated on another plate so that the reproducibility of each assay can later be checked and reported.

Robotics and plate sealing

Although it is possible in theory to fill the wells a 384-well plate with a manual pipette, this requires great concentration and is therefore tiring and inherently inaccurate. Laboratories that regularly use high-throughput methods require robotic liquid-handlers. These deliver more reproducible volumes than manual pipettes and do not tire or get distracted, and so they reduce error rates and improve uniformity across a large study. Robots are not necessarily faster than human pipettors, particularly at the lower end of the sample-size spectrum, and their major disadvantage is that they are less flexible. The variety of liquid-handling robots on the market are changing all the time and at the top end there are some with built-in flexibility, e.g. some are capable of cherry-picking individual samples as well as rapidly pipetting into sets of 8, 12, 96 and even 384 wells at a time. However, most laboratories will find they need more than one robot for different activities (i.e. pre- and post-PCR work) and, when choosing these expensive items, careful thought needs to go into exactly what processes are required, as well as what the laboratory might need in the foreseeable future. A point to bear in mind when choosing a robot is that, although it is desirable to keep sample and reagent volumes as low as possible to save on reagent costs, there is a lower limit to the volumes that are practical. This is a result of inaccuracies generated by evaporation, surface tension effects and the tolerances of the robots themselves. There is nothing to be gained by reducing volumes to the point where inaccuracies and inconsistencies

adversely affect the results of the entire study. Liquid-handling robots will be needed in addition to chip-spotting robots that come as part of the genotype detection process for certain methods, such as Sequenom and Affymetrix.

The other main place where problems of non-uniformity across an entire study set of samples can erupt is in the sealing of plates before thermal cycling or incubation steps. Microtitre plates are usually sealed with some type of adhesive film, which can be applied by hand, but again this manual process is slow and can be rate-limiting. It is also all too easy for human operators to miss sealing the edges or corners properly and this will lead to evaporation form certain wells and loss of data. Investing in an automated plate-sealer is recommended whenever avoidance of evaporation is a priority.

The most effective way of achieving high-throughput rates comes from the ability to prepare arrays of normalized DNA and using this to 'print-out' multiple sets of assay plate in advance. Small volumes (5–10µl) of DNA can be pipetted into microtitre plates, left in a clean place to air-dry and then be stored for many months. When desired, a set of prepared plates can then be quickly picked and have assay reagents added. By far the most time-consuming part of any genotyping operation is the preparation and arraying of DNA into a study set. This can take weeks to do properly and so it is highly desirable to plan carefully exactly which samples are required in any given array, so that the array only needs to be made once and can then be used for many assays, so as to make the preparation time worthwhile.

QC steps

Although good planning reduces the number of times or places in which things can go wrong, errors will inevitably happen, and the best way to minimize their impact is to incorporate QC steps into every assay so that the problem can be quickly diagnosed and corrected. I would suggest the following as minimum QC steps in high-throughput genotyping.

The most devastating error, affecting an entire dataset that occurs all too easily when samples are spread over several microtitre plates, is to mislabel and confuse the plates or to rotate one or more of them. This ultimately results in samples being ascribed the wrong genotypes. The most sophisticated way of avoiding this problem is to label all samples and all plates with a unique barcode. Pre-barcoded plates can already be purchased or users can print their own on adhesive labels. The barcode can only be stuck on one face of a plate and many genotyping systems will read the barcode as the plate is passed into the final detector, so preventing it being rotated and enabling software to put the correct genotype against the appropriate samples. Barcodes are useful, even without a laboratory information management (LIM) system. However, when used within a LIM system they can be used for many purposes, including tracking samples through the process, recording which batch of reagent every sample has been treated with, auditing the rate at which any given DNA sample is being used up, and ensuring that the correct genotype is entered against the appropriate sample within a database of results.

However, smaller laboratories without a LIM system or barcoding can still achieve good QC using a few simple tools. We have already mentioned the need to use empty wells as negative controls on every plate to demonstrate that there is minimal cross-contamination of DNA across the wells in a microtitre plate. Increasingly, when publishing results, it is also necessary to quote the reproducibility rate of the assay; to do this, the same assay needs to be tested more than once on a subset of samples. It is simplest to build the duplicates into the array as it is being made, so that certain samples are put into wells on more than one plate. It is relatively easy to design a pattern of negatives and duplicates that are different on each plate, so that a laboratory scientist reading the results can quickly check that the pattern of data-points corresponds to the pattern of samples on any plate. It is also advisable, when making the arrays, to set up a spreadsheet showing where all the samples are on each plate. The genotype calls, when made, can be pasted directly into the sheet, which can also be set up to check that empty wells give negative results and that results from duplicated samples correspond. Such a spreadsheet can

then be used to calculate the failure rate and error rate for a given assay. Over time it may become clear that certain low-quality DNA samples within an array rarely if ever yield results, and so the spreadsheet can be edited to exclude these unreliable samples from analysis. Although spreadsheets are extremely useful for data checking, the fact that they can be too easily over-written makes them insecure for long-term data-storage, and ultimately results need to be exported into a proper database for archiving, cross-comparison and retrieval.

The final QC test that can be very useful in checking genotyping quality is a test of deviation from Hardy–Weinberg equilibrium (HWE) among control samples from a single population. This test was not designed for this purpose – it was initially a test for evidence of natural selection on organisms with a particular genotype – but the test is even more sensitive to non-random failure-to-call genotypes (as happens when genotype clusters start to merge) and will give extremely significant results if this has occurred – so alerting the laboratory scientist to look again at the data-points.

11.8. AFTER THE ASSOCIATION HAS BEEN PROVED – WHAT NEXT?

Using all the techniques and tools that are now available, in conjunction with good study design, the era of false-positive and irreproducible results that have damaged the reputation of genetic association studies should now be over. Truly significant genetic associations will start to be published on a regular basis and we will soon have a good indication of how often the common disease–common variant model fits better than the multiple rare-variant hypothesis. Successful genetic association studies will rapidly open up other avenues for research that will bring new problems to resolve. The most immediately visible one is that from genetic association studies; unlike in classical epidemiology, it will be possible to identify the variant that causes the trait of increased susceptibility to disease (since genotype is present from conception, the genetic trait always arises later and the order of variant *causing* the trait can be clearly established). The obvious problem is that we may find ourselves with very significant genetic associations of one or more SNP markers within a particular haplotype block, with a specified trait but an inability to determine which variant in that block is causative. The first step towards addressing this problem will be full, deep resequencing to identify all the variants within a given block. Presently this is a difficult, expensive and time-consuming task, which is the reason why there are no properly comprehensive SNP databases. The recent arrival of massively parallel clonal sequencers, such as the Roche GS FLX™ (http://www.454.com/) and the Illumina Genome Analyser™ (http://www.illumina.com/pages.ilmn?ID = 204) systems may soon ease this bottleneck. But it is only when we are able to define the full set of potentially causative variants within a given locus that we will be able to hand the work back over to the functional biologists to test out which of the set are more likely to affect biological function than others.

References

Amundadottir LT, Sulem P, Gudmundsson J *et al.* 2006. A common variant associated with prostate cancer in European and African populations. *Nat Genet* **38**: 652–658.

Corder EH, Saunders AM, Strittmatter WJ *et al.* 1993. Gene dose of apolipoprotein E type 4 allele and the risk of Alzheimer's disease in late onset families. *Science* **261**: 921–923.

Dunning AM, Dowsett M, Healey CS *et al.* 2004. Polymorphisms associated with circulating sex hormone levels in postmenopausal women. *J Natl Cancer Inst* **96**: 936–945.

Freedman ML, Haiman CA, Patterson N *et al.* 2006. Admixture mapping identifies 8q24 as a prostate cancer risk locus in African-American men. *Proc Natl Acad Sci USA* **103**: 14068–14073.

Key T, Appleby P, Barnes I *et al.* 2002. Endogenous sex hormones and breast cancer in postmenopausal women: reanalysis of nine prospective studies. *J Natl Cancer Inst* **94**: 606–616.

Ozaki K, Tanaka T. 2005. Genome-wide association study to identify SNPs conferring risk of myocardial infarction and their functional analyses. *Cell Mol Life Sci* **62**: 1804–1813.

Satagopan JM, Elston RC. 2003. Optimal two-stage genotyping in population-based association studies. *Genet Epidemiol* **25**: 149–157.

Sladek R, Rocheleau G, Rung J *et al.* 2007. A genome-wide association study identifies novel risk loci for type 2 diabetes. *Nature* **445**: 881–885.

Smyth DJ, Cooper JD, Bailey R *et al.* 2006. A genome-wide association study of nonsynonymous SNPs identifies a type 1 diabetes locus in the interferon-induced helicase (IFIH1) region. *Nat Genet* **38**: 617–619.

Stein LD. 2004. Human genome: end of the beginning. *Nature* **431**: 915–916.

Tsai MS, Tangalos EG, Petersen RC *et al.* 1994. Apolipoprotein E: risk factor for Alzheimer disease. *Am J Hum Genet* **54**: 643–649.

12

Proteomics and Molecular Epidemiology

Jeff N. Keen and John B. C. Findlay

LIGHT Laboratories, University of Leeds, UK

12.1. INTRODUCTION

At their most basic, biological systems are integrated expressions of protein function. Thus, any alteration in the activity of a protein can be expected to result in biological changes, although they may range from the inconsequential to the very profound. Elucidation of the genomes of organisms has paved the way for a further '-omics' wave, namely 'proteomics'. Put simply, the 'proteome', a term introduced by Marc Wilkins in 1994 (Wasinger *et al.* 1995), is the protein content of a system under a given set of conditions; a sub-proteome would be a segment of that system. For example, the proteome could be all the proteins in a tissue or cell and a sub-proteome, the glycoproteins, the phosphorylated species or the membrane-bound set, and so on. At its most fundamental, it defines the functionality of the system because proteins are in essence the functional representation of the genome. The immediate consequence is that the proteome: (a) is much more complex that the genome, since all forms of post-translational modification can potentially alter the functional behaviour of the system; and (b) is a moveable feast which varies depending on the prevailing conditions to which the system is subjected. Thus, if we can determine what is there and what is changing, the potential exists for both understanding biological events and predicting effects. Moreover, since changes can be both short- and long-term, one can begin to appreciate how the environment of the system can exert far-distant influences.

As a result, proteomics can be used to: (a) define a system/individual; (b) identify and define components that change form, quantity or location in response to stimuli; (c) detect system changes that arise from altered conditions or exposure; and (d) elucidate key markers defining predisposition, exposure and disease conditions. It is easy to appreciate, therefore, that one will be able to define both the changes that are characteristic of a disease process and, ultimately, interventions that may be able to reverse, counteract or ameliorate the dysfunction.

Whilst the theory seems plausible, the complexity of biological systems and the fragmented state of our knowledge mean that definitive answers and clear solutions require considerable time, effort and resources. However, real progress is being made in the application of this new high-throughput science, largely through developments in three major areas – genome sequences, bioinformatics and mass spectrometry. Before going on to describe the new technologies, however, some points on the importance of this new kind of science are worth making.

12.2. GENERAL CONSIDERATIONS

In the area of epidemiology as applied to disease, the primary application of proteomics lies in the areas of biomarkers, exposure and effect. Changes can be seen in both the short term and in the long term. For example, there may be acute reactions to exposure which subside once the exposure conditions have ended. In the longer term, exposure may engineer

Molecular Epidemiology of Chronic Diseases, Edited by C. P. Wild, P. Vineis, and S. Garte
© 2008 John Wiley & Sons, Ltd

a more permanent alteration in system behaviour (usually being 'fixed' at the genetic level) – indicative of a 'biomarker of effect'. A disease condition always results in the changed behaviour of a cell, be it altered function, i.e. increase or reduction in an important key signal, or aberrant growth, e.g. in cancer, where growth lacks the normal control. The observation so far is that such alterations are usually characterized by a large number of protein changes, both in the amounts present and in their form, i.e. alteration in post-translational modification. At this stage in the development of proteomics, it is very difficult to identify which change is the key one causing the altered behaviour; most changes are likely to be the consequence. Nevertheless, there could well be, in every situation, a variation or set of variations (however caused) that are characteristic of that condition and no others. These changes would then constitute biomarkers of that condition. But for such biomarkers to be reliable they must be statistically significant. Herein lies one of the important features of proteomics – the need to survey sufficiently large numbers of individuals and/or situations to confirm that the variability seen is related only to the dysfunction and not to any other or to any more normal genetic difference. This problem is appreciated well by epidemiologists but the scale is even greater in proteomics, especially if the changes seen are small (down to two-fold). For future therapeutic developments, knowledge of the protein changes that may precipitate dysfunction is clearly extremely valuable, for they will be the targets for intervention and drug discovery. In epidemiology, they may be the key indicators of present or future effects. In order for these changes to be secure, not only must they be statistically significant, they must also be validated as key elements. For this, other techniques, such as transgenic knock-outs, RNAi, etc. should be employed.

Of course, the changes seen in a sample may not be due to a disease condition but to an external influence of some kind. Substances in the environment, for example, may exert direct or indirect effects on the system, either to induce one or more changes, e.g. in protein expression levels, or to cause modifications in one or more components. Where modifications to the nature, as distinct from the amount, of a protein are concerned, there will be at least two versions of the same polypeptide – one the normal form, the others modified ones. Potentially, such markers of exposure could be potent indicators of susceptibility to adverse consequences and could in the long run constitute very important early indicators of dysfunction. They would not necessarily be the cause of the dysfunction themselves – other low-abundance proteins, such as transcription factors, may be the direct cause. However, they are indicators of exposure directly predictive of a future disease process.

The ability to monitor individuals and populations in this way represents a very powerful tool for preventative medicine and early intervention, but huge amounts of work are required, often at the extremes of biochemical detection and quantitation. Assuming validation of the target, the speed of mass spectrometry (MS) can also be the method of choice for screening. Currently techniques such as ELISAs are utilized, but for high-throughput analysis perhaps robotic MS will be the most cost-effective and rapid method.

12.3. SAMPLE SELECTION

Clearly, when monitoring the human individual, strategies that are not invasive are to be preferred. Thus, urine constitutes the best sample for analysis. Blood is the next-best alternative from this perspective, and it constitutes by far the richer source of information. The variety of proteins, protein forms and protein abundances is truly enormous, constituting at least 500 000 species, with molecular weights from a few hundred into six figures and relative abundance ranging over at least nine orders of magnitude. Identifying the appropriate markers in such a mixture represents a formidable challenge, but it is an important one to tackle and overcome, since there is a general, if not fully validated, belief that serum/plasma contains the past history, the present condition and future prediction of the health of an individual (Anderson and Anderson 2002).

For such information to be reliable and valuable, however, it must be robust and reproducible. Herein lies the first of the problems. All biological samples are constantly changing; thus, for reproducibility, they must be treated identically. Therefore, the sampling, storage and analytical processes must be rigorous and identical. Differences in any of these procedures, even as simple as time and temperature, are likely to create irrelevant variations or obscure real ones. Whilst complexity such as with serum/plasma is inherent in the nature of the material, similar problems can arise where tissue samples such as tumours are involved. Tissue is inherently complex, with many different varieties of cells being present. Thus, biopsy material has not only to be treated in a consistent fashion but the nature of the sample has also to be rigorously policed. Thus, every effort, such as the use of microdissection, has to be made to ensure that the sample taken for analysis is as pure or consistent (in cellular terms) as possible. For this reason, the discovery and development of meaningful biomarkers often starts with studies on cultured cell systems. Here the biological material can be much more uniform and the amounts obtained sufficiently large for accurate analysis and clear validation. Then one can turn to the native tissue to authenticate the culture observations.

Whilst these few observations may suggest that the highest aspirations for proteomic-derived reliable biomarker discovery may not be realized, significant progress is being achieved in the methodology, accuracy and sensitivity of both procedures and instrumentation, and in the bioinformatic analysis of the data, so much so that trustworthy indicators are emerging and will grow in numbers inexorably if sufficient determination and resources are applied to the discipline. The current state of the practical science of proteomics is outlined below (and see Figure 12.1), together with what might be a typical gel-based experiment (Figure 12.2).

12.4. PROTEOMICS TECHNOLOGIES

Proteomic analysis requires the separation and analysis of components of complex protein systems. There are two major approaches. The first, 'top-down' proteomics, requires the separation of complex protein mixtures into individual components, followed by analysis of the discrete polypeptides. This typically can be achieved using either electrophoretic or chromatographic methods. The resolved proteins are then analysed, usually following proteolysis into constituent peptides. The second, 'bottom-up' proteomics, involves the initial digestion of the entire unfractionated protein mixture into peptides and subsequent chromatographic separation and analysis of these partially resolved protein fragments. Both approaches depend on the utilization of techniques in mass spectrometry to provide information that allows the identification of proteins (see Figure 12.1).

Protein identification using mass spectrometry

Central to any proteomic investigation is protein identification using MS (Aebersold 2003). MS is almost universally applied to peptides produced by proteolytic (generally tryptic) digestion of proteins. The peptides, generated from individual proteins (top-down) or mixtures of proteins (bottom-up) are ionized using either electrospray ionization (ESI; Yamashita and Fenn 1984) or matrix-assisted laser desorption ionization (MALDI; Karas *et al.* 1987) and the peptide ions are separated by a magnetic or electric field in a mass analyser according to their mass:charge (m/z) ratios. The profile of these masses allows identification of the parent protein (see later). This technology has transformed the subject and the significance of these technological developments is reflected in the award of the Nobel Prize for Chemistry 2002 to John Fenn (ESI) and Koichi Tanaka (MALDI). A further technique related to MALDI, known as surface-enhanced laser desorption ionization (SELDI; Hutchens and Yip 1993), can be used for the analysis of peptide or protein samples, but usually on the basis of profiling samples using mass rather than direct protein identification, although this can be bolted on. The significance of SELDI is that changes occur in the form of an altered component profile due to the absence/presence or smaller/larger amounts of a particular mass. Typically, to effect mass-dependent separation of the peptide/protein species, time-of-flight (TOF), quadrupole (Q), ion trap (IT) or Fourier transform (FT) analysers are employed in conjunction with an appropriate ionization source. Analysers can also be combined to enable tandem MS (MS–MS), in which peptide ions emerging from the first analyser can be subjected to (further) fragmentation, using collision-induced dissociation

(CID) prior to determination of the fragment masses in a second analyser. The combination of these fragment ions will generate amino acid sequence information to aid protein identification.

Commonly-used instruments are MALDI–TOF and liquid chromatography (LC) coupled to ESI MS–MS. In MALDI–TOF–MS, which is typically used to analyse tryptic digests produced from

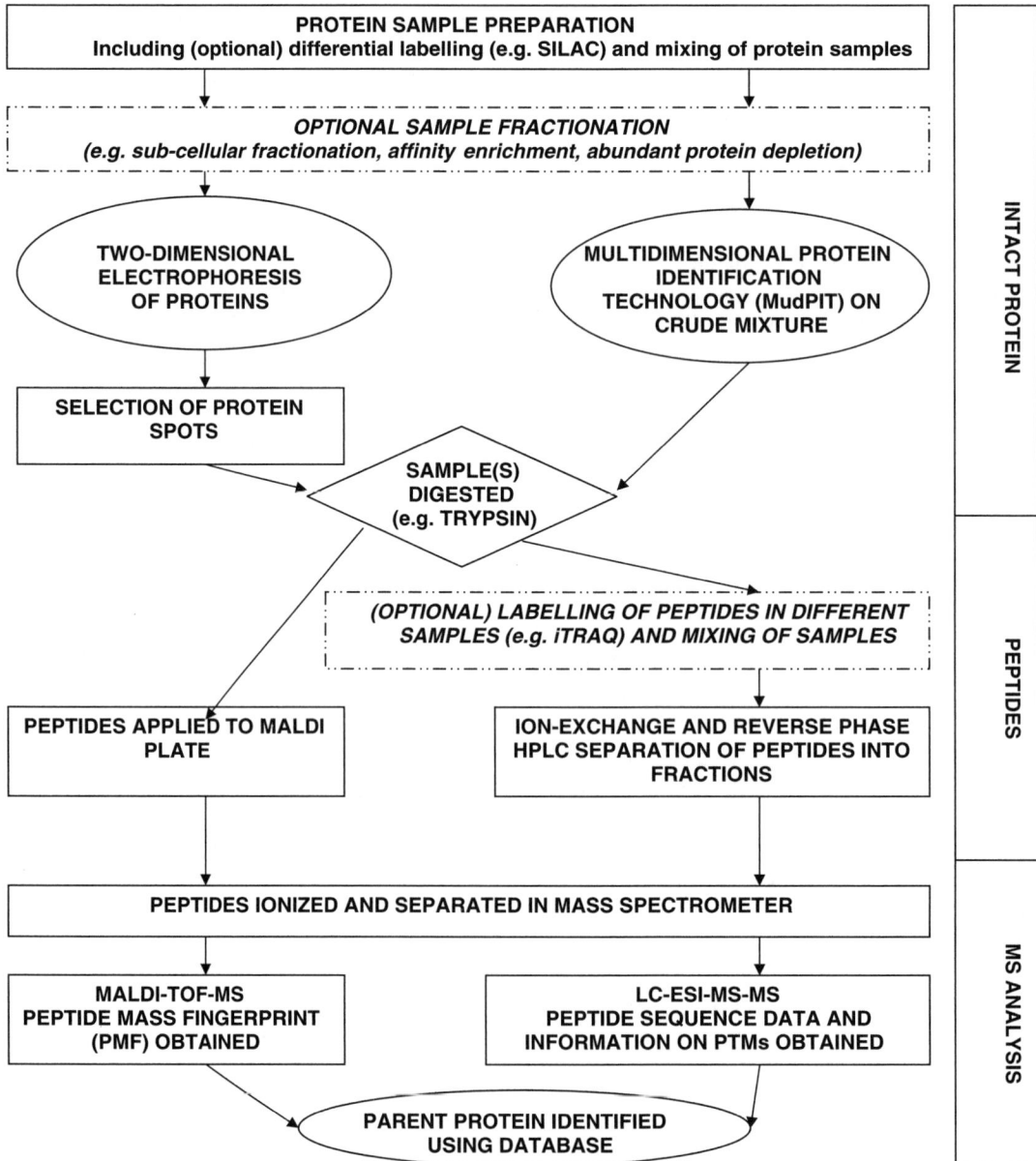

Figure 12.1. Proteomic workflow. Samples are prepared for analysis in either gel-based or MudPIT experiments, including the use of optional labelling and fractionation approaches. Proteins are then separated using 2D-PAGE and spots of interest selected for digestion into peptides for mass spectrometric analysis, or samples are digested into peptides and fractionated using ion-exchange and reverse-phase chromatography for mass spectrometry. Peptide mass fingerprints or sequences obtained are used for database interrogation in order to identify proteins

Phase 1 – Sample Fractionation

Day 1   Plasma sample (200 μl)   Biopsy sample (5000 μg)
              ↓   ↓
         Depletion (500 μg)   Microdissected
                                                              sample (500 μg)
              ↓
Day 2   Sample preparation
              ↓
         Apply sample to IPG strips
              ↓
Day 3   Begin IEF separation (and prepare SDS-gels)
              ↓
Day 4   Apply IPG strip to gel and perform SDS–PAGE
              ↓
Day 5   Staining and imaging

Minimum of 2 samples in duplicate per experiment, minimum of three experiments (12 gels)

              ↓

Phase 2 – Analysis

Day 1-2   Comparative image analysis (three experiments)
                ↓
Day 3   Spot cutting and digestion (100 spots)
                ↓
Day 4   MS analysis and database searching
                ↓
Day 5   Operator validation of bioinformatic readout

Minimum of 20 scientist days per complete experiment

Repeat of whole experiment, including spot identification

Consumables: £4000 (two sets of 12 gels and 100 spots)

Maintenance costs: £3000–4000 (two complete experiments)

£500K capital investment

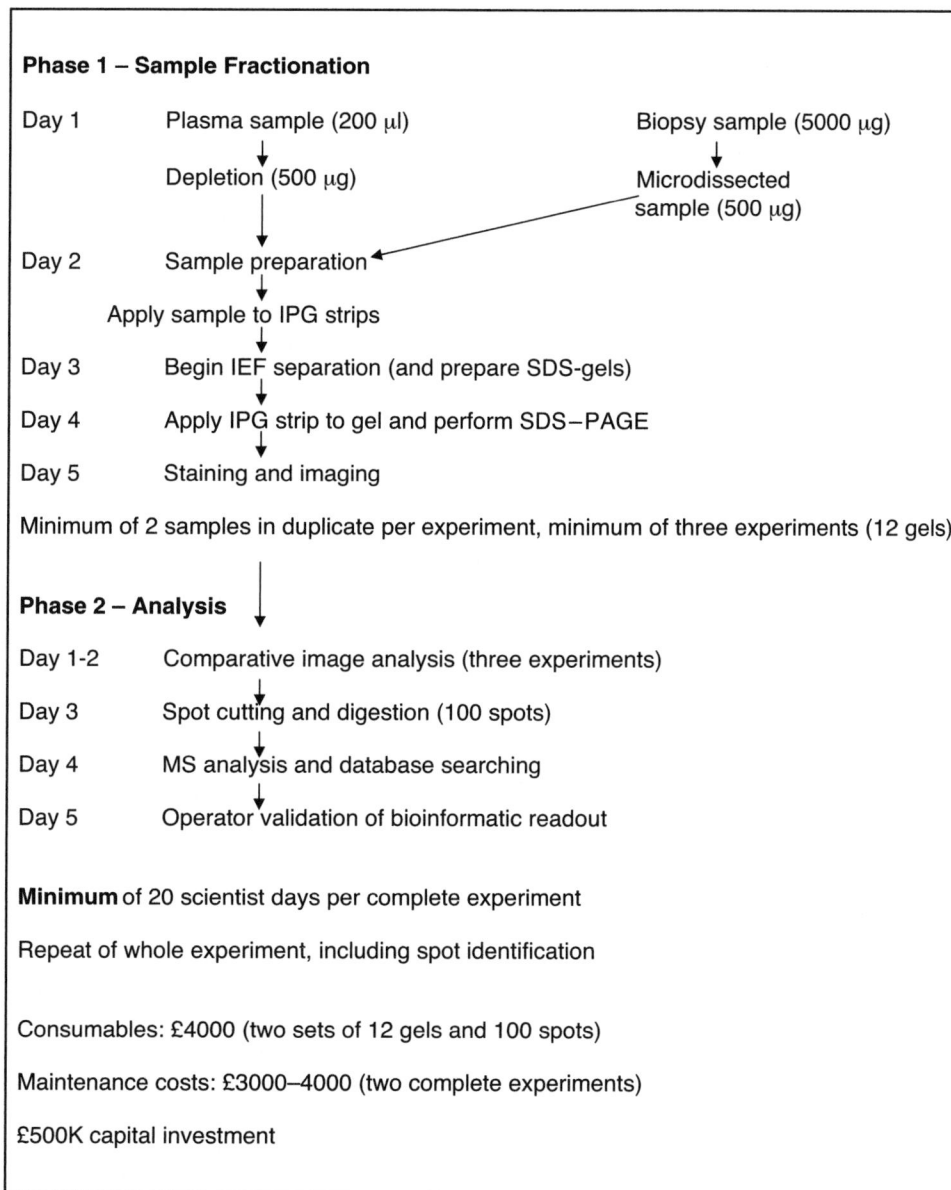

Figure 12.2. Outline of a typical gel-based proteomics experiment. Samples (e.g. patient plasma or tissue biopsies) are prepared for analysis in at least duplicates to allow for experimental variability. Plasma samples may be depleted of abundant proteins using antibody-based affinity removal to enable visualization of the less abundant proteins. Tissue biopsies may be enriched for cancerous cells using laser microdissection. Protein extracts are then applied to immobilized pH gradient (IPG) strips and allowed to absorb overnight. The IPG strips are then transferred to the iso-electric focusing (IEF) equipment and subjected to IEF. The strips containing the separated proteins are transferred to SDS gels and subjected to SDS–PAGE. The gels are then stained to visualize protein patterns and imaged. The experiment is repeated twice more to allow for biological sample variation and then all images can be compared to establish differences in profile. Protein spots of interest are excised, subjected to digestion and the peptides analysed using MS. Database interrogation then enables identification of the proteins of interest and their relevance can be established. Approximate time-scales and costs for equipment, equipment maintenance and consumables are presented

purified proteins generated using a top-down approach, analysis of the set of peptide ions produces a peptide mass fingerprint (PMF), which is unique to the parent protein and can be used to search databases for identification (Pappin *et al.* 1993). The experimental PMF is compared with theoretical digests generated from sequence databases, with the closest match providing the most likely identity for the parent protein. The use of statistics, e.g. what proportion of the protein is covered by the peptides detected and biological knowledge (mass and pI values), provides a level of confidence to the identification. This approach can be fast, sensitive and relatively inexpensive, but is somewhat restricted to almost pure proteins. The addition of a collision cell and a second TOF analyser enables MS–MS, enhancing this approach by enabling sequence generation to support the PMF data. Often six or seven residues of sequence from only one or two peptides is enough to identify the protein. The ambiguities arise when proteins have common domains or where several isoforms can be present.

In LC–ESI–MS–MS, which can be used for both digests of pure proteins and highly complex mixtures, the peptide sample is pre-fractionated, usually using either reverse-phase (RP) chromatography or a combination of strong cation-exchange (SCX) and RP chromatography, into pools which simplify the peptide mixture for analysis. Each pool can then be analysed by the first analyser (e.g. TOF or quadrupole), producing peptide mass data (e.g. PMF from a pure protein sample), but more usually is subjected to ion selection procedures within the mass spectrometer to feed discrete component peptide ions into a collision cell for bombardment with inert gas (e.g. argon). This produces further fragmentation into daughter ions, which are analysed by the second analyser. The combination of these masses can usually reveal a unique sequence. Each peptide within the mixture can be sequentially analysed by this approach, generating sequence tags from each component for protein identification. This latter approach lies at the heart of the major bottom-up approach known as multi-dimensional protein identification technology (MudPIT; Link *et al.* 1999).

Sample fractionation

Top-down proteomics utilizes high-resolution approaches for protein separation, including electrophoretic or multi-dimensional chromatographic procedures.

The approach used most commonly is two-dimensional (2D) polyacrylamide gel electrophoresis (PAGE) (also known as 2-DE; Gorg *et al.* 2004). In 2D PAGE, solubilized denatured proteins in crude mixtures are first separated according to their native charge, using isoelectric focusing (IEF). The denatured protein mixture is applied to an immobilized pH gradient (IPG) strip and allowed to migrate freely under the influence of an applied electrical potential, until each protein reaches the position within the pH gradient at which its native charge is neutralized (the isoelectric point, pI). Each polypeptide thus becomes 'focused', typically as a discrete sharp band, at a single point along the IPG strip. It can readily be seen, therefore, that if the protein becomes, for example, phosphorylated it will focus at a different point from the unphosphorylated version. The strip is then laid onto an SDS-PAGE gel and the protein bands eluted from the strip into the gel and separated in the second dimension on the basis of their mass. This second electrophoretic step separates proteins of very similar pIs but differing masses to produce a 2D array of protein spots. After electrophoresis, the protein spots are visualized using one of a number of general protein stains, such as Coomassie blue, silver (for high sensitivity) or fluorescent stains such as Sypro Ruby (for quantitative detection across a wide dynamic range). Protein patterns of different gels can be compared using specialized software packages in order to detect differences in position or intensity between samples (e.g. disease condition vs. control). The protein spots of interest can then be extracted, digested and identified using MS and their potential involvement in the biological process given due consideration.

Up to several thousand proteins may be separated and visualized using 2D PAGE, enabling detailed qualitative and quantitative differences between samples to be identified. This can be improved further by pre-fractionation of the original sample into pools that can be analysed separately. This may be achieved by dividing the pH range of the IEF separation into sections (e.g. pH 3–6, 5–8 and 7–10) and running samples on the reduced ranges, thus improving loading and separation to achieve a greater overall spot count. The technique does have its limitations, however. The amount of material that may be applied to an IPG strip to generate a well-resolved protein pattern is limited to a few mg at most, so the detection of low abundance proteins may be compromised, particularly where high-abundance proteins may mask areas of the gel and contaminate other spots. Fractionation and enrichment strategies are essential for maximizing the amount of information that may be obtained. Integral membrane proteins present another difficult class, since they are usually not only present in limited amounts but migrate poorly (streak), if at all, in the first dimension. High molecular weight or very basic proteins also resolve poorly.

In some situations, standard SDS–PAGE can prove more suitable than 2D PAGE. This is particularly the case for these hydrophobic membrane proteins, which tend to aggregate, precipitate and smear during IEF. SDS is an excellent solubilizing agent and allows electrophoresis on the basis of size, producing bands that can subsequently be analysed using a version of the MudPIT approach. With complex mixtures, the bands are likely to contain several components and maximal information can only be obtained by subjecting each band on the gel to MS–MS so that the different components can each be identified from sequence information.

The bottom-up approach of MudPIT avoids the problems associated with achieving separation of very complex protein mixtures by converting the sample into more easily managed peptides at the start. The digestion of the sample with trypsin produces peptides with at least one positive charge (usually two), enabling them to adhere to a cation-exchange column. The bound peptides are then eluted stepwise, using increasing concentrations of salt, and each pool is transferred serially to a reverse-phase column to enable further separation on the basis of hydrophobicity. The much simplified fractions from the RP column are then transferred into the ionization source of the mass spectrometer, typically an electrospray instrument. The two-phase (multi-dimensional) chromatography is performed

at low flow rates on capillary columns to enable relatively fast, high resolution separations compatible with solvent flow into the MS. The peptides are fractionated within the MS and peptides fragmented to generate sequence data as outlined above. An experiment can easily analyse tens of thousands of peptides and identify thousands of proteins from the resulting sequences.

However, again with integral membrane proteins, hydrophobic fragments are not easily generated, solubilized and separated, and can therefore 'disappear.' Another potential drawback of this approach is associating the peptide sequences with their parent proteins. Although identification itself is straightforward, it may be more difficult to ascertain whether or not a particular protein isoform or domain is present, since only a relatively limited proportion of the protein is actually sequenced. All isoforms of a protein will produce a number of equivalent peptides that are indistinguishable and will co-migrate through the chromatography stages into the MS. Thus, it will not be possible to determine the origin of common peptide sequences. Differing sequences from isoform variants will only identify the presence of those variants if those particular peptides are analysed. Similarly, with post-translational modifications (PTMs), it is very often not possible to detect or identify that such events have occurred. For these reasons it is often important to identify as many peptides as possible in any experiment, but the cost can be prohibitive!

The MudPIT approach is often combined with a sample labelling method to allow quantitation of peptides (discussed below), which helps to determine the presence of protein variants in different samples.

Sample enrichment

A significant problem for the proteomic analysis of clinical samples is the heterogeneity of tissue sampled from patients. The ability to detect proteins of potential relevance may be hampered by the fact that the diseased cells represent only a small proportion of the total samples, so that differences in proteins in these cells are diluted out by the protein profiles of the quantitatively dominant surrounding cells. A number of approaches may be undertaken to alleviate this problem. These include the use of cell sorting, using specific antibodies targeted against proteins only expressed at the surface of cells of interest, and the use of laser microdissection microscopy, which has become popular (Bonner *et al.* 1997). In this latter technique, the tissue can be viewed under the microscope and specific cells or clumps of cells of interest can be dissected out and combined to produce a much-enriched sample of relevance. Although these approaches can greatly improve the relevance of any study, the problems may be avoided by analysing samples of plasma, which is considered to represent the deepest version of the human proteome, representative of all the physiological and pathological processes present in the body. Its homogeneity and relative ease of sampling are major benefits, but its complexity and vast dynamic range of protein concentrations are major hurdles (Anderson and Anderson 2002).

Complex mixtures of proteins from whole cells or tissues may be simplified to enable more thorough coverage of particular parts of the proteome. Various approaches may be undertaken to pre-fractionate the samples and enrich for particular proteins of interest. Cellular or sub-cellular isolations may be undertaken, e.g. by ultracentrifugation to prepare cellular membranes, or density gradient centrifugation to prepare organelles, but in all cases the procedures must be exhaustively reproducible. More specific approaches include the use of epitope-tagged recombinant proteins to target and isolate protein complexes, immunoaffinity purification, lectin affinity chromatography for glycoproteins, immunoprecipitation of phosphoproteins, and so on.

A particularly useful approach for the investigation of signalling events based on phosphorylation of proteins is immobilized metal affinity chromatography (IMAC; Andersson and Porath 1986). Reversible phosphorylation of proteins on specific serine, threonine or tyrosine side-chains plays a key role in eukaryotic signalling and is at the heart of some disease processes. Identification of key protein kinases or phosphatases, therefore, may assist the development of novel drug intervention strategies. IMAC can be employed with crude protein digests to isolate the phosphopeptides in order to identify proteins undergoing phosphorylation. The IMAC technique exploits the affinity of metal ions, such as Fe^{3+}, Ga^{3+} and Al^{3+}, for acidic phosphate groups. Non-phosphorylated peptides containing numerous acidic residues may also be enriched, causing contamination, but this potential problem can be considerably reduced through methyl esterification of the acidic side-chains (Ficarro *et al.* 2002).

An affinity chip-based approach has been developed to fractionate and enrich samples on the basis of physical, chemical or biological properties. Such a strategy forms the basis of the SELDI–TOF–MS technology for determining protein profiles. The profiles for different samples e.g. biopsies, can then be compared to identify 'fingerprints' diagnostic of disease conditions. The chip surface can be derivatized in various ways to allow binding of subsets of proteins in the sample. The chip is then washed to remove loosely-bound proteins from the surface. The tightly-bound protein fraction on the chip is then analysed using SELDI–TOF–MS, thereby generating a profile or fingerprint. Such an approach is most frequently applied to clinical studies, particularly plasma samples, in an attempt to generate reproducible protein 'fingerprints' diagnostic of particular diseases. Fairly limited resolution and mass accuracy render this approach most suitable for peptides and small proteins (< 20 kDa). Discrete diagnostic patterns of mass peaks have been determined in relation to a number of diseases, but a further limitation of the approach is the lack of progression to protein identification *per se*. Although new instruments may tackle this limitation, it is currently more usual to conduct parallel studies involving protein purification methods to isolate features of interest. The mass of a feature of interest can be used as a marker in order to follow the progress of purification using approaches (e.g. column chromatography) based on the chip-binding characteristics of the molecule of interest. Purification of the protein peak of interest can then be linked to more detailed characterization, using mass spectrometry (fingerprinting or sequencing) in order to determine identity. In general, the approach is an attractive one and allows a rapid throughput of samples. In practice, variability in sample, chip surfaces and procedures have frustrated attempts to make the technology sufficiently robust and reproducible.

Sample depletion

A major problem in the analysis of highly complex proteomes is the dynamic range of protein abundance. Human plasma represents the most extreme example of this, with protein constituents varying in concentration by more than nine orders of magnitude, from albumin at ca. 50 mg/ml to prostate-specific antigen at ca. 20 pg/ml. In order to visualize minor components that may represent potential biomarkers, it is necessary to remove the more abundant proteins to enable mining of the residual proteome. Several commercially-available depletion methods have been developed, ranging from simple pseudo-affinity removal of albumin to antibody-based affinity columns (such as the Agilent Technologies Multiple Affinity Removal System and the Beckman Coulter ProteomeLab IgY-12 Affinity Column), capable of removing several of the most abundant proteins from plasma. Depletion of such components eliminates 'obscuring' effects and allows greater volumes of plasma to be analysed, thus enabling detection of low-abundance proteins (Echan *et al.* 2005). One drawback of such an approach which has exercised the literature is that minor proteins might be interacting with those undergoing depletion (Granger *et al.* 2005).

Quantification of proteins and peptides

Various strategies exist to assist the quantitation of differences between samples in proteomic experiments. In the case of top-down approaches, the separation of proteins allows the direct visualization of individual protein species, so the detection of spots on gels or peaks in column eluates enables relative quantitation of equivalent proteins. Experimental variation can be overcome by labelling samples with dyes of different colours (e.g. the CyDyes™; GE Healthcare) and then co-running the mixed samples on gels or columns to enable direct comparison. Differential in-gel electrophoresis (DIGE; Ünlü *et al.* 1997; Alban *et al.* 2003) is an example of such an approach.

Many of the quantitative methods can be applied to both top-down and bottom-up approaches. They depend on the ability of the mass spectrometer to distinguish isotopically distinct versions of the same peptide generated by treatment of the different samples with labels that are different but very closely related. When the samples are mixed and co-analysed, equivalent proteins or peptides from both samples co-migrate and can be analysed together, as their properties are virtually identical. The mass spectrometer will detect the peptide from both samples at the same time and display the relative amounts of each, distinguishable only by the defined difference in mass of the two 'tags' used to label the peptides.

One such approach is stable isotope labelling in cell culture (SILAC; Ong *et al.* 2002). In this method, one batch of cells is cultured in the presence of an isotopically different amino acid (typically $^{13}C/^{15}N$-Arg), another in the presence of the normal amino acid. The two samples are mixed together and, following either protein separation and digestion or peptide separation, the relative amounts of the distinguishable isoforms can be quantitated.

Another method employs isotopically-labelled affinity tags (ICAT; Gygi *et al.* 1999; Yi *et al.* 2005), whereby different samples are labelled at cysteine residues by isotopically distinguishable ICAT reagents containing a biotin tag. Following proteolysis, peptides containing the modified cysteine residues are isolated using avidin (which binds biotin) affinity chromatography. The relative amounts of the peptide pairs obtained by MS provides quantitative information on their original abundance.

A further method (iTRAQ; Ross *et al.* 2004) uses isobaric tags for relative and absolute quantitation of peptides. A number of different iTRAQ reagents can be used to label different samples, enabling complex comparisons to be undertaken. The amino-terminus of every peptide in the digested protein sample is labelled with the reagent, which contains a specific reporter moiety that is subsequently removed and detected in the mass spectrometer to provide information which reflects the amount of every peptide present. This method is particularly useful in the MudPIT approach, where multiple samples can be prepared, mixed and analysed at the same time.

Validation

The proteomic procedures outlined here are designed to identify changes that may be diagnostic of a particular condition, environment, dysfunction, etc. However, several other important stages need to be carried out before the information can be applied to screening populations or systems.

The first of these is validation, and it too requires several stages. At the start, the protein identification needs to be confirmed. This is most easily carried out by immunological recognition (e.g. Western blotting), using specific antibodies, preferably a panel of antibodies raised against different epitopes of the protein of interest. This approach may enable discrimination of particular protein isoforms. Western blotting can provide quantitative information as well as confirming protein identity. A limitation of this approach might be the availability of a suitable antibody, but if not already available an antibody can be readily generated using a synthetic peptide designed from the known sequence. In addition, the production of antibodies against proteins encoded in the entire human genome is an on-going project.

There can also be functional validation, more usually in cellular systems. Here 'interaction proteomics' or knock-outs/knock-downs can play an important part in establishing the role of the identified protein within the system under investigation.

12.5. ILLUSTRATIVE APPLICATIONS

Two recent applications highlight both the potential and problems in the use of proteomic technologies in biomarker discovery and utilization. Both papers, from the same research group, utilized SELDI–TOF–MS for the identification of 'clusters' or profile signatures in serum samples that are potentially indicative of ovarian or prostate cancer (Petricoin *et al.* 2002a, 2002b). The experiments consisted of the application of serum samples from patients (50 and 38, respectively) with disease and non-diseased controls (66 and 228, respectively) to hydrophobic (C_{16}) interaction ProteinChips™. The chip surface was subjected to repeated laser bombardment and the masses of the released ionized species determined by SELDI–TOF–MS. After each ionization, cluster analysis was used to define

the mass profiles obtained from training sets of diseased and non-diseased subjects. Note that the profiles represented only masses; the components making up the mass patterns were not identified and so, without going a stage further, the proteins predictive of disease and non-diseased states would not have been revealed. Unknown samples were then examined to identify whether they more closely resembled the 'diseased' or 'control' cluster pattern. Positive predictive values for both cancers were reported as > 90%. The studies are interesting in a number of ways. On the positive side, they suggest that discriminating profiles can be obtained. But the question remains – discriminating of what? The cluster patterns obtained are unlikely to define key protein indicators of the disease, which are probably present in amounts too small to be detected (much less identified) by this approach. They are more likely to be indicative of a dysfunction 'of some sort' and in this respect fall into a 'positive' category containing both true and false positives. Thus, when used for general population screening, only one or two of 50 positives out of a sample size of 1000 (5% false-positives) would be true-positives, a positive predictive value of 2–4%, too low for practical use. Thus, the general consensus would be that at the present stage of development, it is more appropriate to identify and then validate individual components that are tightly and specifically linked to particular disease states (Rockhill *et al.* 2002).

But, as the authors point out, it does raise an interesting issue, yet to be resolved one way or another, and that is whether the profiles generated for an altered (e.g. disease) state, composed of whatever components, may nevertheless be sufficiently discriminating to be predictive. Our experience would tend to weigh against this possibility, since we see similar spectra of changes in a variety of experimental situations, which can obscure the relatively small number of specific alterations. Moreover, since it is unlikely that an 'all-or-none' situation will prevail, two factors are probably going to be important – quantitation of the key components, combined with a diagnostic pattern of changes.

The core approaches based on the use of either 2D PAGE or MudPIT technologies determine the actual identities of potential biomarkers, which in the longer term are more likely to be useful. Population screening methods for specific targets can be developed more easily, e.g. by using antibodies against the particular target in high-throughput assays. Thus, these technologies are being utilized in the study of an enormous number of disease states and are beginning to generate meaningful information.

An interesting example of the use of the 2D PAGE approach is reported by Craven *et al.* (2006) for biomarker discovery in renal cell carcinoma (RCC). Short-term cultures of epithelial cell populations derived from RCC tissue and matched normal kidney cortex were compared using 2D PAGE and protein spots displaying > two-fold variation between matched pairs were selected for MS analysis. A number of proteins displaying up- or down-regulation were identified, many of which had been previously reported in RCC, demonstrating the validity of using cell culture to increase the amount of material available from biopsies and overcome problems of tissue heterogeneity and serum contamination. The increased availability of material enabled more detailed analysis and allowed the discovery and validation of new potential biomarkers.

The MudPIT approach has recently been applied to the analysis of urine, a potentially useful source of clinical biomarkers, as it can be easily obtained non-invasively (Chen *et al.* 2007). Over 1000 peptide sequences and sites of post-translational modification could be assigned to 334 unique gene products, over 100 of which had not been detected in urine previously. Some of these proteins could be related to disease states, indicating the potential for routine screening of urine in clinical diagnosis.

Many more examples of the application of the 2D PAGE, MudPIT and SELDI approaches could be selected for discussion, but it is beyond the scope of this chapter to do justice to the extensive amount of clinical research under way.

12.6. FINAL CONSIDERATIONS

Although there is an abundance of literature indicating the identities of potential biomarkers for specific conditions and diseases, it is unlikely that a specific disease will be absolutely defined by a single component. It is much more likely that a set of protein biomarkers (a biomarker profile or signature) will be required in order to define accurately a particular disease condition, with a certain degree of variability within this disease profile. Single components may be fairly diagnostic in themselves, but the statistical probability of disease definition is seen to improve when additional features are encompassed.

An example of the improvement in predictive accuracy in ovarian cancer through the use of multiple features is demonstrated in the work of Moshkovskii *et al.* (2007). They showed that although the use of SELDI–TOF–MS and immunoassay data on serum amyloid A protein in conjunction with cancer antigen CA 125 data did not improve predictive ability, the inclusion of an additional 48 SELDI–TOF–MS features improved prediction of disease from 86.2% (CA 125 alone) to 95.2%. The potential power of feature combination is thus apparent.

There is also considerable debate as to how to screen for a biomarker in the population once it has been identified and validated. Where the biomarker is likely to be a profile, then SELDI–TOF–MS may be the best option, unless the profile can be categorized thoroughly and its components identified and quantitated separately.

Where the biomarker is clearly identified and validated, ELISA assays are a viable option for screening populations. It is debatable whether ELISA assays or an MS screen will prove the most rapid and cost-effective in the long run, once capital and operator costs are taken into account. A further approach would be to use antibody chip arrays, created to detect specific biomarker profiles characteristic of particular diseases, and this ultimately may prove to be the method of choice.

References

Aebersold R. 2003. A mass spectrometric journey into protein and proteome research. *J Am Soc Mass Spectrom* **14**: 685–695.

Alban A, David SO, Bjorkesten L et al. 2003. A novel experimental design for comparative two-dimensional gel analysis: Two-dimensional difference gel electrophoresis incorporating a pooled internal standard. *Proteomics* **3**: 36–44.

Anderson NL, Anderson NG. 2002. The human plasma proteome: history, character, and diagnostic prospects. *Mol Cell Proteom* **1**: 845–867.

Andersson L, Porath J. 1986. Isolation of phosphoproteins by immobilized metal (Fe^{3+}) affinity chromatography. *Anal Biochem* **154**: 250–254.

Bonner RF, Emmert-Buck M, Cole K *et al.* 1997. Laser capture microdissection: molecular analysis of tissue. *Science* **278**: 1481–1483.

Chen Y-T, Tsao C-Y, Li J-M *et al.* 2007. Large-scale protein identification of human urine proteome by multi-dimensional LC and MS–MS. *Proteom Clin Appl* **1**: 577–587.

Craven RA, Stanley AJ, Hanrahan S *et al.* 2006. Proteomic analysis of primary cell lines identifies protein changes present in renal cell carcinoma. *Proteomics* **6**: 2853–2864.

Echan LA, Tang H-Y, Ali-Khan N *et al.* 2005. Depletion of multiple high-abundance proteins improves protein profiling capacities of human serum and plasma. *Proteomics* **5**: 3292–3303.

Ficarro SB, McCleland ML, Stukenberg PT *et al.* 2002. Phosphoproteome analysis by mass spectrometry and its application to *Saccharomyces cerevisiae*. *Nat Biotechnol* **20**: 301–305.

Gorg A, Weiss W, Dunn MJ. 2004. Current two-dimensional electrophoresis technology for proteomics. *Proteomics* **4**: 3665–3685.

Granger J, Siddiqui J, Copeland S *et al.* 2005. Albumin depletion of human plasma also removes low abundance proteins including the cytokines. *Proteomics* **5**: 4713–4718.

Gygi SP, Rist B, Gerber SA *et al.* 1999. Quantitative analysis of complex protein mixtures using isotope-coded affinity tags. *Nat Biotechnol* **17**: 994–999.

Hutchens TW, Yip T-T 1993. New desorption strategies for the mass spectrometric analysis of macromolecules. *Rapid Commun Mass Spectrom* **7**: 576–580.

Karas M, Bachmann D, Bahr U *et al.* 1987. Matrix-assisted ultraviolet laser desorption of non-volatile compounds. *Int J Mass Spectrom Ion Process* **78**: 53–68.

Link AJ, Eng J, Schieltz DM *et al.* 1999. Direct analysis of protein complexes using mass spectrometry. *Nat Biotechnol* **17**: 676–682.

Moshkovskii SA, Vlasova MA, Pyatnitskiy MA *et al.* 2007. Acute phase serum amyloid A in ovarian cancer as an important component of proteome diagnostic profiling. *Proteom Clin Appl* **1**: 107–117.

Ong S-E, Blagoev B, Kratchmarova I *et al.* 2002. Stable isotope labeling by amino acids in cell culture, SILAC, as a simple and accurate approach to expression proteomics. *Mol Cell Proteom* **1**: 376–386.

Pappin DJ, Hojrup P, Bleasby AJ. 1993. Rapid identification of proteins by peptide-mass fingerprinting. *Curr Biol* **1**: 327–332.

Petricoin EF III, Ardekani AM, Hitt BA *et al.* 2002. Use of proteomic patterns in serum to identify ovarian cancer. *Lancet* **359**: 572–577.

Petricoin EF III, Ornstein DK, Paweletz CP *et al.* 2002. Serum proteomic patterns for detection of prostate cancer. *J Natl Cancer Inst* **94**: 1576–1578.

Rockhill B, Pearl DC, Elwood M *et al.* 2002. Correspondence: proteomic patterns in serum and identification of ovarian cancer. *Lancet* **360**: 169–171.

Ross PL, Huang YN, Marchese JN *et al.* 2004. Multiplexed protein quantitation in *Saccharomyces cerevisiae* using amine-reactive isobaric tagging reagents. *Mol Cell Proteom* **3**: 1154–1169.

Ünlü M, Morgan ME, Minden JS. 1997. Difference gel electrophoresis: a single gel method for detecting changes in cell extracts. *Electrophoresis* **18**: 2071–2077.

Wasinger VC, Cordwell SJ, Cerpa-Poljak A *et al.* 1995. Progress with gene-product mapping of the mollicutes: *Mycoplasma genitalium*. *Electrophoresis* **16**: 1090–1094.

Yamashita M, Fenn JB. 1984. Electrospray ion source. Another variation on the free-jet theme. *J Phys Chem* **88**: 4451–4459.

Yi EC, Li X-J, Cooke K *et al.* 2005. Increased quantitative proteome coverage with ^{13}C/^{12}C-based, acid-cleavable isotope-coded affinity tag reagent and modified data acquisition scheme. *Proteomics* **5**: 380–387.

13

Exploring the Contribution of Metabolic Profiling to Epidemiological Studies

M. Bictash,[1] E. Holmes,[1] H. Keun,[1] P. Elliott[2] and J. K. Nicholson[1]

[1]*Biomolecular Medicine, Division of Surgery Oncology Reproductive Bilogy and Anaesthetics (SORA), Faculty of Medicine Imperial Collage London, South Kensington Campus London SW7 2AZ UK and [2]Department of Epidemiological and Public Helth, Imperial Collage London, St. Margs Campus London UK. London, UK.*

13.1. BACKGROUND

The incorporation of post-genomic disciplines such as transcriptomics, proteomics and metabonomics, into large-scale population studies such as Biobank, or the retrospective analysis of samples from sample banks of completed studies such as INERSALT, INTERMAP and EPIC, is increasingly common (Stamler *et al.* 2003; INTERSALT Cooperative Research Group 1988; Riboli *et al.* 2002). Until recently, these post-genomic platforms were not suitable for human population studies, mainly due to their limited capacity for sample throughput and subsequent processing of the vast amounts of data generated. In parallel with the technological developments in these fields came the increasing awareness of biological intricacy and the need to integrate data obtained from multiple biological levels using systems-based approaches capable of accommodating the complexity of biological networks (Jeong *et al.* 2000; Coen *et al.* 2004; Ideker *et al.* 2001; Losoalzo *et al.* 2007). Significant effort was expended in optimizing the technological aspects of analysis and data interpretation, and consequently these disciplines are now reaching a deeper level of maturity that will allow them to fulfil their potential in epidemiological studies. Here we discuss the current role of metabonomic technology and comment on its potential and limitations in exploring associations with disease risk in population studies.

Metabonomics, defined as 'the quantitative measurement of the dynamic multiparametric metabolic response of living systems to pathophysiological stimuli or genetic modification' (Nicholson *et al.* 1999), combines spectroscopic methods with pattern recognition analysis to generate complex high data-density fingerprints of biological tissues and fluids, which can be used to characterize certain physiological or pathological states. Metabonomics can often be confused with the field of metabolomics. Although they both utilize similar analytical tools (typically spectroscopy) with pattern recognition techniques, metabolomics involves the comprehensive analysis of all measurable metabolite concentrations under a given set of conditions. As such, this will always require the addition of a chromatographic separation or fractionation step (Fiehn 2002). The field of metabolomics originated in plant and microbial sciences and favoured gas chromatography (GC) analysis. Generally speaking, metabolomic studies originally focused main on cellular compartments or other contained systems, but more recently the term 'metabolomics' has been used to refer to studies with wider biological scope. In order to fulfil the objectives of epidemiological studies, with their typically large sample sizes and requirements for rapid sample characterization and hence classification, it is the field of metabonomics that is better suited for the epidemiological style study, since the emphasis is on modelling the response of a whole organism or system. Moreover, the field of metabonomics offers a means of identifying candidate biomarkers of disease (Yang *et al.* 2004; Sabatine *et al.* 2005; Barshop 2004; Holmes *et al.* 2006; Nicholson *et al.* 1984a; Brindle *et al.* 2002) and defining response to therapeutic interventions Nicholson *et al.* 1984b; Nicholls *et al.* 2000; Plumb *et al.* 2003; Rezzi *et al.* 2007a; 2007b and it can be used prognostically in some cases to product effects of drags on individuals bused on a pre-wose metabolic profiles (Clayton *et al.* 2006) Metabonomics provides a relatively high throughput and real end-point metrics of the metabolic status of an organism. Any analytical method capable of generating

molecular profiles of multiple molecular species can be used to obtain comprehensive molecular fingerprints of body fluids or tissues, which can then be analysed and compared to a database of spectra, using computer-based pattern recognition algorithms. Techniques that have been adopted for metabolic profiling studies include, but are not limited to: nuclear magnetic resonance (NMR) spectroscopy; mass spectrometry (MS), with or without liquid or gas phase chromatographic separation; chromatography; infra-red spectroscopy; Raman spectroscopy; and capillary electrophoresis (usually with MS detection) (Bonnier *et al.* 2006; Oliveira *et al.* 2006; Lagali *et al.* 2006; Zomer *et al.* 2004; Ullsten *et al.* 2006; Cavaggioni *et al.* 2006; Lenz and Wilson 2007). However, the vast majority of metabonomic studies have been conducted using either NMR spectroscopy or MS methods. Both of these technologies generate high-density data from a single biofluid or tissue sample, and

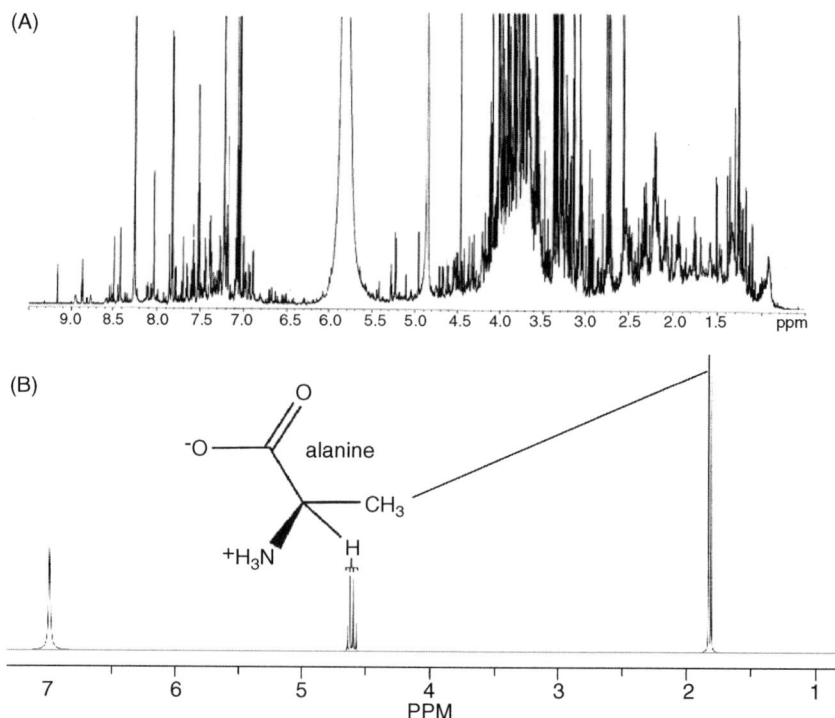

Figure 13.1   (A) 900 MHz one-dimensional ^{1}H-NMR spectrum of a typical urine sample. This spectrum of peaks is a fingerprint of all hydrogen-containing metabolites arising from different chemical types. The area, relative position and splittings of the peaks will provide information on the structure of the molecule and their molar concentration. (B) An example of structural assignment for an individual compound that is present in a complex biological mixture

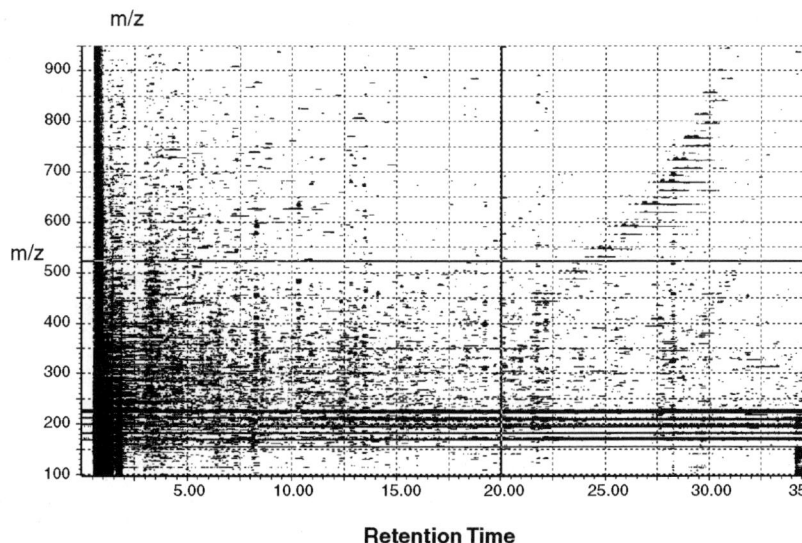

Figure 13.2 The retention time vs. *m/z* plot for a urine sample, showing over 50 000 resolved metabolites from an ultra-performance liquid chromatography with time-of-flight mass spectrometry (UPLC–TOF–MS) experiment

the resulting spectrum typically contains signals from hundreds or even thousands of molecules (Figures 13.1, 13.2). Parallel metabolic profiling technologies were developed in the 1980s, with biofluid NMR being pioneered by Nicholson *et al.* and applied mainly to toxicological problems and small-scale human clinical studies (Nicholson *et al.* 2002; Nicholls *et al.* 2001), whilst MS profiling originated from the plant and microbiology fields (Fiehn 2002). These two approaches have to a large extent converged in objectives and methods, particularly with regard to the mathematical modelling of spectral data. In the case of NMR spectroscopy, the standard experiment is one-dimensional and the registration of peaks on the chemical shift scale, together with the pattern of splitting and the signal intensity provide molecular structural information (Figure 13.1). For MS, typically chromatographic separation is used prior to obtaining a mass:charge ratio (*m/z*), and therefore these data are mostly two-dimensional (2D), consisting of a retention time and *m/z* value (Figure 13.2). Latent information reflecting the 'global' metabolic status of an individual and encoding metabolic responses to genetic and environmental factors, such as nutrition, ageing, gender, stress, disease, etc., can be extracted from the spectral data using multivariate mathematical modelling tools.

Metabonomics was adopted in several of the larger pharmaceutical companies in both the drug discovery and development phases in the 1990s (Robertson *et al.* 2000; Reily *et al.* 2007). At this stage, some automated sample measurement and processing methods were available, but the sample throughput was not sufficient to efficiently analyse specimens collected from large epidemiological studies. Initially, the mathematical algorithms applied to the analysis of spectral data were capable of characterizing metabolic responses to toxicity in animals maintained in a well-controlled laboratory environment, or for generating mathematical models of certain overt human diseases, such as inborn errors of metabolism (Moolenaar *et al.* 2003) or kidney failure (Foxall PS *et al.* 1993. Neild GH *et al.* 1997). However, they were not optimized for analysing human population studies, where the wide range of genetic and environmental variation can obscure the often more subtle metabolic signature of disease and a more sensitive data modelling strategy became a necessity.

Multivariate projection methods such as principal component analysis (PCA) can be used to reduce the dimensionality or complexity of datasets and to visualize inherent patterns within the data. Here, a series of linear orthogonal components are computed with the first component describing the greatest amount of variance in the data, with subsequent components being orthogonal to each other and describing progressively less of the variation. The distribution of samples can be viewed as a 2D or three-dimensional (3D) scores plot, and the metabolites with the greatest influence on the distribution of the coordinates (in this case biological specimens) can be extracted from the coefficients defining the principal components (PC). Where phenotypic data characterizing a specimen are known, more complex modelling methods can be used to maximize the discrimination between disease states and to enhance the potential for biomarker recovery, utilizing the knowledge of class membership for any given set of samples. Numerous multivariate methods have been applied to spectral data (Trygg *et al.* 2007), some of the more common including artificial neural networks (Holmes *et al.* 2001), non-linear mapping (Gartland *et al.* 1990), projection to latent structure discriminant analysis (Eriksson *et al.* 1999), genetic algorithms (Metzger *et al.* 1996), Bayesian probabilistic algorithms (Stoyanova *et al.* 2004) and hierarchical cluster analysis (Eriksson *et al.* 2004). Although the increased sample throughput

has expedited the ability to analyse large populations, it is the development of new mathematical tools for processing spectral data that has had the greatest impact on the implementation of metabonomic technologies in epidemiological studies. Recently we have developed a series of statistical spectroscopic techniques designed to enhance information recovery by analysis of correlative features across several samples, either in NMR spectroscopy, LC–MS or combinations of spectroscopic techniques (Jonsson *et al.* 2005; Shockcor *et al.* 1996; Crockford *et al.* 2006), in particular statistical total correlation spectroscopy (STOCSY; Cloarec *et al.* 2005), which relies on the near-unitary variance in signal intensity between signals from nuclei on the same molecule across several samples. Plotting the correlation matrix as a 2D contour correlation plot of aller port of the spactrum or projected into a 1D spectrum when calculated for a selected peak, allows the assignment of intramolecular correlations as well as inter-molecular correlations for molecules that are in related or co-regulated pathways (Figure 13.3). These spectral

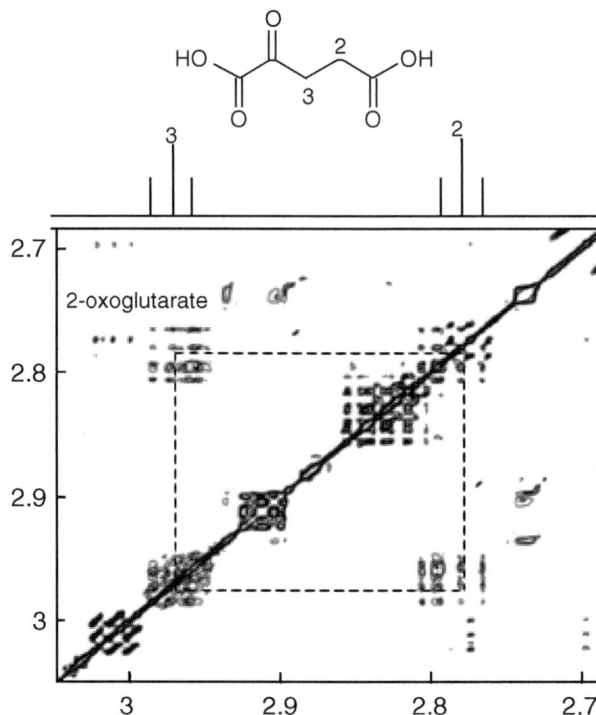

Figure 13.3 Two dimensional STOCSY presented as a contour plot showing the correlation between the two triplets of z-oxoglutorate

reconstruction methods have revolutionized the ability to interpret multivariate analyses by back-transforming significance levels onto a visually recognizable spectrum and have been applied to metabolic profiling of populations we select for example analysic resonances (Holmes *et al.* 2007).

Mathematical tools for data alignment, filtering extraneous variation and statistical correlation spectroscopy algorithms have revolutionized the capacity to extract latent information from spectral datasets. For example, ^{1}H-NMR spectra of urine can suffer from misalignment, due to small variations in pH or complexation with metal ions, although the number of metabolites affected is relatively small, whereas high-performance liquid-chromatography (HPLC)–MS suffers from misalignment in both chromatographic retention time dimension and exact mass. Several solutions have been presented to overcome this problem, including automatic peak detection, alignment, setting of retention time windows, summing in the chromatographic dimension and data compression by means of alternating regression, where the relevant metabolic variation is retained for further modelling using multivariate analysis (Jonsson *et al.* 2005; Bijlsma *et al.* 2006). Differential ionization of analytes in chromatographically overlapped peaks and analytical drift have also been problematic. Ultra-performance liquid-chromatograpy (UPLC)–MS systems use smaller sorbent particle sizes than conventional HPLC, which sustains higher flow rates or higher pressure and enables the use of shorter columns. Consequently, significant analytical improvement over conventional HPLC–MS is achieved with UPLC; it is inherently more sensitive, gives a more rapid separation and improved resolution, and has proved to be more reproducible in terms of inter-run variability in column retention times (Plumb *et al.* 2004).

Appropriats preprocessing of the original spectral data makes the task of identifying candidate biomarkers of disease or condition easier. Spectral regions that systematically differentiate between two or more classes of biological sample can be further examined to identify metabolite(s) that are potentially biomarkers for a particular condition (physiological or pathological state). Tables of metabolites and their properties are available for both NMR and MS data, to aid the identification of such compounds. However, if the metabolite can not be found in these look-up tables, analytical strategies such as LC–MS–MS can be used to obtain further information on the molecular structure of the candidate biomarker. In the case of MS data, the parent ion can be broken down into daughter fragments. For NMR, a series of 2D pulse sequences can be used to uncover further structural information, or STOCSY can be applied.

Epidemiological studies have a markooly different set of requirements to controlled animal or clinical studies. For example, patient recruitment and sample collection on a large scale often requires a substantial amount of time, where samples are generally collected over a period of months or even years. Therefore, it is crucial that the analytical platforms used remain stable over time, and that any analytical drift can be recognized and compensated for to ensure the derivation of valid statistical conclusions. For this reason, as well as good practice, it is important that experimental design be applied to the analysis of samples, and that samples are run in randomized order. Several studies have been conducted to assess the effects of instrumental variation over time (Keun *et al.* 2002; Dumas *et al.* 2006). The coefficient of variation associated with inter- and intra-laboratory NMR experimentation is <5% (Keun *et al.* 2002). With GC–MS, this figure is around 8% for plant extracts (Fiehn *et al.* 2000).

Recent technological advances in automation, and improved robustness in MS systems, have vastly increased sample throughput, allowing the measurement of >100 samples/instrument/day. In line with this, quality control methods have been developed and implemented in large studies. Typically, blank samples (water or buffer) are interspersed in each well plate to ensure that there is no carry-over between samples. In addition, aliquots of quality control (QC) samples, which are prepared at the outset of each study, are included in each run to allow assessment of the instrumental and biological reproducibility over the study duration. These QC samples should be representative of the samples included in the study and are often made up of pooled aliquots of the original samples (Dumas *et al.* 2006). A total of 10 143 urine samples obtained from the International Study of Macronutrients and Blood Pressure (INTERMAP) study were measured by ^{1}H-NMR spectroscopy over 7 months. Analysis of 144 QC samples over this period showed good coefficients of variation <5% (Dumas *et al.* 2006). Another QC assessment was made using duplicate samples, which were split at source and hidden within a > 10-fold bigger population. For MS analysis the issue of quality control is even more important, due to the greater propensity for analytical drift.

Although the potential use of metabolite profiling strategies extends to a vast number of disease areas, we have chosen to expand on key diseases attributed to high mortality and morbidity in the developed world (cardiovascular, cancer and neurodegenerative diseases).

13.2. CANCER

A simple screening test for cancer has been a challenging aspiration. Since many biomarkers widely used to monitor tumour regression during therapy do not have sufficient predictive power (sensitivity and specificity) to justify population screening, there is an urgent need for more biomarkers in the field of oncology. Furthermore, the basic causal factors in the onset of the majority of cancers, and the mechanisms of commonly observed resistance to chemotherapies remain poorly understood, despite vast amounts of research funding. Analytical strategies without an *a priori* selection of target molecules have an obvious advantage over the more conventional single biochemical assay approaches, since they allow the simultaneous generation (and to some degree the testing) of several hypotheses from a single experiment. Many genetic and environmental factors have been linked to specific cancers. There is convincing epidemiological evidence that diet plays a protective role in the development of breast and bowel cancers (Willett 1995), but there is limited information on the efficacy of dietary interventions. Moreover, with no clear understanding of the interaction between these genetic and environmental factors, there can be no good biochemical biomarkers that can be used to identify individuals who are at increased risk and who may benefit from dietary intervention.

Imaging, typically looking at changes in positron emission tomography (PET) and the measurement of protein products by proteomics, have been the most widely applied technologies for biomarker discovery and disease characterization in the field of oncology. With such an obvious genetic basis for the disease, tissue/cell transcriptomics for prognostic markers has also been a popular avenue of research. Metabolic profiling technology has obvious potential in cancer diagnostics. However, screening for biomarkers is limited by numerous difficulties, including sample bias, whereby false discrimination occurs due to sample handling and/or differential analysis between case and control groups. The field of biomarker discovery in oncology has been tainted by several studies succumbing to bias (Baggerly *et al.* 2005; Fossel *et al.* 1986). An early application of NMR spectroscopy in the oncology field resulted in an index measurement, that claimed high diagnostic capability. The "test" was based on the half-height line width of the methyl or methylene plasma lipid resonance in the ^{1}H-NMR spectrum. which was reported to be narrower in patients with malignant tumours however, these could not be validated, as the lipoprotein concentration, contributing to Fossel index measurement, also varies with other diseases (Okunieff *et al.* 1990) and hence was not a unique marker for cancer. It was this past failure in specificity for cancer that has hampered biomarker discovery by metabonomics, and it is only recently that metabolic profiling technologies such as NMR and MS are being reapplied to study a range of cancers (Odunsi *et al.* 2005; Beckonert *et al.* 2003; Beger *et al.* 2006).

Recently a metabonomic study has been successfully applied to epithelial ovarian cancer (EOC) and was able to discriminate women with EOC from healthy controls, based on the ^{1}H-NMR spectroscopic profiles of pre-operative serum specimens (Odunsi *et al.* 2005). In order to achieve this level of discrimination, specific regions of the spectrum were identified as particularly contributing to the sensitivity and specificity of the test. Further metabolic profiling in tumour cells has revealed the presence of elevated choline metabolites, along with an increase in the phosphocholine: glycerophosphocholine ratio; these metabolites have been shown to be a general marker for rapid proliferation (Wheatley 2005; Ackerstaff *et al.* 2003). In conjunction with other metabolites, these may provide some indication of the invasiveness of a tumour that could be of significant utility in the clinic.

Environmental exposures are known to play an important role in the development of some cancers. However, to what extent the metabolic phenotype of cancer is causal or consequential to carcinogenesis and disease progression is not clear. Several key examples show how metabolic profiling has revealed relationships between metabolic perturbation and oncogenic transformation. Ramanathan *et al.* demonstrated changes in a systematic way across fibroblast cell lines progressively transformed from a primary to a cancerous state, using a variety of inhibitors of aerobic and anaerobic metabolism (Rarnanathan 2005). Tumours are frequently hypoxic and consequently the transcription factor hypoxia-inducible factor (HIF) has been shown to be upregulated across a variety of cancers, and is indirectly involved in the modulation of tumour progression, through angiogenic factors such as VEGF (vascular endothelial growth factor) and also glycolytic enzymes and glucose transporters (Jaakkola *et al.* 2001). In fact, unlike healthy cells, cancerous cells preferentially rely on glycolysis for the production of adenosine triphosphate (ATP). This difference in metabolism is known as the Warburg effect, and is exploited in PET imaging, where the presence of a tumour utilizing glucose is highlighted. During conditions of hypoxia, increases in the tricarboxylic acid cycle (TCA) intermediates, fumarate and succinate, may prevent the degradation of HIF, and hence HIF acts as major regulator of cell metabolism, as well as changes in cell metabolism having important effects on HIF response. But is HIF activation a necessary condition for observing the metabolic phenotype of cancer? Metabolic profiling suggests that HIF-1β-deficient Hepa-1 cells continue to upregulate glycolysis in hypoxic conditions, but has revealed unexpected effects, such as lowered purine biosynthesis and ATP (Griffiths *et al.* 2002; Troy *et al.* 2006). It is likely that the latest revelations of how one of the most frequently mutated genes in cancers, *p53*, regulates mitochondrial respiration will also inspire similar interest in the application of metabolic profiling to cancer research (Matoba *et al.* 2006). Collectively, it is these metabolic consequences of altered genomic status that help characterize the neoplastic state and hopefully will allow metabonomics to reliably 'fingerprint' disease status.

A long-awaited promise of the post-genomic era was the use of biomolecular profiling, particularly genetic profiling, to tailor the therapy of each individual to their specific needs and susceptibilities. This is already beginning in the field of oncology, with the assignment of breast cancer treatments based on genomic tests assessing for the tumour expression of marker proteins (Paik *et al.* 2004). The consequence of such diagnostic tests has been the subsequent approval of two cancer drugs, Herceptin and Iressa, for the treatments of breast and lung cancer, respectively, in the patient populations, with tumours shown to overexpress the appropriate marker proteins. Metabonomics has enormous potential in remodelling the field of personalized medicine (Nebert and Vessell 2006), not only because metabolic biomarkers can act as phenotypic indicators for expression of genetic differences, as described above, but also because metabolism is the major interface between an organism and its environment. In this way, metabolic profiling can report on the effect of diet, commensurate and pathogenic microbial populations, exposure to drugs or carcinogens. Since carcinogenesis is initiated through environmental exposures that operate commonly on the metabolic level, metabonomics could have a huge role to play in prevention, as well as cure, by identifying risk factors for disease, such as altered microbial populations, that would otherwise remain undetected. These applications, combined with the ready translation of metabolic markers and the strong evidence for a metabolic phenotype for oncogenic transformation, suggest that metabolic profiling is set to make a strong impact on cancer research in the years to come.

The application of the biomarker technologies is driving a move from the clinic-style study to more population-based studies to enhance our understanding of the roles of genetics, environmental factors and their interactions on disease aetiology. The repository of biological samples from epidemiological studies, such as INTERSALT and INTERMAP, provides a unique opportunity to explore such questions based on high-quality population-based data, collected in standardized fashion from population samples in different countries around the world.

13.3. CARDIOVASCULAR DISEASE

Cardiovascular disease is ranked as the number one cause of deaths in the UK and USA and is rapidly increasing across developing countries. Most cardiovascular-focused metabonomic investigations to date have been pilot studies on a small scale (Brindle *et al.* 2002; Mayr *et al.* 2007). Nevertheless, the technology has shown diagnostic promise in its ability to characterize the presence and even degree of cardiovascular pathology. Several large population-based studies on cardiovascular health have now been initiated, many as part of large European Union (EU)- or National Institute of Health (NIH)-funded consortium grants. Other strategies include the retrospective application of metabolic profiling approaches to sample banks stored from epidemiological studies such as INTERSALT and INTERMAP (Stamler *et al.* 2003; INTERSALT Cooperative Research Group 1988). Although there is much to be learned from these retrospective studies, the original objectives of these studies did not include '-omic' analysis, and so the study design is often suboptimal in terms of sample collection and storage conditions. However, despite this, NMR spectroscopic analysis of urine samples from the INTERMAP study has provided good-quality data, as discussed in the section above considering the issue of quality control.

INTERSALT is an international cooperative cross-sectional study of the relation of electrolytes and other factors to blood pressure (BP) and high BP. It was one of the largest undertakings of its type: 10 079 men and women aged 20–59 were studied in 52 diverse population samples in 32 countries, covering a wide range of cultures and dietary intakes. Extensive material has been published on the multiple implications of the INTERSALT data for the prevention and control of adverse blood pressure levels, prevalent in a majority of the population (Dyer *et al.* 1994, 1997; Stamler 1996, 1997). The potential and limitations of applying MS-based approaches to analyse epidemiological data was assessed using a subset of the INTERSALT samples from populations in

Shanxi (China) and Honolulu (USA; predominantly Japanese ethnicity). Multivariate modelling procedures were developed to remove analytical drift and to align the MS data prior to modelling metabolic differences between the two populations (Jonsson *et al.* 2005). Using this methodology, the two populations were well differentiated and metabolic differences were profiled (Figure 13.4). This strategy is now being extended to further characterize population differences in blood pressure.

INTERMAP is a basic epidemiological investigation designed to clarify and advance the findings from the INTERSALT study. INTERMAP investigated the role of multiple dietary factors (macromicronutrients) and their urinary metabolites in the aetiology of unfavourable BP levels prevailing for a majority of middle-aged and older adults in most populations around the world. INTERMAP has a sample size of 4680 men and women aged 40–59 from 17 diverse population samples but, in contrast to INTERSALT, data collection took place in only four countries, Japan, the People's Republic of China, the UK and the USA (Stamler *et al.* 2003).

Published findings from this study include quantification of the extent that multiple dietary factors help explain the strong inverse relation of education to BP in US participants (Stamler *et al.* 2003) and the higher BP of northern than southern Chinese (Zhao *et al.* 2004). ^{1}H-NMR spectroscopy has been used to profile the urine specimens from male and female participants from three of the population samples, in China, Japan and the USA. The aim of this initial study, aside from the QC investigation described above (Figure 13.4), was to examine for population differences in INTERMAP and to determine whether the metabonomic approach was a suitable method with which to investigate this. The NMR data, together with extensive databases of nutrients extracted from 4×24 h multiple-pass dietary recall questionnaires, eight measurements of systolic BP and diastolic BP, and anthropometric and sociodemographic information for each participant, provide a unique resource for elucidation of metabolic patterns. Multivariate analysis of the spectra for the selected samples revealed discrimination between the three populations and enabled identification of major metabolites responsible for this (Figure 13.5). Elevated levels of trimethylamine-*N*-oxide (TMAO)

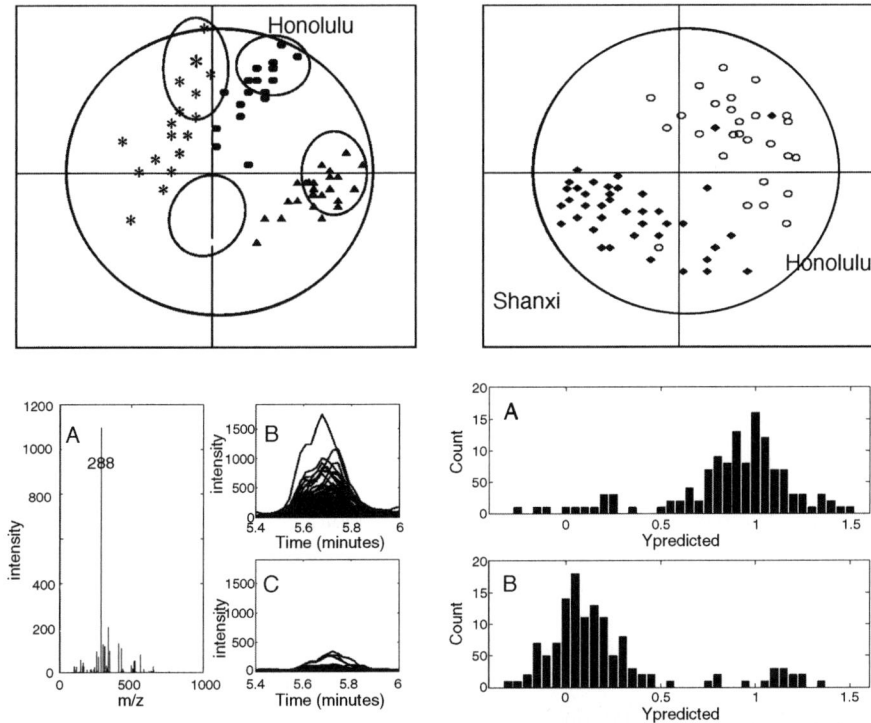

Figure 13.4 LC–MS analysis of urine samples obtained from the Honolulu and Shanxi populations, collected as part of the INTERMAP study (Jonsson *et al.* 2005). The top left-hand PCA scores plot shows each LC–MS measurement for a given urine sample as a single coordinate. The plot shows four clusters which relate to four separate well plates (the samples are coded according to well plate), revealing differences that were attributable to instrumental drift. In the top right-hand plot, the samples are coded according to population and, following analysis using an alternating least squares algorithm to remove these analytical differences, a separation between the Shanxi and Honolulu populations could be observed (top right). The bottom left-hand plot focuses on the identified discriminatory variable, $m/z = 288$ in the time window 250 (5.72 min). (A) Mass spectrum for the differentiating metabolite having the highest intensity for the Honolulu population, as can be observed from the ion chromatogram distributions for the Honolulu (B) and Shanxi (C) samples. The bottom right-hand plot is a histogram representation for the prediction of class for the 253 independent test samples, which gave a correct classification in 221 cases (87.4%). By examining the distribution of prediction results for Honolulu (A) and Shanxi (B), the high predictive ability of the model is revealed. *x* axis, predicted values; *y* axis, count in each interval (interval = 0.05)

were observed in the Japanese populations, consistent with their high dietary intake of fish. The urinary metabolic profiles were also indicative of differences in muscle mass and gut microflora across the populations (Dumas *et al.* 2006).

Metabonomic analysis of the urine samples from two cross-sectional epidemiological studies is beginning to enable the identification of characteristic population-based metabolic phenotypes or metabotypes, defined as 'a probabilistic multiparametric description of an organism in a given physiological state based on analysis of its cell types, biofluids

or tissues' (Gavaghan *et al.* 2000). These population differences in human metabotypes are likely to reflect a combination of genetic and environmental factors, such as diet and life style, that contribute to the observed population diversity. This is an area rich in opportunity for making finer biological distinctions among such phenotypes/metabotypes, and is a challenge for metabonomics to enhance the scientific information relevant to aetiology that can be obtained from epidemiological studies.

The epidemiological studies discussed so far investigated the contribution of dietary and life

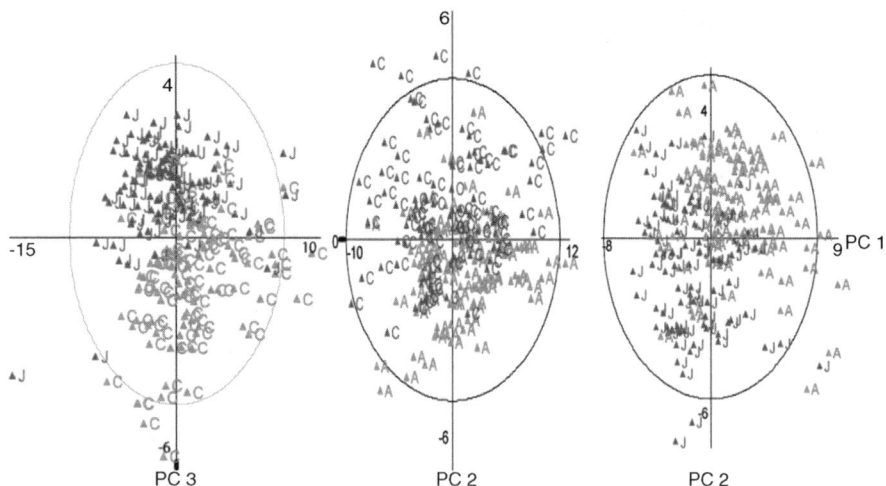

Figure 13.5 Large-scale metabonomic screening of human populations. PCA otherwise scores plot derived from the ^{1}H-NMR spectra of urine samples obtained from INTERMAP American, Japanese and Chinese samples adapted from Teague *et al.*

style factors in the development of high blood pressure, a major risk factor for cardiovascular disease. Because of the ageing of the population in the developed world, the burden of cardiovascular disease is set to rise further (Lakatta and Levy 2003). A longitudinal study investigating high functioning in older adults (the MacArthur Study of Successful Aging) has identified, in this population, biomarker combinations predictive of death, using a recursive portioning technique (Gruenewald *et al.* 2006). However. this study focused on a selected panel of biomarkers; it would be of great interest to further such an investigation using a metabonomic approach to characterize ageing.

13.4. NEURODEGENERATIVE DISORDERS

The disease area perhaps most commonly profiled by NMR technology has been neurodegeneration. Magnetic resonance imaging (MRI) has been routinely applied as an imaging tool to generate anatomical information, via the measurement of water molecules in different physical environments, whilst magnetic resonance spectroscopy (MRS), which provides chemical structural information from *in vivo* tissues, is often applied in conjunction with imaging. MRI

has been established as one of the most important clinical diagnostic tools for diseases such as multiple sclerosis (Frisoni and Fillppi 2005) and Alzheimer's disease (Frisoni and Fillppi 2005), Creutzfeldt–Jakob syndrome (Kleiner-Fisman 2004), Huntington's disease (Muhlau *et al.* 2006) and others. Although MRI is an extremely important clinical tool, MRS and high-resolution NMR spectroscopy of biofluids and tissues may prove more useful in the context of epidemiology, since they generate multivariate information on a wide range of molecules. Yet few epidemiological studies to date have used metabolic profiling methods in the neurodegenerative area. Much work has been done in either genetically modified animal models of neurodegenerative disease (Choi *et al.* 2003; Tsang *et al.* 2006) or by using various neuroactive compounds in toxicological models (Podell M *et al.* 2003; Boska *et al.* 2005). However, the validity of extrapolation from these animal models to man is questionable. Clinical studies have thus far been on a small scale, but preliminary evidence suggests that metabolic profiling methods may be useful for detecting biomarkers of disease progression in Huntington's disease (Underwood *et al.* 2006) and Pick's disease (Cheng *et al.* 1997). Recently a study investigating blood plasma samples from schizophrenic patients showed potential in monitoring response to therapeutic intervention (Holmes *et al.* 2006).

13.5. THE WAY FORWARD

The revolution in '-omics' technology promises substantial capability in achieving a global systems view of physiological and pathological processes and in understanding gene–environment interactions and their role in disease aetiology. Metabonomics, in particular, has a proven ability to generate large metabolic datasets in a reasonably rapid time frame, and is therefore a suitable tool for population screening. The analytical methodologies employed are generally robust and cost-effective. However, careful attention is required to prevent issues such as instrumental variability and the biochemical stability of the biofluid samples from compromising the integrity of population studies, particularly where the sample collection period extends over a number of months or even years. Consideration of experimental design, sampling strategies and the inclusion of quality control measures can help to ensure that the conclusions drawn from epidemiological studies are robust and valid.

One of the most powerful strategies for understanding the impact of gene–environment interactions in disease aetiology lies in combining data from the different '-omics' platforms. Each of the '-omics' technologies has unique data structure and is associated with its own technical limitations and constraints. However, several studies have shown that it is possible to integrate data at the gene, protein and metabolite levels in a meaningful way. Several examples in the literature describe bioinformatic approaches to mining and interpreting multi-omic data, but most operate by extracting lists of the most discriminatory variables and drawing inferences about potential biological significance from the response at the different levels (Griffin *et al.* 2004; Kleno *et al.* 2004; Craig *et al.* 2006; Rantalainen *et al.* 2006). Other modelling strategies seek to combine data from the different '-omics' platforms by exploring covariation across the different data matrices (Rantalainen *et al.* 2006). Modelling strategies

will undoubtedly play a key role in the analysis of epidemiological cohorts and, together with the advances in freely available databases for mapping metabolic responses onto molecular networks, will increase the potential for generating and subsequently testing hypotheses concerning the role of genetic or environmental factors in disease development and risk.

Acknowledgements

The INTERMAP study is supported by Grant No. 2-RO1-HL50490, US NHLBI and by national and local agencies in the four countries. The INTERMAP Metabonomics study was supported by Grant No. 5-RO1-HL71950-2 US NHLBI.

References

Ackerstaff E, Glunde K, Bhujwalla ZM. 2003. Choline phospholipid metabolism: a target in cancer cells? *J Cell Biochem* **90**: 525–533.

Baggerly KA, Morris JS, Edmonson SR *et al.* 2005. Signal in noise: evaluating reported reproducibility of serum proteomic tests for ovarian cancer. *J Natl Cancer Inst* **97**: 307–309.

Barshop BA. 2004. Metabolomic approaches to mitochondrial disease: correlation of urine organic acids. *Mitochondrion* **4**: 521–527.

Beckonert O, MonnerjahnJ, Bonk U *et al.* 2003. Visualizing metabolic changes in breast cancer tissue using ^{1}H-NMR spectroscopy and self organizing maps. *NMR Biomed* **16**: 1–11.

Beger RD, Schnackenberg LK, Holland RD *et al.* 2006. Metabonomic models of human pancreatic cancer using 1D proton NMR spectra of lipids in plasma. *Metabolomics* **2**: 125–134.

Bijlsma S, Bobeldijk I, Verheij ER *et al.* 2006. Large scale metabolomics studies: a strategy for data (pre-) processing and validation. *Anal Chem* **78**: 567–574.

Bonnier F, Rubin S, Venteo L *et al.* 2006. *In vitro* analysis of normal and aneurismal human ascending aortic tissues using FT-IR microspectroscopy. *Biochim Biophys Acta* **1758**: 968–973.

Boska MD, Lewis TB, Destache CJ *et al.* 2005. Quantitative ^{1}H magnetic resonance spectroscopic imaging determines therapeutic immunization efficacy in an animal model of Parkinson's disease. *J Neurosci* **25**: 1691–1700.

Brindle JT, Antti H, Holmes E *et al.* 2002. Rapid and noninvasive diagnosis of the presence and severity of coronary heart disease using ^{1}H-NMR-based metabo-nomics. *Nat Med* **8**: 1439–1444.

Cavaggioni A, Mucignat-Caretta C, Redaelli M *et al.* 2006. The scent of urine spots of male mice, *Mus musculus*: changes in chemical composition over time. *Rapid Commun Mass Spectrum* **20**: 3741–3746.

Cheng LL, Ma MJ, Becerra L *et al.* 1997. Quantitative neuropathology by high resolution magic angle spinning proton magnetic resonance spectrometry. *Proc Natl Acad Sci USA* **94**: 6408–6413.

Choi IY, Lee SP, Guilfoyle DN *et al.* 2003. *In vivo* NMR studies of neurodegenerative diseases in transgenic and rodent models. *Neurochem Res* **28**: 987–1001.

Clayton TA, Lindon JC, Cloarec O *et al.* 2006. Pharmaco-metabonomic phenotyping and personalized drug treatment. *Nature* **440**: 1073–1075.

Cloarec O, Dumas ME, Craig A *et al.* 2005. Statistical total correlation spectroscopy: an exploratory approach for latent biomarker identification from metabolic ^{1}H-NMR datasets. *Anal Chem* **77**: 1282–1289.

Coen M, Reupp SU, Lindon JC *et al.* 2004. Integrated application of transcriptomics and metabonomics yields new insight into the toxicity due to para-cetamol inn the mouse. *J Pharm Biomed Anal* **35**: 93–105.

Craig A, Sidaway J, Holmes E *et al.* 2006. Systems toxicology: integrated genomic, proteomic and metabonomic analysis of methapyrilene induced heptotoxicity in the rat. *J Proteome Res* **5**: 1586–1601.

Crockford DJ, Holmes E, Lindon JC *et al.* 2006. Statistical heteropectroscopy, an approach integrated analysis of NMR and UPLC–MS datasets application in metabonomic toxicology studies. *Anal Chem* **78**: 363–371.

Dumas ME, Maibaum EC, Teague C *et al.* 2006. Assessment of analytical reproducibility of ^{1}H-NMR spectroscopy based metabonomics for large scale epidemiological research: the INTERMAP study. *Anal Chem* **78**: 2199–2208.

Dyer A, Elliott P, Chee D *et al.* 1997. Urinary biochemical markers of dietary intake in the INTERSALT study. *Am J Clin Nutr* **65**: 1246–1235S.

Dyer AR, Elliott P, Shipley M *et al.* 1994. Body mass index and associations of sodium and potassium with blood pressure in Inetrsalt. *Hypertension* **23**: 729–736.

Eriksson L, Antti H, Gottfries J *et al.* 2004. Using chemometrics for navigating in the large datasets of genomics, proteomics and metabonomics (gpm). *Anal Bioanal Chem* 380: 419–429.

Eriksson L, Johansson E, Kettanah-Wold N *et al.* 1999. Discriminant analysis. In *Multi- and Megavariate Data Analysis*. Umetrics Acadame: Umea, Sweden; 193–914.

Fiehn O, Kopka J, Dormann P *et al.* 2000. Metabolite profiling for plant functional genomics. *Nat Biotechnol* 18: 1157–1161.

Fiehn O. 2002. Metabolomics-the link between genotypes and phenotypes. *Plant Mol Biol* **48**: 155–171.

Fossel ET, Carr JM, McDonagh J. 1986. Detection of malignant tumours. Water-suppressed proton nuclear magnetic resonance spectroscopy of plasma. *N Engl J Med* **315**: 1369–1376.

Foxall PJ, Mellothe GJ, Bonding MR *et al.* 1993. NMR spactroscopy as a noval approach to the monitaring of renal transplant function. *Kidney Int* **43**: 234–245.

Frisoni GB, Fillppi M. 2005. Multiple sclerosis and Alzheimer disease through the looking glass of MR imaging. *Am J Neuroradiol* **26**: 2488–2491.

Gartland KP, Sanins SM, Nicholson JK *et al.* 1990. Pattern recognition analysis of high resolution ^{1}H-NMR spectra of urine. A non-linear mapping approach to the classification of toxicological data. *NMR Biomed* **3**: 166–172.

Gavaghan CL, Holmes E, Lenz E *et al.* 2000. An NMR-based metabonomic approach to investigate the biochemical consequences of genetic strain differences: application to the C57BL10J and Alpk:ApfCD mouse. *FEBS Lett* **484**: 169–174.

Griffin JL, Bonney SA, Mann C *et al.* 2004. An integrated reverse functional genomic and metabolic approach to understanding orotic acid-induced fatty acid liver. *Physiol Genom* **17**: 140–149.

Griffiths JR, McSheehy PMJ, Robinson SP *et al.* 2002. Metabolic changes detected by *in vivo* magnetic resonance studies of *HEPA-1* wild-type tumours and tumours deficient in hypoxia-inducible factor-1β (HIF-1β): evidence of an anabolic role for the HIF-1 pathway. *Cancer Res* **62**: 688–695.

Gruenewald TL, Seeman TE, Ryff CD *et al.* 2006. Combinations of biomarkers predictive of later life mortality. *Proc Natl Acad Sci USA* **103**: 14158–14163.

Holmes E, Loo RL, Cloorec O *et al.* 2007. Detection of urinary drug metabolite (xenometabolome) signatures in molecular epidemiology studies via statistical total correlation (NMR) spectrosropy. *Anal Cham* **19**: 2629–2640.

Holmes E, Nicholson JK, Tranter G. 2001. Metabonomic characterization of genetic variations in toxicological and metabolic responses using probabilistic neural networks. *Chem Res Toxicol* **14**: 182–191.

Holmes E, Tsang T, Huang JT *et al.* 2006. Metabolic profiling of CSF: evidence that early intervention may impact on disease progression and outcome in schizophrenia. *PLoS Med* **3**: e327.

Ideker T, Thorsson V, Ranish JA *et al.* 2001. Integrated genomic and proteomic analyses of a systematically perturbed metabolic network. *Science* **292**: 929–934.

INTERSALT co-operative research group. 1988. INTERSALT: an international study of electrolyte excretion and blood pressure. Results for 24 hour urinary sodium and potassium excretion. *Br Med J* **297**: 319–328.

Jaakkola P, Mole DR, Tian YM *et al*. 2001. Targeting of HIF-α to the von Hippel–Lindau ubiquitylation complex by O_2-regulated prolyl hydroxylation. *Science* **292**: 468–472.

Jeong H, Tombor B, Albert R *et al*. 2000. The large-scale organization of metabolic networks. *Nature* **407**: 651–654.

Jonsson P, Bruce SJ, Moritz T *et al*. 2005. Extraction, interpretation and validation of information for comparing samples in metabolic LC/MS datasets. *Analyst* **130**: 701–707.

Jonsson P, Bruce SJ, Moritz T *et al*. 2005. Extraction, interpretation and validation of information comparing samples in metabolic LC–MS datasets. *Analyst* **130**: 701–707.

Keun HC, Ebbels TM, Antti H *et al*. 2002. Analytical reproducibility in ¹H-NMR-based metabonomic urinalysis. *Chem Res Toxicol* **15**: 1380–1386.

Kleiner-Fisman G, Bergeron C, Lang AE. 2004. Presentation of Creutzfeldt–Jakob disease as acute corticobasal degeneration syndrome. *Mov Disord* **19**: 948–949.

Kleno TG, Kiehr B, Baunsgaard D *et al*. 2004. Combination of 'omics' data to investigate the mechanisms of hydrazine induced hepatotoxicity in rats and to identify potential biomarkers. *Biomarkers* **9**: 116–138.

Lagali N, Burns K, Zimmerman D *et al*. 2006. Hemodialysis monitoring in whole blood using transmission and diffusion reflection spectroscopy: a pilot study. *J. Biomed Opt* **11**: 54003.

Lakatta EG, Levy D. 2003. Arterial and cardiac aging: major shareholders in cardiovascular disease enterprises. *Circulation* **107**: 139.

Lenz EM, Wilson ID. 2007. Analytical strategies in metabonomics. *J Proteome Res* **6**: 443–458.

Losralzo J, Kohone I. Barabsi AL. 2007. Human Diseases Clasification in The post-genomic area: A Complex systems approach to human pathobiology. *Molecular Systems Biology* **3**: 124.

Matoba S, Kang JG, Patino WD *et al*. 2006. p53 regulates mitochondrial respiration. *Science* **312**: 1650–1653.

Mayr M, Madhu B, Xu QB. 2007. Proteomic and metabolomics combined in cardiovascular research. *Trends Cardiovasc Med* **17**: 43–48.

Metzger GJ, Patel M, Hu X. 1996. Application of genetic algorithms to spectral quantification. *J Magn Reson B* **110**: 316–320.

Moolenaar SH, Engelke UF, Wevers RA. 2003. Proton nuclear magnetic resonance spectrometry of body fluids in the field of inborn errors of metabolism. *Ann Clin Biochem* **40**: 16–24.

Muhlau M, Weindl A, Wohlschlager AM *et al*. 2007. Voxel-based morphometry indicates relative preservation of the limbic prefrontal cortex in early Huntington disease. *J Neural Transm* **114**: 367–372.

Nebert DW, Vessell ES. 2006. Can personalized drug therapy be achieved? A closer look at pharmacometabonomics. *Trends Pharmacol Sci* **27**: 580–586.

Neild GH, Foxall PJ, Lindon JC *et al*. 1997. Uroscopy in the 21st century: high field NMR spectroscopy. *Nephrol Dial Transplant* **12**: 404–417.

Nicholls AW, Holmes E, Lindon JC *et al*. 2001. Metabonomic investigation into hydrazine toxicity in the rat. *Chem Res Toxicol* **14**: 975–987.

Nicholson JK, Lindon JC, Holmes E. 1999. 'Metabonomics': understanding the metabolic responses of living systems to pathophysiological stimuli via multivariate statistical analysis of biological NMR spectroscopic data. *Xenobiotica* **29**: 1181–1189.

Nicholson JK, O'Flym MP, Sodier PJ, Macleod AF, Jool SM, Sonkson PH. 1989. Proton-nuclear-magnetic-resonance studies of serum, plasma and urine from fastine normal and diabeke subjects. Biochem J **217**: 365–375.

Nicholson JK, Sadler PJ, Bales JR *et al*. 1984a. Monitoring metabolic disease by proton NMR of urine. *Lancet* **2**: 751–752.

Odunsi K, Wollman RM, Ambrosone CB *et al*. 2005. Detection of epithelial ovarian cancer using ¹H-NMR-based metabonomics. *Int J Cancer* **113**: 782–788.

Okunieff P, Zietman A, Khan J *et al*. 1990. Lack of efficacy of water-suppressed proton nuclear magnetic resonance spectroscopy of plasma for the detection of malignant tumours. *N Engl J Med* **322**: 953–958.

Oliveira AP, Bitar RA, Silveira L *et al*. 2006. Near infrared Raman spectroscopy for oral carcinoma diagnosis. *Photomed Laser Surg* **24**: 348–353.

Paik S, Shak S, Tang G *et al*. 2004. A multigene assay to predict recurrence of Tamoxifen-treated, node-negative breast cancer. *N Engl J Med* **351**: 2817–2826.

Plumb R, Castro-Perez J, Granger J *et al*. 2004. Ultra-performance liquid chromatography coupled to quadrupole-orthogonal time of flight spectrometry. *Rapid Commun Mass Spectrom* **18**: 2331–2337.

Plumb RS, Stumpf CL, Granger JH *et al*. 2003. Use of liquid chromatography/time-of-flight mass spectrometry and multivariate statistical analysis shows promise for the detection of drug metabolites in biological fluids. *Rapid Commun Mass Spectrom* **17**: 2632–2638.

Podell M, Hadjiconstantinou M, Smith MA *et al.* 2003. Proton magnetic resonance imaging and spectroscopy identify metabolic changes in the striatum in the MPTP feline model of Parkinsonism. *Exp Neurol* **179**: 159–166.

Rantalainen M, Cloarec O, Beckonert O *et al.* 2006. Statistically integrated metabonomic–proteomic studies on human prostate cancer xenograft model in mice. *J Proteome Res* **5**: 2642–2655.

Rarnanathan A, Wang C, Schreiber SL. 2005. Perturbational profiling of a cell-line model of tumourigenesis by using metabolic measurements. *Proc Natl Acad Sci USA* **102**: 5992–5997.

Razzi S, Ramadan Z, Martin FP. 2007a. Human metabolic phenotypas in directly to specific discovery preferances in healthy individuals. *J Proteome Res* **6**: 4469–4477.

Razzi S, Ramadan Z, Fay LB *et al.* 2007b. Nuhitional metabonouics: applications and parspactives. *J Proteome Res* **6**: 513–525.

Reily MD, Baker JD, Roberston DG. 2007. Metabonomics in pharmaceutical discovery and development. *J Proteome Res* **6**: 526–539.

Riboli E, Hunt KJ, Slimani N *et al.* 2002. European Prospective Investigation into Cancer and Nutrition (EPIC): study population and data collection. *Publ Health Nutr* **5**: 1113–1124.

Robertson DG, Reily MD, Sigler RE *et al.* 2000. Metabonomics: evaluation of nuclear magnetic resonance (NMR) and pattern recognition technology for rapid *in vivo* screening of liver and kidney toxicants. *Toxicol Sci* **57**: 326–337.

Sabatine MS, Liu E, Morrow DA *et al.* 2005. Metabolomic identification of novel biomarkers of myocardial ischemia. *Circulation* **112**: 3868–3875.

Satmler J. 1997. The INTERSALT study: background, methods, findings and implications. *Am J Clin Nutr* **65**: 626–642S.

Shockcor J, Unger S, Wilson ID *et al.* 1996. Combined hyphenation of HPLC, NMR spectroscopy and ion-trap mass spectrometry (HPLC–NMR–MS) with application to the detection and characterization of xenobiotic and endogenous metabolites in human urine. *Anal Chem* **68**: 4431–4435.

Stamler J, Elliott P, Appel L *et al.* 2003. Higher blood pressure in middle-aged American adults with less education – role of multiple dietary factors: the INTERMAP study. *J Hum Hypertens* **9**: 655–775.

Stamler J, Elliott P, Dennis B *et al.* 2003. INTERMAP: background, aims, design, methods and descriptive statistics (non-dietary). *J Hum Hypertens* **17**: 591–608.

Stamler J, Elliott P, Kesteloot H *et al.* 1996. Inverse relation of dietary protein markers with blood pressure. Findings for 10 020 men and women in the Intersalt study. INTERSALT cooperative research group. INTERnational study of SALT and blood pressure. *Circulation* **94**: 1629–1634.

Stoyanova R, Nicholson JK, Lindon JC *et al.* 2004. Sample classification based on Bayesian spectral decomposition of metabonomic NMR datasets. *Anal Chem* **76**: 3666–3674.

Troy H, Chung YL, Manuel M *et al.* 2006. Metabolic profiling of hypoxia-inducible factor-1β-deficient and wild-type Hepa-1 cells: effects of hypoxia measured by ^{1}H magnetic resonance spectroscopy. *Metabolomics* **1**: 293–303.

Trygg J, Holmes E, Lundstedt T. 2007. Chemometrics in metabonomics. *J Proteome Res* **6**: 469–479.

Tsang TM, McGLoughlin G, Woodman B *et al.* 2006. Metabolic characterization of the R6/2 transgenic mouse model of Huntington's disease by high-resolution MAS ^{1}H-NMR spectroscopy. *J. Proteome Res* **5**: 483–492.

Ullsten S, Danielsson R, Backstrom D *et al.* 2006. Urine profiling using capillary electrophoresis-mass spectrometry and multivariate data anlaysis. *J Chromatogr A* **1117**: 87–93.

Underwood BR, Broadhurst D, Dunn WB *et al.* 2006. Huntington disease patients and transgenic mice have similar pro-catabolic serum metabolite profiles. *Brain* **129**: 877–886.

Wheatley DN. 2005. Arginine deprivation and metabolomics: important aspects of intermediary metabolism in relation to the differential sensitivity of normal and tumour cells. *Semin Cancer Biol* **15**: 247–253.

Willett WC. 1995. Diet, nutrition and avoidable cancer. *Environ Health Perspect* **103**: 165–170.

Yang J, Xu G, Hong Q *et al.* 2004. Discrimination of type 2 diabetic patients from healthy controls by using metabonomics method based on their serum fatty acid profiles. *J Chromatogr B Analyt Technol Biomed Life Sci* **813**: 53–58.

Zhao L, Stamler J, Yan LL *et al.* 2004. INTERMAP Research Group. Blood pressure differences between northern and southern Chinese: role of dietary factors: the International Study on Macronutrients and Blood Pressure. *Hypertension* **43**: 1332–1337.

Zomer S, Guillo C, Brereton RG *et al.* 2004. Toxicological classification of urine samples using pattern recognition techniques and capillary electrophoresis. *Anal Bioanal Chem* **378**: 2008–2020.

14

Univariate and Multivariate Data Analysis

Yu-Kang Tu and Mark S. Gilthorpe

Centre for Epidemiology and Biostatistics, University of Leeds, UK

14.1. INTRODUCTION

Overview

The objective of this chapter is to introduce *univariate* and *multivariate* methods from the viewpoint of parametric statistical modelling, to provide some insights to the nuances of modelling observational (non-randomized) data, as typically encountered in epidemiological studies. This chapter does not cover statistical analyses in general, neither does it address non-parametric methods, as there are already several good texts that cover generic statistics, including non-parametric methods (see Bland 2000; Norman and Streiner 2000; Dawson and Trapp 2001; Kirkwood and Sterne 2003). It is also impossible and impractical to discuss in one chapter all commonly-used univariate and multivariate statistical modelling methods relevant to the analysis of epidemiological data. Fortunately, excellent discussions of most methods can be found in standard statistical and epidemiological texts (as well as those previously cited, see also Jewell 2004).

We seek to present univariate and multivariate methods specifically from the perspective of *structural equation modelling* (SEM), with the aid of *path diagrams*, as it is hoped that the reader will become familiar with the use of path diagrams, which we would argue are essential in making good sense of epidemiological data. It is also hoped that the reader gains a good introductory insight to the utility of SEM – a family of modelling methodologies, useful for complex datasets where variables

are measured with error and may also exhibit non-linear inter-relationships with each other, as found within most epidemiological data.

Within a single chapter, we cannot cover entirely the enormous topic of SEM, although what is presented here, especially in the context of molecular epidemiology, should enable readers to realize the potential of SEM and therefore anticipate opportunities open to them with their own research data. Through recognizing the value of path diagrams and associated SEM methodology, it becomes possible for epidemiologists to formulate research hypotheses that accommodate explicit and/or implicit causal relationships within one's data. Acknowledging underlying assumptions of causality often contributes to a much clearer interpretation of results from the statistical analyses.

Throughout examples in this chapter, we use data from a study on the associations between body growth and blood aflatoxin levels in a group of 200 African children (Gong *et al.* 2004). We are very grateful to the contributors of this original study for making these data available.

Terminology and definitions

It is helpful to clarify the terminology surrounding the terms *univariate* and *multivariate*. Statistically speaking, a 'variate' is a variable that is measured or recorded with error. Within statistical models, the term *univariate* indicates there is only one variable that has been measured with error, and this is

Molecular Epidemiology of Chronic Diseases, Edited by C. P. Wild, P. Vineis, and S. Garte

referred to as the *dependent* variable; all other variables, usually called *independent* variables or 'covariates', are assumed to be measured without error. The term *multivariate* indicates that there are multiple variables associated with measurement error and in standard regression this would imply there are multiple dependent variables.

For instance, consider the regression model of a dataset for a group of children, where body weight (dependent variable) is regressed simultaneously on the children's age and aflatoxin blood levels (independent variables). The analysis is univariate because there is only one dependent variable in the model (weight). The analysis is also described as *multiple linear regression*, because there are multiple independent variables in the model (age and aflatoxin blood level). In this example, only weight is permitted to have measurement error; it is assumed that both age and blood levels of aflatoxin are measured precisely, i.e. error-free. These assumptions may not be realistic, although they are nevertheless the assumptions made by standard linear regression. Different statistical processes may adopt different assumptions, as for instance with correlation, which is a *bivariate* process, since both variables assessed for their association to each other are assumed to have measurement error.

To test associations between *multiple* dependent variables and *multiple* covariates, whilst also accommodating potential measurement error and non-linear inter-variable relationships, we need more advanced statistical methods than standard linear regression, such as multivariate analysis of variance, multilevel modelling (also known as random effects modelling or mixed effects modelling), or structural equation modelling. We focus on the latter two modelling methodologies in the remainder of this chapter.

Before we start examining the various modelling strategies, we first define a few statistical terms: *Variance* is one of several indices of variability used to characterize the dispersion of a set of variable values. To calculate the variance, it is necessary first to calculate the mean and then determine the amount that each value deviates from the mean, square this, and sum. The variance equals the average of the squared deviations from the mean,

i.e. 'variance' = 'standard deviation' squared. *Covariance* is the extent to which two variables vary together (co-vary). To calculate the covariance, it is necessary first to calculate the mean of both variables and then determine the amount that each variable value deviates from its respective mean, multiply these deviations together, and sum. *Correlation* is the strength and direction of a relationship between two variables. The Pearson correlation coefficient assesses the strength of a linear relation, whilst the Spearman (rank) correlation is a non-parametric measure of correlation. *Regression coefficient* is the estimated linear rate of change in the dependent variable (variate) as a function of change in the independent variable (covariate) in a linear regression model with only one independent variable (termed 'simple linear regression'). *Partial regression coefficient* is the estimated linear rate of change of the dependent variable as a function of change in one independent variable, conditional on all other independent variables in the model remaining constant. Finally, the formula for the Pearson correlation coefficient (r) between the two variables x and y is given by:

$$r = \frac{Cov(x, y)}{\sqrt{Var(x)Var(y)}} \qquad (1)$$

where $Cov(x,y)$ is the covariance of x and y, $Var(x)$ the variance of x, and $Var(y)$ the variance of y.

A priori model assumptions

Within statistical regression models, assumptions regarding measurement errors are implicit, although possibly quite often unrealistic, and this point is too frequently overlooked. Even modest measurement error in covariates can cause sizeable bias in the estimated relationship between the dependent variable and covariates (Greenwood *et al.* 2006). In molecular epidemiology (as within all epidemiology), if measurement error occurs and is not duly considered within the statistical modelling process, inferences from the model may be biased and therefore potentially erroneous. The extent of bias is context-specific and is generally unknown.

The strategy of multiple linear regression is to test only for *associations* between the dependent variable and covariates. Although the covariates are also sometimes referred to as 'explanatory variables', because we may infer that they 'explain' some of the variation in the outcome, in practice the only sure thing we can established from regression analysis is the degree of association; we cannot establish causality. Whilst researchers may wish to infer causality, additional extraneous information is required to justify this. Causality is too often inferred without such extraneous information.

It is also important to distinguish between *testing* preconceived hypotheses and *generating* them. When *testing* preconceived hypotheses, researchers formulate statistical models in which causal relationships between variables are explicitly defined (and such assumptions may then be represented in path diagrams – see later) and the researchers collect data to test the validity of the model. When *generating* hypotheses, researchers do not have a clear theory to formulate relationships amongst variables. Instead, researchers adopt a trial-and-error modelling strategy to look for statistically significant relationships, which would then require validation in independent datasets. Forward or backward stepwise regression analysis to select variables in multiple regression analysis is a typical example of hypothesis generation. However, it should be noted that the 'optimal' models identified by these methods, i.e. where the majority of the variation within dependent variable is accommodated, may not be replicable, and the variables selected may make little sense scientifically (Thompson 1995; Berk 2004). In general, stepwise procedures should not be relied upon; scientific exploration is better guided by *a priori* knowledge and understanding. In any event, all the statistical methods discussed in this chapter can perform both tasks of testing and generating hypotheses, and the difference between the two approaches lies in the *interpretation* of the results.

Initial data exploration

Before commencing complex data analysis, it is necessary to undertake simple descriptive and exploratory analyses to gain a basic understanding of one's data. For instance, histograms can be used to look at the distributions of outcome variables – whether they are normal, skewed or truncated – and scatter plots

used to observe the relationship between variables – whether they are linear or non-linear. After basic descriptive analyses, one can then examine the data against any number of *a priori* research questions or explore the data in order to generate hypotheses.

14.2. UNIVARIATE ANALYSIS

Simple linear regression

We now introduce the example dataset and test the relationship between body weight (in kg) and age (in months) at baseline examination. Figure 14.1 is the scatter plot which shows a linear relationship between the variables *Weight*1 and *Age*1. We may initially examine the relationship between the two variables using correlation (a bivariate process), or model their relationship using *simple linear regression* (a univariate process). For our example, the correlation coefficient between *Weight*1 and *Age*1 is 0.63, and is highly significant ($p < 0.001$). This is hardly surprising, as older children are expected to have a larger body size. The simple linear model for *Weight*1 regressed on *Age*1 is given as:

$$Weight1 = b_0 + b_1\,Age1 + e \qquad (2)$$

where *Weight*1 and *Age*1 are body weight and age at baseline examination and e the residual error

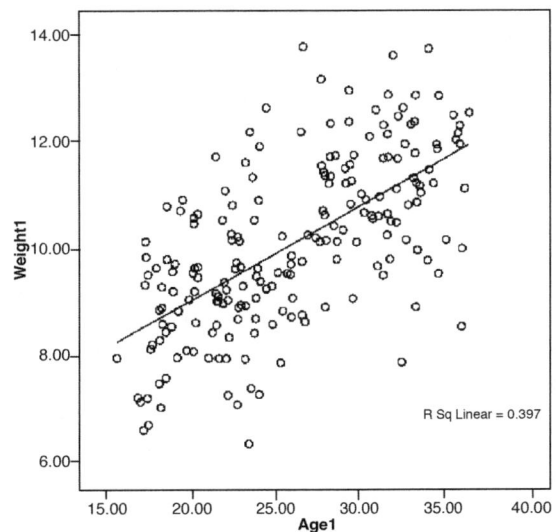

Figure 14.1 Scatter plot for the relationship between body weight (*Weight*1) and ages (*Age*1) at baseline examination for 200 children

term associated with *Weight*1 (note that *Age*1 is assumed to be measured without error). The regression coefficient b_1 is 0.175, indicating that children who were 1 month older than other children were on average 0.175 kg heavier. The square of the correlation coefficient, termed R^2 ('r-squared'), is 0.397 for equation 1, and indicates that almost 40% of the variation in the baseline body weight is 'explained' by the age of these children. That is, 40% of the variation in the dependent variable (*Weight*1) is accommodated in the regression model by including the covariate (*Age*1); this does not show that age is *causally* related to weight, which may or may not be true – such an inference may be made separately from, or in conjunction with, the information provided by the simple regression model. Reality is likely to be far more complex, and one should be cautious at making precise causal inferences.

Multiple linear regression

It is probably important to know the correlation between body weight and aflatoxin, as the research hypothesis is that high levels of aflatoxin would have a negative impact on the body growth of these children. As the distribution of aflatoxin levels was positively skewed, we generate a new variable *Lnafb*1, which is simply the natural logarithm transformation of the baseline aflatoxin blood levels. The scatter plot for *Weight*1 and *Lnafb*1 is shown in Figure 14.2, indicating a positive association with a correlation coefficient of 0.074 ($p = 0.300$). Although this positive correlation is not statistically significant, it is contradictory to expectation, as we hypothesize that aflatoxin derived from food would have a negative impact on the growth of these children, yet a (modest) positive association suggests otherwise.

One plausible explanation is what is termed 'confounding', and here we might consider the age of the child to be a confounder, since older children were both bigger and also tended to have higher levels of aflatoxin. To adjust for the confounding, we perform a *multiple* linear regression model (also described as a *multivariable* linear regression

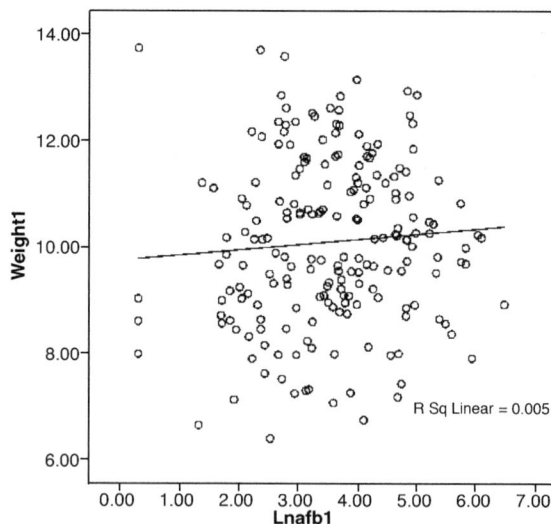

Figure 14.2 Scatter plot for the relationship between body weight (*Weight*1) and aflatoxin levels (*Lnafb*1) at baseline examination for 200 children

model, but not a *multivariate* model, as we still have only one variate, *Weight*1):

$$Weight1 = b_0 + b_1\, Age1 + b_2\, Lnafb1 + e \qquad (3)$$

The results show that b_1 is 0.181 [95% confidence interval (CI) 0.152, 0.215; $p < 0.001$) and b_2 is -0.137 (CI -0.288, 0.013; $p = 0.074$), i.e. after adjusting for their ages at baseline examination, children with higher levels of aflatoxin (on a log scale) tended to have smaller body sizes.

Path diagram

Simple linear regression

An alternative visual presentation of the regression analysis is to draw a path diagram. To illustrate what a path diagram is, we begin with a simple example of linear regression given by equation 2 (Figure 14.3). In a path diagram, observed variables such as *Weight*1 and *Age*1 are within squares, and latent (unobserved) variables, such as residual errors (e.g. e in equation 2), are within circles. Continuous latent variables are referred to as *factors*, whilst categorical latent variables are referred to as *classes*. Here we focus only on continuous latent variables.

The arrow from *Age*1 to *Weight*1 in Figure 14.3 implies that *Age*1 affects *Weight*1, whilst *Weight*1

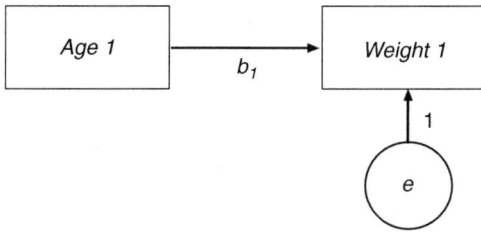

Figure 14.3 Path diagram of simple linear regression in equation 2. The regression weight b_1 for the path from *Age*1 to *Weight*1 in the path diagram is equivalent to the regression coefficient b_1 in equation 2

does not affect *Age*1. In the path diagram, this unidirectional relationship is assumed to imply causality, which is then inferred by the specified statistical regression model. In contrast, a double arrow connecting *Age*1 and *Weight*1 means that these two variables are correlated without specific direction of causality being implied. Whilst this assumption of causal relationships between variables is useful for formulating statistical models to be tested, it does not mean that the assumed causality must exist in reality. When there is no arrow (single or double) between two variables, this implies that they are assumed to be independent, i.e. the underlying (true) correlation between them in the population is assumed to be zero for the specified model – this is not to say that they may not be correlated in the sample data. For instance, *Weight*1 and *e* are assumed to be uncorrelated in all instances; this is in fact one of the assumptions underlying linear regression analysis.

It is important to be aware of the assumptions made within path diagrams, as these diagrams may then inform the interpretation of linear regression models, i.e. an association in one's dataset does not dictate the path diagram, rather the path diagram aids interpretation of the linear regression model. Therefore, it is important to note that the association between *Age*1 and *Weight*1 in the regression model (either equation 2 or equation 3) does not itself imply causality in either direction, and neither does an association in the study sample imply an association in the population that has been sampled.

Finally, it is noted that there is no intercept b_0 in Figure 14.3, because the path diagram only seeks to represent the relationship between the 'change' in *Weight*1 and the 'change' in *Age*1. From a statistical viewpoint, b_0 is the estimated body weight when *Age*1 is zero, i.e. the birth weight, but zero is not in the range of the data (Figure 14.1 shows that the range of these children's ages is 15–37 months). The interpretation of b_0 is therefore not meaningful. Nevertheless, we can still estimate the intercept or means of variables in a path diagram using *structural equation modelling* (SEM), an example of which is discussed in detail in the section on the latent growth curve models.

In general, SEM software such as Amos (Arbuckle 2005), EQS (Bentler 2006) and LISREL (Jöreskog and Sörbom 2006) provides a graphical interface for drawing path diagrams, which are then evaluated as statistical models. The results of each model, such as regression weights for the path (regression coefficients in linear regression, or path coefficients in a path analysis, or factor loadings in a factor analysis), as well as means and variances, are usually displayed on the computer screen. To illustrate this point, Figure 14.4 is the graphical output from Amos (version 6.0) for the path diagram for equation 2.

In the results displayed in Figure 14.4, the mean age at baseline is 26.14 months, with an estimated variance (standard deviation (SD) squared) of 31.65 (i.e. SD of 5.64 months). The intercept for *Weight*1 is 5.54, which is b_0 in equation 2. The variance of the residual error term (*e*) is 1.47.

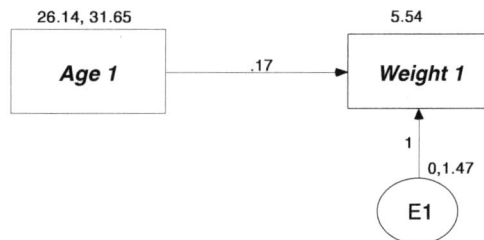

Figure 14.4 Graphical output from Amos for simple linear regression analysis given by equation 2

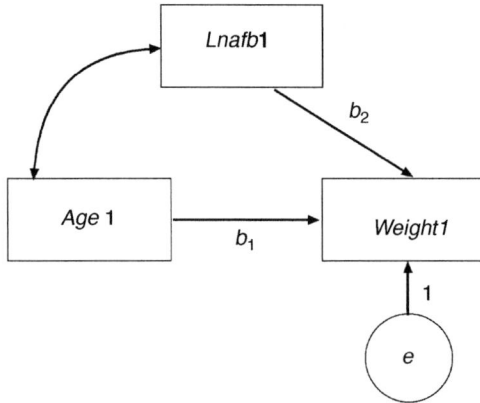

Figure 14.5. Path diagram of multiple linear regression in equation 3. The regression weights b_1 and b_2 for the paths from $Age1$ to $Weight1$ and from $Lnafb1$ to $Weight1$ in the path diagram, respectively, are equivalent to those in equation 2

Multiple linear regression

The path diagram for equation 3 is shown in Figure 14.5. The difference between Figures 14.5 and 14.3 is that $Lnafb1$ is now included in the diagram as an observed variable. It should be noted that in Figure 14.5 we do not specify the direction of the relationship between $Age1$ and $Lnafb1$, but instead we assume that they are correlated and that they both affect $Weight1$. To illustrate this as a path diagram, we examine the graphical output from Amos in Figure 14.6, where the results are identical to those from the multiple regression analysis. The intercept is 5.83, and the mean of $Lnafb1$ is

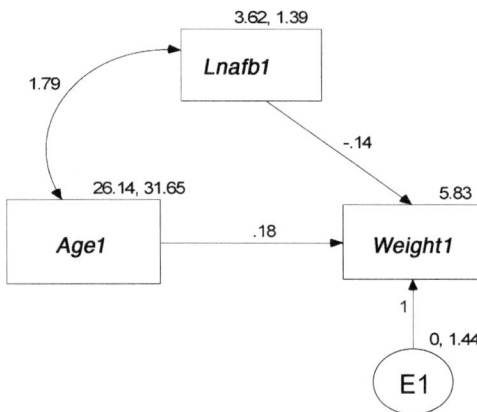

Figure 14.6 Graphical output from Amos for simple linear regression analysis given by equation 3

3.62 (i.e. natural logarithm of the geometric mean of the originally untransformed aflatoxin levels). The covariance between $Age1$ and $Lnafb1$ is 1.79, and hence the correlation coefficient is obtained using equation 1, $1.79/\sqrt{31.65*1.39} = 0.27$, which is indeed the observed bivariate Pearson correlation between these two variables.

14.3. GENERALIZED LINEAR MODELS

The regression models in the previous section are known as generalized linear models. The dependent variables, such as $Weight1$, are continuous variables and assumed to follow a Normal distribution for constant values of all potential covariates. Strictly speaking, it is the residuals of $Weight1$ that are Normally distributed, i.e. for a given age and aflatoxin blood level (and any other variables that are potentially related to body weight) the observed values of weight are expected to follow a Normal distribution. This does not imply that all population (or sample) values of weight are Normally distributed – indeed, this would be most unlikely, since weight increases with age. Often, when it is stated that the dependent variable is 'Normally distributed', this is also contingent on the additional clause that Normality occurs for each fixed value of all related factors (in this instance, age and aflatoxin blood levels). This is identical to suggesting that the regression model residuals (e in equation 3) are Normally distributed (i.e. conditional on the covariates $Age1$ and $Lnafb1$). The general form of a linear model for dependent variable y, and independent variables $x_1 \ldots x_p (1 \ldots p)$ is:

$$y = b_0 + b_1 x_1 + b_2 x_2 + \cdots + b_p x_p^{+e} \varepsilon \qquad (4)$$

where it is assumed that the residuals e have mean zero and are Normally distributed.

When the outcome variable y is binary (e.g. healthy vs. diabetic) or counts (the number of childhood leukaemia cases in a given period), it is not appropriate to use linear regression to analyse these data. Instead, we model the transformation of y, rather than y itself, via a link function. For instance, the link function for continuous variables is $f(y) = y$, which is simply the *identity* function, i.e. there is no transformation at all. For binary variables, as within case–control studies, we often use the *logit* link function, as this permits us to model the log-odds of the binary outcome, and the regression model is then known as logistic regression. Other link functions exist for binary dependent variables, such as the *probit* link function, but these tend to have less appeal to biomedical statisticians because they are not as interpretable as the more common *logit* link function. The odds ratio is directly interpretable and is often used to estimate relative risk. For outcomes recorded as counts, we use the *natural logarithm* transformation to model rates, and the regression model is known as Poisson regression. More details about generalized linear models can be found in excellent texts (e.g. Dupont 2002; Kirkwood and Sterne 2003; Jewell 2004). We do not cover these types of model in detail here.

The interpretation of models for binary and count data remains similar to that of standard linear models for continuous outcomes, except that the effects of covariates operate in an additive manner on the transformed scale of y, so their effects on y directly will be multiplicative, i.e. no longer additive and hence no longer linear (strictly, they are linear models with a non-identify or non-linear link function). The implication of this becomes important when examining interactions, e.g. gene–environment interactions, since the interpretation of an interaction is scale-dependent. For instance, switching between the continuous response model and a related binary response model (where the continuous outcome is dichotomized) changes the estimated effect size (and hence formal significance level) of the coefficient of the interaction term. Hence, the use of interactions within linear models, and especially their interpretation, should be pursued cautiously, with consideration to the model context and the scales of the variables for which inferences are to be made; note that the transformation of variables in a model can generate or nullify significant statistical interactions, irrespective of the underlying biological or physical interactions the model might seek to represent.

14.4. MULTIVARIATE METHODS

In this section we introduce two multivariate methods to analyse longitudinal epidemiological data: *multilevel modelling* (MLM) and *structural equation modelling* (SEM). We first start with multilevel modelling, and then show how multilevel models can be specified within the framework of structural equation modelling. This comparison of modelling methodologies – their similarities and differences – is not merely of technical interest, rather of practical value, since it reveals how flexibly one can be in modelling complex datasets, using various analytical strategies. Readers are thus advised to consider carefully their own research data, their research context, and their associated research questions, before settling on a modelling strategy. For those familiar with either MLM or SEM but not both, the comparison of these two methods will aid comprehension of the lesser-known method.

Multilevel modelling (MLM)

This approach is also known as mixed effects modelling, random effects modelling, or hierarchical linear modelling (Goldstein 1995; Raudenbush and Bryk 2002; Hox 2002; Singer and Willett 2003; West *et al.* 2007). MLM has been used in epidemiological research mainly to deal with hierarchical data structure, e.g. patients nested within doctors' or within geographical areas, or both; complex cross-classified structures may also be modelled in this analytical framework (Leyland and Boddy 1998), but these types of model are not discussed here. Another application of MLM is to analyse longitudinal data, treating the repeated measurements as the lowest level (Gilthorpe *et al.* 2003).

For our example on the relationship between growth in body weights and aflatoxin levels, three measurements of body weight are considered: one made at baseline, then again at 3 and 8 months following baseline. These repeated measurements form the lowest level (level 1) of the multilevel hierarchy, nested within children at the highest level (level 2). Table 14.1 summarizes the variables that are used for all subsequent analyses using multivariate methods.

The basic multilevel model (MLM-1) for the growth trajectory of these children is given as:

$$Weight_{ij} = \beta_{0ij} + \beta_{1j} Month_{ij} \qquad (5)$$

where the $Weight_{ij}$ is body weight measured on occasion i (1, 2, or 3) for subject j, and $Month_{ij}$ is a

Table 14.1. Summary of observed variables used for multivariable analysis in a group of 200 African children

Variable	n	Minimum	Maximum	Mean	SD
*Weight*1	200	6.40	13.76	10.1056	1.56425
*Weight*2	195	7.10	14.41	10.5522	1.48574
*Weight*3	193	6.00	15.04	11.2033	1.64133
*Lnafb*1	197	0.41	6.53	3.6213	1.18169
*Lnafb*2	194	0.41	6.61	3.6562	1.07174
*Lnafb*3	193	1.54	7.36	4.4631	1.22568
*Age*1	200	15.74	36.67	26.1383	5.64024

continuous variable for time (0 at baseline, three on occasion 2, and eight for occasion 3); and multilevel regression coefficients, β_{0ij} and β_{1j}, are given by:

$$\beta_{0ij} = \beta_0 + u_{0j} + e_{0ij}$$
$$\beta_{1j} = \beta_1 + u_{1j} \tag{6}$$

where β_0 is the overall mean intercept (at month zero); β_1 is the overall mean gradient of the growth trajectory (for all children); e_{0ij} is the residual error term at level 1, representing the difference between observed and predicted weight on each occasion (for each child); u_{0j} and u_{1j} are residuals at level 2, representing, respectively, the intercept and slope differences between observed mean weight trajectories for each child and the overall mean weight trajectory for all children. Combined, equations 5 and 6 define a multilevel model referred to as a *random coefficient model*, since the regression coefficients exhibit random variation about their mean (across occasions and across children). More detailed non-technical explanations of MLM can be found elsewhere (Gilthorpe *et al.* 2000; Twisk 2003, 2006).

It should be noted that in this model (MLM-1), the growth in body weight is assumed to be linear throughout the 8 month observation period and hence an individual linear growth trajectory is estimated for each child separately. Figure 14.7 shows the observed and estimated growth trajectories of these 200 children. Figure 14.8 shows the results

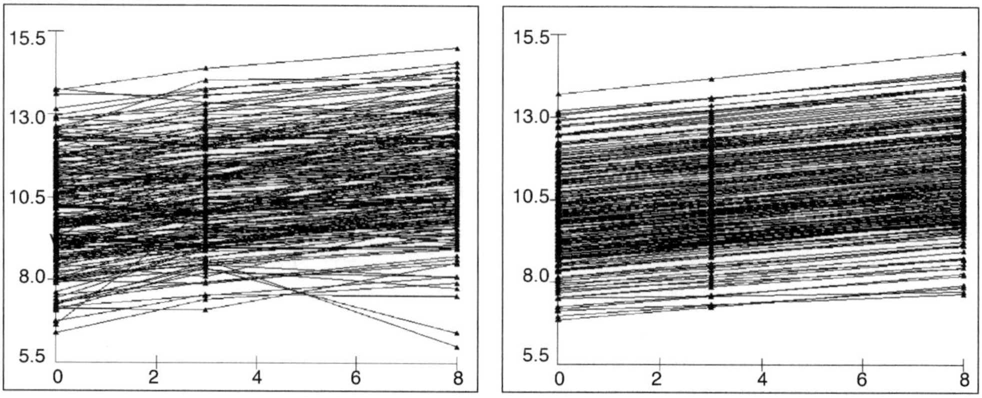

Figure 14.7 Observed (left panel) and estimated (right panel) body weight growth trajectories in 200 African children over an 8 month observation period

$$\text{Weight}_{ij} \sim N(XB, \Omega)$$
$$\text{Weight}_{ij} = \beta_{0ij}\text{cons} + \beta_{1j}\text{Month}_{ij}$$
$$\beta_{0ij} = 10.116(0.107) + u_{0j} + e_{0ij}$$
$$\beta_{1j} = 0.139(0.006) + u_{1j}$$

$$\begin{bmatrix} u_{0j} \\ u_{1j} \end{bmatrix} \sim N(0, \Omega_u) : \Omega_u = \begin{bmatrix} 2.143(0.229) \\ 0.007(0.010) \ 0.002(0.001) \end{bmatrix}$$

$$[e_{0ij}] \sim N(0, \Omega_e) : \Omega_e = [0.184(0.019)]$$

-2*loglikelihood(IGLS Deviance) = 1443.648(588 of 600 cases in use)

Figure 14.8 Output from MLwiN for MLM-1 given by equations 5 and 6 for the growth trajectories of body weight of MLM-1 from the multilevel modelling software MLwiN 2.02 (Rasbash *et al.* 2005).

The regression coefficient for *Month*$_{ij}$, 0.139, suggests that the children experience an average change in weight of 0.139 kg with the passing of each month. The intercept, 10.116, is the estimated mean baseline body weight, which is almost identical to the observed value shown in Table 14.1. The random intercept at level 1 has a variance of 0.184, with a standard error (SE) of 0.019, indicating that variation between the observed and predicted weight trajectory on each occasion for each child is statistically significant (significance at the 5% level occurs when the estimate is approximately twice its standard error). The random intercept at level 2 has a variance of 2.143 with SE of 0.229, indicating that the extent of variation between child-specific (predicted) baseline body weight and overall mean baseline body weight for all children is also statistically significant. Moreover, variation between each child's baseline body weight and that of the mean of all children is much greater than the variation between observed and predicted for each child (the SD of predicted baseline body weight for all children is around 1.46 kg, whereas the SD for measurement error and/or biological variation between observed and predicted linear body weight on each occasion is around 0.43 kg). The random slope has a variance of 0.002 with SE 0.001, which is modest and of borderline statistical significance ($p = 0.055$). This indicates there is modest variation in the slope of the growth trajectories (i.e. the speed of body weight growth) amongst these children. The covariance between the random intercept and the random slope (0.007, with SE of 0.010) is not significant, indicating no relationship between change in growth trajectory and baseline body

weight (see Blance *et al.* 2005 for more details on this type of assessment).

To accommodate possible non-linear body weight growth trajectories, we could include a quadratic term, *Month*$^2_{ij}$ (i.e. the square of *Month*$_{ij}$), in the multilevel model. The results from MLwiN show that the regression coefficient for *Month*$^2_{ij}$ is −0.002 with SE of 0.003, which is not statistically significant, suggesting that the growth in body weight is approximately linear during the 8 month observation period.

Our research objective is to test the relationship between body weight growth and blood aflatoxin levels. However, as discussed in the previous section, the age of these children is an important confounder for this relationship. Therefore, we include *Age*1$_j$ (the baseline ages of children) in our model. Since *Age*1$_j$ is a subject-level variable (hence it has only the subscript j), this allows each child to adopt a different baseline body weight according to the association of body weight with age. In order to model a different slope (growth trajectory), we then need to include an interaction term between *Age*1$_j$ and *Month*$_{ij}$ (MLM-2). The model becomes:

$$Weight_{ij} = \beta_{0ij} + \beta_{1j} Month_{ij} + \beta_2 Age1_j + \beta_3 Age1_j \cdot Month_{ij} \tag{7}$$

where β_{0ij} and β_{1j} are defined as before in equation 6.

It should be noted that, since the body weights of these children are measured at almost identical time intervals, there is no need to adjust for the ages on the second and third occasions (indeed, it is impossible to do so, due to the problems of collinearity; see Tu *et al.* 2004). The results of the model for equation 7 are shown in Figure 14.9, where the regression coefficient for *Month*$_{ij}$ is unaltered,

Weight$_{ij}$ ~ N(*XB*, Ω)

Weight$_{ij}$ = β$_{0ij}$cons + β$_{1j}$Month$_{ij}$ + 0.169(0.015)Age1$_j$ + -0.002(0.001)Age1.Month$_{ij}$

β$_{0ij}$ = 5.704(0.395) + u$_{0j}$ + e$_{0ij}$

β$_{1j}$ = 0.197(0.029) + u$_{1j}$

$$\begin{bmatrix} u_{0j} \\ u_{1j} \end{bmatrix} \sim N(0, \Omega_u) : \Omega_u = \begin{bmatrix} 1.242(0.139) \\ 0.019(0.008) \quad 0.002(0.001) \end{bmatrix}$$

[e$_{0ij}$] ~ N(0, Ω$_e$) : Ω$_e$ = [0.184(0.019)]

-2*loglikelihood(IGLS Deviance) = 1341.118(588 of 600 cases in use)

Figure 14.9 Output from MLwiN for MLM-2 given by equation 7 for the growth trajectories of body weight

$Weight_{ij} \sim N(XB, \Omega)$

$Weight_{ij} = \beta_{0ij}cons + \beta_{1j}Month_{ij} + 0.172(0.015)Age1_j + -0.003(0.001)Age1.Month_{ij} + -0.059(0.029)Lnafb_{ij}$

$\beta_{0ij} = 5.836(0.399) + u_{0j} + e_{0ij}$

$\beta_{1j} = 0.211(0.030) + u_{1j}$

$$\begin{bmatrix} u_{0j} \\ u_{1j} \end{bmatrix} \sim N(0, \Omega_u) : \Omega_u = \begin{bmatrix} 1.236(0.138) \\ 0.018(0.007) \; 0.002(0.001) \end{bmatrix}$$

$[e_{0ij}] \sim N(0, \Omega_e) : \Omega_e = [0.184(0.019)]$

$-2*loglikelihood(IGLS\ Deviance) = 1327.368(582\ of\ 600\ cases\ in\ use)$

Figure 14.10   Output from MLwiN for MLM-3 given by equation 8 for the growth trajectories of body weight

although β_0 has changed to 5.7, since it is now the body weight when $Age1_j$ is zero (which is not directly interpretable, given that these data do not contain age at zero within their range). The variance of the random intercept has been reduced to 1.242, indicating that some variation in the baseline body weight is accounted for by the association between the body weights and the ages of these children. The regression coefficient for $Age1_j$ is 0.169, indicating that older children tended to have higher aflatoxin blood levels at baseline (as shown previously in the univariate analysis). The regression coefficient for the term $Age1_j.Month_{ij}$ (the interaction between $Age1_j$ and $Month_{ij}$, derived by multiplying $Month_{ij}$ by $Age1_j$) is -0.002 ($p = 0.042$), suggesting that older children gained less weight than younger children over the 8 month period.

The next step is to test the relationship between body weight and aflatoxin blood levels by considering the covariate $Lnafb_{ij}$ in the multilevel model. Since $Lnafb_{ij}$ is a level-1 covariate (as it is different on each occasion), there is no need to consider an interaction term between $Lnafb_{ij}$ and $Month_{ij}$ (MLM-3):

$$Weight_{ij} = \beta_0 + \beta_{1j}Month_{ij} + \beta_2 Age1_j \quad (8)$$
$$+ \beta_3 Age1_j \cdot Month_{ij} + \beta_4 Lnafb_{ij}$$

where β_{0ij} and β_{1j} are defined as before in equation 6. The results from MLwiN are shown in Figure 14.10. The regression coefficient for $Lnafb_{ij}$ is 0.059 (SE of 0.029, $p = 0.037$), suggesting that children with higher levels of aflatoxin had a smaller body size after adjusting for their biological age.

It is possible to test whether or not this relationship between body weight and aflatoxin blood levels varies over time. First, we first split $Lnafb_{ij}$ into three distinct level 1 variables, $Lnafb1_{ij}$, $Lnafb2_{ij}$, and $Lnafb3_{ij}$, to represent aflatoxin levels at baseline, 3 months and 8 months, respectively: $Lnafb1_{ij}$ takes the values of $Lnafb_{ij}$ at baseline and zero elsewhere; $Lnafb2_{ij}$ takes the values of $Lnafb_{ij}$ at 3 months and zero elsewhere; and $Lnafb3_{ij}$ takes the values of $Lnafb_{ij}$ at 8 months and zero elsewhere. The revised model then becomes (MLM-4):

$$Weight_{ij} = \beta_{0ij} + \beta_{1ij}Month_{ij} + \beta_{2j}Age1_j$$
$$+ \beta_{3ij}Age1_j \cdot Month_{ij} + \beta_{4ij}Lnafb1_{ij}$$
$$+ \beta_{5ij}Lnafb2_{ij} + \beta_{6ij}Lnafb3_{ij} \quad (9)$$

and in so doing, the association with body weight of aflatoxin blood levels is assessed on each measurement occasion separately, whilst controlling for the potential confounding of age at baseline. The results are shown in Figure 14.11.

$Weight_{ij} \sim N(XB, \Omega)$

$Weight_{ij} = \beta_{0ij}cons + \beta_{1j}Month_{ij} + 0.170(0.015)Age1_j + -0.002(0.001)Age1.Month_{ij} + -0.016(0.035)Lnafb_{ij} +$
         $-0.047(0.031)Lnafb2_{ij} + -0.105(0.037)Lnafb3_{ij}$

$\beta_{0ij} = 5.725(0.404) + u_{0j} + e_{0ij}$

$\beta_{1j} = 0.246(0.034) + u_{1j}$

$$\begin{bmatrix} u_{0j} \\ u_{1j} \end{bmatrix} \sim N(0, \Omega_u) : \Omega_u = \begin{bmatrix} 1.250(0.140) \\ 0.015(0.007) \; 0.002(0.001) \end{bmatrix}$$

$[e_{0ij}] \sim N(0, \Omega_e) : \Omega_e = [0.181(0.019)]$

$-2*loglikelihood(IGLS\ Deviance) = 1322.774(582\ of\ 600\ cases\ in\ use)$

Figure 14.11   Output from MLwiN for MLM-4 given by equation 9 for the growth trajectories of body weight

The negative association of aflatoxin blood levels with body weight became greater as the children grew older, although only the association at the third measurement occasion (β_{6ij}) is statistically significant. To revert to the overall test of whether or not there is a significant inverse association observed between body weight and aflatoxin blood levels across all three occasions, we constrain β_4, β_5 and β_6 to be equal. The results are identical to those in Figure 14.10, in which only $Lnafb_{ij}$ was in the model, and reveals that $\beta_4 = \beta_5 = \beta_6 = -0.059$. The likelihood ratio test (taking deviance to be 4.594 with two degrees of freedom) is not statistically significant ($p = 0.101$) and the null hypothesis, namely that the negative association is equal across all three occasions, cannot be rejected.

Structural equation modelling (SEM)

Structural equation modelling (SEM) is a statistical methodology for testing complex theories. The development of modern SEM theory is generally attributed to the works by Professor Karl Jöreskog in the late 1960s and 1970s. The software LISREL, developed by him and his colleagues (Jöreskog and Sörbom 2006), is almost synonymous with SEM. SEM can be considered a general theoretical framework for all univariate and multivariate linear statistical models, i.e. all the commonly used statistical methods – correlation, linear regression, analysis of variance, multivariate analysis of variance, canonical correlation and factor analysis – are all within the family of SEM. Therefore, all these analyses can also be performed using SEM software, such as LISREL (Jöreskog and Sörbom 2006), EQS (Bentler 2006), Amos (Arbuckle 2005), and Mplus (Muthen and Muthen 2006). The statistical theory of SEM is, however, complex and the formal equations for SEM models are usually written using matrix algebra. An alternative way for epidemiologists to appreciate the concepts of SEM is through understanding the path diagrams of these statistical models (Miles 2003; Kline 2005), especially now that some SEM software provides a graphical interface for users to draw path diagrams for their models, and then the software performs the analysis specified by the path diagram.

The SEM methodology has provided a plethora of additional statistical indices by which one can assess model fit, to help researchers with the modification and selection of their models. However, there might be the temptation to focus on model fit indices that comply with an *a priori* perspective of the 'correct' model. For this reason, the entire philosophy of what is 'correct' must be approached cautiously when using SEM. As George Box once commented, 'all models are wrong' (Box 1976); however, some are more plausible and useful than others.

Latent growth curve modelling (LGCM)

Latent growth curve modelling (LGCM) is a special application of SEM to analyse longitudinal data and repeated measurements (Byrne and Crombie 2003; Bollen and Curran 2005; Duncan *et al.* 2006). There are many similarities between LGCM and MLM, and it can be shown that, in some scenarios, both approaches give rise to identical results (Curran 2003). The random effects in multilevel models are treated as latent variables in LGCM. The three multilevel models described in the previous section can therefore be specified in the framework of LGCM and analysed using SEM software. The advantage of this is that LGCM provides greater flexibility in modelling both non-linear change and the residual variances. The latter is very important where the study data exhibit non-constant variances over time (e.g. blood pressure varies much more amongst adults than children, and hence blood pressure does not have constant variance across age). Data that exhibit non-constant variance are described by statisticians as exhibiting heteroscedasticity, and data that exhibit constant variance are described as exhibiting homoscedasticity (alternative spellings of heteroskedasticity/homoskedasticity).

Whilst the formal presentation of LGCM in statistical equations requires the use of matrix algebra, the use of path diagrams does not require a mathematical background, and is thus more accessible for epidemiologists. Figure 14.12 is the path diagram of the LGCM for MLM-2 discussed in the previous section, where there are four observed variables (*Weight*1, *Weight*2, *Weight*3 and *Age*1) and seven latent variables: *E*1–*E*3 are the residual error terms for the three successive measurements of body weight, and *D*1 and *D*2 are the residual errors for the two latent variables *F*1 and *F*2 (in SEM, 'disturbances' is the technical name for the

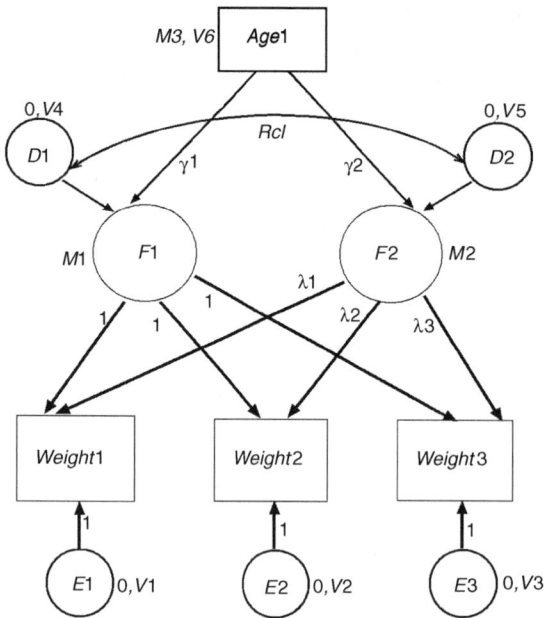

Figure 14.12. Path diagram for the latent growth curve model equivalent to MLM-2

residual errors of latent variables). The numbers associated with the arrows are 'factor loadings', which can be interpreted as a measure of association akin to regression coefficients. All factor loadings for the arrows from $F1$ to the three measurements of body weight are unity, which indicates that the association between $Weight1$ and $F1$ is scaled by unity (see equation 10). Similarly, all the factor loadings for the arrows from the residual errors are fixed to be 1. In contrast, the factor loadings for F2 are undetermined and are given labels (λ_1, λ_2, and λ_3). Hence, the implied regression equations for $Weight1$ through to $Weight3$ in relation to all other variables in Figure 14.12 are:

$$Weight1 = 1F1 + \lambda_1 F2 + 1E1$$
$$Weight2 = 1F1 + \lambda_2 F2 + 1E2 \quad\quad (10)$$
$$Weight3 = 1F1 + \lambda_3 F2 + 1E3$$

and the model in Figure 14.12 is then complete with the additional regression equations:

$$F1 = M1 + \gamma_1 Age1 + 1D1$$
$$F2 = M2 + \gamma_2 Age1 + 1D2 \quad\quad (11)$$

where $M1$ an $M2$ are intercepts, γ_1, γ_2 the slope coefficients, with the covariate $Age1$, which has

mean $M3$. In Figure 14.12, the means of residual errors are fixed to be zero (as with the means of residual errors in standard linear regression). Factor loadings for $Weight1$, $Weight2$ and $Weight3$ are set to unity, since their means become 'absorbed' and hence estimated via the mean values of the latent variables $F1$ (estimated intercept for the trajectory) and $F2$ (estimated slope for the trajectory). The variances ($V1$–$V6$) are estimated, although here we simplify matters by constraining $V1 = V2 = V3$, and thus make the implicit assumption of homoscedasticity (constant variance throughout the study period), as is the default within linear regression. Relaxing this constraint provides far greater flexibility than the usual linear regression approach and this is discussed in the next section.

We thus use multiple equations to define the complex relationships amongst the observed and latent variables. Some model parameters are identified, where factor loadings are set to unity or given predefined values, whilst other model parameters are estimated. In most structural equation models, researchers are interested only in the covariance/correlation structure of their data, whereas in LGCM we are also interested in the means of the observed and latent variables.

To specify a LGCM model equivalent to MLM-2 (as given by equations 6 and 7), we fix the factor loadings λ_1, λ_2, and λ_3 to be 0, 3 and 8, respectively, corresponding to the times when the repeated measurements of body weight were made. Then, the latent variable $F1$ is equivalent to the estimated mean body weight at baseline and the latent variable $F2$ is equivalent to the estimated mean change in body weight over the 8 month period, and the variances of $D1$ and $D2$ ($V4$ and $V5$) are equivalent to the random effects of the intercept (u_{0j}) and slope (u_{1j}) in Figure 14.8, respectively. By constraining the variances of $E1$, $E2$ and $E3$ ($V1$, $V2$ and $V3$) to be equal, these are then equivalent to e_{0ij} in Figure 14.8. The regression coefficient for the interaction term between $Age1$ and $Month$ in Figure 14.8 is 0.002, and this is equivalent to γ_2 in the LGCM in Figure 14.12. The intercept (5.704), regression coefficient for $Month$ (0.197) and the regression coefficient for $Age1$ (0.169) in Figure 14.8 are $M1$, $M2$ and γ_1, respectively, in the LGCM in Figure 14.12.

The results of the LGCM model in Amos (Figure 14.12) are almost identical to those for the MLM model in MLwiN (Figure 14.8). This is to be expected, since the estimation procedure in MLwiN was iterative generalized least squares (IGLS), which gives maximum likelihood estimates, and Amos (and other SEM software) often uses maximum likelihood estimation as the default. However, if within MLwiN one used restricted iterative generalized least squares (RIGLS; equivalent to restricted maximum likelihood), the results would be slightly different. Another cause of modest differences between the two analytical strategies can be down to incomplete data due to missing values. Different software packages use different approaches to deal with missing values in the dataset, and this may yield slight differences in the results. Amos uses full information maximum likelihood (FIML) estimations to account for the missing values in the data, as long as they are missing at random. Details about the properties of FIML can be found in Arbuckle (1996). MLwiN uses cases with data available at the lower-level and therefore accommodates missing values at the lower level. These distinctions can be important, since different assumptions regarding missing data can affect the outcomes of the model and the researcher is advised to consider fully the context of their data and the assumptions they need to make regarding any missing values.

It is also possible to specify the multilevel model given by MLM-3 in the framework of LGCM, and the path diagram is shown in Figure 14.13, in which the three measurements of aflatoxin blood levels are included as covariates (exogenous variables to body weight). It is noted that the paths from *Aflatoxin* to *Weight* on each occasion need to be constrained to be equal, and this can be easily done in SEM. To specify MLM-4 within LGCM, we simply remove the constraint of equality for γ_1, γ_2, and γ_3.

Flexibility of LGCM

LGCM provides great flexibility in modelling longitudinal data, achieved by freeing the fixed or constrained factor loadings and model parameters. For instance, in the previous models (equations 6

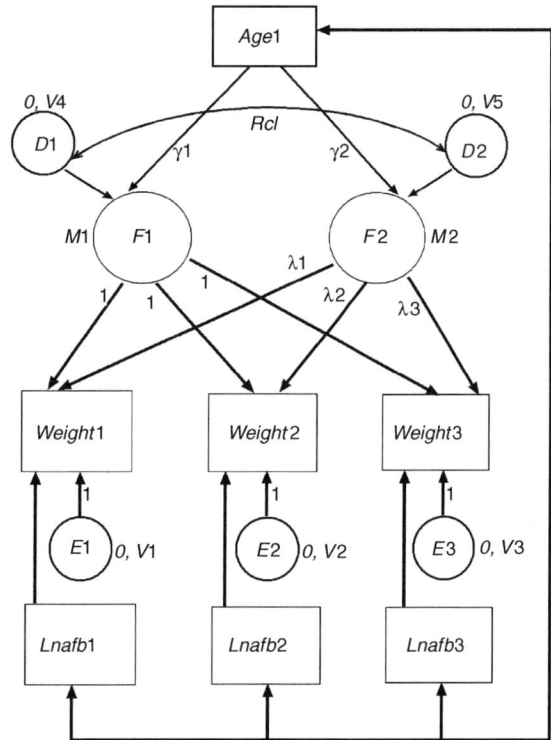

Figure 14.13 Path diagram for the latent growth curve model equivalent to MLM-3 and MLM-4

and 10), the variances of the residual errors (e_{0ij}, $E1$, $E2$, and $E3$) are forced to be identical across the three occasions. This constraint could be removed, and the error variances on each occasion would be estimated freely, allowing heteroscedasticity. This can be modelled in all SEM software but only some MLM software.

Another parameter that could be freed is λ_2. In the original specification, we assumed a linear growth trajectory and tested a quadratic growth curve model by considering a quadratic term in the multilevel model. In LGCM, by removing the constraint on λ_2, we estimate a possible nonlinear growth curve for body weights. For instance, when the constraint is removed and λ_2 freely estimated for the LGCM in Figure 14.12 (MLM-3) and Figure 14.13 (MLM-4), its value is 3.16 and 3.08, respectively. These estimates are both very close to 3, indicating that the growth curves are indeed very close to linear changes over time, as shown in the MLM analyses.

One important advantage of using LGCM for longitudinal data analysis is that it is straightforward to model the growth curves or change patterns for multiple outcomes simultaneously, and it is also possible to specify the relationship between the growth processes of these variables. For instance, our previous analyses treated aflatoxin levels measured on the three occasions as covariates and showed that there was an overall inverse association between body weight and aflatoxin blood levels. Another important research question is whether or not the *growth* in body weight is affected by the *change* in aflatoxin blood levels. There are several alternative ways to specify a latent growth curve model for this research question, and different models imply different *a priori* assumptions in the causal relationships amongst the considered variables. Figure 14.14 shows one possible latent growth model, in which we model the growth trajectories of both body weight and aflatoxin blood levels, and assess the association of the latter with the former.

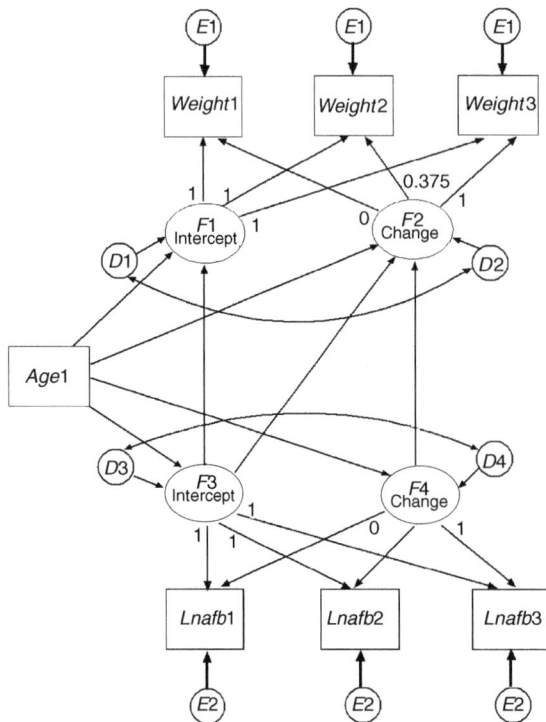

Figure 14.14 Path diagram for structural equation models (SEM-1)

In the model shown in Figure 14.14, it is hypothesized that the growth curves of body weight are affected by the *changes* in aflatoxin blood levels. The latent variable $F1$ is the baseline body weight (i.e. the intercept in multilevel models, as discussed in the previous section), and $F2$ is the *linear* growth curve for body weight, since we set the factor loadings from $F2$ to *Weight*1, *Weight*2 and *Weight*3 to be 0, 0.375 and 1, respectively (recall that these variables are measured at months 0, 3 and 8, so by treating 8 months as one unit, 3 months becomes $3/8 = 0.375$). The latent variable $F3$ is the baseline aflatoxin blood levels, and $F4$ is the potentially non-linear change in aflatoxin blood levels; non-linear since the factor loadings from $F4$ to *Lnafb*1 and *Lnafb*3 are fixed to be 0 and 1 (as with *Weight*1 and *Weight*3), yet the factor loading from $F4$ to *Lnafb*2 is a free parameter, to be estimated from the data. The reason for setting it free is because we know from Table 14.1 that the changes in aflatoxin blood levels are far from linear. It is also hypothesized that baseline body weight ($F1$) is associated with the baseline aflatoxin levels ($F3$), and this is represented by the arrow from $F3$ to $F1$. Furthermore, in this model, it is assumed that change in the growth trajectories of body weight ($F2$) is associated with both the baseline aflatoxin blood levels ($F3$) and changes in these levels throughout the study period. The baseline ages of these children is again considered a covariate in the model to accommodate the confounding relationship it has with both body weight and aflatoxin blood levels. In this instance, the residual errors for the three measurements of body weight are constrained to be equal ($E1$), as too are the residual errors for the aflatoxin blood levels ($E2$), thereby assuming homoscedasticity for both measures.

The results from Amos (using FIML to account for missing values) are shown in the column under the heading SEM-1 in Table 14.2. The χ^2 value for the overall model fit to the data is 18.932 with 13 degrees of freedom ($p = 0.125$). In SEM, the null hypothesis for the χ^2 test is that there is no difference in the covariance structures between the proposed model and the data, and a p value > 0.05 suggests that we cannot reject the null hypothesis. This is different from the usual null hypothesis testing, where researchers try to reject the null hypothesis. For the χ^2 test in SEM, we seek *insufficient*

Table 14.2. Results of latent growth curve models

			SEM-1				SEM-2			
			Estimate	SE	CR	p Value	Estimate	SE	CR	p Value
Regression weights										
F3	<—	*Age*1	0.047	0.013	3.721	< 0.001	0.047	0.013	3.728	< 0.001
F4	<—	*Age*1	−0.030	0.013	−2.311	0.021	−0.030	0.013	−2.314	0.021
F1	<—	*Age*1	0.179	0.016	11.490	< 0.001	0.179	0.016	11.494	< 0.001
F2	<—	*Age*1	−0.034	0.015	−2.198	0.028	−0.035	0.013	−2.633	0.008
F1	<—	F3	−0.215	0.110	−1.958	0.050	−0.217	0.110	−1.972	0.049
F2	<—	F4	−0.553	0.307	−1.803	0.071	−0.577	0.283	−2.037	0.042
F2	<—	F3	−0.012	0.085	−0.147	0.883				
*Weight*2	<—	F1	1				1			
*Weight*1	<—	F2	0				0			
*Weight*2	<—	F2	0.375				0.375			
*Weight*3	<—	F2	1				1			
*Weight*1	<—	F1	1				1			
*Weight*3	<—	F1	1				1			
*Lnafb*1	<—	F3	1				1			
*Lnafb*2	<—	F3	1				1			
*Lnafb*3	<—	F3	1				1			
*Lnafb*1	<—	F4	0				0			
*Lnafb*2	<—	F4	0.060	0.075	0.803	0.422	0.059	0.075	0.791	0.429
*Lnafb*3	<—	F4	1				1			
Means										
*Age*1			26.138	0.399	65.538	< 0.001	26.138	0.399	65.538	< 0.001
Intercept										
F3			2.389	0.340	7.033	< 0.001	2.390	0.339	7.047	< 0.001
F4			1.639	0.353	4.651	< 0.001	1.639	0.352	4.658	< 0.001
F1			6.218	0.472	13.161	< 0.001	6.222	0.473	13.166	< 0.001
F2			2.510	0.521	4.814	< 0.001	2.519	0.538	4.679	< 0.001
Covariance										
D3	<–>	D4	0.014	0.080	0.176	0.860	0.022	0.063	0.344	0.731
D1	<–>	D2	0.135	0.059	2.297	0.022	0.135	0.059	2.302	0.021
Correlation										
D3	<–>	D4	0.033				0.051			
D1	<–>	D2	0.701				0.753			
Variance										
*Age*1			31.653	3.173	9.975	< 0.001	31.653	3.173	9.975	< 0.001
D3			0.720	0.102	7.041	< 0.001	0.715	0.096	7.438	< 0.001
D4			0.257	0.132	1.943	0.052	0.249	0.122	2.039	0.041
D1			1.209	0.137	8.847	< 0.001	1.209	0.137	8.846	< 0.001
D2			0.031	0.077	0.397	0.691	0.027	0.074	0.359	0.720
E1			0.184	0.019	9.810	< 0.001	0.184	0.019	9.810	< 0.001
E2			0.483	0.049	9.789	< 0.001	0.486	0.047	10.317	< 0.001

SE, standard error; CR, critical ratio.

evidence to reject the null hypothesis. Another commonly used model fit index is root mean-square error of approximation (RMSEA) and this is 0.048 (90% CI 0, 0.092) for our model in Figure 14.14. A small RMSEA (0.06 is usually used as the cut-off value) suggests that the proposed model fits the data reasonably (Kline 2005). Both model fit indices indicate an acceable model.

Although our proposed model seems to fit the data relatively well, it is noted that the regression weight for the path from $F3$ to $F2$ is small (–0.012) and non-significant, i.e. baseline aflatoxin blood levels seem to have little association with the growth in body weight compared to the association between changes in aflatoxin blood levels and the growth in body weight. For model parsimony, we remove the path from $F3$ to $F2$, and the results for this modified model are shown in Table 14.2 in the column under the heading of SEM-2. The χ^2 value for the overall model fit to the data is 18.951 with 14 degrees of freedom ($p = 0.167$) and the RMSEA is 0.042 (90% CI 0, 0.086). Both indices indicate that the modified model provides an acceptable fit whilst being more 'economic' (i.e. parsimonious), due to estimating one less parameter and having one more degrees of freedom. For other parameters in the model, the differences between the original and modified models are small.

Baseline aflatoxin blood levels thus had an inverse association with baseline body weight, and children with a greater increase in aflatoxin blood levels experienced less growth in their body weight over the 8 month observation period. Older children had greater body weights and higher levels of aflatoxin, yet they showed smaller associations with increases in body weight and with aflatoxin blood levels than did younger children.

14.5. CONCLUSIONS

In this chapter, we discuss and demonstrate that commonly used univariate and multivariate methods for longitudinal data analysis can be considered within the family of *structural equation modelling* (SEM). These methods, if applied appropriately, can be very useful and powerful statistical tools for epidemiologists. SEM illustrated in this chapter can also be used to accommodate multiple change proc-

esses and, therefore, provide a statistical framework for epidemiologists to evaluate complex causal relationships in observational and experimental studies.

Although there is dedicated software for the analysis of *multilevel models* (MLM), there are a few advantages of using SEM software to conduct these analyses. For instance, SEM provides additional indices of model fit (such as RMSEA), latent variables can be easily incorporated in SEM but not in MLM (thereby accommodating measurement error more appropriately) and longitudinal data analyses can be more flexible (in terms of modelling both non-linearity and heteroscedasticity).

It is particularly straightforward to model complex variance and covariance terms for level-1 residuals in SEM, whereas it usually requires a working knowledge of programming to model the same using statistical software for MLM. The statistical package SAS® PROC MIXED, for random effects models, provides several commands to model covariance structures of level-1 residuals (Littell *et al.* 2006), and epidemiologists can obtain greater understanding of these commands from the perspective of SEM.

LGCM also has some limitations. In our examples on growth, the intervals between the measurements of body weight were approximately identical for all children. If, for example, body weight was measured at baseline 2 and 7 months for some children, but at baseline 4 and 9 months for others, this does not pose any problem in the analyses using MLM but will be a problem using LGCM. Nevertheless, this is not a methodological weakness of LGCM *per se*, rather a limitation in the currently available SEM software. These problems will most likely be resolved in the future, as these methods are constantly being developed.

Researchers should always choose the methods best suited for their research questions and study design and keep abreast of new analytical developments.

Acknowledgements

The authors were funded by the UK government's Higher Education Funding Council for England (HEFCE).

References

Arbuckle JL. 2005. Amos 6.0 [computer software]. SPSS: Chicago, IL.

Arbuckle JL. 1996. Full information estimation in the presence of incomplete data. In *Advanced Structural equation Modelling: Issues and Techniques*, Marcoulides GA, Schmacker RE (eds). Laurence Erlbaum: Mahwah, NJ.

Bentler PM. 2006. EQS 6.1 [computer software]. Multivariate Software: Encino, CA.

Berk RA. 2004. *Regression Analysis: A Constructive Critique*. Sage: Thousand Oaks, CA.

Blance A, Tu Y-K, Gilthorpe MS. 2005. A multilevel modelling solution to mathematical coupling. *Statist Methods Med Res* **14**: 553–565.

Bland M. 2000. *An Introduction to Medical Statistics*. Oxford University Press: Oxford, UK.

Bollen KA, Curran PJ. 2006. *Latent Curve Models*. Wiley: Hoboken, NJ.

Box GEP. 1976. Science and statistics. *J Am Statist Assoc* **71**: 791–799.

Byrne BM, Crombie G. 2003. Modeling and testing change: an introduction to the latent growth curve model. *Understanding Statist* **2**: 177–203.

Curran P. 2003. Have multilevel models been structural equation models all along? *Multivar Behav Res* **38**: 529–569.

Dawson B, Trapp RG. 2001. *Basic and Clinical Biostatistics*, 3rd edn. McGraw-Hill: New York.

Duncan TE, Duncan SC, Strycker LA. 2006. *An Introduction to Latent Variable Growth Curve Modeling*, 2nd edn. Laurence Erlbaum: Mahwah, NJ.

Dupont WD. 2002. *Statistical Modeling for Biomedical Researchers: A Simple Introduction to the Analysis of Complex Data*. Cambridge University Press: Cambridge, MA.

Gilthorpe MS, Maddick IH, Petrie A. 2000. Introduction to multilevel modelling in dental research. *Commun Dent Health* **17**: 222–226.

Gilthorpe MS, Zamzuri AT, Griffiths GS *et al.* 2003. Unification of the 'burst' and 'linear' theories of periodontal disease progression: a multilevel manifestation of the same phenomenon. *J Dent Res* **82**: 200–205.

Goldstein H. 1995. *Multilevel Statistical Models*, 2nd edn. Wiley: New York.

Gong Y, Hounsa A, Egal S *et al.* 2004. Postweaning exposure to aflatoxin results in impaired child growth: a longitudinal study in Benin, West Africa. *Environ Health Perspect* **112**: 1334–1338.

Greenwood DG, Gilthorpe MS, Cade JE. 2006. The impact of imprecisely measured covariates on estimating gene–environment interactions. *BMC Med Res Methodol* **6**: 21.

Hox J. 2002. *Multilevel Analysis*. Laurence Erlbaum: Mahwah, NJ.

Jewell NP. 2004. *Statistics for Epidemiology*. Chapman and Hall/CRC: Boca Raton, FL.

Jöreskog K, Sörbom D. 2006. LISREL 8.80 [computer software]. Scientific Software International: Lincolnwood, IL.

Kirkwood B, Sterne JAC. 2003. *Essential Medical Statistics*. Blackwell Science: Oxford, UK.

Kline RB. 2005. *Principles and Practice of Structural equation Modeling*, 2nd edn. Guilford Press: New York.

Leyland AH, Boddy FA. 1998. League tables and acute myocardial infarction. *Lancet* **351**: 555–558.

Littell RC, Milliken GA, Stroup WW *et al.* 2006. *SAS® for Mixed Models*, 2nd edn. SAS Institute: Cary, NC.

Miles J. 2003. A framework for power analysis using a structural equation modelling procedure. *BMC Med Res Methodol* **3**: 27.

Muthén LK, Muthén B. 2006. *Mplus User's Guide*, 4th edn. Muthén & Muthén: Los Angeles, CA.

Norman GR, Streiner DL. 2000. *Biostatistics*, 2nd edn. B. C. Decker: Hamilton, Ontario, Canada.

Rasbash J, Browne W, Healy M *et al.* 2005. MLwiN 2.02. Multilevel Project, Institute of Education: London.

Raudenbush SW, Bryk AS. 2002. *Hierarchical Linear Models*, 2nd edn. Sage: Thousand Oaks, CA.

Singer JB, Willett JD. 2003. *Applied Longitudinal Data Analysis*. Oxford University Press: New York.

Thompson B. 1995. Stepwise regression and stepwise discriminant analysis need not apply here: a guidelines editorial. *Educ Psychol Measurem* **55**: 525–534.

Tu YK, Clerehugh V, Gilthorpe MS. 2004. Collinearity in linear regression is a serious problem in oral health research. *Eur J Oral Sci* **112**: 389–397.

Twisk JWR. 2003. *Applied Longitudinal Data Analysis for Epidemiology*. Cambridge University Press: Cambridge, UK.

Twisk JWR. 2006. *Applied Multilevel Analysis*. Cambridge University Press: Cambridge, UK.

Vickers AJ. 2001. The use of percentage changes from baseline as an outcome in a controlled trial is statistically inefficient: a simulation study. *BMC Med Res Methodol* **1**: 6.

West BT, Welch KB, Gałecki AT. 2007. *Linear Mixed Models*. Chapman and Hall/CRC: Boca Raton, FL.

15

Meta-Analysis and Pooled Analysis – Genetic and Environmental Data

Camille Ragin and Emanuela Taioli

University of Pittsburgh Cancer Institute and School of Public Health, Pittsburgh, PA, USA

15.1. INTRODUCTION

A large amount of data have been produced and published on biological markers measured in human subjects, but a common obstacle in reaching definite conclusions has been the lack of statistical power of the individual studies. To overcome this problem, it is customary to perform summaries of the published studies, while new, larger studies are completed. Reviews of the evidence can be performed in two main ways: by completing a meta-analysis of published data, or a pooled analysis of individual data (both published and unpublished).

15.2. META-ANALYSIS

Meta-analysis provides summary estimates by combining the individual results published by independent scientists. This approach increases power, produces a more accurate estimate of the risk while reducing the possibility of false-negative results (Greenland 1987). Some limitations are the impossibility of performing more refined analyses, such as dose–response and stratified analyses. While several guidelines and methodological papers have been published for clinical trials, the organization and summarization of data for observational studies (Blettner *et al.* 1999; Stroup *et al.* 2000) and for molecular genetic research (Bogardus *et al.* 1999) has been addressed less frequently.

Four steps can be identified when performing a meta-analysis (See Box 15.1)

1. Identification of the relevant studies.
2. Setting of the eligibility criteria for the inclusion and exclusion in the meta-analysis of the identified studies.
3. Abstracting the relevant data.
4. Analysis of the data, including formal statistical testing for heterogeneity and investigation of the reasons for heterogeneity if it exists.

Database searching, eligibility criteria and data extraction

A bibliographic search (e.g. in MEDLINE or EMBASE) should be conducted to identify the studies of interest. Potentially relevant publications may be identified from the abstracts, and full-text versions should be obtained for a review. The next step is to define eligibility criteria for the meta-analysis, since not all studies can or should be included in a meta-analysis. This is to ensure reproducibility of the meta-analysis and minimize bias. If the eligibility criteria are well defined, any person should be able to reproduce the path followed for identifying studies to be included. Bias will be reduced by the systematic selection of studies, which should not be influenced by the knowledge of the study results or aspects of the study conduct. The basic considerations in defining

Molecular Epidemiology of Chronic Diseases, Edited by C. P. Wild, P. Vineis, and S. Garte
© 2008 John Wiley & Sons, Ltd

Box 15.1   Four steps to performing a meta-analysis

1. Identify the relevant studies.
2. Set the eligibility criteria for the inclusion and exclusion.
3. Abstract the relevant data.
4. Analyse the data:
   - (a) Generate summary estimates.
   - (b) Testing for heterogeneity.
   - (c) Assess publication bias.

---

eligibility criteria for a meta-analysis should include study design, years of publication, completeness of information in the publication, similarities of treatment and/or exposure, languages, and the choice of studies that have overlapping datasets.

Data extraction should include data on the relevant outcomes and the general characteristics of each study such as samples size, source of the control population, race or ethnicity of the study population.

Graphical summaries

The results for each of the studies included in the meta-analysis can be graphically displayed with the odds ratios (ODs) or relative risk ratios (RRs) and confidence intervals (CIs), using a Forest plot (Figure 15.1). Each study is represented by a square and a solid line. The square corresponds to the OR or RR. The size of each square corresponds to the contribution or weight of that particular study in the meta-analysis. Larger studies tend to contribute more to the meta-analyses than smaller studies. The solid horizontal line drawn through each square represents the study's 95% CI. Note that the 95% CI shows the true underlying effect of the measured outcome 95% of the times if the study were repeated again and again. The solid vertical line (at OR = 1) represents no association. If the study's 95% CI crosses this line, then the effect of the measured outcome is not statistically significant (i.e. $p > 0.05$). The diamond corresponds to the summary estimate or combined OR for all the studies in the analysis.

Summary estimates and assessment of heterogeneity

The summary estimate provides the overall effect of the measured outcome by combining the data from all the studies included in the meta-analysis. A weighted average of the results of each study is used to calculate this summary estimate; the simple arithmetic average would be misleading. The size of the study must be taken into consideration when calculating the summary estimate. Larger studies have more weight than smaller studies because their results are less subject to chance. Fixed effects and random effects are two types of models used for calculating the summary estimate. *Fixed effects models* consider that the variability between studies is due to random variation because of the size of the study. This means that if all the studies were large, they would yield the same results (i.e. the studies are homogeneous). Statistical methods which calculate the summary estimates based on the assumption of fixed effects include the Mantel–Haenszel method (Mantel and Haenszel 1959), the Peto method (Yusuf *et al.* 1985), general variance-based methods (Wolf 1986) and the CI methods (Greenland 1987; Prentice and Thomas 1987). *Random effects models* consider that the variability between studies is due to distinct differences between the studies (i.e. the studies are heterogeneous). Heterogeneity can arise when there are differences in study design, lengths of follow-up or inclusion criteria of the study participants. The statistical methods described by DerSimonian and Laird (1986) calculate the summary estimates based on the assumption of random effects.

Although meta-analyses provide the opportunity to generate summary estimates of published studies, it is important to note that this summary estimate may not be appropriate when the included studies are heterogeneous. To establish whether the results are consistent between studies, reports of meta-analyses commonly present a statistical test of heterogeneity. The classical measure of heterogeneity between studies is the Q-statistic, where heterogeneity exists when $p < 0.05$. Another statistical test, the I^2 statistic, describes the percentage of variation across studies that are due to heterogeneity rather than chance (Higgins *et al.* 2003; Higgins and Thompson 2002). The I^2 ranges from 0% to 100%,

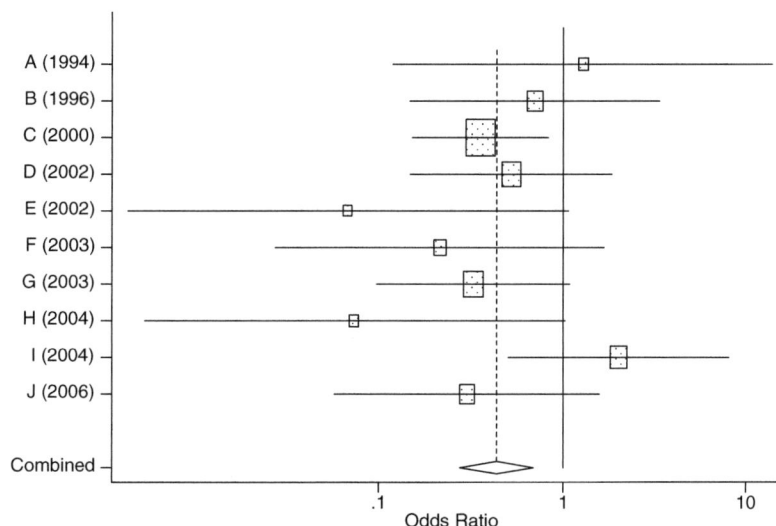

Figure 15.1   Forest plot of studies included in the meta-analysis, and summary odds ratio for the combined studies

where a value of 0% indicates that there is no heterogeneity between the studies in the meta-analysis.

A fixed effects model should be used to calculate the summary estimate when no heterogeneity is observed between studies, while a random-effects model should be used was when heterogeneity across studies is observed (Normand 1999). When differences among studies exist, it is the task of the meta-analyst to determine the sources of these differences. The reporting of the random effects estimate does not remedy the problem and can sometimes conceal the fact that the summary estimates or fitted model is a poor summary of the data (due to heterogeneity). Sources of heterogeneity can be examined by stratification of the studies by ethnicity or control source (Raimondi *et al.* 2006), or detection method (Hobbs *et al.* 2006), but in some cases this may not be possible because the relevant information is missing in the publication.

Assessing publication bias

Publication bias arises when studies with statistically significant results are more likely to be published and cited, and are preferentially published by scientific journals. A formal test for publication bias can be done by performing the test for funnel plot asymmetry proposed by Egger *et al.* (1997). In the absence

of bias, the plot will resemble a symmetrical inverted funnel (Figure 15.2) and the Egger's test value will be $p > 0.05$. Conversely, if there is bias, funnel plots will often be skewed and asymmetrical. In this case, the Egger's test value would be $p < 0.05$.

15.3. POOLED ANALYSIS

Another way to summarize results from observational studies is to pool individual records and reanalyse the data (Fenech *et al.* 1999; Friedenreich 2002; Taioli 1999). This approach (see Box 15.2) allows for the performance of statistical interaction tests, sub-group analyses, and refined dose–response curves (Friedenreich 1993). Guidelines and methods for pooling data from molecular epidemiological research have been published (Taioli and Bonassi 2002).

15.4. ISSUES IN POOLED ANALYSIS OF EPIDEMIOLOGICAL STUDIES INVOLVING MOLECULAR MARKERS

Choice of study design

There are two commonly used study designs: case–control and cross-sectional studies. Case–control studies are more popular when genetic

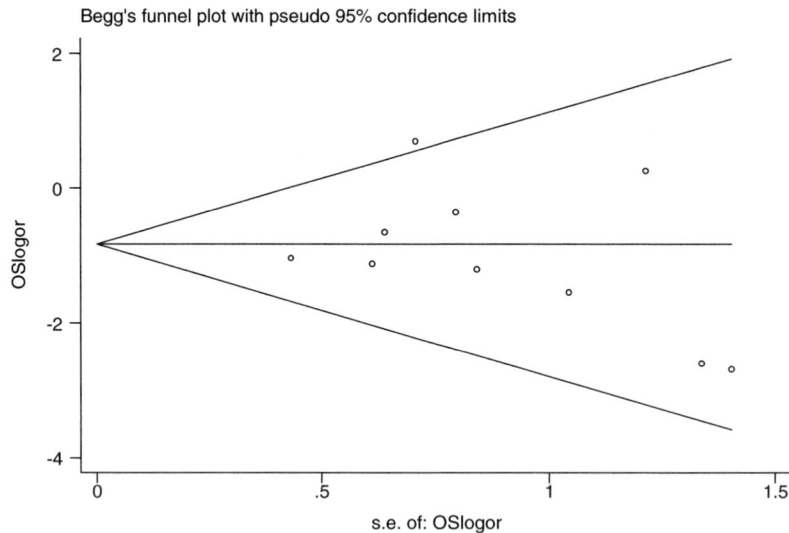

Figure 15.2 Funnel plot for the evaluation of publication bias

markers are involved, while cross-sectional studies are preferred when markers of exposure in otherwise healthy populations are explored. Cohort studies may also be included in pooled analyses. Pooling data from cohorts has its advantages as well as limitations. For example, pooling cohort studies that involve diet assessed prior to development of the disease would limit recall and selection biases. A limitation that these studies, shared with the case–control studies, is that they may have been designed and implemented differently, thus leading to heterogeneity between studies and in many cases to the inability to standardize data before pooling.

Planning of the study

A pooled analysis of studies involving molecular markers in human populations needs to take into account all the aspects related to biological modelling or laboratory practice. The first step is locating and collecting a list of relevant studies, followed by the tabulation of the study design, laboratory methods, and analysis of the data. Human studies using biomarkers often represent the joint effort of researchers from different areas; therefore an advisory group of recognized experts in several disciplines, usually epidemiology, molecular biology and biostatistics, would greatly benefit the success of the pooled analysis.

Box 15.2 Steps to performing a pooled analysis

1. Select the studies:
    (a) Develop clear inclusion and exclusion criteria.
    (b) Request the data.
    (c) Validate the data.
2. Standardize the data.
3. Analyse the data:
    (a) Generate pooled estimates.
    (b) Test for heterogeneity.
    (c) Assess publication bias – stratify according to published/unpublished datasets.
    (d) Conduct subgroup and stratified analyses.

Selection of studies

A further step is the selection of studies to be included in the pooled analysis. At this point, clear inclusion and exclusion criteria need to be set and agreed upon, e.g. it may be important to exclude convenience samples of healthy subjects, or subjects with precursor lesions, e.g. colon polyposis, or with specific life-style habits, e.g. biological markers measured in alcoholic subjects, or with specific exposure, e.g. workers exposed to chemicals. For genotype studies, it must be evaluated whether phenotype studies can be used as a surrogate of genotype.

Data request

The collection of the original databases from the investigators can be performed through a communication network based on e-mail. Pre-compiled worksheets reporting the variables to be included in the study should be sent along with the invitation to participate in the pooled analysis. This will encourage the investigators to contribute data according to a common framework. All the data received and included in the dataset must be anonymous.

Evaluation of the validity of the study

An *ad hoc* questionnaire collected from all the participants, in order to characterize each study, is very useful. This information is essential to evaluate how reliable is the evidence provided by single studies. A minimum epidemiological information should be collected, e.g. type of controls in case–control studies, incident vs. prevalent cases, availability of histological diagnosis, response rate for both cases and controls, percentage of subjects in which the biological marker was measured. Information dealing with the sensitivity and reproducibility of the assay collected, such as the time of sample collection and storage, the length of storage, the reagents used, the scoring criteria, as well as data on inter- and intra-subject and - laboratory variability, should also be collected. It is possible to apply a quality scoring system to studies included in the analysis. This method, used in some meta-analyses of randomized clinical trials, has many potential advantages and could be tentatively applied to biomarkers, where an agreement on the quality of technical procedures seems easier to reach.

Data standardization

An important issue when pooling data from different studies is the degree of standardization of the biomarker measure across studies. Studies involving genetic markers suffer fewer problems of standardization of the laboratory technique, since the most common method (PCR) is fairly universal. However, new techniques are now used that require comparison with common PCR methods. The outcome of genetic testing is also commonly defined as either binary or in three levels (wild-type, heterozygous, homozygous variant). Comparison of levels of biomarkers of exposure is more difficult, since the protocols, equipment and methods may be different among studies. An example of this issue can be found in a re-evaluation of 25 cytogenetic studies (Bonassi *et al.* 1996). A possible solution is to obtain a post-hoc data standardization by asking the participating investigators to test their method against a common gold standard, so that some correcting factor can be produced. Often the laboratories involved are not available for further testing; an alternative approach is to ask them to provide detailed information on the method used, so that differences and similarities across laboratories can be better evaluated. Another aspect to consider is the standardization of the questionnaires to collect risk factors. In a pooled analysis on micronuclei (the HUMN study; Bonassi *et al.* 2001), few variables were included in the overall statistical analysis, i.e. age, sex, country of origin, because several other factors, such as alcohol drinking, medical treatments and occupational exposure, were collected but not standardized and were not comparable across studies categories. A common approach in pooled analysis is to collect a minimum set of epidemiological variables, for which a standard can be set and compliance to the standard can be expected from most of the investigators. The standardization of the outcome variable is also relevant. A requirement is that comparable measurements of the biomarker reflects the same range of biological events, but this assumption is quite often not verified (Bonassi *et al.* 2001). Two approaches are possible. One involves the categorization of data within each laboratory before data are assembled. This approach was used by the European Study Group on Cytogenetic Biomarkers and Health (ESCH) (Bonassi *et al.* 2000; Hagmar *et al.* 1998) and it is generally used in cohort studies linking the frequency of cytogenetic biomarkers to cancer (Bonassi *et al.* 2000; Hagmar *et al.* 1998; Smerhovsky *et al.* 2001, 2002). Another solution is to apply statistical modelling for correlated data. Each laboratory is considered as a cluster of events, and the within-cluster correlation is taken into account when the association biomarker/outcome is evaluated (Bonassi *et al.* 1999; Golgstein 1995; Rasbash *et al.* 1999).

Heterogeneity among studies

The presence of heterogeneity among studies, and the distinction between heterogeneity due to the different distribution of risk factors in the study population and heterogeneity attributable to external variables (study design, laboratory protocols, etc.) needs to be addressed. Some factors that determine heterogeneity can be efficiently studied with a pooled analysis, where major sources of variability in protocols and in scoring can be identified (Bonassi *et al.* 2001).

The first step is to run a univariate analysis of the biomarker distribution by study or laboratory, to assess the distribution of relevant variables and identify outliers. This approach should provide information on whether to include a dataset, and in certain circumstances to rank the datasets by quality. Information on the time of sample collection and storage, length of storage, reagents used, scoring criteria, as well as inter- and intra-subject and laboratory variability are useful elements to judge the possible source of variability. Inter-study variability can be accounted for in the statistical analysis; however, the removal of outliers with no obvious possible explanation should be considered.

Publication bias

Pooled analysis allows for the collection of unpublished studies, of unpublished data from published studies or of pilot studies. Sometimes, epidemiological studies involving biomarkers remain unpublished for reasons that do not necessarily relate to less rigorous study design, analysis or laboratory work, but because of the quick decrease in interest for markers that are less fashionable, lack of funding or difficulty in recruiting subjects, etc. Pooling these data gives the opportunity to collect data on rare cancers or on small, geographically isolated populations, with the potential of raising new research questions. Differences between published and unpublished data should be always tested by comparing the frequency of relevant variables in the two groups before pooling data. In the Genetic Susceptibility to Environmental Carcinogens (GSEC) study, for example, we assessed the frequency of the polymorphisms in the control population coming from published studies vs. unpublished studies, and no differences were observed (Garte *et al.* 2001).

Any study that pools original data should establish inclusion criteria beforehand. For example, the GSEC study uses the following criteria to include unpublished studies: (a) score the quality of the study according to a pre-defined list of variables including study design, choice of controls, participation rate of cases and controls, laboratory methods, reasons for nonpublishing the data; and (b) check the frequency of both epidemiological and genetic variables to assess any unusual pattern, such as genotype frequencies not in Hardy–Weinberg equilibrium. In addition, a table of studies included in the analysis and studies not included because of lack of interest of the original investigator should always be generated, to assess inclusion bias.

Another option is to identify a committee of experts to evaluate the quality of the assays and the reliability of the information. Among basic criteria that should be considered for including a database, size is one of the most relevant. Often biomarker studies are small because of costs and working time required by technical procedures; however, a certain number of samples processed are necessary to achieve good practical skill. An important issue is the use of the same population by several different laboratories, with the possibility of duplicating the same subjects when pooling. This is typical of studies involving biomarkers, where a population of interest, e.g. a case–control study on lung cancer, is shared by laboratories performing different assays, or different genotypes. This has to be checked carefully by cross-checking IDs, ages, dates of birth if available, or the names of the authors on related publications.

Ethical issues

Ethical considerations should be made in all phases of pooling studies. A suitable strategy is to request the original investigators to send data to the coordinating centre without personal identifiers but with a numerical ID. A second ID can be centrally assigned to each individual when the data are received, and this ID will then be used in the analysis. This will make it impossible for the scientist who pools the data to identify and/or contact the subjects included in the dataset.

Usually pooled analysis involves studies conducted in different countries, which implies a variety of ethical regulations (see Chapter 22). The investigator should strictly adhere to all existing ethical and safety provisions applicable in the countries in which the research is carried out, and in no circumstances should data with personal identifications be sent from one country to another.

References

Blettner M *et al*. 1999 Traditional reviews, meta-analyses and pooled analyses in epidemiology. *Int J Epidemiol* **28**: 1–9.

Bogardus ST Jr, Concato J, Feinstein AR. 1999. Clinical epidemiological quality in molecular genetic research: the need for methodological standards. *J Am Med Assoc* **281**: 1919–1926.

Bonassi S *et al*. 2001. HUman MicroNucleus project: international database comparison for results with the cytokinesis-block micronucleus assay in human lymphocytes: I. Effect of laboratory protocol, scoring criteria, and host factors on the frequency of micronuclei. *Environ Mol Mutagen* **37**: 31–45.

Bonassi S *et al*. 1999. Analysis of correlated data in human biomonitoring studies. The case of high sister chromatid exchange frequency cells. *Mutat Res* **438**: 13–21.

Bonassi S *et al*. 2000. Chromosomal aberrations in lymphocytes predict human cancer independently of exposure to carcinogens. European Study Group on Cytogenetic Biomarkers and Health. *Cancer Res* **60**: 1619–1625.

Bonassi S *et al*. 1996. Is human exposure to styrene a cause of cytogenetic damage? Re-analysis of the available evidence. *Biomarkers* **1**: 217–225.

DerSimonian R, Laird N. 1986. Meta-analysis in clinical trials. *Control Clin Trials* **7**: 177–188.

Egger M *et al*. 1997. Bias in meta-analysis detected by a simple, graphical test. *Br Med J* **315**: 629–634.

Fenech M *et al*. 1999. The HUman MicroNucleus Project – an international collaborative study on the use of the micronucleus technique for measuring DNA damage in humans. *Mutat Res* **428**: 271–283.

Friedenreich CM. 1993. Methods for pooled analyses of epidemiologic studies. *Epidemiology* **4**: 295–302.

Friedenreich CM. 2002. Commentary: improving pooled analyses in epidemiology. *Int J Epidemiol* **31**: 86–87.

Garte S *et al*. 2001. Metabolic gene polymorphism frequencies in control populations. *Cancer Epidemiol Biomarkers Prevent* **10**: 1239–1248.

Golgstein H. 1995. *Multilevel Statistical Models*. Halsted: New York.

Greenland S. 1987. Quantitative methods in the review of epidemiologic literature. *Epidemiol Rev* **9**: 1–30.

Hagmar L *et al*. 1998. Chromosomal aberrations in lymphocytes predict human cancer: a report from the European Study Group on Cytogenetic Biomarkers and Health (ESCH). *Cancer Res* **58**: 4117–4121.

Higgins JP, Thompson SG. 2002. Quantifying heterogeneity in a meta-analysis. *Stat Med* **21**: 1539–1558.

Higgins JP *et al*. 2003. Measuring inconsistency in meta-analyses. *Br Med J* **327**: 557–560.

Hobbs CG *et al*. 2006. Human papillomavirus and head and neck cancer: a systematic review and meta-analysis. *Clin Otolaryngol* **31**: 259–266.

Mantel N, Haenszel W. 1959. Statistical aspects of the analysis of data from retrospective studies of disease. *J Natl Cancer Inst* **22**: 719–748.

Normand SL. 1999. Meta-analysis: formulating, evaluating, combining, and reporting. *Stat Med* **18**: 321–359.

Prentice RL, Thomas DB. 1987. On the epidemiology of oral contraceptives and disease. *Adv Cancer Res* **49**: 285–401.

Raimondi S *et al*. 2006. Meta- and pooled analysis of GSTT1 and lung cancer: a HUGE-GSEC review. *Am J Epidemiol* **164**: 1027–1042.

Rasbash J *et al*. 1999. MLwiN version 2.0, Multilevel Models Project. Institute of Education, University of London.

Smerhovsky Z *et al*. 2001. Risk of cancer in an occupationally exposed cohort with increased level of chromosomal aberrations. *Environ Health Perspect* **109**: 41–45.

Smerhovsky Z *et al*. 2002. Increased risk of cancer in radon-exposed miners with elevated frequency of chromosomal aberrations. *Mutat Res* **514**: 165–176.

Stroup DF *et al*. 2000. Meta-analysis of observational studies in epidemiology: a proposal for reporting. Meta-analysis Of Observational Studies in Epidemiology (MOOSE) group. *J Am Med Assoc* **283**: 2008–2012.

Taioli E. 1999. International collaborative study on genetic susceptibility to environmental carcinogens. *Cancer Epidemiol Biomarkers Prevent* **8**: 727–728.

Taioli E, Bonassi S. 2002. Methodological issues in pooled analysis of biomarker studies. *Mutat Res* **512**: 85–92.

Wolf F. 1986. *Meta-Analysis: Quantitative Methods for Research Synthesis*. Sage: Newbury Park, CA.

Yusuf S *et al*. 1985. Beta blockade during and after myocardial infarction: an overview of the randomized trials. *Prog Cardiovasc Dis* **27**: 335–371.

16

Analysis of Complex Datasets

Jason H. Moore,[1,2] Margaret R. Karagas[1] and Angeline S. Andrew[1]

[1]*Dartmouth Medical School, Lebanon, NH,* [2]*University of New Hampshire, Durham, NH, and* [2]*University of Vermont, Burlington, VT, USA*

16.1. INTRODUCTION

Molecular epidemiology is an interdisciplinary field that combines traditional epidemiology with pieces of other disciplines, such as human genetics and genomics. An important goal of molecular epidemiology is to understand the mapping relationship between inter-individual variation in DNA sequences (i.e. the genome), variation in environmental exposure (i.e. ecology) and variation in disease susceptibility (i.e. the phenotype). Stated another way, how do one or more changes in an individual's DNA sequence increase or decrease their risk of developing a common disease, such as cancer or cardiovascular disease, through complex networks of biomolecules that are hierarchically organized, highly interactive and dependent on ecology? Understanding the role of DNA sequences in disease susceptibility is likely to improve diagnosis, prevention and treatment. Success in this important public health endeavour will depend critically on the degree of non-linearity in the mapping between genotype to phenotype, i.e. how complex is the transfer of information from the genome to the phenotype of interest? Non-linearities can arise from phenomena such as locus heterogeneity (i.e. different DNA sequence variations leading to the same phenotype), phenocopy (i.e. environmentally determined phenotypes that do not have a genetic basis) and the dependence of genotypic effects on environmental factors (i.e. gene–environment interactions or plastic reaction norms) and genotypes at other loci (i.e. gene–gene interactions or epistasis).

All of these phenomena have been recently reviewed and discussed by Thornton-Wells *et al.* (2004), who call for an analytical retooling to address these complexities. We direct the reader elsewhere for recent work on locus heterogeneity (Thornton-Wells *et al.* 2004, 2006). We focus here on non-linearities due to interactions between multiple genetic and environmental factors.

16.2. GENE—ENVIRONMENT INTERACTION

The role of the environment in genetics has had a long history. The German researcher Woltereck (1909) coined the term 'reaction norm' to refer to the set of phenotypes that can be produced by a genotype in different environmental contexts. Reaction norms or gene–environment interactions were revived by Schmalhausen (1949) and recently reviewed in books by Schlicting and Pigliucci (1998) and Pigliucci (2001). An excellent basic science example of gene–environment interactions can be found in a study of *E. coli* by Remold and Lenski (2004). In this study, 18 random insertion mutations were introduced in *E. coli* on five different genetic backgrounds exposed to two different resource environments (glucose or maltose). The authors of the study found no examples of an environmental effect on fitness. However, six of the 18 mutations had an effect on fitness that was dependent on both genetic background and environmental context demonstrating a plastic reaction norm. These functional studies in model organisms lay

Molecular Epidemiology of Chronic Diseases, Edited by C. P. Wild, P. Vineis, and S. Garte

an important foundation for understanding the role of the environment in modulating genetic effects in humans. The importance of gene–environment interactions in human disease has been recently reviewed by Hunter (2005).

16.3. GENE—GENE INTERACTION

Gene–gene interaction or epistasis has been recognized for many years as a deviation from the simple inheritance patterns observed by Mendel (Bateson 1909) or deviations from additivity in a linear statistical model (Fisher 1918) and is likely due, in part, to canalization or mechanisms of stabilizing selection that evolve robust (i.e. redundant) gene networks (Waddington 1942, 1957; Gibson and Wagner 2000; Proulx and Phillips 2005).

Epistasis has been defined in multiple different ways (e.g. Hollander 1955; Philips 1998; Brodie 2000). We have reviewed two types of epistasis, biological and statistical (Moore 2005; Moore and Williams 2005). *Biological epistasis* results from physical interactions between biomolecules (e.g. DNA, RNA, proteins, enzymes, etc.) and occurs at the cellular level in an individual. This type of epistasis is what Bateson (1909) had in mind when he coined the term. *Statistical epistasis*, on the other hand, occurs at the population level and is realized when there is inter-individual variation in DNA sequences. The statistical phenomenon of epistasis is what Fisher (1918) had in mind. The relationship between biological and statistical epistasis is often confusing but will be important to understand if we are to make biological inferences from statistical results (Moore 2005; Moore and Williams 2005). The focus of this chapter is the detection, characterization and interpretation of statistical patterns of interaction in human populations. since interaction or synergy among predictors in a dataset is one of the primary sources of complexity.

16.4. STATISTICAL INTERACTION

As discussed above, interactions among biomolecules and environmental agents occurs at the cellular level in an individual. The focus of this chapter is detecting statistical patterns of interaction in human populations. Consider the following simple example of statistical epistasis in the form

Table 16.1 Penetrance values for genotypes from two SNPs

SNP B	SNP A		
	AA (0.25)	*Aa* (0.50)	*aa* (0.25)
BB (0.25)	0	0.1	0
Bb (0.50)	0.1	0	0.1
bb (0.25)	0	0.1	0

of a penetrance function. Penetrance is simply the probability (p) of disease (D) given a particular combination of genotypes (G) that was inherited (i.e. $p[D|G]$). The model illustrated in Table 16.1 is an extreme example of epistasis among two single nucleotide polymorphisms (SNPs), A and B. Let's assume that genotypes *AA*, *aa*, *BB*, and *bb* have population frequencies of 0.25, while genotypes *Aa* and *Bb* have frequencies of 0.5 (values in parentheses in Table 16.1). What makes this model interesting is that disease risk is entirely dependent on the particular *combination* of genotypes inherited. Individuals have a very high risk of disease if they inherit *Aa* or *Bb* but not both (i.e. the exclusive OR function). The penetrance for each individual genotype in this model is 0.5 and is computed by summing the products of the genotype frequencies and penetrance values. Thus, in this model there is no difference in disease risk for each single genotype as specified by the single-genotype penetrance values (all 0.5). This model is labelled M170 by Li and Reich (2000) in their categorization of genetic models involving two SNPs and is an example of a pattern that is not linearly separable. Heritability or the size of the genetic effect is a function of these penetrance values (e.g. Culverhouse *et al.* 2002). The model specified in Table 16.1 has a heritability of 0.053, which represents a relatively small genetic effect size. This model is a special case where all of the heritability is due to epistasis.

Detecting statistical patterns of interaction

As discussed above, one of the early definitions of epistasis was deviation from additivity in a linear model (Fisher 1918). The linear model plays a very important role in modern epidemiology because it has solid theoretical foundation, is easy to implement using a wide-range of different software packages, and is easy to interpret. Despite these good reasons to use linear models, they do have limitations for detecting non-linear patterns of interaction (Moore and Williams 2002). The first problem is that modelling interactions requires looking at combinations of variables. Considering multiple variables simultaneously is challenging because the available data gets spread thinly across multiple combinations of genotypes, for example. Estimation of parameters in a linear model can be problematic when the data are sparse. The second problem is that linear models are often implemented such that interaction effects are only considered after independent main effects are identified. This certainly makes model fitting easier but assumes that the important predictors will have main effects. For example, the FITF approach of Millstein *et al.* (2006) provides a powerful logistic regression approach to detecting interactions but conditions on main effects. Moore (2003) argues that this is an unrealistic assumption for common human diseases. The limitations of the linear model and other parametric statistical approaches has motivated the development of computational approaches, such as those from machine learning and data mining (McKinney *et al.* 2006), that make fewer assumptions about the functional form of the model and the effects being modelled. We review below two different data-mining approaches, random forests and multifactor dimensionality reduction (MDR), that can be applied to detecting gene–gene and gene–environment interactions in molecular epidemiology studies. Since the focus of this review is interaction as a source of complexity, the reader is directed elsewhere for reviews of methods for detecting independent main effects (see Chapter 14). A recent series of seven reviews summarize many of the basics of genetic and epidemiological association studies in human populations, thus providing a starting point for those needing to learn more about basic analytical methods such as logistic regression (see Burton *et al.* 2005; Teare and Barrett 2005; Cordell and Clayton 2005; Palmer and Cardon 2005; Hattersley and McCarthy 2005; Hopper *et al.* 2005; Smith *et al.* 2005). Several other recent reviews also provide some basic concepts (e.g. Balding 2006).

Decision trees, classification trees and random forests

Decision trees (sometimes called classification trees) are one of the most widely used methods for modelling the relationship between one of more variables and a discrete endpoint, such as case–control status (Mitchell 1997). A decision tree classifies subjects as case or control, for example, by sorting them through a tree from node to node, where each node is a variable with a decision rule that guides that subject through different branches of the tree to a leaf that provides its classification. The primary advantage of this approach is that it is simple and the resulting tree can be interpreted as IF–THEN rules that are easy to understand. For example, a dominant genetic model of genotype data coded *{AA = 0, Aa = 1, aa = 2}* might look like IF genotype at SNP1 < 2 then case ELSE control. In this simple model, the root node of the tree would be SNP1 with decision rule < 2 and leaves equal to case and control. Additional nodes or variables below the root node allows hierarchical dependencies (i.e. interactions) to be modelled. Random forests (RF) build on the decision tree idea and have been used to detect gene–gene and gene–environment interactions in molecular epidemiology studies.

An RF is a collection of individual decision tree classifiers, where each tree in the forest has been trained using a bootstrap sample of instances (i.e. subjects) from the data, and each variable or attribute in the tree is chosen from among a random subset of attributes (Breiman 2001). Classification of instances is based upon aggregate voting over all trees in the forest.

Individual trees are constructed as follows from data having *n* samples and *M* attributes:

1. Choose a training set by selecting *n* samples, with replacement, from the data.
2. At each node in the tree, randomly select *m* attributes from the entire set of *M* attributes in the data (the magnitude of *m* is constant throughout the forest building).
3. Choose the best split at that node from among the *m* attributes.
4. Iterate the second and third steps until the tree is fully grown (no pruning).

Repetition of this algorithm yields a forest of trees, each of which have been trained on bootstrap samples of instances. Thus, for a given tree, certain instances will have been left out during training. Prediction error is estimated from these 'out-of-bag' instances. The out-of-bag instances are also used to estimate the importance of particular attributes via permutation testing. If randomly permuting values of a particular attribute does not affect the predictive ability of trees on out-of-bag samples, that attribute is assigned a low importance score (Bureau et al. 2005).

The decision trees comprising a RF provide an explicit representation of attribute interaction that is readily applicable to the study of gene–gene or gene–environment interactions. These models may uncover interactions among genes and/or environmental factors that do not exhibit strong marginal effects (Cook et al. 2004). Additionally, tree methods are suited to dealing with certain types of genetic heterogeneity, since early splits in the tree define separate model subsets in the data (Lunetta et al. 2004). Random forests capitalize on the benefits of decision trees and have demonstrated excellent predictive performance when the forest is diverse (i.e. trees are not highly correlated with each other) and composed of individually strong classifier trees (Breiman 2001). The RF method is a natural approach for studying gene–gene or gene–environment interactions because importance scores for particular attributes take interactions into account without demanding a pre-specified model (Lunetta et al. 2004). However, most current implementations of the importance score are calculated in the context of all other attributes in the model. Therefore, assessing the interactions between particular sets of attributes must be done through careful model interpretation, although there has been preliminary success in jointly permuting explicit sets of attributes to capture their interactive effects (Bureau et al. 2005).

In selecting functional SNP attributes from simulated case–control data, RFs outperform traditional methods, such as the Fisher exact test, when the 'risk' SNPs interact, and the relative superiority of the RF method increases as more interacting SNPs are added to the model (Lunetta et al. 2004). Random forests have also shown to be more robust in the presence of noise SNPs relative to methods that rely on main effects, such as Fisher's exact test (Bureau et al. 2005). Initial results of RF applications to genetic data in studies of asthma (Bureau et al. 2005), breast cancer (Schwender et al. 2004), DNA repair capacity (Jones et al. 2007) and drug response (Sabbah and Darlu 2006) are encouraging, and it is anticipated that random forests will prove a useful tool for detecting gene–gene interactions. They may also be useful when multiple different data types are present (Reif et al. 2006). The primary limitation of tree-based methods is that the standard implementations condition on main effects. That is, the algorithm finds the best single variable for the root node before adding additional variables as nodes in the model. Random forests can be implemented using the open-source Weka data mining software package (Witten and Frank 2005).

Multifactor dimensionality reduction (MDR)

Multifactor dimensionality reduction (MDR) was developed as a non-parametric (i.e. no parameters are estimated) and genetic model-free (i.e. no genetic model is assumed) data-mining strategy for identifying combinations of discrete genetics and environmental factors that are predictive of a discrete clinical endpoint (Ritchie et al. 2001, 2003; Hahn et al. 2003; Hahn and Moore 2004; Moore 2004, 2007; Moore et al. 2006). Unlike most other methods, MDR was designed to detect interactions in the absence of detectable main effects and thus complements approaches such as logistic regression and random forests. At the heart of the MDR approach is a feature or attribute construction algorithm that creates a new variable or attribute by pooling, for example, genotypes from multiple SNPs. The process of defining a new attribute as a function of two or more other attributes is referred to as constructive induction or attribute construction and was first developed by Michalski (1983). Constructive induction using the MDR kernel is accomplished in the following way. Given a threshold T, a multilocus genotype combination is considered high-risk if the ratio of cases (subjects with disease) to controls (healthy subjects) exceeds T, else it is considered low-risk. Genotype combinations considered to be high-risk are labelled G_1, while those considered low-risk are labelled G_0.

This process constructs a new one-dimensional attribute with levels G_0 and G_1. It is this new single variable that is assessed using any classification method. The MDR method is based on the idea that changing the representation space of the data will make it easier for a classifier, such as a decision tree or a naive Bayes learner, to detect attribute dependencies. Open-source software in Java and C are freely available from www.epistasis.org.

Consider the simple example presented above and in Table 16.1. This penetrance function was used to simulate a dataset with 200 cases (diseased subjects) and 200 controls (healthy subjects) for a total of 400 instances. The list of attributes included the two functional interacting SNPs (SNP1 and SNP2) in addition to three randomly generated SNPs (SNP3–SNP5). All attributes in these datasets are categorical. The SNPs each have three levels (0, 1, 2), while the class has two levels (0, 1) that code controls and cases. Figure 16.1A illustrates the distribution of cases (left bars) and controls (right bars) for each of the three genotypes of

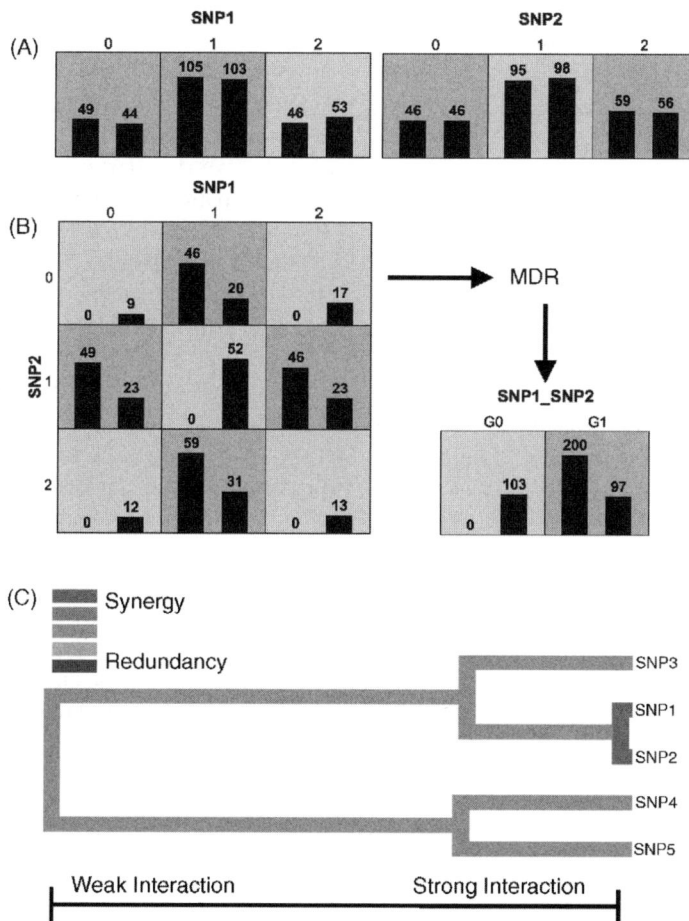

Figure 16.1 (A) Distribution of cases (left bars) and controls (right bars) across three genotypes (0, 1, 2) for two simulated interacting SNPs. Note that the ratios of cases to controls for these two SNPs are nearly identical. The dark-shaded cells signify 'high-risk' genotypes. (B) Distribution of cases and controls across nine two-locus genotype combinations. Note that considering the two SNPs jointly reveals larger case–control ratios. Also illustrated is the use of the MDR attribute construction function that produces a single attribute (SNP1_SNP2) from the two SNPs. (C) An interaction dendrogram summarizing the information gain associated with constructing pairs of attributes using MDR. The length of the connection between two SNPs is inversely related to the strength of the information gain. Red lines indicate a positive information gain that can be interpreted as synergistic interaction. Brown lines indicate no information gain

SNP1 and SNP2. The dark-shaded cells have been labelled 'high-risk' using a threshold of $T = 1$. The light-shaded cells have been labelled 'low-risk'. Note that when considered individually, the ratio of cases to controls is close to 1 for each single genotype. Figure 16.1B illustrates the distribution of cases and controls when the two functional SNPs are considered jointly. Note the larger ratios that are consistent with the genetic model in Table 16.1. Also illustrated in Figure 16.1B is the distribution of cases and controls for the new single attribute constructed using MDR. This new single attribute captures much of the information from the interaction and could be assessed using logistic regression, for example.

The MDR method has been successfully applied to detecting gene–gene and gene–environment interactions for a variety of common human diseases and clinical endpoints including, for example, antiretroviral therapy (Haas et al. 2006), asthma (Chan et al. 2006; Millstein et al. 2006), atrial fibrillation (Tsai et al. 2004; Asselbergs et al. 2006; Moore et al. 2006), autism (Coutinho et al. 2007), bladder cancer (Andrew et al. 2006), cervical cancer (Chung et al. 2006), coronary calcification (Bastone et al. 2004), coronary artery disease (Tsai et al. 2007; Agirbasli et al. 2006), diabetic nephropathy (Hsieh et al. 2006), drug metabolism (Sabbah et al. 2006), essential hypertension (Williams et al. 2004), familial amyloid polyneuropathy (Soares et al. 2005), multiple sclerosis (Motsinger et al. 2007; Brassat et al. 2006), myocardial infarction (Coffey et al. 2004; Mannila et al. 2006), osteoporosis (Xiong et al. 2006), preterm birth (Menon et al. 2006), prostate cancer (Xu et al. 2005), schizophrenia (Qin et al. 2005; Yasuno et al. 2006), sporadic breast cancer (Ritchie et al. 2001; Oestergaard et al. 2006; Nordgard et al. 2007) and type II diabetes (Cho et al. 2004). The MDR method has also been proposed for pharmacogenetics and toxicogenetics (e.g. Wilke et al. 2005).

Statistical interpretation of interaction models

Random forests and MDR are powerful methods for detecting gene–gene and gene–environment interactions in epidemiologic studies of common human diseases. The models that these methods produce are by nature multidimensional and thus difficult to interpret. For example, an interaction model with four SNPs, each with three genotypes, summarizes 81 different genotype (i.e. level) combinations (i.e. 3^4). How do each of these level combinations relate back to biological processes in a cell? Why are some combinations associated with high risk for disease and some associated with low risk for disease? Moore et al. (2006) have proposed using information-theoretic approaches with graph-based models to provide both a statistical and a visual interpretation of models from MDR and other methods, such as symbolic discriminant analysis (Moore et al. 2007). Statistical interpretation should facilitate biological interpretation because it provides a deeper understanding of the relationship between the attributes and the class variable. We describe next the concept of interaction information and how it can be used to facilitate statistical interpretation.

Jakulin and Bratko (2003) have provided a metric for determining the gain in information about a class variable (e.g. case–control status) from merging two attributes into one (i.e. attribute construction) over that provided by the attributes independently. This measure of *information gain* allows us to gauge the benefit of considering two (or more) attributes as one unit. While the concept of information gain is not new (McGill 1954), its application to the study of attribute interactions has been the focus of several recent studies (Jakulin and Bratko 2003; Jakulin et al. 2003). Consider two attributes, A and B, and a class label C. Let $H(X)$ be the Shannon entropy (see Pierce 1980) of X. The information gain (IG) of A, B and C can be written as equation (1) and defined in terms of Shannon entropy (equations 2 and 3):

$$IG(ABC) = I(A;B|C) - I(A;B) \qquad (1)$$

$$I(A;B|C) = H(A|C) + H(B|C) \\ - H(A, B|C) \qquad (2)$$

$$I(A;B) = H(A) + H(B) - H(A,B) \qquad (3)$$

The first term in (1), $I(A;B|C)$, measures the *interaction* of A and B. The second term, $I(A;B)$, measures the *dependency* or correlation between A and B. If this difference is positive, then there is evidence for an attribute interaction that cannot be linearly decomposed. If the difference is negative, then the information between *A* and *B* is redundant. If the difference is zero, then there is evidence of conditional independence or a mixture of synergy and redundancy. These measures of interaction information can be used to construct interaction graphs (i.e. network diagrams) and an interaction dendrograms using the entropy estimates from the algorithms described first by Jakulin and Bratko (2003) and more recently in the context of genetic analysis by Moore *et al.* (2006). Interaction graphs are comprised of a node for each attribute with pairwise connections between them. The percentage of entropy removed (i.e. information gain) by each attribute is visualized for each node. The percentage of entropy removed for each pairwise MDR product of attributes is visualized for each connection. Thus, the independent main effects of each polymorphism can be quickly compared to the interaction effect. Additive and non-additive interactions can be quickly assessed and used to interpret the MDR model which consists of distributions of cases and controls for each genotype combination. Positive entropy values indicate synergistic interaction while negative entropy values indicate redundancy.

Interaction dendrograms are also a useful way to visualize interaction (Jakulin and Bratko 2003; Moore *et al.* 2006). Here, hierarchical clustering is used to build a dendrogram that places strongly interacting attributes close together at the leaves of the tree. Jakulin and Bratko (2003) define the following dissimilarity measure, *D* (equation 5), that is used by a hierarchical clustering algorithm to build a dendrogram. The value of 1000 is used as an upper bound to scale the dendrograms:

$$D(A,B) = |I(A;B;C)|^{-1} \text{ if } |I(A;B;C)|^{-1} < 1000$$
$$1000 \text{ otherwise} \tag{5}$$

Using this measure, a dissimilarity matrix can be estimated and used with hierarchical cluster analysis to build an interaction dendrogram. This facilitates rapid identification and interpretation of pairs of interactions. The algorithms for the entropy-based measures of information gain are implemented in the open-source MDR software package available from www.epistasis.org. Output in the form of interaction dendrograms is provided. Figure 16.1C illustrates an interaction dendrogram

for the simple simulated dataset described above. Note the strong synergistic relationship between SNP1 and SNP2. All other SNPs are independent, which is consistent with the simulation model.

16.5. CASE STUDY: BLADDER CANCER

Consider the following case study. Andrew *et al.* (2006) carried out an epidemiological study to identify genetic and environmental predictors of bladder cancer susceptibility in a large sample of Caucasians (*n* = 914) from New Hampshire. This study focused specifically on genes that play an important role in the repair of DNA sequences that have been damaged by chemical compounds (e.g. carcinogens). Seven SNPs were measured, including two from the *X-ray repair cross-complementing group 1* gene (*XRCC1*), one from the *XRCC3* gene, two from the *xeroderma pigmentosum group D (XPD)* gene, one from the *nucleotide excision repair* gene (*XPC*) and one from the *AP endonuclease 1* gene (*APE1*). Each of these genes plays an important role in DNA repair. Smoking is a known risk factor for bladder cancer and was included in the analysis, along with gender and age, for a total of 10 attributes. Age was discretized to > or ≤ 50 years.

A parametric statistical analysis of each attribute individually revealed a significant independent main effect of smoking, as expected. However, none of the measured SNPs were significant predictors of bladder cancer individually. Andrew *et al.* (2006) used MDR to exhaustively evaluate all possible two-, three- and four-way interactions among the genetic and environmental variables. For each combination of attributes, a single constructed attribute was evaluated using a naïve Bayes classifier. Training and testing accuracy were estimated using 10-fold cross-validation. A best model was selected that maximized the testing accuracy. The best model had a testing accuracy of 0.66 and included two SNPs from the *XPD* gene and smoking. The distribution of cases and controls with each genotype/smoking combination is illustrated in Figure 16.2A. Also illustrated in Figure 16.2A is the single variable constructed by MDR. The empirical *p* value of this model was < 0.001, suggesting that a testing accuracy of ≥ 0.66 is unlikely under the null hypothesis of no association, as assessed using a 1000-fold permutation test.

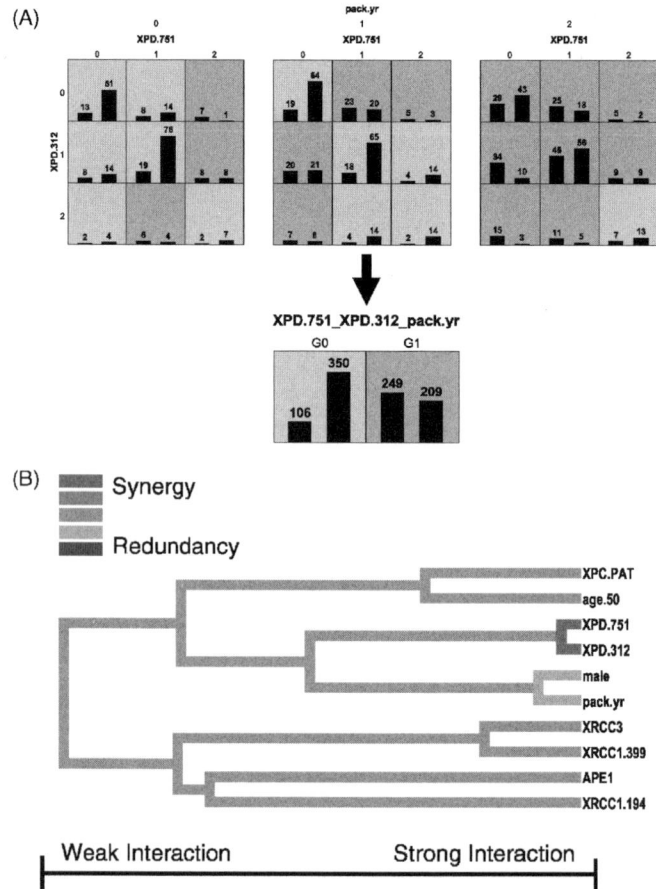

Figure 16.2 (A) Distribution of cases (left bars) and controls (right bars) for each genotype-smoking combination in the bladder cancer example. (B) An interaction dendrogram summarizing the information gain associated with constructing pairs of attributes using MDR. Note that the *XPD* polymorphisms are connected by a short red line, indicating a strong synergistic interaction. Pack years of smoking is connected to the *XPD* polymorphisms by a brown line, suggesting that the effects are independent or additive. It is also interesting to note that gender is connected to smoking with a short green line, indicating that these two attributes are correlated or redundant

Decomposition of this model using the measures of information gain described above demonstrated that the effects of the two *XPD* SNPs were non-additive or synergistic, suggestive of non-linear interaction. The same analysis indicated that the effect of smoking was mostly independent from the gene–gene interaction effect. Figure 16.2B illustrates an interaction dendrogram summarizing these measures of information gain. Variables connected by short lines have stronger interactions than variables connected by longer lines. The colour of the line indicates the type of interaction (i.e.

synergistic or redundant). Note that the two *XPD* polymorphisms are connected by a short red line, indicating a strong synergistic interaction. In other words, combining these two polymorphisms into a single variable using MDR provides more information about case–control status than considering them additively. The short green line connecting pack years of smoking to gender suggests that these two variables are correlated or redundant. In other words, information about case–control status is lost when putting these two variables together. The long yellow line between smoking and the two

XPD polymorphisms indicates independence. It is important to note that parametric logistic regression was unable to model this three-attribute interaction, owing to lack of convergence. This study illustrates the power of MDR to identify complex relationships between genes, environmental factors, such as smoking, and susceptibility to a common disease, such as bladder cancer. The MDR approach works well in the context of a candidate gene study, but how does it scale to genome-wide analysis of thousands of attributes?

16.6. GENOME-WIDE ANALYSIS

Biological and biomedical sciences are undergoing an information explosion and an understanding implosion. That is, our ability to generate data is far outpacing our ability to interpret it. This is especially true in the domain of human genetics, where it is now technically and economically feasible to measure thousands of SNPs from across the human genome. It is anticipated that at least one SNP occurs approximately every 100 nucleotides across the 3×10^9 nucleotide human genome. An important goal in human genetics is to determine which of the many thousands of SNPs are useful for predicting who is at risk for common diseases. This 'genome-wide' approach is expected to revolutionize the genetic analysis of common human diseases (Hirschhorn and Daly 2005; Wang *et al.* 2005) and is quickly replacing the traditional 'candidate gene' approach that focuses on several genes selected by their known or suspected function.

Moore and Ritchie (2004) have outlined three significant challenges that must be overcome if we are to successfully identify genetic predictors of health and disease using a genome-wide approach. First, powerful data-mining and machine-learning methods will need to be developed to statistically model the relationship between combinations of DNA sequence variations and disease susceptibility. Traditional methods, such as logistic regression, have limited power for modelling high-order non-linear interactions (Moore and Williams 2002). The MDR approach was discussed above as an alternative to logistic regression. A second challenge is the selection of genetic features or attributes that should be included for analysis. If interactions

between genes explain most of the heritability of common diseases, then combinations of DNA sequence variations will need to be evaluated from a list of thousands of candidates. Filter and wrapper methods will play an important role because there are more combinations than can be exhaustively evaluated. A third challenge is the interpretation of gene–gene interaction models. Although a statistical model can be used to identify DNA sequence variations that confer risk for disease, this approach cannot be translated into specific prevention and treatment strategies without interpreting the results in the context of human biology. Making aetiological inferences from computational models may be the most important and the most difficult challenge of all (Moore and Williams 2005).

Combining the concept of interaction described above with the challenge of variable selection yields what Goldberg (2002) calls a *needle-in-a-haystack* problem. That is, there may be a particular combination of SNPs that, together with the right non-linear function, are a significant predictor of disease susceptibility. However, individually they may not look any different than thousands of other SNPs that are not involved in the disease process and are thus noisy. Under these models, the learning algorithm is truly looking for a genetic needle in a genomic haystack. A recent report from the International HapMap Consortium (Altshuler *et al.* 2005) suggests that approximately 300 000 carefully selected SNPs may be necessary to capture all of the relevant variation across the Caucasian human genome. Assuming that this is true (it is probably a lower bound), we would need to scan 4.5×10^{10} pairwise combinations of SNPs to find a genetic needle. The number of higher-order combinations is astronomical. What is the optimal approach to this problem?

There are two general approaches to selecting attributes for predictive models. The filter approach pre-processes the data by algorithmically or statistically assessing the quality or relevance of each variable and then using that information to select a subset for classification. The wrapper approach iteratively selects subsets of attributes for classification using either a deterministic or stochastic algorithm. The key difference between the two approaches is that the classifier plays no role in selecting which attributes to consider in the filter approach. As Freitas (2002) reviews, the advantage of the filter is speed, while the wrapper approach has the potential to do a better job of classification. We discuss each of these general approaches in turn for the specific problem of detecting epistasis or gene–gene interactions on a genome-wide scale.

A filter strategy for genome-wide analysis

There are many different statistical and computational methods for determining the quality of attributes. A standard strategy in human genetics and epidemiology is to assess the quality of each SNP, using a χ^2 test of independence, followed by a correction of the significance level that takes into account an increased false-positive (i.e. type I error) rate due to multiple tests. This is a very efficient filtering method but it ignores the dependencies or interactions between genes. Kira and Rendell (1992) developed an algorithm called Relief that is capable of detecting attribute dependencies. Relief estimates the quality of attributes through a type of nearest-neighbour algorithm that selects neighbours (instances) from the same class and from the different class, based on the vector of values across attributes. Weights (W) or quality estimates for each attribute (A) are estimated based on whether the nearest neighbour (nearest hit, H) of a randomly selected instance (R) from the same class and the nearest neighbour from the other class (nearest miss, M) have the same or different values. This process of adjusting weights is repeated for m instances. The algorithm produces weights for each attribute, ranging from -1 (worst) to $+1$ (best). The Relief pseudocode is outlined below:

```
set all weights W[A] = 0
for i = 1 to m do begin
        randomly select an instance R_i
        find nearest hit H and nearest miss M
        for A = 1 to a do
                W[A] = W[A] − diff(A, R_i, H)/m
                        + diff(A, R_i, M)/m
    end
```

The function diff(A, I_1, I_2) calculates the difference between the values of the attribute A for two

instances, I_1 and I_2. For nominal attributes such as SNPs, it is defined as:

diff(A, I_1, I_2) = 0 if genotype(A, I_1)
              = genotype (A, I_2), 1 otherwise

The time complexity of Relief is $O(m \times n \times a)$, where m is the number of instances randomly sampled from a dataset with n total instances and a attributes. Kononenko (1994) improved upon Relief by choosing n nearest neighbours instead of just one. This new ReliefF algorithm has been shown to be more robust to noisy attributes (Kononenko 1994; Robnik-Šikonja and Kononenko 2001, 2003) and is widely used in data-mining applications.

ReliefF is able to capture attribute interactions because it selects nearest neighbours using the entire vector of values across all attributes. However, this advantage is also a disadvantage, because the presence of many noisy attributes can reduce the signal the algorithm is trying to capture. Moore and White (2007a) proposed a 'tuned' ReliefF algorithm (TuRF) that systematically removes attributes that have low-quality estimates, so that the ReliefF values of the remaining attributes can be re-estimated. The pseudocode for TuRF is outlined below:

```
let a be the number of attributes
for i = 1 to n do begin
        estimate ReliefF
        sort attributes
        remove worst n/a attributes
    end
return last ReliefF estimate for each attribute
```

The motivation behind this algorithm is that the ReliefF estimates of the true functional attributes will improve as the noisy attributes are removed from the dataset.

Moore and White (2007a) carried out a simulation study to evaluate the power of ReliefF, TuRF, and a naïve χ^2 test of independence for selecting functional attributes in a filtered subset. Five genetic models, in the form of penetrance functions (e.g. Table 16.1), were generated. Each model consisted of two SNPs that define a non-linear relationship with disease susceptibility. The heritability of each model was 0.1, which reflects a moderate to small genetic effect size. Each of the five models was used to generate 100 replicate datasets with sample sizes of 200, 400, 800, 1600, 3200 and 6400. This range of sample sizes represents a spectrum that is consistent with small- to medium-sized genetic studies. Each dataset consisted of an equal number of case (disease) and control (no disease) subjects. Each pair of functional SNPs was combined within a genome-wide set of 998 randomly generated SNPs for a total of 1000 attributes. A total of 600 datasets were generated and analysed.

ReliefF, TuRF and the univariate χ^2 test of independence were applied to each of the datasets. The 1000 SNPs were sorted according to their quality, using each method, and the top 50, 100, 150, 200, 250, 300, 350, 400, 450 and 500 SNPs out of 1000 were selected. From each subset we counted the number of times the two functional SNPs were selected out of each set of 100 replicates. This proportion is an estimate of the power, or how likely we are to find the true SNPs if they exist in the dataset. The number of times each method found the correct two SNPs was statistically compared. A difference in counts (i.e. power) was considered statistically significant at a type I error rate of 0.05. Moore and White (2007a) found that the power of ReliefF to pick (filter) the correct two functional attributes was consistently better ($p \leq 0.05$) than a naïve χ^2 test of independence across subset sizes and models when the sample size was 800 or larger. These results suggest that ReliefF is capable of identifying interacting SNPs with a moderate genetic effect size (heritability = 0.1) in moderate sample sizes. Next, Moore and White (2007a) compared the power of TuRF to the power of ReliefF. They found that the TuRF algorithm was consistently better ($p \leq 0.05$) than ReliefF across small SNP subset sizes (50, 100 and 150) and across all five models when the sample size was 1600 or larger. These results suggest that algorithms based on ReliefF show promise for filtering interacting attributes in this domain. The disadvantage of the filter approach is that important attributes might be discarded prior to analysis. Stochastic search or wrapper methods provide a flexible alternative. The ReliefF algorithm has been included in the open-source MDR software package (available from: www.epistasis. org/mdr.html).

A wrapper strategy for genome-wide analysis

Stochastic search or wrapper methods may be more powerful than filter approaches because no attributes are discarded in the process. As a result, every attribute retains some probability of being selected for evaluation by the classifier. There are many different stochastic wrapper algorithms that can be applied to this problem. Moore and White (2007b) have explored the use of genetic programming (GP), which is an automated computational discovery tool that is inspired by Darwinian evolution and natural selection (Koza 1992, 1994; Koza et al. 1999, 2003; Banzhaf et al. 1998; Langdon 1998; Langdon and Poli 2002). The goal of GP is evolve computer programs to solve problems. This is accomplished by first generating random computer programs that are composed of the basic building blocks needed to solve or approximate a solution to the problem. Each randomly generated program is evaluated and the good programs are selected and recombined to form new computer programs. This process of selection based on fitness and recombination to generate variability is repeated until a best program or set of programs is identified. Genetic programming and its many variations have been applied successfully in a wide range of different problem domains, including data mining and knowledge discovery (e.g. Freitas 2002), electrical engineering (e.g. Koza et al. 2003) and bioinformatics (e.g. Fogel and Corne 2003).

Moore and White (2007b) developed and evaluated a simple GP wrapper for attribute selection in the context of an MDR analysis. The goal of this study was to develop a stochastic wrapper method that is able to select attributes that interact in the absence of independent main effects. At face value, there is no reason to expect that a GP or any other wrapper method would perform better than a random attribute selector, because there are no 'building blocks' for this problem when accuracy is used as the fitness measure. That is, the fitness of any given classifier would look no better than any other with just one of the correct SNPs in the MDR model. Preliminary studies by White et al. (2005) support this idea. For GP or any other wrapper to work, there need to be recognizable building blocks. Moore and White (2007b) specifically evaluated whether including pre-processed attribute quality estimates using TuRF (see above) in a multi-objective fitness function improved attribute selection over a random search or just using accuracy as the fitness. Using a wide variety of simulated data, Moore and White (2007b) demonstrated that including TuRF scores in addition to accuracy in the fitness function significantly improved the power of GP to pick the correct two functional SNPs out of 1000 total attributes. A subsequent study showed that using TuRF scores to select trees for recombination and reproduction performed significantly better than using TuRF in a multiobjective fitness function (Moore and White 2006).

This study presents preliminary evidence suggesting that GP might be useful for the genome-wide genetic analysis of common human diseases that have a complex genetic architecture. The results raise numerous questions. How well does GP do when faced with finding three, four or more SNPs that interact in a non-linear manner to predict disease susceptibility? How does extending the function set to additional attribute construction functions impact performance? How does extending the attribute set impact performance? Is using GP better than filter approaches? To what extent can GP theory help formulate an optimal GP approach to this problem? Does GP outperform other evolutionary or non-evolutionary search methods? Does the computational expense of a stochastic wrapper such as GP outweigh the potential for increased power? These studies provide a starting point to begin addressing some of these questions.

16.7. SUMMARY

We have reviewed several phenomena, including gene–gene interactions (i.e. epistasis) and gene–environment interactions (i.e. plastic reaction norms), that are an important part of the genetic architecture of common human diseases and that partly explain the non-linear mapping relationship between genotype and phenotype. The success of any molecular epidemiology study will depend on whether an analytical strategy is employed that embraces, rather than ignores, these complexities. We have introduced random forests (RF) and multifactor dimensionality reduction (MDR) as examples of data-mining meth-

ods that can be used to complement traditional parametric methods, such as logistic regression. Random forests build on the very popular decision tree-based methods, while MDR was designed specifically for the purpose of detecting, characterizing and interpreting non-linear interaction among multiple factors in the absence of detectable main effects. Both methods are available in easy-to-use and freely available software packages and come with tools for visualizing and interpreting interactions. We have also reviewed a filter method using ReliefF and a stochastic wrapper method using genetic programming (GP) for the analysis of interactions on a genome-wide scale with thousands of attributes. These data-mining and knowledge-discovery methods and others will play an increasingly important role in molecular epidemiology as the field moves away from the candidate-gene approach, that focuses on a few targeted genes, to the genome-wide approach that measures DNA sequence variations from across the genome.

Acknowledgements

This work was supported by National Institutes of Health (USA) Grants CA102327 LM009012, AI59694, HD047447, RR018787 and HL65234.

References

Agirbasli D, Agirbasli M, Williams SM *et al.* 2006. Interaction among 5,10 methylenetetrahydrofolate reductase, plasminogen activator inhibitor and endothelial nitric oxide synthase gene polymorphisms predicts the severity of coronary artery disease in Turkish patients. *Coron Artery Dis* **17**: 413–417.

Altshuler D, Brooks LD, Chakravarti A *et al.* 2005. International HapMap Consortium. A haplotype map of the human genome. *Nature* **437**: 1299–1320.

Andrew AS, Nelson HH, Kelsey KT *et al.* 2006. Concordance of multiple analytical approaches demonstrates a complex relationship between DNA repair gene SNPs, smoking, and bladder cancer susceptibility. *Carcinogenesis* **27**: 1030–1037.

Asselbergs FW, Moore JH, van den Berg MP *et al.* 2006. A role for CETP TaqB polymorphism in determining susceptibility to atrial fibrillation: a nested case control study. *BMC Med Genet* **7**: 39.

Balding DJ. 2006. A tutorial on statistical methods for population association studies. *Nat Rev Genet* **7**: 781–791.

Banzhaf W, Nordin P, Keller RE *et al.* 1998. *Genetic Programming: An Introduction: On the Automatic Evolution of Computer Programs and Its Applications.* Morgan Kaufmann: San Francisco, CA.

Bastone L, Reilly M, Rader DJ *et al.* 2004. MDR and PRP: a comparison of methods for high-order genotype-phenotype associations. *Hum Hered* **58**: 82–92.

Bateson W. 1909. *Mendel's Principles of Heredity.* Cambridge University Press: Cambridge, UK.

Brassat D, Motsinger AA, Caillier SJ *et al.* 2006. Multifactor dimensionality reduction reveals gene–gene interactions associated with multiple sclerosis susceptibility in African Americans. *Genes Immun* **7**: 310–315.

Breiman L. 2001. Random Forests. *Machine Learning* **45**: 5–32.

Brodie ED III. 2000. Why evolutionary genetics does not always add up. In *Epistasis and the Evolutionary Process*, Wolf J, Brodie B III, Wade M (eds). Oxford University Press: New York; 3–19.

Bureau A, Dupuis J, Falls K *et al.* 2005. Identifying SNPs predictive of phenotype using random forests. *Genet Epidemiol* **28**: 171–182.

Burton PR, Tobin MD, Hopper JL. 2005. Key concepts in genetic epidemiology. *Lancet* **366**: 941–951.

Chan IH, Leung TF, Tang NL *et al.* 2006. Gene–gene interactions for asthma and plasma total IgE concentration in Chinese children. *J Allergy Clin Immunol* **117**: 127–133.

Cho YM, Ritchie MD, Moore JH *et al.* 2004. Multifactor-dimensionality reduction shows a two-locus interaction associated with type 2 diabetes mellitus. *Diabetologia* **47**: 549–554.

Chung HH, Kim MK, Kim JW *et al.* 2006. *XRCC1 R399Q* polymorphism is associated with response to platinum-based neoadjuvant chemotherapy in bulky cervical cancer. *Gynecol Oncol* **103**: 1031–1037.

Coffey CS, Hebert PR, Ritchie MD *et al.* 2004. An application of conditional logistic regression and multifactor dimensionality reduction for detecting gene–gene interactions on risk of myocardial infarction: the importance of model validation. *BMC Bioinf* **4**: 49.

Cook NR, Zee RY, Ridker PM. 2004. Tree and spline based association analysis of gene–gene interaction models for ischemic stroke. *Stat Med* **23**: 1439–1453.

Cordell HJ, Clayton DG. Genetic association studies. *Lancet* **366**: 1121–1131.

Coutinho AM, Sousa I, Martins M *et al.* 2007. Evidence for epistasis between *SLC6A4* and *ITGB3* in autism etiology and in the determination of platelet serotonin levels. *Hum Genet* **121**: 243–256.

Culverhouse R, Suarez BK, Lin J *et al.* 2002. A perspective on epistasis: limits of models displaying no main effect. *Am J Hum Genet* **70**: 461–471.

Fisher RA. 1918. The correlations between relatives on the supposition of Mendelian inheritance. *Trans R Soc Edinb* **52**: 399–433.

Fogel GB, Corne DW. 2003. *Evolutionary Computation in Bioinformatics.* Morgan Kaufmann: San Francisco, CA.

Freitas A. 2001. Understanding the crucial role of attribute interactions. *Artif Intell Rev* **16**: 177–199.

Freitas A. 2002. *Data Mining and Knowledge Discovery with Evolutionary Algorithms.* Springer: New York.

Gibson G, Wagner G. 2000. Canalization in evolutionary genetics: a stabilizing theory? *BioEssays* **22**: 372–380.

Goldberg DE. 2002. *The Design of Innovation.* Kluwer: Boston, MA.

Haas DW, Geraghty DE, Andersen J *et al.* 2006. Immunogenetics of CD4 lymphocyte count recovery during antiretroviral therapy: an AIDS clinical trials group study. *J Infect Dis* **194**: 1098–1107.

Hahn LW, Moore JH. 2004. Ideal discrimination of discrete clinical endpoints using multilocus genotypes. *In Silico Biol* **4**: 183–194.

Hahn LW, Ritchie MD, Moore JH. 2003. Multifactor dimensionality reduction software for detecting gene–gene and gene–environment interactions. *Bioinformatics* **19**: 376–382.

Hattersley AT, McCarthy MI. 2005. What makes a good genetic association study? *Lancet* **366**: 1315–1323.

Hirschhorn JN, Daly MJ. 2005. Genome-wide association studies for common diseases and complex traits. *Nat Rev Genet* **6**: 95–108.

Hollander WF. 1955. Epistasis and hypostasis. *J Heredity* **46**: 222–225.

Hopper JL, Bishop DT, Easton DF. 2005. Population-based family studies in genetic epidemiology. *Lancet* **366**: 1397–1406.

Hsieh CH, Liang KH, Hung YJ *et al.* 2006. Analysis of epistasis for diabetic nephropathy among type 2 diabetic patients. *Hum Mol Genet* **15**: 2701–2708.

Hu Y-J. 1998. Constructive induction: covering attribute spectrum. In *Feature Extraction, Construction and Selection: A Data Mining Perspective*, Liu H, Motoda H (eds). Kluwer: Boston, MA; 257–272.

Hunter DJ. 2005. Gene–environment interactions in human diseases. *Nat Rev Genet* **6**: 287–298.

Jakulin A, Bratko I. 2003. Analyzing attribute interactions. *Lect Notes Artif Intell* **2838**: 229–240.

Jones IM, Thomas CB, Xi T *et al.* 2007. Exploration of methods to identify polymorphisms associated with variation in DNA repair capacity phenotypes. *Mutat Res* **616**: 213–220.

Kira K, Rendell LA. 1992. A practical approach to feature selection. In Sleeman DH, Edwards P (ed.). *Proceedings of the Ninth International Workshop on Machine Learning*. Morgan Kaufmann: San Francisco, CA; 249–256.

Kononenko I. 1994. Estimating attributes: analysis and extension of relief. *Proceedings of the European Conference on Machine Learning*. Springer: New York, NY; 171–182.

Koza JR. 1992. *Genetic Programming: On the Programming of Computers by Means of Natural Selection*. MIT Press: Cambridge, MA.

Koza JR. 1994. *Genetic Programming II: Automatic Discovery of Reusable Programs*. MIT Press: Cambridge, MA.

Koza JR, Bennett FH III, Andre D *et al.* 1999. *Genetic Programming III: Darwinian Invention and Problem Solving*. Morgan Kaufmann: San Francisco, CA.

Koza JR, Keane MA, Streeter MJ *et al.* 2003. *Genetic Programming IV: Routine Human-Competitive Machine Intelligence*. Springer: New York.

Langdon WB. 1998. *Genetic Programming and Data Structures: Genetic Programming + Data Structures = Automatic Programming!* Kluwer: Boston, MA.

Langdon WB, Poli R. 2002. *Foundations of Genetic Programming*. Springer: New York.

Li W, Reich J. 2000. A complete enumeration and classification of two-locus disease models. *Hum Hered* **50**: 334–349.

Lunetta KL, Hayward LB, Segal J *et al.* 2004. Screening large-scale association study data: exploiting interactions using random forests. *BMC Genet* **5**: 32.

Ma DQ, Whitehead PL, Menold MM *et al.* 2005. Identification of significant association and gene–gene interaction of GABA receptor subunit genes in autism. *Am J Hum Genet* **77**: 377–388.

Mannila MN, Eriksson P, Ericsson CG *et al.* 2006. Epistatic and pleiotropic effects of polymorphisms in the fibrinogen and coagulation factor XIII genes on plasma fibrinogen concentration, fibrin gel structure and risk of myocardial infarction. *Thromb Haemost* **95**: 420–427.

Martin ER, Hahn LW, Bass M *et al.* 2006. A combined multifactor dimensionality reduction and pedigree disequilibrium test (MDR-PDT. approach for detecting gene–gene interactions in pedigrees. *Genet Epidemiol* **30**: 111–123.

McGill WJ. 1954. Multivariate information transmission. *Psychometrica* **19**: 97–116.

McKinney BA, Reif DM, Ritchie MD *et al.* 2006. Machine learning for detecting gene–gene interactions: a review. *Appl Bioinf* **5**: 77–88.

Menon R, Velez DR, Simhan H *et al.* 2006. Multilocus interactions at maternal tumor necrosis factor-alpha, tumor necrosis factor receptors, interleukin-6 and interleukin-6 receptor genes predict spontaneous preterm labor in European-American women. *Am J Obstet Gynecol* **194**: 1616–1624.

Michalski RS. 1983. A theory and methodology of inductive learning. *Artif Intell* **20**: 111–161.

Michalewicz Z, Fogel DB. 2000. *How to Solve It: Modern Heuristics*. Springer: New York.

Millstein J, Conti DV, Gilliland FD *et al.* 2006. A testing framework for identifying susceptibility genes in the presence of epistasis. *Am J Hum Genet* **78**: 15–27.

Mitchell TM. *Machine Learning*. MacGraw-Hill: Boston, MA.

Moore JH. 2003. The ubiquitous nature of epistasis in determining susceptibility to common human diseases. *Hum Hered* **56**: 73–82.

Moore JH. 2004. Computational analysis of gene–gene interactions in common human diseases using multifactor dimensionality reduction. *Exp Rev Mol Diagn* **4**: 795–803.

Moore JH. 2005. A global view of epistasis. *Nat Genet* **37**: 13–14.

Moore JH. 2007. Genome-wide analysis of epistasis using multifactor dimensionality reduction: feature selection and construction in the domain of human genetics. In *Knowledge Discovery and Data Mining: Challenges and Realities with Real World Data*, Zhu X, Davidson I (eds). IGI Global, 17–30.

Moore JH, Barney N, Tsai CT *et al.* 2007. Symbolic modelling of epistasis. *Hum Hered* **63**: 120–133.

Moore JH, Gilbert JC, Tsai C-T *et al.* 2006. A flexible computational framework for detecting, characterizing, and interpreting statistical patterns of epistasis in genetic studies of human disease susceptibility. *J Theoret Biol* **241**: 252–261.

Moore JH, Ritchie MD. 2004. The challenges of whole-genome approaches to common diseases. *J Am Med Assoc* **291**: 1642–1643.

Moore JH, White BC. 2007a. Tuning ReliefF for genome-wide genetic analysis. *Lect Notes Comput Sci* **4447**: 166–175.

Moore JH, White BC. 2007b. Genome-wide genetic analysis using genetic programming: The critical need for expert knowledge. In *Genetic Programming Theory and Practice IV*. Riolo R, Soule T, Worzel B (eds). Springer: New York; 11–28.

Moore JH, White BC. 2006. Exploiting expert knowledge in genetic programming for genome-wide genetic analysis. *Lect Notes Comput Sci* **4193**: 696–977.

Moore JH, Williams SW. 2002. New strategies for identifying gene–gene interactions in hypertension. *Ann Med* **34**: 88–95.

Moore JH, Williams SW. 2005. Traversing the conceptual divide between biological and statistical epistasis: systems biology and a more modern synthesis. *BioEssays* **27**: 637–646.

Motsinger AA, Ritchie MD, Shafer RW *et al.* 2006. Multilocus genetic interactions and response to efavirenz-containing regimens: an Adult AIDS Clinical Trials Group study. *Pharmacogenet Genom* **16**, 837–845.

Motsinger AA, Brassat D, Caillier SJ *et al.* 2007. Complex gene–gene interactions in multiple sclerosis: a multifactorial approach reveals associations with inflammatory genes. *Neurogenetics* **8**: 11–20.

Nordgard SH, Ritchie MD, Jensrud SD *et al.* 2007. *ABCB1* and *GST* polymorphisms associated with TP53 status in breast cancer. *Pharmacogenet Genom* **17**: 127–136.

Oestergaard MZ, Tyrer J, Cebrian A *et al.* 2006. Interactions between genes involved in the antioxidant defence system and breast cancer risk. *Br J Cancer* **95**: 525–531.

Palmer LJ, Cardon LR. 2005. Shaking the tree: mapping complex disease genes with linkage disequilibrium. *Lancet* **366**: 1223–1234.

Phillips PC. 1998. The language of gene interaction. *Genetics* **149**: 1167–1171.

Pierce JR. 1980. *An Introduction to Information Theory: Symbols, Signals, and Noise.* Docer: New York.

Picliucci M. 2001. *Phenotypic Plasticity: Beyond Nature and Nurture.* Johns Hopkins Press: Baltimore, MD.

Proulx SR, Phillips PC. 2005. The opportunity for canalization and the evolution of genetic networks. *Am Naturalist* **165**: 147–162.

Qin S, Zhao X, Pan Y *et al.* 2005. An association study of the *N*-methyl-D-aspartate receptor NR1 subunit gene (*GRIN1*) and NR2B subunit gene (*GRIN2B*) in schizophrenia with universal DNA microarray. *Eur J Hum Genet* **13**: 807–814.

Reif DM, Motsinger AA, McKinney BA *et al.* 2006. Feature selection using random forests for the integrated analysis of multiple data types. *Proceedings of the 2006 IEEE Symposium on Computational Intelligence in Bioinformatics and Computational Biology.* IEEE Press: New York; 171–178.

Remold SK, Lenski RE. 2004. Pervasive joint influence of epistasis and plasticity on mutational effects in *Escherichia coli. Nat Genet* **36**: 423–426.

Ritchie MD, Hahn LW, Moore JH. 2003. Power of multifactor dimensionality reduction for detecting gene–gene interactions in the presence of genotyping error, phenocopy, and genetic heterogeneity. *Genet Epidemiol* **24**: 150–157.

Ritchie MD, Hahn LW, Roodi N *et al.* 2001. Multifactor dimensionality reduction reveals high-order interactions among estrogen metabolism genes in sporadic breast cancer. *Am J Hum Genet* **69**: 138–147.

Robnik-Sikonja M, Kononenko I. 2001. Comprehensible interpretation of Relief's Estimates. In *Proceedings of the Eighteenth International Conference on Machine Learning.* Morgan Kaufmann: San Francisco, CA; 433–440.

Robnik-Siknja M, Kononenko I. 2003. Theoretical and empirical analysis of ReliefF and RReliefF. *Machine Learning* **53**: 23–69.

Sabbagh A, Darlu P. 2006. SNP selection at the *NAT2* locus for an accurate prediction of the acetylation phenotype. *Genet Med* **8**: 76–85.

Sabbagh A, Darlu P. 2006. Data-mining methods as useful tools for predicting individual drug response: application to CYP2D6 data. *Hum Hered* **62**: 119–134.

Schlichting CD, Pigliucci M. 1998. *Phenotypic Evolution: A Reaction Norm Perspective.* Sinauer: Sunderland, MA.

Schmalhausen II. 1949. *Factors of Evolution: The Theory of Stabilizing Selection.* University of Chicago Press: Chicago, IL.

Schwender H, Zucknick M, Ickstadt K *et al.* 2004. A pilot study on the application of statistical classification procedures to molecular epidemiological data. *Toxicol Lett* **151**: 291–299.

Smith G, Ebrahim S, Lewis S *et al.* 2005. Genetic epidemiology and public health: hope, hype, and future prospects. *Lancet* **366**: 1484–1498.

Soares ML, Coelho T, Sousa A *et al.* 2005. Susceptibility and modifier genes in Portuguese transthyretin V30M amyloid polyneuropathy: complexity in a single-gene disease. *Hum Mol Genet* **14**: 543–553.

Teare MD, Barrett JH. 2005. Genetic linkage studies. *Lancet* **366**: 1036–1044.

Templeton AR. 2000. Epistasis and complex traits. In *Epistasis and the Evolutionary Process*, Wolf J, Brodie B III, Wade M (eds). Oxford University Press: New York; 41–57.

Thornton-Wells TA, Moore JH, Haines JL. 2004. Genetics, statistics and human disease: analytical retooling for complexity. *Trends Genet* **20**: 640–647.

Thornton-Wells TA, Moore JH, Haines JL. 2006. Dissecting trait heterogeneity: a comparison of three clustering methods applied to genotypic data. *BMC Bioinf* **7**: 204.

Tsai CT, Lai LP, Lin JL *et al*. 2004. Renin–angiotensin system gene polymorphisms and atrial fibrillation. *Circulation* **109**: 1640–1646.

Tsai C-T, Hwang J-J, Ritchie MD *et al*. 2007. Renin–angiotensin system gene polymorphisms and coronary artery disease in a large angiographic cohort: detection of gene–gene interactions. *Atherosclerosis* **195**: 172–180.

Waddington CH. 1942. Canalization of development and the inheritance of acquired characters. *Nature* **150**: 563–565.

Waddington CH. 1957. *The Strategy of the Genes*. Macmillan: New York.

Wang WY, Barratt BJ, Clayton DG *et al*. 2005. Genome-wide association studies: theoretical and practical concerns. *Nat Rev Genet* **6**: 109–118.

White BC, Gilbert JC, Reif DM *et al*. 2005. A statistical comparison of grammatical evolution strategies in the domain of human genetics. *Proceedings of the IEEE Congress on Evolutionary Computing*. IEEE Press: New York; 676–682.

Whitten IH, Frank E. 2005. *Data Mining*. Elsevier: Boston, MA.

Wilke RA, Reif DM, Moore JH. 2005. Combinatorial pharmacogenetics. *Nat Rev Drug Discov* **4**: 911–918.

Williams SM, Ritchie MD, Phillips JA III *et al*. 2004. Multilocus analysis of hypertension: a hierarchical approach. *Hum Hered* **57**: 28–38.

Woltereck R. 1909. Weitere experimentelle Untersuchungen über Artveränderung, speziell über das Wesen quantitativer Artunterschiede bei Daphnien. *Verhandl Deutsch Zool Gesellsch* **19**: 110–173.

Xu J, Lowery J, Wiklund F *et al*. 2005. The interaction of four inflammatory genes significantly predicts prostate cancer risk. *Cancer Epidemiol Biomarkers Prevent* **14**: 2563–2568.

17

Some Implications of Random Exposure Measurement Errors in Occupational and Environmental Epidemiology

S. M. Rappaport[1] and L. L. Kupper[2]

[1]*University of California, Berkeley, CA, USA, and* [2]*University of North Carolina, Chapel Hill, NC, USA*

17.1. INTRODUCTION

A major goal of research studies in occupational and environmental epidemiology is to validly and precisely estimate relationships between levels of exposures to toxic substances and levels of health effects in human populations. If exposure levels are poorly characterized, the estimated exposure–response relationships often underestimate risk for a given exposure (known as attenuation bias). Various terms are used to describe the reason for this biasing effect of inaccurate exposure assessment, namely *measurement error* (when exposure is treated as a continuous variable) and *misclassification error* (when exposure is treated as a categorical variable, e.g. as high, medium and low exposure). In this chapter, we will show how measurement error effects can be related to the underlying random variation in exposure levels within persons, between persons and across groups. It is assumed that individual exposure measurements randomly vary around the true exposure levels at the times of measurement, i.e. there are no systematic errors in measurements of exposure. For example, personal measurements of air contaminants should provide unbiased estimates of true air levels during the periods of measurement.

The effects of measurement error can be elucidated by comparing a true and an estimated exposure–response relationship. For simplicity, we will illustrate the detrimental effects of measurement error using a very simple model, in which a continuous health outcome is linearly related to a continuous exposure variable on the (natural) log scale. Examples of continuous health outcomes include quantitative measures of pulmonary function (arising from exposures to fibrogenic dusts or irritants), amounts of particular urinary proteins (from exposures to heavy metals) or levels of chromosome aberrations (from exposures to genotoxic chemicals). Such a linear relationship is illustrated in Figure 17.1A, where β_1 represents the straight-line slope of the true log-scale relationship relating the expected health outcome in a population to the true exposure level. Although this log-scale linear relationship is simple, it is able to capture much of the non-linear behaviour observed in exposure–response relationships in the natural scale. This is illustrated in Figure 17.1B, where the three straight lines from Figure 17.1A are plotted in the natural scale. When $\beta_1 = 0.5$ in the log scale, the relationship is concave downward (supra-linear) in the natural scale, as might be observed when bioactivation of a chemical to a toxic metabolite is saturable.

Molecular Epidemiology of Chronic Diseases, Edited by C. P. Wild, P. Vineis, and S. Garte
© 2008 John Wiley & Sons, Ltd

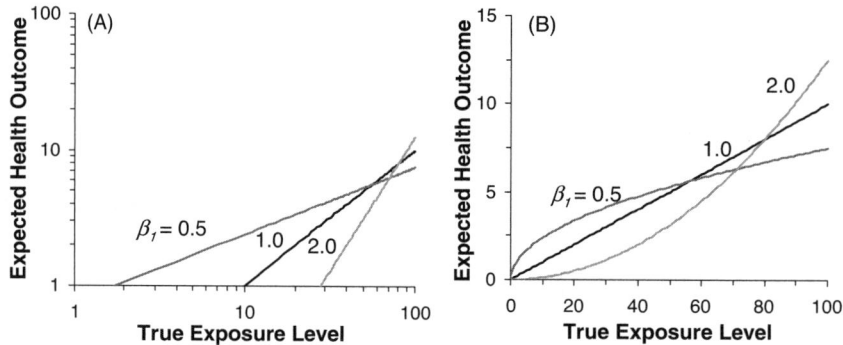

Figure 17.1 Examples of exposure–response relationships. (A) Straight-line relationships between the expected value of the logged continuous health outcome and the true mean logged exposure level, where β_1 is the slope. (B) Natural-scale relationships corresponding to the straight lines in (A): when $\beta_1 < 1$, the curve is concave downward; when $\beta_1 = 1$, the relationship is linear; and when $\beta_1 > 1$, the curve is concave upward

When $\beta_1 = 1.0$ in the log scale, the relationship is linear in the natural scale. And when $\beta_1 = 2.0$ in the log scale, the relationship is concave upward (super-linear) in the natural scale, indicative of saturable detoxification or repair processes.

Because the magnitudes and effects of exposure measurement error differ with the type of study design, two types of epidemiological studies will be considered. First, we will examine an *individual-based study*, in which exposure levels and health outcomes are measured for all d persons in a random sample. Then, we will consider a *group-based study*, in which random samples of k persons are measured in each of several groups. In the group-based study, group-specific mean values of both the logged health outcome and logged exposure levels are used to estimate the exposure–response relationship. In these comparisons we are considering classical measurement error models and not Berkson error models, which have different underlying assumptions (Armstrong 1998).

17.2. INDIVIDUAL-BASED STUDY

For an individual-based study, we will examine the straight-line relationship between the true subject-specific mean logged exposure and the expected logged health response in a population. Because we consider a linear log-scale relationship, the true individual exposure level is interpreted as the true mean value of logged exposure levels experienced

by that person over time; we designate the true mean logged exposure level for the ith person as μ_{Yi}. Then, the health outcome model for an individual-based study can be written as:

$$R_i = \beta_0 + \beta_1 \mu_{Yi} + u_i \quad \text{for } i$$
$$= 1, 2, \ldots, d \text{ persons} \quad (1)$$

where R_i represents the natural logarithm of the continuous health outcome; β_0 is the background value of R_i when $\mu_{Yi} = 0$; β_1 is the true slope relating μ_{Yi} to R_i; $\mu_{Yi} = (\mu_Y + b_i)$ is the true unobservable mean of the logged exposure for the ith person [where $b_i \sim N(0,\sigma_b^2)$]; and, u_i is the error term, with $u_i \sim N(0,\sigma_u^2)$. We assume that the jth logged exposure measurement Y_{ij} for the ith person is modelled as $Y_{ij} = \ln(X_{ij}) = \mu_{Yi} + e_{ij} = \mu_Y + b_i + e_{ij}$, where $e_{ij} \sim N(0,\sigma_w^2)$. Finally, we assume that all the $\{b_i\}$, $\{e_{ij}\}$ and $\{u_i\}$ are mutually independent random variables. The above assumptions imply that the expected value and variance of R_i for the ith person are $\beta_0 + \beta_1\mu_{Yi}$ and σ_u^2, respectively, and that the expected value and variance of R_i for the entire population are $\beta_0 + \beta_1\mu_Y$ and $\beta_1^2 \sigma_b^2 + \sigma_u^2$, respectively.

In the context of the log-scale relationship depicted in Figure 17.1A, the x axis would be μ_{Yi} and the y axis would be the expected value of R_i for the ith person. The corresponding x and y axes in the natural scale (Figure 17.1B) would be $e^{\mu_{Yi}}$ (the true geometric mean exposure for the ith person) and the expected value of e^{R_i} for the ith person, respectively.

Regression analysis

Now, consider a dataset where we have n repeated exposure measurements from each of d persons. Since μ_{Yi} is unobservable, it is reasonable to use $\bar{Y}_i = \frac{1}{n}\sum_{j=1}^{n} Y_{ij}$ as a surrogate measure of μ_{Yi} and to use the data pairs $(\bar{Y}_i, R_i)$, $i = 1, 2, \ldots, d$, to obtain the unweighted least-squares estimator $\hat{\beta}_1$ of β_1. It follows that the expected value of $\hat{\beta}_1$ is given by:

$$E(\hat{\beta}_1) = \frac{\beta_1 \sigma_b^2}{\sigma_b^2 + \frac{\sigma_w^2}{n}} = \beta_1 \left(\frac{1}{1 + \frac{\lambda}{n}} \right), \quad (2)$$

where $\lambda = \frac{\sigma_w^2}{\sigma_b^2}$ is the *variance ratio*. Equation (2) is a well-known result (see e.g. Cochran 1968) that has been discussed in the context of occupational and environmental epidemiology (Brunekreef et al. 1987; Lin et al. 2005; Kromhout et al. 1996; Tielemans et al. 1998). The standard error (SE) of $\hat{\beta}_1$ is given by:

$$SE(\hat{\beta}_1) = \sqrt{\frac{n \left[\beta_1^2 \left(\frac{\sigma_b^2 \sigma_w^2}{n\sigma_b^2 + \sigma_w^2} \right) + \sigma_u^2 \right]}{(d-3)(n\sigma_b^2 + \sigma_w^2)}} \quad (3)$$

(see Tielemans et al. 1998), which indicates that at least four subjects ($d \geq 4$) would be required in an individual-based study to allow SE to be estimated using expression (3).

From Equation (2), we see that $E(\hat{\beta}_1)$ is attenuated (or suppressed) toward zero by the quantity $\left(1 + \frac{\sigma_w^2}{n\sigma_b^2}\right)^{-1}$ referred to as the *reliability coefficient* (Armstrong 1998). The amount of attenuation increases with σ_w^2 and decreases with both σ_b^2 and n. Since attenuation decreases as σ_b^2 increases, it is wise to select subjects covering the widest possible range of exposures.

Estimating sample sizes for a specified bias

Let $B = \frac{E(\hat{\beta}_1)}{\beta_1}$; then the (relative) bias of $\hat{\beta}_1$ as an estimator of β_1 is simply $(B - 1)$. For example, if $B = \frac{E(\hat{\beta}_1)}{\beta_1} = 0.8$, then the relative bias in $E(\hat{\beta}_1)$ is

$(0.8 - 1) = -0.20$, indicating that $\hat{\beta}_1$ tends, on average, to underestimate β_1 by 20%. Since $B = \frac{E(\hat{\beta}_1)}{\beta_1} = \frac{1}{1 + \frac{\lambda}{n}}$, sample sizes can be estimated from the following relationship for given values of B and λ: $n = \left(\frac{B}{1-B}\right)\lambda$. From Lin et al. (2005), the median value of $\hat{\lambda}$ was 1.1 for 17 occupational studies and was 4.7 for 33 environmental studies. Thus, in order to limit the bias in $\hat{\beta}_1$ to at most 20% (on average), one would choose $n = \left(\frac{0.8}{1-0.8}\right)(1.1) = 4.4$, meaning at least five measurements per person for a typical occupational study, or $\left[\text{since } \left(\frac{0.8}{1-0.8}\right)(4.7) = 18.8\right]$ at least 19 measurements per person for a typical environmental study. Assuming that at least 10 subjects would be sampled in such studies, typical total sample sizes would be at least 50 measurements for an occupational study and at least 190 measurements for an environmental study.

Since sample sizes with many repeated measurements per subject are difficult to achieve in most realistic situations, the above calculations suggest that individual-based studies will generally have non-negligible bias (particularly those conducted in environmental settings), assuming that the simple model structure and attendant assumptions used here are reasonable.

17.3. GROUP-BASED STUDIES

Exposure model with a random group effect

Suppose that there are H observational groups ($h = 1, 2, \ldots, H$) of persons to be investigated for exposure-related health effects. To gain some insight about attenuation bias resulting from studies of this type, Kromhout et al. considered the following model for $Y_{hij} = \ln(X_{hij})$, where X_{hij} is the jth exposure level for the ith person in the hth group ($j = 1, 2, \ldots, n; i = 1, 2, \ldots, k; h = 1, 2, \ldots, H$) (Kromhout et al. 1996):

$$Y_{hij} = \ln(X_{hij}) = \mu_Y + a_h + b_{hi} + e_{hij} = \mu_{Yhi} + e_{hij}. \quad (4)$$

In model (4), $\mu_Y = E(Y_{hij})$ is the overall mean (log-scale) exposure, $a_h \sim N(0, \sigma_{bg}^2)$ is the random effect for Group h, $b_{hi} \sim N(0, \sigma_{b|h}^2)$ is the random

Table 17.1   Variance components estimated under model (4) for levels of air contaminants among workers in several industries

Industry	Contaminant	m	N	$\hat{\sigma}_{bg}^2$	$\hat{\sigma}_{b/h}^2$	$\hat{\sigma}_w^2$
Rubber manufacturing	Inhalable dust	231	617	0.18	1.30	0.48
Rubber manufacturing	Solvents	54	111	2.07	1.59	0.63
Animal feed production	Inhalable dust	157	569	0.56	0.84	1.17
Dry cleaning	Perchloroethylene	23	113	0.00	0.88	0.10
Brick manufacturing	Respirable dust	38	150	0.07	0.29	0.32
Reinforced plastics	Styrene	85	258	0.35	0.22	0.46
Petrochemicals	Benzene	418	1949	0.36	0.30	1.56
Bakeries	Flour dust	212	488	0.53	0.42	0.60
Furniture manufacturing	Solvents	16	42	0.55	0.59	0.20
			Median	0.36	0.59	0.48

m, total number of persons sampled; N, total number of air measurements; $\hat{\sigma}_{bg}^2$, estimated between-group variance component; $\sigma_{b/h}^2$, estimated group-specific between-person variance component; $\hat{\sigma}_w^2$, estimated within-person variance component. Reproduced with permission from Kromhout *et al.* (1996).

effect for the ith person in Group h (note that the between-subject variance component $\sigma_{b/h}^2$ can also be regarded as the *within-group variance component*, reflecting variation across subjects in group h), and $e_{hij} \sim N(0,\sigma_w^2)$ is the random error of the jth exposure level for the ith person in the hth group. Note that in model (4), a_h is the random effect of the hth group. Thus, there is a between-group variance component σ_{bg}^2 associated with model (4). We assume that the $\{a_h\}$, $\{b_{hi}\}$ and $\{e_{hij}\}$ are mutually independent random variables. It then follows that $\mu_{Yhi} = \mu_Y + a_h + b_{hi} \sim N(\mu_Y, \sigma_{bg}^2 + \sigma_{b/h}^2)$, and $Y_{hij} \sim N(\mu_Y, \sigma_{bg}^2 + \sigma_{b/h}^2 + \sigma_w^2)$.

In epidemiological studies, it is common to investigate exposures and health outcomes among groups defined by factory, job, city, etc. Here, it is reasonable to assume that a collection of factories under study, for example, represents a random sample of all possible factories that could be chosen (Kromhout *et al.* 1996), and that the particular factories selected are not of unique interest. Thus, model (4) can be more useful in characterizing exposures for epidemiological studies than a model with fixed effects delineating different groups. However, under model (4), it is important to realize that σ_w^2 and $\sigma_{b/h}^2$ are assumed to be common to all groups. Based upon our experience, this assumption (of homogeneous σ_w^2 and $\sigma_{b/h}^2$ across groups) can be problematic. In fact, Weaver *et al.* showed that it was inappropriate to assume common σ_w^2 and $\sigma_{b/h}^2$ in 43 of 117 cases (37%), based upon likelihood

ratio tests at a significance level of 0.01 (Weaver *et al.* 2001). This assumption of homogeneous variance should be kept in mind when applying model (4) to characterize exposures across groups.

Relatively few studies have applied model (4) to estimate variance components across groups. Tables 17.1 and 17.2 summarize results of two such studies that applied model (4) to air contaminants in several factories (Kromhout *et al.* 1996) and in several US cities (Rappaport and Kupper 2004). The median values of these estimated variance components (listed at the bottom of these tables) will be used in subsequent sample size calculations.

Health-outcome model

We will assume the balanced case, where there are H groups, each containing k persons with n measurements per person. Let R_{hi} be the natural logarithm of a continuous health response for the ith person in the hth group. The health outcome model we will consider is:

$$R_{hi} = \beta_0 + \beta_1 \mu_{Yhi} + u_{hi}, \tag{5}$$

which has the same general structure as equation (1), where $\mu_{Yhi} = \mu_Y + a_h + b_{hi}$ is now the true (but unobservable) mean logged exposure level for subject i in group h. Under model (5), the expected value and variance of R_{hi} for the ith person in the hth group are $\beta_0 + \beta_1 \mu_{Yhi}$ and σ_u^2, respectively, while those for the entire population are $\beta_0 + \beta_1 \mu_Y$ and $\beta_1^2 (\sigma_{bg}^2 + \sigma_{b/h}^2) + \sigma_u^2$, respectively.

Table 17.2 Variance components estimated under model (4) for levels of air contaminants among the general population in five US cities

Contaminant	m	N	$\hat{\sigma}^2_{bg}$	$\sigma^2_{b|h}$	$\hat{\sigma}^2_w$
Benzene	421	523	0.21	0.53	0.71
Chloroform	443	553	0.57	0.00	1.26
Ethyl benzene	445	555	0.18	0.20	0.93
Methyl chloroform	447	558	0.07	0.89	1.17
p-Dichlorobenzene	356	398	0.00	1.06	2.00
Perchloroethylene	447	556	0.06	0.07	1.37
Styrene	532	532	0.10	0.07	1.08
Trichloroethylene	444	553	0.29	0.06	2.20
o-Xylene	444	553	0.07	0.00	1.07
		Median	0.10	0.07	1.17

m, total number of persons sampled; N, total number of air measurements; $\hat{\sigma}^2_{bg}$, estimated between-group variance component; $\sigma^2_{b|h}$, estimated group-specific between-person variance component; $\hat{\sigma}^2_w$ estimated within-person variance component.
Reproduced with permission from Rappaport and Kupper (2004).

Regression analysis

We will use the estimated mean exposure for the hth group, $\bar{Y}_h = \frac{1}{kn}\sum_{i=1}^{k}\sum_{j=1}^{n} Y_{hij} \sim N[\mu_{Y,Z} \; \hat{\sigma}^2_{bg} + (\hat{\sigma}^2_{b|h}/k) + (\sigma^2_w/kn]$, as a surrogate for μ_{Yhi} (basically assigning each subject in group h the estimated mean logged exposure level $\bar{Y}_h$ for that group). Then, the h pairs $(\bar{Y}_h, R_{hi})$, $h = 1, 2, \ldots, H$ and $i = 1, 2, \ldots, k$, can be used to obtain the unweighted least-squares estimator $\hat{\beta}^*_1$ of β_1 in model (5). This leads to the result, shown in (Kromhout et al. 1996; Tielemans et al. 1998), that:

$$E\,(\hat{\beta}^*_1) = \frac{\beta_1}{1 + \dfrac{\sigma^2_w}{kn\sigma^2_{bg} + n\sigma^2_{b|h}}}, \tag{6}$$

which is the measure of expected attenuation in $\hat{\beta}^*_1$ for group-based studies. Also, the SE of $\hat{\beta}^*_1$ is given by:

$$SE(\hat{\beta}^*_1) = \sqrt{\frac{\dfrac{\sigma^2_u}{k}\left(\sigma^2_{bg} + \dfrac{\sigma^2_{b|h}}{k} + \dfrac{\sigma^2_w}{kn}\right) + \dfrac{\beta^2_1\sigma^2_w}{kn}\left(\sigma^2_{bg} + \dfrac{\sigma^2_{b|h}}{k}\right)}{(H-3)\left(\sigma^2_{bg} + \dfrac{\sigma^2_{b|h}}{k} + \dfrac{\sigma^2_w}{kn}\right)^2}} \tag{7}$$

(see Tielemans et al. 1998), indicating that at least four groups ($H \geq 4$) would be required in a group-based study, so that SE can be estimated using

expression (7). From equation (6), we see that $E(\hat{\beta}^*_1)$ is attenuated (or suppressed) toward zero by the reliability coefficient $\left(1 + \dfrac{\sigma^2_w}{kn\sigma^2_{bg} + n\sigma^2_{b|h}}\right)^{-1}$. Attenuation increases directly with σ^2_w and decreases with $(kn\sigma^2_{bg} + n\sigma^2_{b|h})$. Attenuation is decreased by increasing σ^2_{bg} and/or $\sigma^2_{b|h}$, as well as k and n, and these quantities are impacted by the study design. In particular, it is important to select groups, and persons within groups, covering the widest possible ranges of exposures. When each group consists of one person ($k = 1$), equations (6) and (7) reduce to equations (2) and (3), provided earlier for individual-based studies [when $k = 1$, each person represents a group, so that σ^2_b in equations (2) and (3) is equal to $(\sigma^2_{bg} + \sigma^2_{b|h})$ in equations (6) and (7)].

Estimating sample sizes

To investigate the sample sizes needed to estimate β_1 with no more than a specified level of bias, let $B^* = \dfrac{E(\hat{\beta}^*_1)}{\beta_1}$, so that the (relative) bias is $(B^* - 1)$. Since $\dfrac{B^*}{1 - B^*} = \dfrac{kn\sigma^2_{bg} + n\sigma^2_{b|h}}{\sigma^2_w}$, then the number of subjects needed to be sampled per group can be derived from the following relationship for given values of B^*, n, σ^2_{bg}, $\sigma^2_{b|h}$, and σ^2_w:

$$k = \left(\frac{B^*}{1 - B^*}\right)\left(\frac{\sigma^2_w}{n\sigma^2_{bg}}\right) - \left(\frac{\sigma^2_{b|h}}{\sigma^2_{bg}}\right),$$ where the total number of measurements needed per group is equal to kn.

Table 17.3 Number of measurements per group kn (for k subjects per group and n measurements per subject) that would be required to maintain attenuation bias at no more than 20% of the true β_1 value for typical occupational and environmental studies using a group-based study design.

Type of study	σ_{bg}^2	$\sigma_{b|h}^2$	σ_w^2	n	k	kn	$N\ (H = 4)$
Occupational	0.36	0.59	0.48	1	4	4	16
Occupational	0.36	0.59	0.48	2	2	4	16
Occupational	0.36	0.59	0.48	3	1	3	12
Environmental	0.10	0.07	1.17	1	47	47	188
Environmental	0.10	0.07	1.17	2	23	46	184
Environmental	0.10	0.07	1.17	3	15	45	180
Environmental	0.10	0.07	1.17	4	11	44	176
Environmental	0.10	0.07	1.17	5	9	45	180
Environmental	0.10	0.07	1.17	6	8	48	192
Environmental	0.10	0.07	1.17	7	6	42	168
Environmental	0.10	0.07	1.17	8	6	48	192
Environmental	0.10	0.07	1.17	9	5	45	180
Environmental	0.10	0.07	1.17	10	4	40	160
Environmental	0.10	0.07	1.17	11	4	44	176
Environmental	0.10	0.07	1.17	12	4	48	192
Environmental	0.10	0.07	1.17	13	3	39	156

n, number of measurements per person; k, number of persons sampled per group; kn, number of measurements per group; N, total number of measurements, assuming $H = 4$ groups; $\hat{\sigma}_{bg}^2$, estimated between-group variance component; $\hat{\sigma}_{b|h}^2$, estimated group-specific between-person variance component; $\hat{\sigma}_w^2$, estimated within-person variance component.

Some sample size calculations employing the above relationship for k are shown in Table 17.3, using median values from Tables 17.1 and 17.2 for the needed values of variance components (occupational, $\sigma_{bg}^2 = 0.36$, $\sigma_{b|h}^2 = 0.59$, $\sigma_w^2 = 0.48$; environmental, $\sigma_{bg}^2 = 0.10$, $\sigma_{b|h}^2 = 0.07$, $\sigma_w^2 = 1.17$). Table 17.3 gives the numbers of subjects per group that would be required to limit attenuation bias to at most 20% of β_1 (i.e. $B^* = 0.8$), along with the total number $N = Hkn$ of measurements required assuming $H = 4$ groups. In order to achieve the same level of bias control, the sample size for a typical environmental study ($N \approx 150$–200 for the illustrations in Table 17.3) would need to be roughly 10 times that of a typical occupational study ($N \approx 15$–20).

Adjusting estimated regression coefficients for attenuation bias

Equations (2) and (6) can be used to adjust estimated slopes for the attenuation biases due to random errors in exposure measurements. That is, the true value of β_1 can be estimated from either $\hat{\beta}_1$ or $\hat{\beta}_1^*$ and the operative estimated reliability coefficient

(Armstrong 1998). Let $\hat{\beta}_{1,adj.}$ represent the estimated slope after adjustment for measurement error, i.e.

$\hat{\beta}_{1,adj.} = \hat{\beta}_1 \left(1 + \dfrac{\hat{\lambda}}{n} \right)$ from equation (2) for an individual-based study, where $\hat{\lambda} = \dfrac{\hat{\sigma}_w^2}{\hat{\sigma}_b^2}$. Analogously,

$\hat{\beta}_{1,adj.} = \hat{\beta}_1^* \left(1 + \dfrac{\hat{\sigma}_w^2}{kn\hat{\sigma}_{bg}^2 + n\hat{\sigma}_{b|h}^2} \right)$ from equation (6) for a group-based study. Armstrong states that these simple adjustments do not affect the p values associated with testing whether the estimated slopes are significantly different from zero. These corrections can also be applied to compute measurement error-corrected confidence intervals for the true slopes (Armstrong 1998). However, if the true reliability coefficients are estimated using small samples, the precision of such adjustments will undoubtedly suffer.

17.4. COMPARING BIASES FOR INDIVIDUAL-BASED AND GROUP-BASED STUDIES

Comparing the relationships that define the amounts of attenuation bias in individual-based studies [equation (2)] and group-based studies [equation (6)], the latter design should lead to less attenuation, owing to the moderating effect of the group-specific sample size (kn) on σ_w^2. This reduction in bias has been mentioned by several investigators as a motivation for employing group-based studies in epidemiology (Armstrong 1998; Kromhout *et al.* 1996; Seixas and Sheppard 1996; Tielemans *et al.* 1998). However, since group-based studies employ larger sample sizes than individual-based studies, the advantage of reduced bias should be weighed against the need for larger numbers of measurements.

In order to determine whether the added costs of a group-based study can be justified by the reduction in attenuation bias relative to an individual-based study, it is helpful to consider the ratio of the expected values of the estimators of β_1 for the two types of studies. This ratio, designated $\frac{E(\hat{\beta}_1)}{E(\hat{\beta}_1^*)}$, is derived from Equations (2) and (6) as:

$$\frac{E(\hat{\beta}_1)}{E(\hat{\beta}_1^*)} = \frac{(\sigma_{bg}^2 + \sigma_{b|h}^2)(\sigma_{bg}^2 + \frac{\sigma_{b|h}^2}{k} + \frac{\sigma_w^2}{kn})}{(\sigma_{bg}^2 + \sigma_{b|h}^2 + \frac{\sigma_w^2}{n})(\sigma_{bg}^2 + \frac{\sigma_{b|h}^2}{k})}. \tag{8}$$

Table 17.4 illustrates calculations, which employ equation (8), to gauge the reduction in attenuation bias that is possible using a group-based study compared to an individual-based study. Again, these calculations employ the median values of the variance components for occupational and environmental studies identified in Tables 17.2 and 17.3. The results indicate that values of $\frac{E(\hat{\beta}_1)}{E(\hat{\beta}_1^*)}$ are close to 1 for the typical occupational study, as long as k is at least 2. This suggests that the potential reduction in attenuation bias that would be conferred by a group-based occupational study [with a total of at least $4kn$ measurements, since $H \geq 4$ to allow the SE to be estimated using expression (7)] relative to an individual-based occupational study [involving a total of at least kn measurements with $k \geq 4$, since at least four persons must be sampled to allow estimation of the SE via expression (3)] would probably not be justified by the increased costs involved with collecting at least four times as many measurements. On the other hand, the typical environmental

Table 17.4 Reduction in attenuation bias for a group-based study compared to an individual-based study

Type of study	σ_{bg}^2	$\sigma_{b|h}^2$	σ_w^2	k	n	$\frac{E(\hat{\beta}_1)}{E(\hat{\beta}_1^*)}$
Occupational	0.36	0.59	0.48	2	2	0.94
Occupational	0.36	0.59	0.48	2	4	0.97
Occupational	0.36	0.59	0.48	2	8	0.98
Occupational	0.36	0.59	0.48	4	2	0.89
Occupational	0.36	0.59	0.48	4	4	0.94
Occupational	0.36	0.59	0.48	4	8	0.97
Occupational	0.36	0.59	0.48	8	2	0.85
Occupational	0.36	0.59	0.48	8	4	0.92
Occupational	0.36	0.59	0.48	8	8	0.96
Environmental	0.10	0.07	1.17	2	2	0.71
Environmental	0.10	0.07	1.17	2	4	0.77
Environmental	0.10	0.07	1.17	2	8	0.83
Environmental	0.10	0.07	1.17	4	2	0.51
Environmental	0.10	0.07	1.17	4	4	0.60
Environmental	0.10	0.07	1.17	4	8	0.70
Environmental	0.10	0.07	1.17	8	2	0.38
Environmental	0.10	0.07	1.17	8	4	0.49
Environmental	0.10	0.07	1.17	8	8	0.63

σ_{bg}^2, between-group variance component; $\sigma_{b|h}^2$, group-specific between-person variance component; σ_w^2, within-person variance component; k, number of persons sampled per group; n, number of measurements per person; $\frac{E(\hat{\beta}_1)}{E(\hat{\beta}_1^*)}$, ratio of the expected values of the estimated regression coefficients obtained from individual-based and group-based studies [equation (8)].

study would probably benefit greatly from a group-based design, since this would reduce attenuation bias by at least 50%, so long as $k \geq 4$. Note that the metric used in these calculations, i.e. $\frac{E(\hat{\beta}_1)}{E(\hat{\beta}_1^*)}$, is a *relative* measure of the reduction in bias conferred by a group-based design compared to an individual-based design; in fact, both individual-based and group-based studies could suffer from considerable attenuation bias in a particular application.

17.5. CONCLUSIONS

In this chapter we used simple straight-line models to evaluate the biasing effects of random exposure measurement errors. The models for exposure assumed random effects for group, person and error, and we considered a continuous health outcome variable. Although the results strictly apply only to continuous health outcomes, the underlying attenuation relationships are roughly appropriate for other types of health outcomes, including odds ratios estimated using logistic regression (Armstrong 1998; Armstrong *et al.* 1992; Rosner *et al.* 1989, 1992).

For both individual- and group-based studies, relationships were derived for the attenuation bias in estimated straight-line regression slopes (in log scale) as a function of sample sizes and the variance components representing exposure variability within persons, between persons and across groups. The results point to several ways to reduce the biasing effects of exposure measurement errors in epidemiological studies. First, it is important to seek the widest possible ranges of exposures. If the ranges of exposure are modest, then increases in the numbers of subjects and/or measurements per subject will be required. For both study designs, the relatively large within-subject variability in occupational or environmental exposures puts a premium on obtaining replicate exposure measurements on each subject. Overall biases tend to be smaller for group-based designs (due to the larger sample sizes collected in such studies) and thus point to grouping as a means of reducing attenuation bias when justified, despite the increased costs of additional measurements.

When comparing occupational and environmental sources of exposure, it appears that the typical environmental population will have larger within-subject variability, and smaller between-person and between-group variability, than the typical occupational population. This suggests that considerably larger sample sizes are required to investigate health effects arising from environmental exposures than from occupational exposures. It also suggests that group-based designs will be more useful than individual-based designs for investigating exposure–response relationships in the general population, but not so for occupational populations.

Finally, we note that this chapter has focused almost exclusively on considerations of bias (i.e. validity) rather than variability (i.e. precision). Although we think that validity issues should take precedence over precision issues in the design of research studies, we realize that both validity and precision can be considered simultaneously when designing an investigation. Also, as mentioned earlier, we have considered a relatively simple regression model measurement error scenario. For treatment of more complicated measurement error situations applicable to investigations of environmental and occupational health, see Lyles and Kupper (1997, 2000).

Acknowledgement

This work was supported in part by the National Institute for Environmental Health Sciences through Grants P42ES05948 and P30ES10126.

References

Armstrong BG. 1998. Effect of measurement error on epidemiological studies of environmental and occupational exposures. *Occup Environ Med* **55**: 651–656.

Armstrong BK, White E, Saracci G. 1992. *Principles of Exposure Measurement in Epidemiology*. Oxford University Press: New York.

Brunekreef B, Noy D, Clausing P. 1987. Variability of exposure measurements in environmental epidemiology. *Am J Epidemiol* **125**: 892–898.

Cochran WG. 1968. Errors of measurement in statistics. *Technometrics* **10**: 637–666.

Kromhout H, Tielemans E, Preller E *et al.* 1996. Estimates of individual dose from current measurements of exposure. *Occup Hyg* **3**: 23–29.

Lin YS, Kupper LL, Rappaport SM. 2005. Air samples vs. biomarkers for epidemiology. *Occup Environ Med* **62**: 750–760.

Lyles RH, Kupper LL. 1997. A detailed evaluation of adjustment methods for multiplicative measurement error in linear regression with applications in occupational epidemiology. *Biometrics* **53**: 1008–1025.

Lyles RH, Kupper LL. 2000. Measurement error models for environmental and occupational health applications. In *Handbook of Statistics, vol 18: Bio-environmental and Public Health Statistics*, Rao CR, Sen PK (eds). Elsevier Science: Amsterdam.

Rappaport SM, Kupper LL. 2004. Variability of environmental exposures to volatile organic compounds. *J Expo Anal Environ Epidemiol* **14**: 92–107.

Rosner B, Spiegelman D, Willett WC. 1992. Correction of logistic regression relative risk estimates and confidence intervals for random within-person measurement error. *Am J Epidemiol* **136**: 1400–1413.

Rosner B, Willett WC, Spiegelman D. 1989. Correction of logistic regression relative risk estimates and confidence intervals for systematic within-person measurement error. *Stat Med* **8**: 1051–1069; discussion 1071–1073.

Seixas NS, Sheppard L. 1996. Maximizing accuracy and precision using individual and grouped exposure assessments. *Scand J Work Environ Health* 22: 94–101.

Tielemans E, Kupper LL, Kromhout H *et al.* 1998. Individual-based and group-based occupational exposure assessment: some equations to evaluate different strategies. *Ann Occup Hyg* **42**: 115–119.

Weaver MA, Kupper LL, Taylor D *et al.* 2001. Simultaneous assessment of occupational exposures from multiple worker groups. *Ann Occup Hyg* **45**: 525–542.

18

Bioinformatics

Jason H. Moore

Dartmouth Medical School, Lebanon, NH, University of New Hampshire, Durham, NH, and University of Vermont, Burlington, VT, USA

18.1. INTRODUCTION

Bioinformatics is an interdisciplinary field that blends computer science and biostatistics with biomedical sciences, such as epidemiology, genetics and genomics. Bioinformatics emerged as an important discipline shortly after the development of high-throughput DNA sequencing technologies in the 1970s (Boguski 1994). It was the momentum of the Human Genome Project that spurred the rapid rise of bioinformatics as a formal discipline. The word 'bioinformatics' did not start appearing in the biomedical literature until around 1990, but quickly caught on as the descriptor of this important new field. An important goal of bioinformatics is to facilitate the management, analysis and interpretation of data from biological experiments and observational studies. Thus, much of bioinformatics can be categorized as database development and implementation, data analysis and data mining, and biological interpretation and inference. The goal of this chapter is to review each of these three areas and to provide some guidance on getting started with a bioinformatics approach to molecular epidemiology investigations of disease susceptibility.

The need to interpret information from whole-genome sequencing projects in the context of biological information acquired in decades of research studies prompted the establishment of the National Center for Biotechnology Information (NCBI) as a division of the National Library of Medicine (NLM) at the National Institutes of Health (NIH) in the USA in November of 1988. When the NCBI was established, it was charged with:

1. Creating automated systems for storing and analysing knowledge about molecular biology, biochemistry and genetics.
2. Performing research into advanced methods of computer-based information processing for analysing the structure and function of biologically important molecules and compounds.
3. Facilitating the use of databases and software by biotechnology researchers and medical care personnel.
4. Coordination of efforts to gather biotechnology information worldwide (Benson *et al.* 1990).

Since 1988, the NCBI has fulfilled many of these goals and has delivered a set of databases and computational tools that are essential for modern biomedical research in a wide range of different disciplines, including molecular epidemiology. The NCBI and other international efforts, such as the European Bioinformatics Institute (EBI), which was established in 1992 (Robinson 1994), have played a very important role in inspiring and motivating the establishment of research groups and centres around the world that are dedicated to providing bioinformatics tools and expertise. Some of these tools and resources are reviewed here.

Molecular Epidemiology of Chronic Diseases, Edited by C. P. Wild, P. Vineis, and S. Garte
© 2008 John Wiley & Sons, Ltd

18.2. DATABASE RESOURCES

One of the most important pre-study activities is the design and development of one or more databases that can accept, store and manage molecular epidemiology data. Haynes and Blach (2006) list eight steps for establishing an information management system for genetic studies:

1. *Develop the experimental plan* for the clinical, demographic, sample and molecular/laboratory information that will be collected – what are the specific needs for the database?
2. *Establish the information flow* – how does the information find its way from the clinic or laboratory to the database?
3. *Create a model for information storage* – how are the data related?
4. *Determine the hardware and software requirements* – how many data need to be stored? – how quickly will investigators need to access the data? – what operating system will be used? – will a freely available database such as mySQL (http://www.mysql.com) serve the needs of the project, or will a commercial dataset solution such as Oracle (http://www.oracle.com) be needed?
5. *Implement the database* – the important consideration here is to define the database structure so that data integrity is maintained.
6. *Choose the user interface to the database* – is a web page portal to the data sufficient?
7. *Determine the security requirements* – do HIPAA regulations (http://www.hhs.gov/ocr/hipaa) need to be followed? Most databases need to be password-protected at a minimum.
8. *Select the software tools* that will interface with the data for summary and analysis. Some of these tools are reviewed below.

Although most investigators choose to develop and manage their own database for security and confidentiality reasons, there are an increasing number of public databases for depositing data and making it widely available to other investigators. The tradition of making data publicly available soon after it has been analysed and published can largely be attributed to the community of investigators using gene expression microarrays.

Microarrays (Shena *et al.* 1995) represent one of the most revolutionary applications that derived from the knowledge of whole genome sequences. The extensive use of this technology has led to the need to store and search expression data for all the genes in the genome acquired in different genetic backgrounds or in different environmental conditions. This has resulted in a number of public databases, such as the Stanford Microarray Database (Sherlock *et al.* 2001; http://genome-www5. stanford.edu), the Gene Expression Omnibus (Barrett *et al.* 2005, 2006; http://www.ncbi.nlm. nih.gov/geo), ArrayExpress (Brazma *et al.* 2003, 2006; http://www.ebi.ac.uk/arrayexpress) and others, from which anyone can download data. The nearly universal acceptance of the data-sharing culture in this area has yielded a number of useful tools that might not otherwise have been developed. The need to define standards for the ontology and annotation of microarray experiments has led to proposals such as the Minimum Information About a Microarray Experiment (MIAME; Brazma *et al.* 2001; Ball and Brazma 2006; http://www. mged.org/Workgroups/MIAME/miame.html), which provided a standard that greatly facilitates the storage, retrieval and sharing of data from microarray experiments. The MIAME standards provide an example for other types of data, such as SNPs and protein mass spectrometry spectra (for a comprehensive review of standards for data sharing in genetics, genomics, proteomics and systems biology, see Brazma *et al.* 2006). The success of the different databases then depends on the availability of methods for easily depositing data and tools for searching the databases often after data normalization.

Despite the acceptance of data sharing in the genomics community, the same culture does not yet exist in molecular epidemiology and genetics. One of the few such examples is the Pharmacogenomics Knowledge Base of PharmGKB (Hewett *et al.* 2002; http://www.pharmgkb.org). PharmGKB was established with funding from the NIH to store, manage and make available molecular data in addition to phenotype data from pharmacogenetic and pharmacogenomic experiments and clinical studies. It is anticipated that similar databases for molecular epidemiology will appear and gain acceptance over the next few years, as the NIH and various journals start to require data from public research be made available to the public.

In addition to the need for a database to store and manage molecular epidemiology data collected from experimental or observational studies, there a number of database resources that can be very helpful for planning a study. A good starting point for database resources are those maintained at the NCBI (Wheeler *et al.* 2006; http://www.ncbi.nlm.nih.gov). Perhaps the most useful resource when planning a molecular epidemiology study is the Online Mendelian Inheritance in Man (OMIM) database (Hamosh *et al.* 2000, 2005; http://www.ncbi.nlm.nih.gov/omim). OMIM is a catalogue of human genes and genetic disorders, with detailed summaries of the literature. The NCBI also maintains the PubMed literature database, with more than 15 million indexed abstracts from published papers in more than 4700 life science journals. The PubMed Central database (http://www.pubmedcentral.nih.gov) is quickly becoming an indispensable tool, with more than 400 000 full text papers from over 200 different journals. Rapid and free access to the complete text of published papers significantly enhances the planning, execution and interpretation phases of any scientific study. The new Books database (http://www.ncbi.nlm.nih.gov/books) provides free access for the first time to electronic versions of many textbooks and other resources, such as the *NCBI Handbook*, which serves as a guide to the resources that NCBI has to offer. This is a particularly important resource for students and investigators who need to learn a new discipline such as genomics. One of the oldest databases provided by the NCBI is the GenBank DNA sequence resource (Benson *et al.* 1993, 2006; http://www.ncbi.nlm.nih.gov/Genbank). DNA sequence data for many different organisms have been deposited in GenBank for more than two decades now, totalling more than 100 gigabases of data. GenBank is a common starting point for the design of PCR primers and other molecular assays that require specific knowledge of gene sequences. Curated information about genes, their chromosomal locations, their functions, their pathways, etc., can be accessed through the Entrez Gene database (http://www.ncbi.nlm.nih.gov/entrez/query.fcgi?db=gene), for example.

Important emerging databases include those that store and summarize DNA sequence variations. NCBI maintains the dbSNP (Sherry *et al.* 1999, 2001; http://www.ncbi.nlm.nih.gov/projects/SNP/) database for single-nucleotide polymorphisms (SNPs). dbSNP provides a wide range of different information about SNPs, including the flanking sequence primers, the position, the validation methods and the frequency of the alleles in different populations. As with all NCBI databases, it is possible to link to a number of other datasets, such as PubMed and OMIM. The recently completed International Haplotype Map (HapMap) project documents genetic similarities and differences among different populations (International HapMap Consortium 2005). Understanding the variability of SNPs and the linkage disequilibrium structure plays an important role in determining which SNPs to measure when planning a molecular epidemiology study. The International HapMap Consortium maintains an online database with all the data from HapMap project (http://www.hapmap.org/thehapmap.html). Another useful database is the Allele Frequency Database (ALFRED; Cheung *et al.* 2000; http://alfred.med.yale.edu/alfred/index.asp), which currently stores information on more than 3700 polymorphisms across 518 populations.

In addition to databases for storing raw data, there are a number of databases that retrieve and store knowledge in an accessible form. For example, the Kyoto Encyclopedia of Genes and Genomes (KEGG) database stores knowledge on genes and their pathways (Kanehisa 1997; Ogata 1999; http://www.genome.jp/keg). The Pathway component of KEGG currently stores knowledge on 42 937 pathways generated from 307 reference pathways. While the Pathway component documents molecular interaction in pathways, the Brite database stores knowledge on higher-order biological functions. One of the most useful knowledge sources is the Gene Ontology (GO) project, which has created a controlled vocabulary to describe genes and gene products in any organism in terms of their biological processes, cellular components and molecular functions (Ashburner *et al.* 2000; Gene Ontology Consortium 2006; http://www.geneontology.org). GO descriptions and KEGG pathways are both captured and summarized in the NCBI databases. For example, the description of *p53* in Entrez Gene includes KEGG pathways, such as cell cycle and apoptosis; it also includes GO descriptions, such as protein binding and cell proliferation.

In general, a good place to start for information about available databases is the annual Database issue and the annual Web Server issue of the journal *Nucleic Acids Research*. These special issues include annual reports from many of the commonly used databases.

18.3. DATA ANALYSIS

Once the data are collected and stored in a database, an important goal of molecular epidemiology is to identify biomarkers or molecular/environmental predictors of disease susceptibility or severity. A recent series of seven reviews summarize many of the basics of genetic and epidemiological association studies in human populations, thus providing a starting point for those needing to learn more about basic analytical methods (see Burton *et al.* 2005; Teare and Barrett 2005; Cordell and Clayton 2005; Palmer and Cardon 2005; Hattersley and McCarthy 2005; Hopper *et al.* 2005; Smith *et al.* 2005). Several other recent reviews also provide some basic concepts (e.g. Balding 2006). In addition to the availability of numerous statistical methods for association studies, there are a number of software packages that can be used for data analysis. A long list of more than 350 genetic software packages can be found at Genetic Analysis Software website (http://www.nslij-genetics.org/soft). A recent comprehensive review by Excoffier and Heckel (2006) takes a critical look at more than 20 such software packages for computing linkage disequilibrium, genetic diversity and association, for example. These resources provide an excellent starting point for deciding which software packages are available for molecular epidemiology analysis. However, the tools available through these resources are not the only ones that a molecular epidemiologist might find useful. We review below several data-mining/machine-learning tools for identifying more complex patterns of association that might be predictive of disease susceptibility.

It is increasingly recognized that the mapping relationship between genotype and phenotype is highly complex, involving phenomena such as locus heterogeneity, gene–environment interactions and gene–gene interactions or epistasis (Templeton 2000; Moore 2003; Sing *et al.* 2003; Thornton-Wells *et al.* 2004; Moore and Williams 2005). As such, modern molecular epidemiology must take these non-linearities into account during the data analysis and modelling process. Traditional methods, such as logistic regression, are not always well suited for modelling high-dimensional datasets (Moore and Williams 2002). Machine learning and data mining are primarily computer science disciplines that play a very important role in bioinformatics, especially for detecting non-linear interactions among variables (Freitas 2001). Methods such as neural networks and classification trees are necessary to complement the traditional statistical methods, such as those based on the linear model (McKinney *et al.* 2006). Machine-learning and data-mining methods have the advantage of having more power to detect non-linear patterns in high-dimensional data. Although power can be gained, these methods are typically more difficult to implement because they require some knowledge of computer science and often require more computing power than is available using a single processor. An important development in the last few years is the availability of open-source and user-friendly software that brings many of these advanced computer science methods to the molecular epidemiologist. We briefly review several of these software packages below. For those looking for a good introduction to machine-learning and data-mining methods, we recommend Hastie *et al.* (2001) and Whitten and Frank (2005).

Data mining using R

R is perhaps the one software package that everyone should have in their bioinformatics arsenal. R is an open-source and freely-available programming language and data analysis and visualization environment that can be downloaded from http://www.r-project.org. According to the web page, R includes: (a) an effective data handling and storage facility; (b) a suite of operators for calculations on arrays, in particular matrices; (c) a large, coherent, integrated collection of intermediate tools for data analysis; (d) graphical facilities for data analysis and display, either on-screen or on hard copy; and (e) a well-developed, simple and effective programming language, which includes conditionals, loops, user-defined recursive functions and input and output facilities. A major strength of R is the enormous community of developers and users, which ensures that just about any analysis method needed is available. This includes analysis packages such as Rgenetics (http://rgenetics.org) for basic genetic and epidemiological analysis, such as testing for deviations from Hardy–Weinberg equilibrium or haplotype estimation, epitools for basic epidemiology analysis (http://www.epitools.net), geneland for spatial genetic analysis (http://www.inapg.inra.fr/ens_rech/mathinfo/personnel/guillot/Geneland.html), and popgen for population genetics (http://cran.r-project.org/src/contrib/Descriptions/popgen.html). Perhaps the most useful contribution to R is the Bioconductor project (Reimers and Carey 2006; http://www.bioconductor.org). According to the Bioconductor web page, the goals of the project are to: (a) provide access to a wide range of powerful statistical and graphical methods for the analysis of genomic data; (b) facilitate the integration of biological metadata (e.g. PubMed, GO) in the analysis of experimental data; (c) allow the rapid development of extensible, scalable and interoperable software; (d) promote high-quality and reproducible research; and (e) provide training in computational and statistical methods for the analysis of genomic data. Many of the available tools are specific to microarray data but some are more generally applicable to genetics and epidemiology, e.g. there are tools in Bioconductor for accessing data from the KEGG database mentioned above.

There are numerous packages for machine learning and data mining that are either part of the base R software or can be easily added. For example, the neural package includes routines for neural network analysis (http://cran.r-project.org/src/contrib/Descriptions/neural.html). Others include 'arules' for association rule mining (http://cran.r-project.org/src/contrib/Descriptions/arules.html), 'cluster' for cluster analysis (http://cran.r-project.org/src/contrib/Descriptions/cluster.html), 'genalg' for genetic algorithms (http://cran.r-project.org/src/contrib/Descriptions/genalg.html), 'som' for self-organizing maps (http://cran.r-project.org/src/contrib/Descriptions/som.html) and 'tree' for classification and regression trees (http://cran.r-project.org/src/contrib/Descriptions/tree.html). Many others are available. A full list of contributed packages for R can be found at: http://cran.r-project.org/src/contrib/PACKAGES.html. The primary advantage of using R as your data-mining software package is its power. However, the learning curve can be challenging at first. Fortunately, there is plenty of documentation available on the web and in published books. Several essential R books include those by Gentleman *et al.* (2005) and Venables and Ripley (2002).

Data mining using Weka

One of the most mature open-source and freely available data-mining software packages is Weka (Whitten and Frank 2005; http://www.cs.waikato.ac.nz/ml/weka). Weka is written in Java and thus will run in any operating system (e.g. Linux, Mac, Sun, Windows). Weka contains a comprehensive list of tools and methods for data processing, unsupervised and supervised classification, regression, clustering, association rule mining and data visualization. Machine-learning methods include classification trees, *k*-means cluster analysis, *k*-nearest neighbours, logistic regression, naïve Bayes, neural networks, self-organizing maps, and support vector machines, for example. Weka includes a number of additional tools, e.g. search algorithms and analysis tools such as cross-validation and bootstrapping. A nice feature of Weka is that it can be run from the command line, making it possible to run the software from Perl or even R (see http://cran.r-project.org/src/contrib/Descriptions/RWeka.html). Weka includes an experimenter module that facilitates comparison of algorithms. It also includes a knowledge flow environment for visual layout of an analysis pipeline. This is a very powerful analysis package that is relatively easy to use. Further, there is a published book that explains many of the methods and the software (Whitten and Frank 2005).

Data mining using Orange

Orange is another open-source and freely-available data mining software package (Demsar and Zupan 2004; http://www.ailab.si/orange) that provides a number of data-processing, data-mining and data-visualization tools. What makes Orange different and in some ways preferable to other packages, such as R, is its intuitive visual programming interface. With orange, methods and tools are represented as icons that are selected and dropped into a window called the canvas. For example, an icon for loading a dataset can be selected along with an icon for visualizing the data table. The file load icon is then 'wired' to the data table icon by drawing a line between them. Double-clicking on the file load icon allows the user to select a data file. Once loaded, the file is then automatically transferred by the 'wire' to the data table icon. Double-clicking on the data table icon brings up a visual display of the data. Similarly, a classifier such as a classification tree can be selected and wired to the file icon. Double-clicking on the classification tree icon allows the user to select the settings for the analysis. Wiring the tree viewer icon then allows the user to view a graphical image of the classification tree inferred from the data. Orange facilitates high-level data mining with minimal knowledge of computer programming. A wide range of different data analysis tools are available. A strength of Orange is its visualization tools for multivariate data (e.g. Leban *et al.* 2005; for example applications of Orange and its interaction graph tool for visualizing gene–gene interactions, see e.g. Angeline *et al.* 2006; Moore *et al.* 2006). Recent additions to Orange include tools for microarray analysis and genomics, such as heat maps and GO analysis (Curk *et al.* 2005).

Data mining using multifactor dimensionality reduction

Multifactor dimensionality reduction (MDR) was developed as a non-parametric (i.e. no parameters are estimated) and genetic model-free (i.e. no genetic model is assumed) data-mining strategy for identifying combinations of SNPs that are predictive of a discrete clinical endpoint (Ritchie *et al.* 2001, 2003; Hahn *et al.* 2003; Hahn and Moore 2004; Moore 2004, 2007; Moore *et al.* 2006). Technical details of the MDR method, along with a summary of MDR applications, can be found in the previous citations or in Chapter 16. The original software package for MDR was programmed

in C and was executed from a simple command-line interface (Hahn *et al.* 2003). This software is no longer being used. MDR is now available in a mature software package that is open-source, freely available and comes with a user-friendly graphic user interface (GUI) for point-and-click analysis (http://www.epistasis.org). As of early 2007, MDR is available in version 1.0 and is the result of more than 2 years of development and testing. The MDR 1.0 software is programmed entirely in Java and is thus compatible with any operating system (e.g. Linux, Mac, Sun, Windows). MDR can be run using the visual interface or can be run from the command line for scripting in Perl, for example. We describe here some of the features of the MDR software.

Figure 18.1 illustrates the MDR GUI. Version 1.0 has four tabs that each shows a different window for carrying out different MDR functions. The analysis tab allows the user to load a dataset, visualize the dataset and run an MDR analysis. The summary table reports statistics, such as the training and testing accuracy for best MDR models as estimated using n-fold cross-validation. The tabs at the bottom of the window provide graphical output illustrating the distribution of cases and controls for each genotype combination (Graphical Model tab), an interaction dendrogram for interpreting interaction results (Dendrogram tab) and the landscape of all models considered along with their training accuracy (Landscape tab). The other tabs report additional detailed statistics about the MDR models. An important feature of the MDR software is the ability to save an MDR analysis (Save Analysis button). This feature allows the user to save all the details about the MDR analysis that was run so that it can be loaded later for inspection of the results without the need to rerun an analysis that might take hours of computing time.

The Configuration tab allows the user to set analysis options, such as the random seed, the order of the models to be considered, the cross-validation count and the search method to be used (e.g. exhaustive, random, etc.). The filter tab allows the user to pre-select a list of SNPs or other discrete variables for MDR analysis, using statistics such as the χ^2 test of independence, odds ratios and the ReliefF algorithm. This feature is useful when the

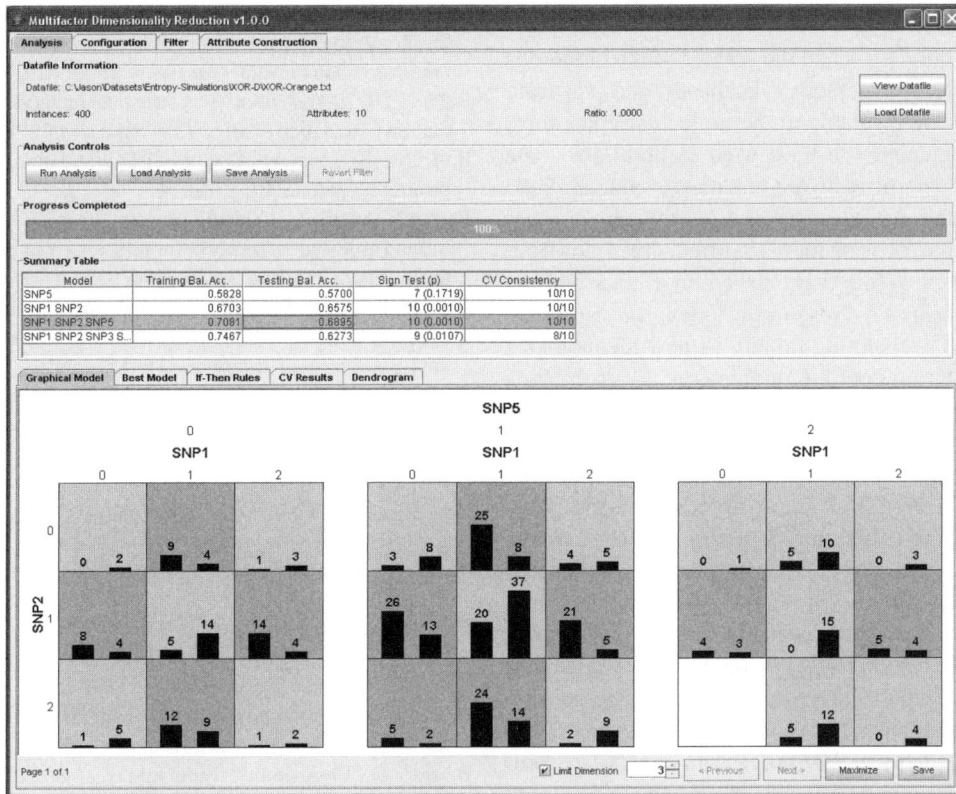

Figure 18.1 Graphic-user interface (GUI) for the multifactor dimensionality reduction (MDR) software package (available from http://epistasis.org). Illustrated in the Graphical Model tab is the distribution of cases (right bars) and controls (left bars) for each three-locus genotype combination

list of SNPs is large and thus not practical for an exhaustive MDR analysis. The fourth tab in the MDR software is the Attribute Construction tab, which allows the user to construct new attributes of variables using the MDR function. These new variables can be added back to the dataset for hierarchical analysis. New features are being added to the MDR software. The software can be downloaded from http://www.epistasis.org.

Interpreting data mining results

Perhaps the greatest challenge of any statistical analysis or data mining exercise is interpreting the results. How does a high-dimensional statistical pattern derived from population-level data relate to biological processes that occur at the cellular level

(Moore and Williams 2005)? This is an important question that is difficult to answer without a close working relationship between epidemiologists, for example, and statisticians and computer scientists. Fortunately, there are a number of emerging software packages that are designed with this in mind, e.g. GenePattern (http://www.broad.mit.edu/cancer/software/genepattern/) provides an integrated set of analysis tools and knowledge sources that facilitates this process (Reich *et al.* 2006). Other tools, such as the Exploratory Visual Analysis (EVA) database and software (http://www.exploratoryvisualanalysis.org/), are designed specifically for integrating research results with biological knowledge from public databases in a framework designed for biologists and epidemiologists (Reif *et al.* 2005). These tools and others will facilitate interpretation.

18.4. THE FUTURE

We have only scratched the surface of the numerous bioinformatics methods, databases and software tools that are available to the molecular epidemiology community. We have tried to highlight some of the important software resources, such as Weka and Orange, which might not be covered in other reviews that focus on more traditional methods from biostatistics. While there are an enormous number of bioinformatics resources today, the software landscape is changing rapidly as new technologies for high-throughput biology emerge. Over the next few years we will witness an explosion of novel bioinformatics tools for the analysis of genome-wide association data and, more importantly, the joint analysis of SNP data with other types of data, such as gene expression data and proteomics data. Each of these new data types and their associated research questions will require special bioinformatics tools and perhaps special hardware, such as faster computers with bigger storage capacity and more memory. Some of these datasets will easily require 1–2 Gb of memory or more for analysis and could require as many as 100 processors or more to complete a data-mining analysis in a reasonable amount of time. The challenge will be to scale our bioinformatics tools and hardware such that a genome-wide SNP dataset can be processed as efficiently as we can process a candidate gene dataset with perhaps 20 SNPs today. Only then can molecular epidemiology truly arrive in the genomics age.

Acknowledgements

This work was supported by National Institutes of Health (USA) Grants LM009012, AI59694, HD047447, RR018787 and HL65234.

References

Altshuler D, Brooks LD, Chakravarti A *et al.* 2005. International HapMap Consortium. A haplotype map of the human genome. *Nature* **437**: 1299–1320.

Andrew AS, Nelson HH, Kelsey KT *et al.* 2006. Concordance of multiple analytical approaches demonstrates a complex relationship between DNA repair gene SNPs, smoking, and bladder cancer susceptibility. *Carcinogenesis* **27**: 1030–1037.

Ashburner M, Ball CA, Blake JA *et al.* 2000. Gene ontology: tool for the unification of biology. The Gene Ontology Consortium. *Nat Genet* **25**: 25–29.

Balding DJ. 2006. A tutorial on statistical methods for population association studies. *Nat Rev Genet* **7**: 781–791.

Ball CA, Brazma A. 2006. MGED standards: work in progress. *OMICS* **10**: 138–144.

Barrett T, Suzek TO, Troup DB *et al.* 2005. NCBI GEO: mining millions of expression profiles – database and tools. *Nucleic Acids Res* **33**: D562–566.

Barrett T, Edgar R. 2006. Gene expression omnibus: microarray data storage, submission, retrieval, and analysis. *Methods Enzymol* **411**: 352–369.

Brazma A, Hingamp P, Quackenbush J *et al.* 2001. Minimum information about a microarray experiment (MIAME) – toward standards for microarray data. *Nat Genet* **29**: 365–371.

Brazma A, Parkinson H, Sarkans U *et al.* 2003. ArrayExpress – a public repository for microarray gene expression data at the EBI. *Nucleic Acids Res* **31**: 68–71.

Brazma A, Kapushesky M, Parkinson H *et al.* 2006. Data storage and analysis in ArrayExpress. *Methods Enzymol* **411**: 370–386.

Brazma A, Krestyaninova M, Sarkans U. 2006. Standards for systems biology. *Nat Rev Genet* **7**: 593–605.

Benson D, Boguski M, Lipman DJ *et al.* 1990. The National Center for Biotechnology Information. *Genomics* **6**: 389–391.

Benson D, Lipman DJ, Ostell J. 1993. GenBank. *Nucleic Acids Res* **21**: 2963–2965.

Benson DA, Karsch-Mizrachi I, Lipman DJ *et al.* 2006. GenBank. *Nucleic Acids Res* **34**: D16–20.

Boguski MS. 1994. Bioinformatics. *Curr Opin Genet Dev* **4**: 383–388.

Burton PR, Tobin MD, Hopper JL. 2005. Key concepts in genetic epidemiology. *Lancet* **366**: 941–951.

Cheung KH, Osier MV, Kidd JR *et al.* 2000. ALFRED: an allele frequency database for diverse populations and DNA polymorphisms. *Nucleic Acids Res* **28**: 361–363.

Cordell HJ, Clayton DG. Genetic association studies. *Lancet* **366**: 1121–1131.

Curk T, Demsar J, Xu Q *et al.* 2005. Microarray data mining with visual programming. *Bioinformatics* **21**: 396–398.

Demsar J, Zupan B, Leban G. 2004. Orange: from experimental machine learning to interactive data mining; white paper (www.ailab.si/orange). Faculty of Computer and Information Science, University of Ljubljana.

Excoffier L, Heckel G. 2006. Computer programs for population genetics data analysis: a survival guide. *Nat Rev Genet* **7**: 745–758.

Freitas A. 2001. Understanding the crucial role of attribute interactions. *Artif Intell Rev* **16**: 177–199.

Gene Ontology Consortium. 2006. The Gene Ontology (GO) project in 2006. *Nucleic Acids Res* **34**: D322–326.

Gentleman R, Carey VJ, Huber W *et al.* 2005. *Bioinformatics and Computational Biology Solutions using R and Bioconductor.* Springer: New York.

Hahn LW, Moore JH. 2004. Ideal discrimination of discrete clinical endpoints using multilocus genotypes. *In Silico Biol* **4**: 183–194.

Hahn LW, Ritchie MD, Moore JH. 2003. Multifactor dimensionality reduction software for detecting gene–gene and gene–environment interactions. *Bioinformatics* **19**: 376–382.

Hamosh A, Scott AF, Amberger J *et al.* 2000. Online Mendelian Inheritance in Man (OMIM). *Hum Mutat* **15**: 57–61.

Hamosh A, Scott AF, Amberger J *et al.* 2005. Online Mendelian Inheritance in Man (OMIM), a knowledge base of human genes and genetic disorders. *Nucleic Acids Res* **33**: D514–517.

Hastie T, Tibshirani R, Friedman J. 2001. *The Elements of Statistical Learning.* Springer: New York.

Hattersley AT, McCarthy MI. 2005. What makes a good genetic association study? *Lancet* **366**: 1315–1323.

Haynes C, Blach C. 2006. Information management. In *Genetic Analysis of Complex Disease*, Haines JL, Pericak-Vance MA (eds). Wiley: Hoboken, NJ.

Hewett M, Oliver DE, Rubin DL *et al.* 2002. PharmGKB: the Pharmacogenetics Knowledge Base. *Nucleic Acids Res* **30**: 163–165.

Hopper JL, Bishop DT, Easton DF. 2005. Population-based family studies in genetic epidemiology. *Lancet* **366**: 1397–1406.

International HapMap Consortium. 2005. A haplotype map of the human genome. *Nature* **437**: 1299–1320.

Kanehisa M. 1997. A database for post-genome analysis. *Trends Genet* **13**: 375–376.

Leban G, Bratko I, Petrovic U *et al.* 2005. VizRank: finding informative data projections in functional genomics by machine learning. *Bioinformatics* **21**: 413–414.

McKinney BA, Reif DM, Ritchie MD *et al.* 2006. Machine learning for detecting gene–gene interactions: a review. *Appl Bioinf* **5**: 77–88.

Moore JH. 2003. The ubiquitous nature of epistasis in determining susceptibility to common human diseases. *Hum Heredity* **56**: 73–82.

Moore JH. 2004. Computational analysis of gene–gene interactions in common human diseases using multifactor dimensionality reduction. *Exp Rev Mol Diagn* **4**: 795–803.

Moore JH. 2007. Genome-wide analysis of epistasis using multifactor dimensionality reduction: feature selection and construction in the domain of human genetics. In *Knowledge Discovery and Data Mining: Challenges and Realities with Real World Data*, Zhu X, Davidson I (eds). IGI Global, 17–30.

Moore JH, Karagas MR, Andrew AS. 2007. Analysis of complex data. In *Molecular Epidemiology of Chronic Diseases: New Techniques and Approaches*, Wild C, Vineis P, Garte S (eds). Wiley: New York.

Moore JH, Gilbert JC, Tsai C-T *et al.* 2006. A flexible computational framework for detecting, characterizing, and interpreting statistical patterns of epistasis in genetic studies of human disease susceptibility. *J Theoret Biol* **241**: 252–261.

Moore JH, Williams SW. 2002. New strategies for identifying gene–gene interactions in hypertension. *Ann Med* **34**: 88–95.

Moore JH, Williams SW. 2005. Traversing the conceptual divide between biological and statistical epistasis: systems biology and a more modern synthesis. *BioEssays* **27**: 637–646.

Ogata H, Goto S, Sato K *et al.* 1999. KEGG: Kyoto Encyclopedia of Genes and Genomes. *Nucleic Acids Res* **27**: 29–34.

Palmer LJ, Cardon LR. 2005. Shaking the tree: mapping complex disease genes with linkage disequilibrium. *Lancet* **366**: 1223–1234.

Reich M, Liefeld T, Gould J *et al.* 2006. GenePattern 2.0. *Nat Genet* **38**: 500–501.

Reif DM, Dudek SM, Shaffer CM *et al.* 2005. Exploratory visual analysis of pharmacogenomic results. *Pac Symp Biocomput*: 296–307.

Reimers M, Carey VJ. 2006. Bioconductor: an open source framework for bioinformatics and computational biology. *Methods Enzymol* **411**: 119–134.

Ritchie MD, Hahn LW, Moore JH. 2003. Power of multifactor dimensionality reduction for detecting gene–gene interactions in the presence of genotyping error, phenocopy, and genetic heterogeneity. *Genet Epidemiol* **24**: 150–157.

Ritchie MD, Hahn LW, Roodi N *et al.* 2001. Multifactor dimensionality reduction reveals high-order interactions among estrogen metabolism genes in sporadic breast cancer. *Am J Hum Genet* **69**: 138–147.

Robinson C. 1994. The European Bioinformatics Institute (EBI) – open for business. *Trends Biotechnol* **12**: 391–392.

Schena M, Shalon D, Davis RW *et al.* 1995. Quantitative monitoring of gene expression patterns with a complementary DNA microarray. *Science* **270**: 467–470.

Sherlock G, Hernandez-Boussard T, Kasarskis A *et al.* 2001. The Stanford Microarray Database. *Nucleic Acids Res* **29**: 152–155.

Sherry ST, Ward M, Sirotkin K. 1999. dbSNP-database for single nucleotide polymorphisms and other classes of minor genetic variation. *Genome Res* **9**: 677–679.

Sherry ST, Ward MH, Kholodov M *et al.* 2001. dbSNP: the NCBI database of genetic variation. *Nucleic Acids Res* **29**: 308–311.

Sing CF, Stengard JH, Kardia SL. 2003. Genes, environment, and cardiovascular disease. *Arterioscler Thromb Vasc Biol* **23**: 1190–1196.

Smith G, Ebrahim S, Lewis S *et al.* 2005. Genetic epidemiology and public health: hope, hype, and future prospects. *Lancet* **366**: 1484–1498.

Teare MD, Barrett JH. 2005. Genetic linkage studies. *Lancet* **366**: 1036–1044.

Templeton AR. 2000. Epistasis and complex traits. In *Epistasis and the Evolutionary Process*, Wolf J, Brodie B III, Wade M (eds). Oxford University Press: New York; 41–57.

Thornton-Wells TA, Moore JH, Haines JL. 2004. Genetics, statistics and human disease: analytical retooling for complexity. *Trends Genet* **20**: 640–647.

Venebles WN, Ripley BD. 2002. *Modern Applied Statistics with S*. Springer: New York.

Wheeler DL, Barrett T, Benson DA *et al.* 2006. Database resources of the National Center for Biotechnology Information. *Nucleic Acids Res* **34**: D173–180.

Whitten IH, Frank E. 2005. *Data Mining*. Elsevier: Boston, MA.

19

Biomarkers, Disease Mechanisms and their Role in Regulatory Decisions

Pier Alberto Bertazzi[1] and Antonio Mutti[2]

[1]*Università degli Studi and IRCCS Maggiore Hospital Policlinico, Milan,*
and [2]*Università degli Studi, Parma, Italy*

19.1. INTRODUCTION

Regulatory decisions are based on and justified according to a set of criteria – including scientific evidence, technical feasibility, people's perceptions and societal sustainability – that are subject to change in time and space. For example, early regulations in occupational and environmental health were aimed at reducing the risk of diseases that were usually diagnosed in their overt, clinical stage. At that time, societal interest in the health conditions of exposed workers was scarce and scientific capabilities were limited. Today, public awareness of environmental risks and concern about their health effects are widespread, and there are strict regulations regarding environmental and occupational exposures in many countries. New techniques accurately measure exposure even at trace concentrations, and early functional, biochemical or genetic changes can be recognized at the cellular and molecular level. Exposed populations have also undergone changes, particularly at work, in terms of ethnicity, gender composition, age structure and education (hence, susceptibility, culture and values).

The contribution of science to the regulatory decision-making process is becoming more and more relevant, thanks especially to its greater capability to identify hazards, assess exposure types and levels and characterize risks. When valid and complete information exists, the regulatory process is based on scientific evidence weighed against other relevant criteria, e.g. economical and technical considerations; instead, when scientific evidence is missing or incomplete, other criteria, e.g. precautionary considerations, may drive the balance (Grandjean 2004).

19.2. HAZARD IDENTIFICATION AND STANDARD SETTING

Defence measures depend upon the nature of the hazard. Hence, identifying the hazard is the primary step in the process of making regulatory decisions. The International Agency for Research on Cancer (IARC) 'Monographs on the Evaluation of Carcinogenic Risk to Humans' programme represents a major example of hazard identification procedure (http://monographs.iarc.fr.). Since programme inception (1971), increasing attention has been devoted to the action mode or 'intermediate biology' (Potter 2005) of suspected agents as a result of greater scientific knowledge and analytical capabilities. To identify hazards, an increasing number of mechanistic markers are becoming available, including biomarkers of internal dose and biologically effective dose, of damage to DNA, proteins, and other chemical and structural components of the cell; biomarkers of early effects, such as mutations, chromosomal

Molecular Epidemiology of Chronic Diseases, Edited by C. P. Wild, P. Vineis, and S. Garte
© 2008 John Wiley & Sons, Ltd

aberrations, genetic and genomic instability, or epigenetic modifications; biomarkers of cellular, tissue or organism responses, such as hormonal, inflammatory or immunological responses; and biomarkers of genetic susceptibility, such as genetic variations affecting gene–environment interactions, which can contribute to supporting the plausibility of suspected causal pathways. In at least two cases (2,3,7,8-tetrachlorodibenzo-*p*-dioxin and ethylene oxide), chemical agents have been categorized as carcinogenic to humans on the basis of mechanistic considerations, even though the epidemiological evidence was considered limited (IARC 1994, 1997), whereas in other cases, e.g. di(2-ethylhexyl)phthalate (DEHP), the compound was downgraded as to its carcinogenicity because the mechanism of carcinogenicity in experimental animals was not considered relevant to humans (IARC 2000).

The extent to which identification of a hazard (e.g. carcinogenic potential) and the action mode of a compound/agent may impact on the regulatory process is nicely illustrated by the work performed by the Scientific Committee on Occupational Exposure Limits (SCOEL), an interdisciplinary advisory group of experts established in 1990 by the European Commission with the mission of identifying and proposing 'health-based' exposure limits for occupational settings (OEL). This implies some knowledge of the function linking exposure–dose and effect–response markers, the identification of a non-observed or lowest-observed adverse effect level (NOAEL/LOAEL) and its weighting against a safety or uncertainty factor. The methods and criteria used to derive exposure limits clearly differ depending upon the nature of the hazard and its mechanism of action (EU 1999). A core issue is to establish whether the substance acts via a non-threshold mechanism or whether a deterministic model with a toxicological threshold can be adopted. To this end, key components of a relevant dataset are likely to include the following: non-threshold effects; effects of long-term and repeated exposure, including dose–response (D-R) relationships; effects resulting from short-term or single exposures; nature of the effects (either local or systemic) and target organs; and methods of measurement of exposure levels. When the avail-

able dataset is satisfactory, it becomes possible to identify: (a) the critical effect(s) on which the limits should be based; (b) the exposure level at which no such critical adverse effect is observable (NOAEL); or (c) the lowest level at which the effect is observable (LOAEL). The OEL is almost regularly set at a level lower than the NOAEL or LOAEL by applying a weighting factor (also called uncertainty factor) that is a function of the degree of confidence in the available database (the higher the uncertainty, the higher the weight). It comprises all adjustment aspects that are related to health (e.g. route-to-route, inter- and intra-species extrapolation). Uncertainty factors must be established on a case-by-case basis and cannot be forecast or established in advance. Other mechanistic information affecting the OEL definition process includes possible short-term effects and dermal absorption. Health effects that may arise from short-term exposure (e.g. irritation, CNS depression, cardiac sensitization) are not adequately controlled by compliance with an 8 h time-weighted average (TWA) value, given the inherent variations in exposure even when compliance with this type of limit is met. In such cases, a short-term exposure limit (STEL) is proposed, usually related to a 15 min reference period, above which exposure should not occur. In order to effectively control total systemic dose, it is further necessary to consider dermal exposure, and not only exposure by inhalation route, which may lead to skin penetration and increase the total body burden of a chemical. In the case of significant uptake through the skin, a notation 'Skin' is associated to the OEL value to further specify the type of rules to be followed for effective health protection.

For recognized carcinogens, no attempt is made by SCOEL to identify health-based limits, under the assumption that the dose–response relationship is linear and no threshold exists. Instead, whenever existing data permit, a risk assessment procedure is implemented in order to estimate the different risks entailed by different degrees of exposure (Zocchetti *et al.* 2004). The results are then forwarded to regulators who eventually come to a decision by weighting the evidence of the estimated risks at different exposure levels against complementary criteria, including risk 'acceptability'. Those limits should not then be considered 'health-based', but rather 'pragmatic'.

Although the assumption of a linear dose–response relationship without threshold has been plausibly supported for many carcinogenic agents, other carcinogens may behave differently. SCOEL is presently discussing a proposal which takes into consideration the ascertained or presumed different mechanisms of carcinogenic action (Bolt et al. 2004a, 2004b). An array of possible modes of action have been proposed (Streffer et al. 2004). A first category includes genotoxic carcinogens. In this case, a linear no-threshold (LNT) model appears appropriate. If agent elimination is not feasible, regulations may then be based on the principle 'as-low-as-reasonably-achievable' (ALARA), on technical feasibility and social considerations. Examples of this type are ionizing radiation, vinyl chloride, 1,3-butadiene and diethyl-nitrosamine (DEN). There are genotoxic carcinogens for which the existence of a threshold can be speculated upon but cannot at present be sufficiently supported by data (Kirsch-Volders et al. 2000; Hengstler et al. 2003). In this case, precautionary considerations

will mostly lead to applying the conservative approach of a linear dose–response extrapolation, and the LNT model is used as a default assumption. Relevant examples being considered by SCOEL include acrylonitrile, benzene, naphthalene and wood dust.

For some genotoxic carcinogens the existence of a 'practical' threshold is supported by studies on mechanisms and/or toxicokinetics. Examples are formaldehyde (Morgan 1997) and vinyl acetate (Bogdanffy and Valentine 2003). In this case, a NOAEL may be established from which to derive a health-based exposure limit. Non-genotoxic carcinogens and non-DNA-reactive carcinogens are characterized by a conventional dose–response relationship with a 'true' threshold that allows establishing a NOAEL. Insertion of an uncertainty (safety) factor permits the derivation of health-based occupational exposure limits. Examples are tumour promoters and hormones (Setzer and Kimmel 2003). SCOEL is also discussing chemicals such as chloroform and carbon tetrachloride. Figure 19.1 summarizes the proposal: different decision outcomes in terms of health-based occupational limit values correspond

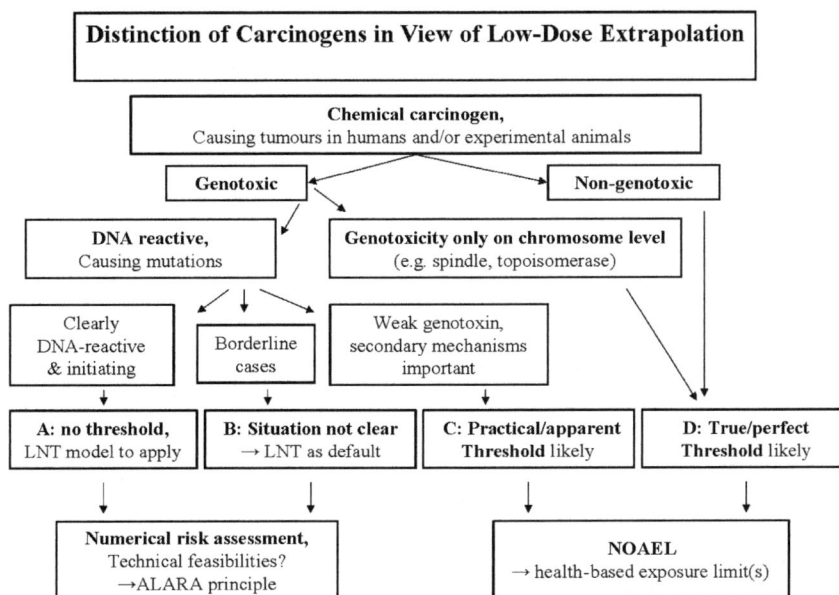

Figure 19.1 Proposal for distinguishing among groups of carcinogens (A–D) for the purpose of risk assessment and standard setting. Reproduced with permission from Bolt and Pegen (2004).

to specific action modes of different categories of chemical carcinogens.

The SCOEL approach to the definition of limit values is one among several possible. For example, alternatives do exist to the use of NOAEL and LOAEL which themselves have limitations: they require modifying and uncertainty factors; they do not consider underlying sample size; they may be inconsistent from study to study, also because they are constrained to be an experimental or empirical dose (Mutti 2001). The benchmark dose (BMD) approach (Allen *et al.* 1994) could be a valid alternative to the NOAEL. The BMD method uses the available biological information better, since the mathematical model fits all the dose–response data in the observable range, rather than simply the lowest dose level at which effects are observed. The model is then used to interpolate a dose estimate that corresponds to a particular level of response. In the BMD approach, information about variability within the dataset and the uncertainty regarding a BMD are accounted for by the use of the lower confidence limit on the BMD (BMDL). As a result, the BMD is sensitive to sample size, as a larger study will give narrower confidence limits on the BMD and thus a higher BMDL (Filipsson *et al.* 2003). In the BMD approach, the probability of adverse effects may be derived from the model describing the relationship between the prevalence of abnormalities and biomarker levels. The BMD (LED_{10}) is defined as the statistical lower boundary on a dose corresponding to a specified risk level (10% risk or 5% excess risk) on the basis of the logistic regression function describing the dose–response curve and is obtained from the upper confidence limit (CL; 95%) on the dose–response curve.

Finally, a quantitative low-dose extrapolation of dose–response data from animal bioassays is often used in setting standards, although it raises numerous scientific uncertainties related to the selection of mathematical models and extrapolation down to levels of human exposure. The margin of exposure (MOE) approach is currently preferred to the application of default uncertainty factors because it is based on the available animal dose–response data, without extrapolation, and on human exposures (Barlow *et al.* 2006).

19.3. RISK CHARACTERIZATION: INDIVIDUALS AND POPULATIONS

To be effective, regulations should take into consideration the risk, rather than simply hazard. Risk may vary among different populations and across individuals exposed to the same hazard. Determinants of inter-individual variability are numerous, and all are relevant when providing evidence on which to base regulatory decisions. Inter-individual differences in uptake, biotransformation, mechanism of action, susceptibility to damage and DNA repair capacity can result in different dose–response relationships for different groups of individuals. The contribution of intra-individual variability in exposure (e.g. day-to-day variation) may also lead to attenuation bias in dose–response relationships when estimates of workers' exposures relying on single measurements are used to evaluate effects resulting from chronic exposure (Symanski *et al.* 2001). The bias in the observed slope coefficient can be reduced by increasing the number of repeated individual measurements (Liljelind *et al.* 2003). In the research context, biological variance should not be reduced, but should be explained by the knowledge of toxicokinetics and toxicodynamics. Differences in physical activity, respiratory rate and body mass index are known to influence the absorption and excretion of chemicals. It is known that different workloads may affect exposure–dose relationships, and interpretation of biological monitoring results should take into consideration that a given biomarker could overestimate exposure in the case of heavy tasks and underestimate it in the case of subjects at rest. Similar interferences in absorption and metabolism of chemicals have been described for other factors, such as sex, fat intake, alcohol consumption, medications and co-exposures to complex mixtures of substances (Viau 2002).

Genetic components of inter-individual variability are of special interest today (Miller *et al.* 2001). They represent an unquestionable advancement in the understanding of disease mechanism, in particular gene–environment interactions. It is a common belief that by enhancing our understanding of environmental disease genetics might also suggest new strategies to prevent and control them.

In risk characterization, the special relevance is recognized of genetic polymorphisms, i.e. sequence variations (inherited and induced) in genes encoding xenobiotic metabolism enzymes or DNA repair enzymes, which may modify individual response to chemical hazards (Kelada *et al.* 2003). Although the number of reports on the impact of genetic polymorphisms on environmental exposure (carcinogens, especially) is steadily growing (Vineis 2004; Garte 2001), examples of strong effect modification are limited by the completeness of the data. In a recent report, the degree of evidence of metabolic polymorphisms as effect modifiers of environmental exposure was systematically reviewed. In only six instances were laboratory evidence and supporting epidemiological data available, whereas laboratory evidence plus suggestive epidemiological data were obtained for 34 'gene–environment relations' and 19 were derived from basic science laboratory reports only (Kelada *et al.* 2003). A consistent, positive association of DNA repair deficiency and increased risk was recently shown by an extended review of inter-individual variability in DNA repair systems and cancer risk (Berwick and Vineis 2000).

Many environmental and occupational chemicals, toxicants and carcinogens require metabolic activation to exert their action. Metabolic polymorphisms can thus modulate individual response, not by producing qualitatively different responses but rather by inducing a shift in the dose–response curve. A further portion of the mechanism linking exposure to response thus becomes understood, with the possible suggestion of new clues for health protection. Figure 19.2 schematically depicts the phenomenon: The central curve denotes the D-R relationship for people with wild-type genotype and the other curves the same relationship in carriers of mutant genotypes. In the example, at the same exposure level, the response (proportion of exposed presenting the effect) is quite different for different genotypes. A shift to the right modifies susceptibility by increasing the response level to the exogenous exposure (thus conferring protection). A shift to the left elicits the same response at a lower level of exposure. The example suggests that, among several plausible models described for the relation between genotypes and environmental

exposures in terms of their effects on disease risk, the focus here is on the so called type B gene–environment interaction (Ottman 1996). It is the exposure itself that increases the risk of disease, even in the absence of the genetic risk factor, and only when the exposure is present does the genetic risk factor have any role. If there is no exposure, then the presence or absence of the genetic risk factor is irrelevant for disease causation. In point of fact, metabolic polymorphisms represent common, low-penetrance conditions of altered metabolic function which become relevant to the disease process only when interacting with the exogenous (or endogenous) chemicals (Vineis *et al.* 2001). The identification of carriers of mutant metabolic genotypes (or of any hyper-susceptibility status, either genetically or otherwise determined) in the exposed population has immediate implications for health protection regulations. The question arises whether the exposure limits, surveillance measures and safety devices adopted are appropriate and effective in protecting the 'susceptible' portion of the exposed population.

This is explicitly recognized, for example, by the American Conference of Governmental Industrial Hygienists (ACGIH®), which every year updates a

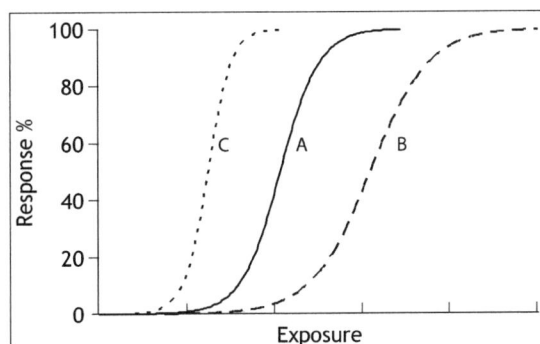

Figure 19.2 Modulation of the exposure–response curve (A) by modifying factors (either genetically determined or acquired). Shifting to the right (B) corresponds to a 'resistant' population: to obtain the same response as in the central typical distribution, a larger exposure is needed. Shifting to the left (C) corresponds to a hyper-susceptible population: the same response as in the typical distribution occurs at a much lower exposure

list of occupational exposure limits called 'threshold limit values'. Their definition is as follows:

> Threshold limit values (TLVs®) refer to airborne concentrations of chemical substances and represent conditions under which it is believed that *nearly all* workers may be repeatedly exposed, day after day, over a working lifetime, without adverse health effects. TLVs® are developed to protect workers who are normal, healthy adults.

'Normal, healthy adults' is a definition that, in fact, includes great variability determined by genetic traits and acquired conditions. Consistently:

> ACGIH® recognizes that there is considerable variation in the level of biological responses to a particular chemical substance, regardless of the airborne concentration.

> Some individuals may experience discomfort or even more serious adverse health effects when exposed to a chemical substance at the TLV® or even at concentrations below the TLVs®...There are numerous possible reasons for increased susceptibility to a chemical substance, including age, gender, ethnicity, genetic factors ..., lifestyle choices ..., medications and pre-existing medical conditions ... (ACGIH 2005).

If we were in the position to include susceptibility variables in the limit value definition process, we could, at least in principle, be more effective in preventing the occurrence of adverse effects by protecting even particularly susceptible individuals.

Consider the example of exposure to lead (Pb), using lead in blood and lead in plasma (the biologically active portion) as exposure markers; δ-aminolevulinic acid dehydratase (ALA-D), involved in the haem synthesis as early effect marker; ALA-D genotypes, which show polymorphic distribution in the population (wild-type 1:1 present in 81.7% of the population; mutant heterozygous 1:2 in 16.8%; mutant homozygous 2:2 in 1.5%) as susceptibility marker; and diastolic blood pressure as the outcome (after De Palma 2005). ALA-D is central in lead toxicity and 80% of lead in blood is bound to it. Polymorphism can result in an increase in the affinity

of ALAD to Pb and may modify its kinetic dynamics and hence its toxicodynamics. Lead *per se* increases the average value of diastolic blood pressure (Table 19.1). Using a cut-off of 300 μg lead/litre of blood, the odds ratio for high diastolic blood pressure (> 90 mmHg) was 1.91 (95% CI 1.20–3.04). After stratifying according to genotype, the odds ratio among the carriers of the wild genotype was 1.68 (95% CI 0.97–2.90), whereas among those with mutant allele it was 6.82 (95% CI 1.68–27.67). In this study, the mutant genotype status evidently modifies the high diastolic blood pressure risk after exposure to lead. An additional result of this study might have immediate impact on regulation. It was shown that the no-effect level was actually around 300 μg/l of lead in blood, whereas the current occupational exposure limit was set at a level twice as high (600 μg/l). The result was simply obtained thanks to a refined investigation of the existing early effect markers of the haem synthesis pathway: ALA-D, ALA-U aminolaevulinic acid in urine, and zinc protoporphyrin in red cells. There probably is an important lesson here: sometimes, for regulatory purposes, it may be more rewarding to perform better studies with existing markers than to adopt brand new markers whose validity has not been fully tested and whose significance not fully understood.

The identification of groups at risk can thus rely on the shift in the distribution of sensitive markers of effect. Whether such a shift is sufficient to trigger action will always be debatable. In fact, the definition of biological limits or health-based criteria is hampered by: (a) the arbitrary nature of attempts to distinguish acceptable from unacceptable effects; (b) the difficulties in assessing the prognostic value of observed changes; and (c) the influence of the study 'power' on statistically defined thresholds. Whatever the approach, translating evidence into standards will always be open to question. The opposite is also true: there is little doubt about the possibility that the many thousands of chemicals currently used may affect human health. This justifies a prudent attitude in interpreting early biological effects as a warning signal, requiring further support from mechanistic studies aimed at understanding their meaning and follow-up studies to confirm the existence of an increased risk for long-term outcomes (Mutti 1995).

Table 19.1 Diastolic blood pressure and blood concentration of lead in a group of 371 workers taken as a whole, and stratified according to δ-aminolevulinic acid dehydratase (ALAD) genotype: wild-type (ALAD1-1, No. 272) and mutant (ALAD1-2/2-2, No: 61)

Diastolic blood pressure (mmHg)	Lead in blood (µg/l)		
	> 300	≤ 300	
> 90	57	43	100
< 90	111	160	271
	168	203	371

OR (95% CI) = 1.91 (1.20–3.04)

ALAD 1-1

Diastolic blood pressure (mmHg)	Lead in blood (µg/l)		
	> 300	≤ 300	
> 90	37	34	71
< 90	79	122	201
	116	156	272

OR (95% CI) = 1.68 (0.97–2.90)

ALAD 1-2/2-2

Diastolic blood pressure (mmHg)	Lead in blood (µg/l)		
	> 300	≤ 300	
> 90	12	3	15
< 90	17	29	46
	29	32	61

OR (95% CI) = 6.82 (1.68–27.67)

Reproduced with permission from De Palma *et al.* (2005).

19.4. MONITORING AND SURVEILLANCE

Regulations in environmental and occupational health require compliance with limit values of pollutants and/or reference values of biological markers of exposure (dose) and/or of effect (early and reversible). To assure compliance with existing standards aimed at preventing adverse health effects in exposed subjects, the monitoring of airborne concentrations or ambient monitoring (AM) is supplemented by biological monitoring (BM) and health surveillance (HS), as depicted in Figure 19.3 (Foà and Alessio 1998). BM is aimed at quantifying the amount of chemical absorbed, transformed into an active metabolite or accumulated in deposits or target organs, tissues or cell as a consequence of exposure to environmental agents. HS is aimed at discovering early, reversible biochemical/functional changes related to the exposure which might call for further action to protect the individual and the

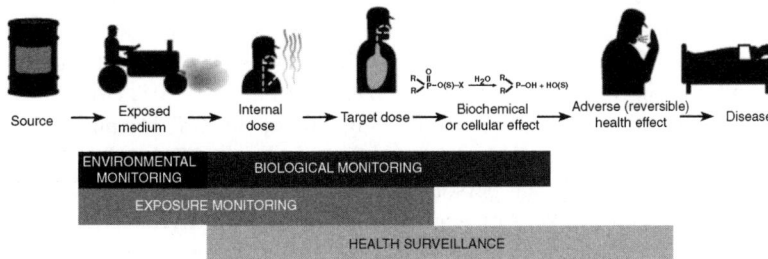

Figure 19.3 The relationship between environmental, biological and exposure monitoring and health surveillance in the prevention and control of effects of environmental toxicants. Reproduced with permission from Foà and Alessio (1998). Copyright © International Labour Organization, 1998.

community. BM gives the opportunity to verify the effectiveness of the adopted environmental exposure limits against excessive intake. HS further verifies that the adopted protective measures do actually prevent even minor but potentially harmful changes in the exposed organism. Should such protective measures prove insufficient, HS results set the need for further action.

Mechanistic studies are essential to develop pertinent, valid, and predictive markers of low-level exposure and early effect. The focus of the investigation is the lower tail of the exposure–response relationship, given that:

- Current exposures are much lower than those at which the majority of the known environmental adverse effects have been observed: The need exists to explore whether at those lower concentrations any residual (similar or different) effect might occur.
- That portion of the D-R curve is where the majority of the susceptible people (i.e. subjects responding to doses lower than those necessary to elicit the same effect in the majority of the population) belong: are the limits in force protective of those people as well?

In addressing these questions, the role of genetic traits, and particularly metabolic polymorphisms, is of utmost relevance. For example, occupational levels of benzene have been steeply decreasing over the years, whereas concentrations in urban and residential areas, although low, raise public concern. Recent studies examined the performance of the available bio-markers in detecting extremely low exposures, as well as the possible role of genetically determined or life style- and personal history-related modifying factors. One study included bus drivers, traffic policemen, gas station attendants and a group of referent subjects working in a reportedly non-polluted environment (Fustinoni 2005). Environmental levels of airborne benzene were determined via personal passive sampler, and biological markers of exposure included *trans,trans*-muconic acid (t,t-MA) in urine, S-phenylmercapturic acid (S-PMA) in urine, and non-metabolized benzene (U-benzene) excreted in urine. The genetic polymorphism of two metabolic enzymes central to benzene biotransformation, CYP2E1- and NADPH-dependent quinone oxidoreductase-1 (NQO1) were considered. The variant *CYP2E1* genotype significantly modified the concentration of t,t-MA at the beginning of the work shift and of U-benzene at the end of it, whereas the NQO1 variant genotypes were not associated with significant differences in levels of exposure markers. The influence of genetic polymorphisms on the variability of biomarkers appeared to be minor in comparison with other factors, in particular tobacco smoking and individual exposure level. The study also showed that among the biomarkers of benzene exposure, U-benzene was performing best at those low levels of exposure. Overall, the results suggest that at the currently prevailing low-level environmental benzene exposure, life style and exposure conditions are the main determinants of individual exposure, not genetic traits.

Another study of taxi drivers exposed to benzene levels close to the recommended air quality parameter of 10 μg/m^3 confirmed the major role played by tobacco smoking in benzene uptake, but identified S-PMA as the most reliable exposure biomarker. In addition, it was observed that subjects bearing the *GSTM1*-positive genotype, expressing GSTM1-1 enzyme activity, excreted significantly higher S-PMA concentrations than *GSTM1*-null subjects (Manini *et al.* 2006).

In workers exposed to styrene, the urine concentrations of styrene-derived mercapturic acids (PHEMAs) were characterized by using tandem liquid chromatography coupled to mass spectrometry (LC–MS–MS) (Manini *et al.* 2000). The excretion profile of PHEMAs was strongly influenced by the GSTM1 status, with *GSTM1*-positive subjects excreting about five-fold higher concentrations of mercapturic acids than *GSTM1*-null subjects (Figure 19.4). On the basis of these data, De Palma *et al.* (2001) concluded that the modifying role of the *GSTM1*-null polymorphism, which affects about 50% of the Caucasian population, limits the practical use of PHEMAs as biomarkers of exposure to styrene. However, the alternative conclusion, i.e. the need for two different limits for biomarkers measured respectively in *GSTM1*-positive and -null subjects, has also been proposed (Haufroid *et al.* 2001).

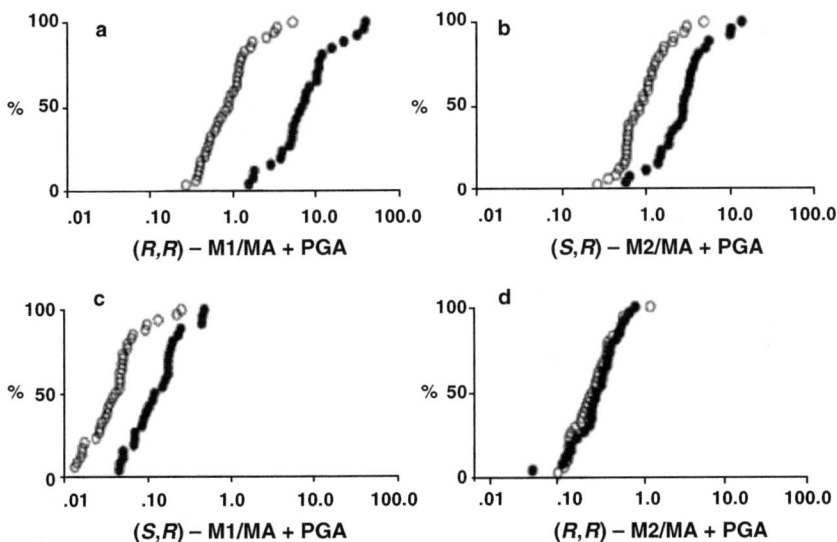

Figure 19.4 Cumulative frequency distribution of the concentration of each diastereoisomer of phenylhydroxyethyl mercapturic acids (PHEMAs) in subgroups of workers classified according to *GSTM1* genotype and adjusted for internal dose (i.e. the sum of mandelic and phenylglyoxylic acids, MA + PGA). Both MA + PGA and PHEMAs were determined in 'next-morning' spot urine samples. A modifying role of the *GSTM1* genotype on the excretion of (*R,R*)-M1, (*S,R*)-M1 and (*S,R*)-M2 is apparent. *GSTM1*-positive people excrete R,R-M1 diastereoisomer at values of at least 1 order of magnitude higher than *GSTM1*-null subjects. Black bullets, GSTM1-positive; white bullets, GSTM1-null. Reproduced with permission from De Palma *et al.* (2001)

Once confirmed in further and larger investigations, the results of this type of study might affect regulations, by pointing out what to measure and by which method in order to be appropriate for monitoring low-level environmental exposures.

19.5. WHAT TO REGULATE: EXPOSURES OR PEOPLE'S ACCESS TO THEM?

A potential concern in applying mechanistic knowledge to exposure monitoring and surveillance of exposed individuals is to divert attention from exposure control and to place the focus on screening susceptible individuals. This raises ethical, social and legal issues which may ultimately have an impact on the entire practice of occupational and environmental health. For example, inexpensive DNA technology and the increasing commercial availability of DNA typing would permit widespread predictive testing and risk stratification at pre-employment and pre-placement assessment. Indeed, genetic testing to identify workplace susceptibility

and predisposition to disease in essentially healthy people is already occurring. In 1999 the American Management Association made a survey with 1054 respondents (Brandt Rauf and Brandt Rauf 2004). The results revealed that 16.7% had performed genetic testing to determine susceptibility to workplace hazards and that 10.3% were using test results in making employment decisions (hiring, job assignment, retention or dismissal). In the UK, the Human Genetic Advisory Commission in 2002 confirmed that this is a real issue by stating that:

> ... people should not be *required* to take a genetic test for employment purposes ... employers should *offer* a genetic test if it is known that a specific working environment or practice, while meeting health and safety requirement, might pose specific risks to individuals with particular genetic constitutions (Palmer *et al.* 2004).

This appears to be a fairly balanced statement suggesting the opportunity, at least in specified circumstances, to complement the *population approach*, which tends to reduce the overall occurrence of a risk factor by meeting health and safety requirement, with the *high-risk approach*, which detects individuals at special risk for whom additional, targeted interventions should be implemented.

Understandably, employers and employees tend to have opposing views (Rawbone 1999). The former tend to support pre-employment genetic screening but generally oppose genetic monitoring, whereas the latter oppose screening as a way of selecting employees but tend to favour genetic monitoring in health surveillance programme. Employers may wish to perform genetic tests to exclude susceptible workers and those predicted to have ill-health in order to reduce costs related to occupational health and safety with reassignment or ill-health retirement, and costs incurred through liability and legal action for compensation. For individual employees, knowing whether they have an increased susceptibility to occupational hazards can be of benefit, warning them of the need for special preventive measures. The difference in perception may also be a reflection of the fact that, while screening is a programme aimed at a population without a one-to-one relationship, surveillance is seen as a programme performed within the context of a doctor–patient relationship.

The concern that genetic testing could open the door to discrimination as well as detracting from attention paid to overall safety in the workplace is reflected in the litigations and controversies that are already taking place. For example, in 2001 in the USA, Burlington Northern Railway Co. was ordered to end its policy of requiring all union members who claimed work-related carpal tunnel syndrome (CTS) to provide blood samples for DNA analysis for a genetic marker that may be related to some forms of CTS. In fact, it was shown that peripheral myelin protein-22 (*PMP-22*) gene deletion contributes negligibly to the incidence of CTS, and therefore there was no scientific basis to the testing programme. In addition, no informed consent was requested, and no genetic counselling offered. On the other hand, Dow Chemical Co. was sued in 2000 by the wife of a deceased worker for failure to include the employee in an experimental cytogenetics testing programme of acquired susceptibility to cancer after exposure to benzene and epichlorohydrin (Brandt Rauf and Brandt Rauf 2004).

19.6. CONCLUSION

Occupational and environmental agents represent components of a usually wide spectrum of risk factors that may combine differently at the individual level in causing ill-health. There is no doubt that, in principle, incorporating genetic and molecular markers into environmental health research may allow more accurate exposure assessment, improved understanding of intermediate endpoints and enhanced risk prediction. Prevention, even at a 'personalized' level, may be fostered by gaining insight into disease mechanisms. However, mechanistic studies do not suggest *per se* effective strategies to reduce risk and, in practice, examples of application of results in this field are few.

This type of research should not be seen as alternative to research on control technology and population intervention strategies. Rather, it may complement that research in several ways, including the following: (a) by identifying and characterizing portions of the population needing special protection; (b) by testing the effectiveness of measures already in place; (c) by suggesting realistically effective health-based limits for agents that cannot be banned; and (d) by pinpointing gaps in knowledge whose bridging might enhance our ability to protect individuals and populations. The time of 'molecular-based primary prevention' (Schpilberg *et al.* 1997), instead, does not yet seem to have come in environmental health.

References

Allen BC, Kavlock RJ, Kimmel CA *et al.* 1994. Dose–response assessment for developmental toxicity. II Comparison of generic benchmark dose estimates with NOAELs. *Fundam Appl Toxicol* **23**: 487–495.

American Conference of Governmental Industrial Hygienists. 2005. *2005 TLVs® and BEIs® Based on the Documentation of the Threshold Limits Values for Chemical Substances and Physical Agents and Biological Exposures Indices.* ACGIH: Cincinnati, OH.

Barlow S, Renwick AG, Kleiner J *et al.* 2006. Risk assessment of substances that are both genotoxic and carcinogenic – report of an International Conference organized by EFSA and WHO with support of ILSI Europe. *Food Chem Toxicol* **44**: 1636–1650.

Berwick M, Vineis P. 2000. Markers of DNA repair and susceptibility to cancer in humans: an epidemiologic review. *J Natl Cancer Inst* **92**: 847–897.

Bogdanffy MS, Valentine R. 2003. Differentiating between local cytotoxicity, mitogenesis, and carcinogenesis in carcinogenic risk assessment: the case of vinyl acetate. *Toxicol Lett* **40–141**: 83–98.

Bolt H, Degen GH. 2004a. Human carcinogenic risk evaluation: contributions of the EUROTOX specialty section for carcinogenesis. *Toxicol Sci* **81**: 3–6.

Bolt H, Foth H, Hengstler JG *et al.* 2004b. Carcinogenicity categorization of chemicals: new aspects to be considered in a European perspective. *Toxicol Lett* **151**: 29–41.

Brandt-Rauf PW, Brandt-Rauf S. 2004. Genetic testing in the workplace: ethical, legal and social implications. *Ann Rev Public Health* **5**: 139–153.

De Palma G, Manini P, Mozzoni P *et al.* 2001. Polymorphism of xenobiotic-metabolizing enzymes and excretion of styrene-specific mercapturic acids. *Chem Res Toxicol* **14**: 1393–1400.

De Palma G, Scotti E, Mozzoni P *et al.* 2005. ALAD polymorphism and indicators of dose and effects of occupational exposure to inorganic lead. *G Ital Med Lav Ergon* **27**: 39–42.

European Commission. 1999. *Methodology for the Derivation of Occupational Exposure Limits: Key Documentation.* EUR 19253 EN. Office for Official Publications of the European Communities: Luxembourg.

Filipsson AF, Sand S, Nilsson J *et al.* 2003. The benchmark dose method – review of available models, and recommendations for application in health risk assessment. *Crit Rev Toxicol* **33**: 505–542.

Foà V, Alessio L. 1998. General principles of biological monitoring. In *Encyclopaedia of Occupational Health and Safety*, 4th edn, vol 1, Stelmann JM (ed.). International Labour Office: Geneva; 27.2–27.6.

Fustinoni S, Consonni D, Campo *et al.* 2005. Monitoring low benzene exposure: comparative evaluation of urinary biomarkers, influence of cigarette smoking, and genetic polymorphisms. *Cancer Epidemiol Biomarkers Prevent* **14**: 2237–2244.

Garte S. 2001. Metabolic susceptibility genes as cancer risk factors: time for reassessment? *Cancer Epidemiol Biomarkers Prevent* **10**: 1233–1237.

Grandjean P. 2004. Implications of the precautionary principle for primary prevention research. *Ann Rev Public Health* **25**: 199–223.

Haufroid V, Buchet J-P, Gardinal S. 2001. Importance of genetic polymorphisms of drug-metabolizing enzymes

for the interpretation of biomarkers of exposure to styrene. *Biomarkers* **6**: 236–249.

Hengstler JG, Bogdanffy MS, Bolt HM *et al.* 2003. Challenging dogma: thresholds for genotoxic carcinogens? The case of vinyl acetate. *Ann Rev Pharmacol Toxicol* **43**: 485–520.

IARC (International Agency for Research on Cancer). 1994. *Some Industrial Chemicals*. IARC Monographs on the Evaluation of Carcinogenic Risk to Humans, vol 60. IARC: Lyon;73–159.

IARC (International Agency for Research on Cancer). 1997. *Polychlorinated Dibenzo-para-dioxins and Polychlorinated Dibenzofurans*. IARC Monographs on the Evaluation of Carcinogenic Risk to Humans, vol 69. IARC: Lyon; 33–343.

IARC (International Agency for Research on Cancer). 2000. *Some Industrial Chemicals*. IARC Monographs on the Evaluation of Carcinogenic Risk to Humans, vol 77. IARC: Lyon; 41–148.

Kelada SN, Eaton DL, Wang SS *et al.* 2003. The role of genetic polymorphisms in environmental health. *Environ Health Perspect* **111**: 1055–1064.

Kirsch-Volders M, Vanhauwaert A, Eichenlaub-Ritter U *et al.* 2003. Indirect mechanism of genotoxicity. *Toxicol Lett* **140–141**: 63–74.

Liljelind I, Rappaport S, Eriksson K *et al.* 2003. Exposure assessment of monoterpenes and styrene: a comparison of air sampling and biomonitoring. *Occup Environ Med* **60**: 599–603.

Manini P, Andreoli R, Bergamaschi E *et al.* 2000. A new method for the analysis of styrene mercapturic acids by liquid chromatography/electrospray tandem mass spectrometry. *Rapid Commun Mass Spectrom* **14**: 2055–2060.

Manini P, De Palma G, Andreoli R *et al.* 2006. Environmental and biological monitoring of benzene exposure in a cohort of Italian taxi drivers. *Toxicol Lett* **167**: 142–151.

Miller MC III, Mohrenweiser HW, Bell DA. 2001. Genetic variability in susceptibility and response to toxicants. *Toxicol Lett* **120**: 269–280.

Morgan KT. 1997. A brief review of formaldehyde carcinogenesis in relation to nasal pathology and human risk assessment. *Toxicol Pathol* **25**: 291–307.

Mutti A. 2001. Biomarkers of exposure and effect for non-carcinogenic end-points. In *International Programme on Chemical Safety. Environmental Health Criteria 222, Biomarkers in Risk Assessment: Validity and Validation*. World Health Organization: Geneva; 130–136.

Mutti A. 1995. Use of intermediate end-points to prevent long-term outcomes. *Toxicol Lett* **77**: 121–125.

Ottman R. 1996. Gene–environment interaction: definitions and study designs. *Prevent Med* **25**: 764–770.

Palmer KT, Poole J, Rawbone RG *et al.* 2004. Quantifying the advantages and disadvantages of pre-placement genetic screening. *Occup Environ Med* **61**: 448–453.

Potter JD. 2002. Epidemiology informing clinical practice: from bills of mortality to population laboratories. *Nat Clin Pract* **2**: 625–633.

Rowbone RG. 1999. Future impact of genetic screening in occupational and environmental medicine. *Occup Environ Med* **56**: 721–724.

Setzer RW, Kimmel CA. 2003. Use of NOAEL, benchmark dose, and other models for human risk assessment of hormonally active substances. *Pure Appl Chem* **75**: 2151–2158.

Schpilberg O, Dorman JS, Ferrell RE *et al.* 1997. The next stage: molecular epidemiology. *J Clin Epidemiol* **50**: 633–638.

Streffer C, Bolt HM, Føllesdal D *et al.* 2004. *Environmental Standards – Dose Effect Relation in the Low Dose Range and Risk Evaluation.* Springer: Berlin.

Symanski E, Bergamaschi E, Mutti A. 2001. Inter- and intra-individual sources of variation in levels of urinary styrene metabolites. *Int Arch Occup Environ Health* **74**: 336–344.

Viau C. 2002. Biological monitoring of exposure to mixtures. *Toxicol Lett* **134**: 9–16.

Vineis P. 2004. Individual susceptibility to carcinogens. *Oncogene* **23**: 6477–6483.

Vineis P, Schulte P, McMichael AJ. 2001. Misconceptions about the use of genetic tests in populations. *Lancet* **357**: 709–712.

Zocchetti C, Pesatori AC, Bertazzi PA. 2004. A simple method for risk assessment and its application to 1,3-butadiene. *Med Lav* **95**: 392–409.

20

Biomarkers as Endpoints in Intervention Studies

Lynnette R. Ferguson

University of Auckland, New Zealand

20.1. INTRODUCTION: WHY ARE BIOMARKERS NEEDED IN INTERVENTION STUDIES?

The methods of molecular epidemiology can be applied in proving the efficacy of preventing hazard exposure, or intervening in the development or progression of a disease. A public health intervention would involve reducing exposure to a potentially hazardous material, often through identifying points of unintentional exposure through occupational monitoring of the environment or personnel. Where such exposures have occurred, chemoprevention approaches may provide an intervention with chemical agents, including minor dietary constituents and/or pharmaceuticals, to halt or slow the disease process. It is essential to prove the efficacy of either type of intervention.

The most desirable intervention study design is a double-blinded, randomized, placebo-controlled clinical trial that directly measures protection against the disease of interest. The definitive proof that any dietary or pharmaceutical intervention prevents disease and/or enhances some physiological process is provided by feeding or administering the test substance at a given level for sufficient time to enable the disease process to occur in the absence of the substance. In practice, however, such a model is almost impossible to attain in most intervention studies, in part because of the length of time that most diseases take to develop. For example, cancer is known to show a 20–30 year lag between disease initiation and development, and the possibility of attaining compliance in a dietary or pharmaceutical trial over such a long period is completely unrealistic. Phase III trials, with cancer reduction as the endpoint, require tens of thousands of subjects, followed-up for times of 5 years or more (Kelloff *et al.* 1992). For this reason, the Chemoprevention Branch of the National Cancer Institute (NCI) have focused a significant part of their efforts on finding surrogate endpoints (biomarkers) with high reliability and predictive value for cancer. The NCI and other US agencies convened a workshop on 'Biomarkers as indicators of cancer risk reduction following dietary manipulation' (Kavanaugh *et al.* 2006; Prentice 2006; Schatzkin 2006; Ferguson *et al.* 2006) that summarizes current views on many of the issues associated with using biomarkers as endpoints of intervention studies.

A *biomarker* has been defined as 'a measurable, biological parameter that predicts the risk of human disease, disorders or conditions, but is not a measure of the disease, disorder or condition itself' (Rothman *et al.* 1995), i.e. biomarkers serve as early indicators of disease in asymptomatic people. They have been used to assess exposure to potential environmental hazards, to gain insight into disease mechanisms and to understand acquired or inherited susceptibility. Thus, biomarkers cover the whole spectrum from hazard exposure to disease pathogenesis. Their successful application to intervention studies requires an understanding of disease natural

Molecular Epidemiology of Chronic Diseases, Edited by C. P. Wild, P. Vineis, and S. Garte
© 2008 John Wiley & Sons, Ltd

history, the mechanism of the intervention and the characteristics and limitations of the biomarker. Schatzkin *et al.* (1993) have distinguished biomarkers from surrogate disease endpoints as follows:

> Whereas any biologic phenomenon can be considered a biomarker, an intermediate end-point is defined as being on the causal pathway between exposure and disease. An intermediate end-point is a valid surrogate for a disease in relation to a given exposure if, and only if, that exposure causes a similar change in the occurrence of both the intermediate end-point and the disease.

The acceptable endpoint of an intervention study might be a single biomarker, or batteries of different markers. In the case of cancer, the NCI's stated criteria (Kelloff *et al.* 1992) include:

- Differential expression in normal and high-risk tissue.
- Appearance early in carcinogenesis (the earlier a reliable biomarker appears, the greater is the chance for successful intervention with a chemopreventive agent).
- High sensitivity, specificity, and accuracy relative to cancer.
- Ease of measurement (use of non-invasive techniques and small tissue samples is preferable).
- Demonstration of modulation by chemopreventive agents.
- Correlation of modulation with decreased cancer incidence.

The same principles also apply to the use of biomarkers in proving efficacy of protection against other diseases.

Convenience is not the only reason for using biomarkers as the endpoint of a clinical trial. Taking a study to a disease endpoint may be unethical, since no hypothesis is foolproof, and there is a small but finite possibility that it could be wrong. This is illustrated vividly by the β-carotene trials (Box 20.1). The rationale

---

Box 20.1   Lessons from the β-carotene trials

It is instructive to consider what has been learned from largely negative trials with disease endpoints, such as those with β-carotene (Altmann *et al.* 1996; Omenn *et al.* 1996). In these examples, the investigators chose high-risk populations to increase the probability of disease occurrence or recurrence in the study group during a short intervention. There appeared to be several advantages to this. Not only did such an approach reduce study numbers and time of study (and therefore costs), it was thought to enhance the quality of the study by making it more likely that there would be continuity of personnel involved in the trial. However, because there was little understanding of the mechanism by which β-carotene was likely to act, biomarkers could not be rationally selected as intermediate endpoints. It was considered sufficient to show the efficacy of the intervention by proving a reduction in incidence of, or mortality from, chronic disease without needing to know the mechanism of action. Indeed, it was suggested that trials would help elucidate possible mechanisms. The strongest argument for early intervention trials was that if such studies were delayed until the major research questions were answered, it would be either unethical to deny the control group the intervention or difficult to find a control group because of widespread antioxidant consumption.

In the event, these trials generally had negative endpoints. While a large number of possible explanations have been attempted, the observations generally provide some lessons. Although the use of high-risk groups appeared justified, it may have been part of the problem. The population groups chosen were either heavy smokers or asbestos-exposed workers, i.e. they were groups that had already been exposed to high risk and likely to have already been experiencing some early disease processes. It is plausible that β-carotene or other antioxidants could have been beneficial in other populations in which disease had not already been initiated, i.e. at a very early stage of the disease process (the normal population). If biomarkers had been used, the studies could have been aborted at an earlier stage.

Kavanaugh *et al.* (2006) detail a research agenda for optimizing the use of biomarkers in subsequent cancer prevention trials. Similar considerations also apply to other disease endpoints.

for suggesting that dietary supplementation with β-carotene would benefit population health was mainly based upon epidemiological studies that related high plasma β-carotene levels in individuals with low disease risk in the same individuals. However, the endpoint of two major intervention studies based in Europe (Albanes *et al.* 1996) or the USA (Omenn *et al.* 1996) was that the interventions enhanced, not reduced, the risk of the diseases that they had been designed to protect against. Box 20.1 discusses the issues uncovered in this trial, and its aftermath.

In practice, experimental studies with disease endpoints are less common than experimental studies linking the test intervention to surrogate disease outcomes.

20.2. IDENTIFICATION AND VALIDATION OF BIOMARKERS

If biomarkers are to provide an effective index of efficacy, they must provide a quantifiable index of reduction in hazard exposure, disease initiation or progression. This implies their being able to measure responses across the range of intakes being studied, being able to be measured with precision and high sensitivity, and being applicable to the population group being studied. For example, Alberts *et al.* (1994) considered rectal mucosal proliferation indices as biomarkers for interventions, but the protocols they developed are applicable to other biomarkers. They emphasized the importance of ongoing quality control/quality assurance programmes in order to achieve high accuracy and reproducibility with minimal variability, and outlined a series of steps that can be followed in the validation process. Validation of a given biomarker is essential if it is to provide meaningful information in intervention studies. A valid biomarker in this context is one that allows the correct inference to be drawn regarding the effect of an intervention on the true clinical endpoint of interest.

Schatzkin *et al.* (1990) developed protocols for establishing whether a given biomarker is a valid intermediate endpoint between exposure to a hazard and disease incidence. They recommend using prospective cohort or, if these are not available, case–control studies, in order to quantify the strength of the biomarker–disease association. The biomarker provides a valid disease surrogate if the attributable proportion is close to 1.0, but not if it is close to 0. In most examples where the attributable proportion is in between 1.0 and 0, the intermediate endpoint must reflect an established exposure–disease relationship.

The importance of biomarker validation is illustrated by the failure of a putative colorectal cancer-predictive endpoint, the aberrant crypt (Hardman *et al.* 1991). This measure had been used as an endpoint of a number of animal studies, especially in the 1980s, and it appeared to be true that dietary components that increased aberrant crypts also enhanced colon cancer incidence in longer-term studies. However, it did not provide an accurate measure of intervention. Hardman and co-workers compared seven different dietary regimes for their ability to modulate colon carcinogenesis in Sprague–Dawley rats by the known colon carcinogen 1,2-dimethylhydrazine. If the studies had only progressed as far as aberrant crypts, they would have falsely predicted the ability of the different diets to protect against colon carcinogenesis.

20.3. USE OF BIOMARKERS IN MAKING HEALTH CLAIMS

An appropriately designed intervention study can be used as the basis for a health claim, and this is often the main reason that such studies are done. While the exact basis for such claims differs between geographical regions, there are some general principles that govern the weight attached to individual studies. For example, the Scientific Advisory group of the Australia/New Zealand Food Safety Authority (FSANZ 2006) considers health claims for foods or dietary regimes, and has stated:

Assessment of the quality of primary evidence includes (but may not be limited to) assessment of the following elements:

- The completeness and appropriateness of the described methodology.
- Appropriate and accurate description and quantification of exposure to the diet, food or food component.

- Appropriate and accurate quantification of the health related outcome.
- Sample size.
- Sample and measurement bias.
- Potential confounding variables.
- Inclusion of appropriate controls.
- Study duration.
- Appropriate statistical methods.

The second and third of these may involve surrogate outcomes or biomarkers. Biomarkers used in studies relating diet, food or food components to a health outcome include serum lipid levels, blood pressure, bone mineral density and adenomatous polyps. While either a disease endpoint or a surrogate outcome is adequate as a study endpoint, it is essential that the outcome measured is one relevant to the study hypothesis. Studies using biomarkers are only useful where there is a well-accepted and validated predictive relationship between the biomarker and the health or disease outcome, and the biomarker is biologically plausible.

20.4. BIOMARKERS OF STUDY COMPLIANCE

In considering the quality of evidence provided by a given trial, one of the most important questions to be asked is whether the study participants adhered to the intervention throughout. There is good evidence that simply asking the participants this question will not necessarily lead to the correct answer (Eliassen *et al.* 2006), stressing the need for biomarker validation of compliance to interventions, especially when comparing populations of different ethnicity. In pharmaceutical trials, compliance may be assessed by checking medicine cabinets, etc., to see whether there is remaining medication. In dietary trials, compliance may be more difficult to assess, and biomarker approaches become important. Compliance may need to be measured at several different stages in the study, especially in longer-term studies where dietary patterns may change during the duration of the trial.

Biomarker-mediated indices of exposure are commonly performed in industrial monitoring to provide indices of risk. For example, Waidyanatha

et al. (2001), used headspace solid-phase microextraction of 0.5 ml urine specimens, followed by gas chromatography–mass spectrometry, to provide an accurate measure of urinary benzene to serve as a biomarker of exposure among benzene-exposed workers and unexposed subjects in Shanghai, China. However, studies of this nature imply a large amount of prior information on the chemistry of the material to which the group is exposed, dose–response relationships, and the level of risk provided by the material in question, i.e. benzene. In the context of an intervention study, it is not risk *per se* but reduction of risk that is important, and the quality of the data is dependent not only upon study compliance but also upon the material reaching its biological target.

Designing a package of biomarkers may be especially important when considering a dietary regime, where more than one component may be mechanistically involved in the desired biological effects. For example, Fowke *et al.* (2006), were looking to show protection against cancer by *Brassica* vegetables. As they pointed out that, not only do these vegetables contain micronutrients that may scavenge free radicals, they also contain isothiocyanates (ITC) and indoles (e.g. indole-3-carbinol) that induce Phase I and Phase II enzymes, leading to the oxidation, reduction and metabolism of endogenous and exogenous carcinogens. The authors performed a randomized cross-over trial of 20 subjects, comparing the effects of a *Brassica* vegetable intervention with those of a micronutrient and dietary fibre supplementation. They considered the effects of the interventions on urinary excretion of F2-isoprostanes (F2-iP), as a stable biomarker of systemic oxidative stress. They not only estimated *Brassica* intake by repeated 24 h dietary recalls and food frequency questionnaire methodology, but also checked the accuracy of the information they were given by comparing the values thus obtained with dietary intakes estimated through urinary ITC levels. In this example, they were able to show that the group on the *Brassica* intervention arm were excreting more ITC in the urine than the other experimental group, from which they implied (as required) a higher *Brassica* intake.

Martini *et al.* (1995) did controlled feeding studies that validated plasma α-carotene, β-carotene, and lutein as biomarkers of intake of carotenoid-rich foods in general, and lutein as an intake biomarker of commonly consumed vegetables in the *Brassica* family. Their methods allowed them not only to monitor intake of some plant foods, but also to distinguish among plant food groups. This general approach has been optimized to distinguish multivitamin intake, as reflected by significantly higher concentrations of plasma retinol, β-carotene and α-tocopherol, from high fruit and vegetable intake, resulting in higher lutein and zeaxanthin and β-cryptoxanthin concentrations (Eliassen *et al.* 2006). It should be noted, however, that high concentrations of carotenoids in plasma do not necessarily provide an indication of tissue levels, since some tissues will selectively accumulate certain carotenoids. For example, lutein and zeaxanthin accumulate at high levels in the retina of the eye (Handelman *et al.* 1988).

Other biomarkers that enable an objective assessment of nutrient consumption include energy expenditure, urinary nitrogen, selected blood fatty acid measurements and various blood micronutrient concentrations (Prentice *et al.* 2002). A measure of levels of the material in either urine or plasma may not be as useful as an index of a biological effect. For example, Baylin and Campos (2006) compared the various types of fatty acid biomarkers that have been used to assess compliance in dietary lipid intervention trials. They concluded that fatty acids in adipose tissue provide a biomarker of choice for long-term intake studies, but there are practical limitations to the tissue's accessibility and measurement, and the measure does not reflect a short term outcome. They were unable to distinguish the desirability of plasma vs. erythrocytes on the basis of the data available to date. The type of study (short- or long-term), the metabolic characteristics of the population group being studied and the probable variability in the fatty acids of interest will help in the selection of the most appropriate tissue to reflect the true intake. If the study design is comparing groups with differing isocaloric intakes, certain fatty acids may not be appropriate for use, especially when trying to assess small differences. Serum cholesterol ester may be the most appropriate serum fraction for measuring short-term but not long-term dietary compliance.

20.5. BIOMARKERS THAT PREDICT THE RISK OF DISEASE

In designing biomarker endpoints to test an intervention in a disease process, it is essential to understand both the mechanisms of development and progression of the disease, and the mechanism(s) by which the intervention is likely to work. Animal studies may provide an essential pilot. Box 20.2 describes the steps that may be desirable in establishing a biomarker methodology. The example used comes from an elegant series of studies that have used biomarkers to question whether Oltipraz might prevent the development of liver cancer in Southern China (Kensler *et al.* 2005; El-Nezami *et al.* 2006; see Chapter 25 for further details).

The ideal design of a chemoprevention study monitors indices of study compliance and biomarkers that relate to disease prevention at the same time. For example, (Luo *et al.* 2006) considered whether green tea polyphenols (GTP) could reduce the risk of liver cancer in high-risk individuals. Their study design entailed volunteers taking GTP capsules daily at doses of 500 mg, 1000 mg or a placebo, for 3 months. Twenty-four hour urine samples, collected before the intervention and at the first and third month of the study, were used to monitor levels of the polyphenols epigallocatechin and epicatechin, as well as of the oxidative stress biomarker, 8-oxo-7,8-dihydrodeoxyguanosine (8-OHdG). By the end of the 3 month intervention, 8-OHdG levels decreased significantly in both GTP-treated groups, which the authors considered as evidence that chemoprevention with GTP was effective in reducing oxidative DNA damage.

20.6. BIOMARKERS RELEVANT TO MORE THAN ONE DISEASE

Oxidative stress

The notion of disease prevention through antioxidant intervention partly stems from the fact that fruits and vegetables contain antioxidants, and are linked to lower disease rates in those who consume high levels of them. Protection against DNA damage by plant food products can be demonstrated *in vitro*. Dalle-Donne *et al.* (2006) claim

---

Box 20.2   Steps in designing appropriate biomarkers for intervention studies

The examples are taken from a series of studies on the possibility of Oltipraz protection against liver cancer in Qidong, People's Republic of China (Kensler *et al.* 2005; El-Nezami *et al.* 2006; see also Chapter 25 for further details).

1. *Establish what type of disease an intervention might protect against, and what population would be most responsive.* In the examples provided here, liver cancer is known to be a significant risk factor for the population, and there is good evidence linking it to aflatoxin B exposure (Kenzler *et al.* 1998).
2. *If the causal factor in high disease risk is known, establish how this is working and whether there is a good animal model on which preventive hypotheses can be tested.* Aflatoxin B1 is known to act as a tumour initiator through the formation of DNA adducts in both animals and humans. Cancer risk in male F344 rats has been shown to relate to the formation and disappearance of aflatoxin–albumin adducts of rats chronically exposed to aflatoxin B1 (Egner *et al.* 1995).
3. *Check what tissue is easily collected and reflective of the disease process in the animal model.* The urinary excretion of aflatoxin B(1)–N(7)–guanine (AFB-N(7)-guanine) reflects tissue levels of this important DNA adduct in the rodent (Egner *et al.* 1995). Additionally, serum aflatoxin adduct biomarkers complement this urinary information.
4. *Ensure that this same biomarker will be also appropriate in humans.* Both serum aflatoxin–albumin adducts and urinary aflatoxin metabolites have been associated with increased liver cancer risk in prospective studies (Wild and Turner 2002).
5. *Consider what type of intervention might reduce the formation of this lesion.* In the Oltipraz trials, there was good preliminary data that this chemical modulated enzyme induction. In animal models, it reduced the risk of cancer through aflatoxin B1 exposure, and this chemoprevention effect was mimicked by enhanced excretion of urinary aflatoxin metabolites or by serum aflatoxin–albumin adduct biomarkers (Egner *et al.* 1995).
6. *Once the scientific basis for an intervention is established and a biomarker validated, human studies can proceed.* Kenzler and co-workers (2005) have extensively described the results of their Oltipraz chemoprevention trials in this high-risk human population in China.

---

that, because oxidative stress markers can be measured accurately and objectively, they provide valuable indicators of responses to therapeutic interventions. Sies (1997) defined oxidative stress as a 'disturbance in prooxidant-antioxidant balance in favor of the former, leading to potential damage'. High levels of oxidative stress contribute significantly to the age-related development of some cancers through DNA damage, while lipid peroxidation plays a role in the development of cardiovascular disease. In diabetes, hyperglycaemia leads to the autoxidation of glucose, glycation of proteins and the activation of polyol metabolism. These changes accelerate oxidative stress, which may play an important role in the development of complications in diabetes, such as lens cataracts, nephropathy and neuropathy (Osawa and Kato, 2005). Protein carbonyl content has been used as a general biomarker of severe oxidative protein damage. High levels of protein carbonylation are associated with various human diseases, including Alzheimer's disease, chronic lung disease, chronic renal failure, diabetes and sepsis (Dalle-Donne *et al.* 2006).

More generally, biomarkers of oxidative stress have been used as an index of susceptibility to degenerative disease, as a means of studying mechanisms of action for certain chemopreventive agents and of defining the optimal intake of those antioxidants (Ferguson *et al.* 2006). (Halliwell 2002) points to mass spectrometric measurements of various families of isoprostanes

(F2-, F3- and F4-isoprostanes) and of multiple DNA base oxidation products as among the most promising biomarkers for use in human nutritional intervention studies. For example, Fowke *et al.* (2006) used urinary F2-iP levels to provide a stable biomarker of systemic oxidative stress in studies comparing fruit and vegetables with micronutrient/dietary fibre interventions. First-morning urine samples were collected at baseline and after each intervention, and the changes in natural log transformed urinary F2-iP levels, analysed using repeated measures regression. Sustainable changes in F2-iP levels were used as evidence that such interventions were likely to protect against cancer and other diseases.

There have been specific concerns as to the validity of biochemical markers of mechanism of action for antioxidants. For example, plasma and tissue malondialdehyde (MDA) concentrations have been used as an index of risk of cardiovascular disease. MDA is a volatile carbonyl resulting from the oxidative breakdown of polyunsaturated fatty acids with three or more double bonds, which has been considered as a measure of lipid peroxidation. Bowen and Mobarhan (1995) have pointed to the problems with this measure. Since MDA has a tendency to bind to nucleic acids, proteins and amino acids, there may be more bound than free MDA in tissues and plasma. The most common assay for MDA is not specific for it, the compound tends to react with other products of biologic oxidations, and it forms chromophores with substances such as ribose, 2-aminopyrimidine or sialic acid. Such technical artifacts have led several investigators to modify the assay according to the particular design of their individual study, leading to a lack of standardization that prevents comparisons across intervention studies with reduction of MDA as an endpoint.

Estimates of background levels of DNA oxidation in human cells range over three orders of magnitude, depending on the method used. In part, this is because oxidation occurs readily during sample preparation, creating a serious potential artefact (Collins 2005). Nevertheless, using validated, reliable biomarker assays for DNA oxidation, it is possible to demonstrate a decrease in oxidative damage after supplementation with isolated antioxidants or whole plant foods in humans. A number of bioactive compounds that function in disease prevention commonly act as antioxidants or show ability to quench singlet-oxygen. It should be recognized, however, that this does not necessarily mean that they protect against disease through their antioxidant properties, since many such compounds may enhance immune response, promote gap junction communications, modify nuclear DNA signalling and modify production of eicosanoids (Ferguson *et al.* 2006).

Inflammation

Inflammation is an important contributor to atherothrombosis. The C-reactive protein (CRP) is not only a biomarker of inflammation, but it is also a direct participant in atherogenesis. CRP consistently predicts new coronary events, including myocardial infarction and death, in patients with ischaemic heart disease. CRP and other markers of inflammation may also be useful in the diagnosis and management of childhood asthma (Li *et al.* 2005).

Cancer

Various biomarkers have been suggested to relate to the risk of certain cancers. As with other diseases, the closer to the formation of a frank tumour, the more likely is the marker to be informative. For example, the likely number of new cases of colorectal cancer has been estimated through measurement of numbers of adenomatous polyps as a biomarker. However, even estimating such polyps is invasive, labour-intensive and costly. Measurements of recurrence of adenomatous polyps over short time periods can be inaccurate because of a large potential rate of missing small adenomas. More importantly, they may not provide an accurate biomarker, since not all polyps progress to form cancers (Wargovich 2006). It should also be recognized that a chemopreventive intervention may be affecting adenoma development, rather than adenoma formation, and a negative answer in a chemoprevention trial that estimates numbers of new polyps could be a false-negative result.

In 1992, Rozen considered measurement of rectal epithelial proliferation as a biomarker of risk for colorectal neoplasia, and response in intervention studies. The justification for using this marker is that a phase of epithelial hyperproliferation typically precedes frank colorectal carcinogenesis. Although the marker seemed of some value in experimental studies, it has been less conclusive in humans. Part of the reason is technical, including difficulties in obtaining tissue samples and sampling error, the type of labelling technique used, lack of an objective method for quantifying proliferation and confounding factors influencing the degree of proliferation. Schatzkin (2006) also pointed out that rectal epithelial proliferation is not necessary for colorectal cancer development, and there may be alternative pathways that bypass this, such as an apoptotic pathway. An agent that reduces hyperproliferation will only be likely to reduce cancer, provided that it does not also reduce apoptosis.

Environmentally induced lung cancer has been identified as a serious problem in industrialized nations, and polycyclic aromatic hydrocarbons (PAHs) and related compounds recognized as lung carcinogens. Because the disease is invariably fatal, lung cancer prevention has been seen as an important target for intervention. Talaska et al. (1996) describe optimization of methods for collecting cells from bronchial-alveolar lavage or sputum, and validation of measures of carcinogen–DNA adduct levels in the lung-derived cells thus available. In practice, however, it has been more common and less invasive to determine levels of the marker in surrogate molecules, such as protein, or markers from surrogate tissues, such as lymphocyte DNA, to assess the risk to the target organ, the lung.

Prostate-specific antigen (PSA) is the most commonly used biomarker in prostate cancer, although other measures have been considered for use. Collette et al. (2005) questioned whether PSA could be considered a valid surrogate endpoint for survival in hormonally treated patients with metastatic prostate cancer. Their meta-analysis used data from 2161 patients with advanced prostate cancer, treated either with bicalutamide or with castration. Endpoints studied were PSA response,

PSA normalization, time to PSA progression and longitudinal PSA measurements. They were able to show that PSA was useful at the individual patient level, but that the correlation between the effect of intervention on any PSA endpoint and overall survival was generally low. They thus concluded that the effect of hormonal treatment on survival cannot be accurately predicted from observed treatment effects on PSA endpoints. This emphasizes the fact that, although a marker is commonly used as a study endpoint, this does not necessarily imply that it has been validated.

DNA adducts are well recognized as a biomarker of exposure to various environmental, life style or occupational chemical carcinogens that have been shown to be modified through various interventions (Santella 1997). For example, administration of chlorophyllin three times a day led to a 50% reduction in the median level of urinary excretion of aflatoxin-N(7)-guanine compared to placebo (Egner et al. 2003). This excreted DNA adduct biomarker is derived from the ultimate carcinogenic metabolite of aflatoxin B(1), aflatoxin-8,9-epoxide, and is associated with increased risk of developing liver cancer in prospective epidemiological studies.

While a decrease in urinary DNA adducts would seem to be a desirable endpoint of intervention studies, implying reduction of exposure, it is also important to realize that urinary DNA adducts are a product of a DNA repair process. While a reduction in these levels may indeed imply an effective reduction in carcinogen exposure, the same result could be achieved by the undesirable endpoint of similar exposure levels, coupled with reduced DNA repair activity. It may be more appropriate to measure adducts levels in cells. For example, DNA adducts of various industrial carcinogens and oxidative DNA damage can be measured in blood mononuclear cells and exfoliated oral and bladder cells from human subjects (Egner et al. 2003). Such DNA damage may also be implied by a measure of DNA breakage, such as the single cell gel electrophoresis or COMET assay (Karunasinghe et al. 2004).

Other biomarkers suggested for cancer include genomic markers, such as micronuclei and specific chromosomal alterations (Bonassi et al. 2007).

Specific genetic markers include oncogenes, growth factors and their receptors, and tumour suppressor genes, such as the *ras* gene family, the *myc* family, *erb B1*, *int-2/hst-1* and the *p53* tumor suppressor gene. Squamous cell differentiation markers include keratins, involucrin and transglutaminase 1, while rectal epithelial proliferation has been implied by proliferating cell nuclei antigen (PCNA).

Cardiovascular disease

A clinician's manual of clinical tests would accept total, LDL- and HDL-cholesterol and blood pressure as biomarkers for cardiovascular disease (CVD) risk. However, an increasing number of other biomarkers are being used, including plasma triglycerides (TG), heart rate (HR), heart rate variability (HRV), atrial fibrillation (AF), arterial compliance, endothelial vasodilator function, intima-media thickness (IMT) and plaque stability.

A meta-analysis of 17 population-based prospective studies, including a total of 46 413 men and 10 864 women, confirmed a strong relationship between plasma TG and CVD, with every 1 mmol/l increase in plasma TG associated with a 30% increase in relative risk of CVD in men and 75% in women (Hokanson and Austin 1996). The Prospective Cardiovascular Münster (PROCAM) study enrolled 19 698 people aged 16–65 years, examined their lipid profile and known or suggested CVD risk factors at study entry, and then followed them for 8 years in order to relate CVD risk factors to the occurrence of fatal or non-fatal myocardial infarct and sudden cardiac death (Assman *et al.* 1998). Again, elevated TG appeared to be the most predictive risk factor for CVD, after controlling for all other important factors. Similarly with the Copenhagen Male study, which followed 3000 middle-aged and elderly Danish men, free of CVD at enrolment, for 8 years (Jeppesen *et al.* 1998). Several other studies (Dyer *et al.* 1980; Kannel *et al.* 1987; Shaper *et al.* 1993; Wannamethee *et al.* 1995; Palatini *et al.* 1999, 2002; Jouven *et al.* 2001), suggest that increased HR is an independent risk factor for SVD and SCD. Hartikainen *et al.* (1996) found that decreased HR variability (HRV) predicted

SCD and arrhythmic events in patients who had survived MI. Vulnerability of the atherosclerotic plaque to rupture (atherosclerotic plaque stability), arterial dysfunction (as decreased compliance or elasticity) or altered vasomotor reactivity were all suggested to be important early markers of CVD risk (Plutzky 1999; Blacher *et al.* 1999; Simons *et al.* 1999).

The endothelium is important in maintaining arterial vasomotor tone and modulating vasoconstrictor, inflammatory, chemotactic and proliferative processes in the artery wall. Abnormalities of endothelial function have been associated with other CVD risk factors. Flow-mediated dilatation (FMD) in the brachial artery is strongly correlated with coronary artery FMD, and has been suggested as a non-invasive method for assessing the extent of coronary artery disease (Takase *et al.* 1998; Celermajer *et al.* 1994; Enderle *et al.* 1998).

It is important to recognize that many of these suggestive relationships have only been established through correlations in observational studies. While RCTs would provide useful proof, it is difficult to affect one of these above measures without having a concomitant effect on the others. Nevertheless, questions are being raised about their increasing use. For example, the primary biological mechanism suggesting that omega-3 polyunsaturated fatty acids (PUFA) would protect against CVD is their anti-arrhythmic effects, but this endpoint has not translated to reduced CVD events when investigated further. One randomized controlled trial (Raitt *et al.* 2005) observed a negative effect of omega-3 PUFA supplementation and arrhythmia, leading to questions about the relevance and reproducibility of this biomarker.

Diabetes

Biomarkers are important for both the diagnosis and management of type II diabetes mellitus (DM). Serum glucose levels, before and after a glucose challenge, is the most common marker of impaired glucose tolerance and pre-diabetic state. Management focuses primarily on such peripheral blood biomarkers as daily serum glucose levels, in association with predictors of long-term morbidity.

20.7. BIOMARKERS THAT PREDICT THE OPTIMIZATION OF HEALTH OR PERFORMANCE

Traditional nutritional science has focused on providing nutrients to nourish populations, and prevent deficiency diseases. However, a more modern approach focuses on maintaining or improving health and optimized performance of individuals through tailored dietary regimes. A considerable challenge is in how to define a 'healthy' phenotype. Rational personalization of diet utilizes the approaches of 'nutrigenetics' and 'nutrigenomics' (Ferguson 2006). Nutrigenetics considers the implications of human variation, including single nucleotide polymorphisms, copy-number polymorphisms and epigenetic phenomena, on dietary requirements and disease susceptibility, while nutrigenomics considers how diet influences gene transcription, protein expression and metabolism. Both sciences are defined by systems biology approaches that integrate several of the '–omics' technologies (transcriptomics, proteomics, metabolomics and metabonomics) with more classic pathological and nutritional science. These approaches have the potential to enable discovery of early indicators for disease disposition, differentiate dietary responders from non-responders, and to lead to the discovery of beneficial bioactive food components.

20.8. CONCLUSIONS

The concept of disease intervention, either by dietary or by pharmaceutical means, has considerable attractions, and the use of biomarkers permits a rational and affordable approach. However, there are some lessons to be learned from studies to date. There needs to be a very strong case for more than just an association in observational studies, in order to justify testing a disease intervention. It is essential to begin with a strong experimental basis, to ensure that the starting hypothesis is based firmly on an understanding of mechanism. It is essential to know what type of disease an intervention might protect against, what population would be most responsive, what the optimal study duration should be, where along the continuum of disease initiation to progression the intervention should be started,

and what levels or concentrations should be tested. In selection of a biomarker, there is an inevitable conflict between the degree of invasiveness of tissue sampling and the necessity to get close to disease and affected tissue for optimal predictivity. Tissues such as peripheral blood lymphocytes are commonly used because of the ease of harvesting. However, there are questions as to whether a systemic effect can predict changes in disease risk in a different tissue. It is necessary to understand the potential mechanism if a biomarker is to provide the maximum amount of information. Biomarkers of oxidative stress or of inflammatory response may have utility in predicting interventions against a range of different events. There may be value in more specific biomarkers, especially for chronic diseases such as cancer, cardiovascular disease or diabetes. Population subgroups at high risk of disease have been used in intervention studies, but answers provided by these groups may not be representative of the general population. The optimal design of an intervention trial using biomarkers might stratify volunteers according to genotype.

Preliminary intervention studies utilizing rationally designed and well-validated biomarkers may well provide many of the answers that are needed to reduce disease risk in human populations.

References

Albanes D, Heinonen OP, Taylor PR *et al.* 1996. α-Tocopherol and β-carotene supplements and lung cancer incidence in the α-Tocopherol, β-Carotene Cancer Prevention Study: effects of base-line characteristics and study compliance [see comment]. *J Natl Cancer Inst* **88**: 1560–1570.

Alberts DS, Einspahr J, Aickin M *et al.* 1994. Validation of proliferation indices as surrogate endpoint biomarkers. *J Cell Biochem Suppl* **19**: 76–83.

Batlin A, Campos H. 2006. The use of fatty acid biomarkers to reflect dietary intake. *Curr Opin Lipidol* **17**: 22–27.

Blacher J, Asmar R, Djane S *et al.* 1999. Aortic pulse wave velocity as a marker of cardiovascular risk in hypertensive patients. *Hypertension* **33**: 1111–1117.

Bonassi S, Znaor A, Ceppi M *et al.* 2007. An increased micronucleus frequency in peripheral blood lymphocytes predicts the risk of cancer in humans. *Carcinogenesis* **28**: 625–631.

Bowen PE, Mobarhan S. 1995. Evidence from cancer intervention and biomarker studies and the development of biochemical markers. *Am J Clin Nutrit* **62**: 1403–1409S.

Celermajer DS, Sorensen KE, Bull C *et al.* 1994. Endothelium-dependent dilation in the systemic arteries of asymptomatic subjects relates to coronary risk factors and their interaction. *J Am Coll Cardiol* **24**: 1468–1474.

Collette L, Burzykowski T, Carroll KJ *et al.* 2005. Is prostate-specific antigen a valid surrogate endpoint for survival in hormonally treated patients with metastatic prostate cancer? Joint research of the European Organisation for Research and Treatment of Cancer, the Limburgs Universitair Centrum, and AstraZeneca Pharmaceuticals. *J Clin Oncol* **23**: 6139–6148.

Collins AR. 2005. Antioxidant intervention as a route to cancer prevention. *Eur J Cancer* **41**: 1923–1930.

Dalle-Donne I, Rossi R, Colombo R *et al.* 2006. Biomarkers of oxidative damage in human disease. *Clin Chem* **52**: 601–623.

Dyer AR, Persky V, Stamler J *et al.* 1980. Heart rate as a prognostic factor for coronary heart disease and mortality: findings in three Chicago epidemiologic studies. *Am J Epidemiol* **112**: 736–749.

Egner PA, Munoz A, Kensler TW. 2003. Chemoprevention with chlorophyllin in individuals exposed to dietary aflatoxin. *Mutat Res* **523–524**: 209–216.

El-Nezami HS, Polychronaki NN, Ma J *et al.* 2006. Probiotic supplementation reduces a biomarker for increased risk of liver cancer in young men from Southern China. *Am J Clin Nutrit* **83**: 1199–1203.

Eliassen AH, Colditz GA, Peterson KE *et al.* 2006. Biomarker validation of dietary intervention in two multiethnic populations. *Prevent Chronic Dis* **3**: A44.

Enderle MD, Schroeder S, Ossen R *et al.* 1998. Comparison of peripheral endothelial dysfunction and intimal media thickness in patients with suspected coronary artery disease. *Heart* **80**: 349–354.

Ferguson LR. 2006. Nutrigenomics: integrating genomic approaches into nutrition research. *Mol Diagn Ther* **10**: 101–108.

Ferguson LR, Philpott M, Karunasinghe N. 2006. Oxidative DNA damage and repair: significance and biomarkers. *J Nutrit* **136**: 2687–2689S.

Fowke JH, Morrow JD, Motley S *et al.* 2006. Brassica vegetable consumption reduces urinary F2-isoprostane levels independent of micronutrient intake. *Carcinogenesis* **27**: 2096–3102.

Halliwell B. 2002. Effect of diet on cancer development: is oxidative DNA damage a biomarker? *Free Rad Biol Med* **32**: 968–974.

Handelman GJ, Dratz EA, Reay CC *et al.* 1988. Carotenoids in the human macula and whole retina. *Invest Ophthalmol Vis Sci* **29**: 850–855.

Hardman WE, Cameron IL, Heitman DW *et al.* 1991. Demonstration of the need for endpoint validation of putative biomarkers: failure of aberrant crypt foci to predict colon cancer incidence. *Cancer Res* **51**: 6388–6392.

Hartikainen J, Malik M, Staunton A *et al.* 1996. Distinction between arrhythmic and nonarrhythmic death after acute myocardial infarction based on heart rate variability, signal-averaged electrocardiogram, ventricular arrhythmias and left ventricular ejection fraction. *J Am Coll Cardiol* **28**: 296–304.

Hokanson JE, Austin MA. 1996. Plasma triglyceride level is a risk factor for cardiovascular disease independent of high-density lipoprotein cholesterol level: a meta-analysis of population-based prospective studies. *J Cardiovasc Risk* **3**: 213–219.

Jeppesen J, Hein HO, Suadicani P *et al.* 1998. Triglyceride concentration and ischemic heart disease: an eight-year follow-up in the Copenhagen Male Study. *Circulation* **97**: 1029–1036.

Jouven X, Zureik M, Desnos M *et al.* 2001. Resting heart rate as a predictive risk factor for sudden death in middle-aged men. *Cardiovasc Res* **50**: 373–378.

Kannel WB, Kannel C, Paffenbarger RS Jr *et al.* 1987. Heart rate and cardiovascular mortality: the Framingham Study. *Am Heart J* **113**: 1489–1494.

Karunasinghe N, Ryan J, Tuckey J *et al.* 2004. DNA stability and serum selenium levels in a high-risk group for prostate cancer. *Cancer Epidemiol Biomarkers Prevent* **13**: 391–397.

Kavanaugh C, Seifried H, Ellwood K *et al.* 2006. A research agenda for biomarkers as indicators of cancer risk reduction following dietary manipulation. *J Nutrit* **136**: 2666–2667S.

Kelloff GJ, Malone WF, Boone CW *et al.* 1992. Intermediate biomarkers of precancer and their application in chemoprevention. *J Cell Biochem Suppl* **16G**: 15–21.

Kensler TW, Chen J-G, Egner PA *et al.* 2005. Effects of glucosinolate-rich broccoli sprouts on urinary levels of aflatoxin–DNA adducts and phenanthrene tetraols in a randomized clinical trial in He Zuo township, Qidong, People's Republic of China. *Cancer Epidemiol Biomarkers Prevent* **14**: 2605–2613.

Li J, Zhao J, Yu X *et al.* 2005. Identification of biomarkers for breast cancer in nipple aspiration and ductal lavage fluid. *Clin Cancer Res* **11**: 8312–8320.

Luo H, Tang L, Tang M *et al.* 2006. Phase IIa chemoprevention trial of green tea polyphenols in high-risk individuals of liver cancer: modulation of urinary excretion of green tea polyphenols and 8-hydroxyde-oxyguanosine. *Carcinogenesis* **27**: 262–268.

Martini MC, Campbell DR, Gross MD *et al.* 1995. Plasma carotenoids as biomarkers of vegetable intake: the University of Minnesota Cancer Prevention Research Unit Feeding Studies. *Cancer Epidemiol Biomarkers Prevent* **4**: 491–496.

Omenn GS, Goodman G, Thornquist M *et al.* 1996. *Chemoprevention of Lung Cancer: the β-Carotene and Retinol Efficacy Trial (CARET) in High-risk Smokers and Asbestos-exposed Workers.* IARC Scientific Publications. IARC: Lyon, France; 67–85.

Osawa T, Kato Y. 2005. Protective role of antioxidative food factors in oxidative stress caused by hyperglycemia. *Ann N Y Acad Sci* **1043**: 440–451.

Palatini P, Casiglia E, Julius S *et al.* 1999. High heart rate: a risk factor for cardiovascular death in elderly men. *Arch Intern Med* **159**: 585–592.

Palatini P, Thijs L, Staessen JA *et al.* 2002. Predictive value of clinic and ambulatory heart rate for mortality in elderly subjects with systolic hypertension. *Arch Intern Med* **162**: 2313–2321.

Plutzky J. 1999. Atherosclerotic plaque rupture: emerging insights and opportunities. *Am J Cardiol* **84**: 15–20J.

Prentice RL. 2006. Research opportunities and needs in the study of dietary modification and cancer risk reduction: the role of biomarkers. *J Nutrit* **136**: 2668–2670S.

Prentice RL, Sugar E, Wang CY *et al.* 2002. Research strategies and the use of nutrient biomarkers in studies of diet and chronic disease. *Publ Health Nutrit* **5**: 977–984.

Raitt MH, Connor WE, Morris C *et al.* 2005. Fish oil supplementation and risk of ventricular tachycardia and ventricular fibrillation in patients with implantable defibrillators: a randomized controlled trial. *J Am Med Assoc* **293**: 2884–2891.

Rothman N, Stewart WF, Schulte PA. 1995. Incorporating biomarkers into cancer epidemiology: a matrix of biomarker and study design categories. *Cancer Epidemiol Biomarkers Prevent* **4**: 301–311.

Santella RM. 1997. DNA damage as an intermediate biomarker in intervention studies. *Proc Soc Exp Biol Med* **216**: 166–171.

Schatzkin A. 2006. Promises and perils of validating biomarkers for cancer risk. *J Nutrit* **136**: 2671–2672S.

Schatzkin A, Freedman L, Schiffman M. 1993. An epidemiologic perspective on biomarkers. *J Intern Med* **233**: 75–79.

Schatzkin A, Freedman LS, Schiffman MH *et al.* 1990. Validation of intermediate endpoints in cancer research. *J Natl Cancer Inst* **82**: 1746–1752.

Shaper AG, Wannamethee G, Macfarlane PW *et al.* 1993. Heart rate, ischaemic heart disease, and sudden cardiac death in middle-aged British men. *Br Heart J* **70**: 49–55.

Sies H. 1997. Oxidative stress: oxidants and antioxidants. *Exp Physiol* **82**: 291–295.

Simons PC, Algra A, Bots ML *et al.* 1999. Common carotid intima–media thickness and arterial stiffness: indicators of cardiovascular risk in high-risk patients. The SMART Study (Second Manifestations of ARTerial disease). *Circulation* **100**: 951–957.

Takase B, Uehata A, Akima T *et al.* 1998. Endothelium-dependent flow-mediated vasodilation in coronary and brachial arteries in suspected coronary artery disease. *Am J Cardiol* **82**: 1535–1539.

Talaska G, Roh J, Zhou Q. 1996. Molecular biomarkers of occupational lung cancer. *Yonsei Med J* **37**: 1–18.

Waidyanatha S, Rothman N, Fustinoni S *et al.* 2001. Urinary benzene as a biomarker of exposure among occupationally exposed and unexposed subjects. *Carcinogenesis* **22**: 279–286.

Wannamethee G, Shaper AG, Macfarlane PW *et al.* 1995. Risk factors for sudden cardiac death in middle-aged British men. *Circulation* **91**: 1749–1756.

Wargovich MJ. 2006. What do diet-induced alterations in colorectal polyps and aberrant crypts indicate for risk? *J Nutrit* **136**: 2679–2680S.

Wild CP, Turner PC. 2002. The toxicology of aflatoxins as a basis for public health decisions. *Mutagenesis* **17**: 471–481.

21

Biological Resource Centres in Molecular Epidemiology: Collecting, Storing and Analysing Biospecimens

Elodie Caboux[1], Pierre Hainaut[1] and Emmanuelle Gormally[1,2]

[1]IARC, Lyon, France and [2]Université Catholique de Lyon France

21.1. INTRODUCTION

Molecular epidemiology consists of the integration of molecular approaches using biospecimens within the context of studies that follow the methods of epidemiological study design. In turn, this integration has considerable impact on the logistics of such studies, as well as on many aspects of study design (Schulte and Perera 1997). Collecting biospecimens and identifying biomarkers are the cornerstones of this approach. Biomarkers are used to measure exposure, susceptibility, immediate biological effects or disease progression, and thus provide intermediate endpoints to understand the relationships between exposure and outcomes. Biomarkers cover the whole range of biological materials used as indicators of physiological or pathological processes, including DNA, RNA, proteins, lipids, cellular and tissue materials, biochemical building blocks of macromolecules, metabolites, hormones, markers of infections, inflammation, immune responses, as well as their modifications. This biological complexity raises considerable difficulties for biospecimen collection, processing and storage. At present, there is no general collection and storage standard applicable for all types of biomarkers in a generic manner. To allow the proper use of biomarkers and prevent bias, the amount and type of biospecimens to be collected has to be defined, and the collection procedure, labelling, processing, transport and storage has to be organized in a specific manner that depends upon the nature and stability of the biomarkers under investigation. For example, in a case-control study, important biases may occur if biospecimens from controls are collected in a different place, at a different time, with a different protocol or stored in different conditions to those of the cases. Making correct decisions in biomarker selection and storage procedures has considerable budgetary and logistical implications.

Biological Resource Centres (BRCs), as defined by the Organization for Economic Co-operation and Development (OECD), are service providers and repositories of living cells, replicable parts of these (genomes, plasmids, viruses, cDNA), viable cells and tissues and information relating to these materials stored in databases (http://wdcm.nig.ac.jp/brc.pdf). The main purpose of BRCs is not only storage but also traceability of all steps between specimen collection and final laboratory analysis. In this context, BRC play a crucial role in providing recommendations and all the practical and technical information for the collection procedures essential to the design, initiation and progression of the study.

Collections of biospecimens have always existed in medical and biological research and have been used for studying biological, physiological and biochemical processes and diseases. Collections have varied from containing a few biospecimens to millions, such as the European Prospective Investigation into Cancer and nutrition (EPIC; Bingham and Riboli 2004). BRCs have had a central role in organizing, managing and ensuring proper labelling, use and storage of these biospecimens. Two main types of BRC can be distinguished. First, a BRC can be developed to provide a logistical and managerial umbrella for diverse biospecimen collections which were collected at different times and places, using different protocols and to answer different scientific hypotheses. These BRCs are typically found in institutes or hospitals conducting many different studies but providing a centralized system for maintaining the collections associated to these studies. The second type of BRC is purpose-built and corresponds to one particular study, e.g. a prospective study in which many biospecimens are collected using a unique protocol. In this type of BRC, the specific study design constitutes the main constraint for the structure, management and exploitation of biospecimen collections.

Networking between BRCs is becoming a strategic need, in particular in multi-centric studies aimed at comparing diseases between populations, or in analysing rare diseases for which sufficient numbers of cases cannot be accrued in any particular location (Ioannidis *et al.* 2005). Networking requires the implementation of standardized procedures and protocols in BRCs. Development of high-quality and accessible BRCs will facilitate biospecimen exchanges but will also raise problems of rights to use biospecimens and of intellectual property. Furthermore, networking opens opportunities for accessing biospecimen collections for studies that have a different purpose to those in which biospecimens were collected in the first place. Such 're-assembly of collections' poses specific ethical and practical problems. Thus, BRCs have to work in close cooperation with all actors involved in the decision chain for biospecimen collection and use, as well as under the monitoring of relevant, institutional or legal ethical issues (see also Chapter 22).

Figure 21.1 shows a time sequence and flow of instructions, information, data and biospecimens, from the study design to final laboratory analyses. This scheme underlines the central role of BRCs as the transfer structure between biospecimen collection and laboratory analysis. It also underlines the fact that, in developing a study protocol, each step in this sequence of events must be clearly defined. This includes in particular the relationships between the five main actors, the principal investigator and his working group, the participant, the medical staff, the BRC and the laboratory in which biomarker analysis will be conducted. The flow of information and biospecimens, as defined by protocols and procedures, will ensure the constitution of a collection containing traceable biospecimens yielding interpretable results. In this complex picture the BRC is an essential source of information and recommendations for the collection of biospecimens, their annotation, storage, processing and flow from the participant to the laboratory where it will be analysed.

In this review, we describe the key steps underlying the different aspects of biospecimen collection, processing, labelling, transport, storage and retrieval for laboratory analysis in molecular epidemiological studies. The layout of this chapter follows the outline of Figure 21.1. Documents with recommendations for the development of BRCs are given in Table 21.1.

21.2. OBTAINING AND COLLECTING BIOSPECIMENS

Obtaining and collecting biospecimens are not, as such, within the scope of the activities run by a BRC. However, the careful planning and implementation of these activities is fundamental to determine the type of facilities for biospecimen storage and processing that the BRC will be required to operate. The BRC should be capable of providing advice on these matters to make sure that they are compatible with appropriate standards for biospecimen storage.

Planning a collection

Collection of biospecimens must be performed according to consistent ethical standards. The

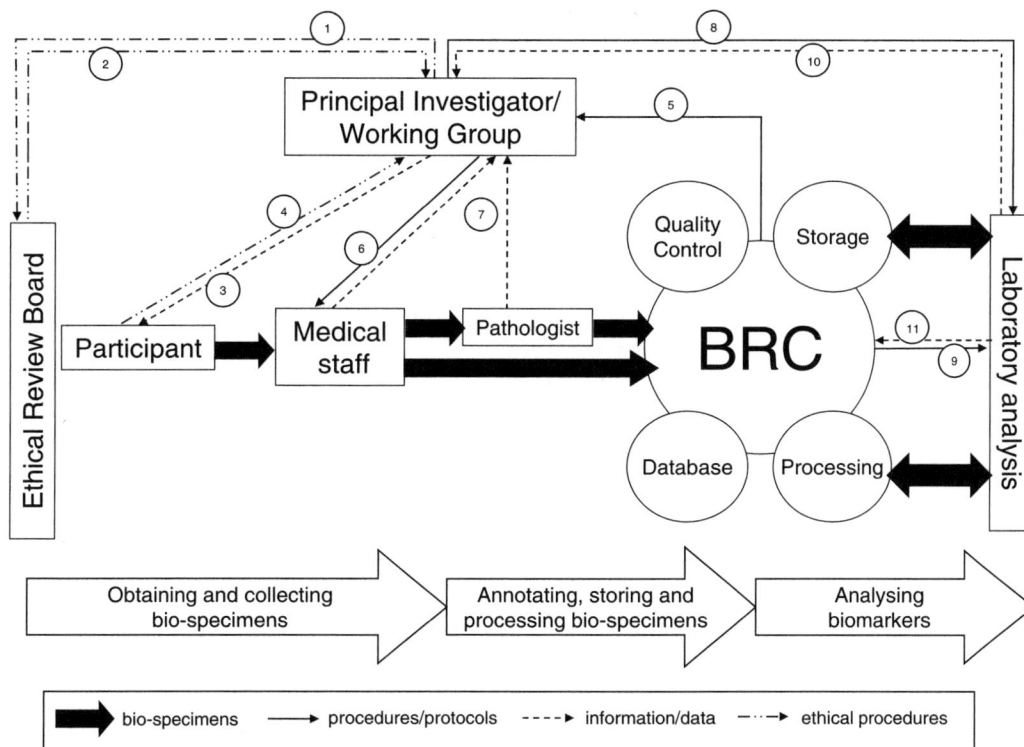

Figure 21.1 Flow of biospecimens, information, procedures/protocols and data between the participants, through the BRC until laboratory analysis. The procedures/protocols, information/data and ethical procedures are described by the following documents: 1, the proposal describing the study design; 2, the approbation of the proposal by the Ethical Review Board (ERB); 3, information to the participants; 4, the informed consent; 5, guidelines for biospecimens collection; 6, collection protocol; 7, biospecimen form containing the information on biospecimen sampling and participant; 8, biomarker selection; 9, guidelines for biospecimen handling; 10, laboratory results; 11, biospecimen and by-product history information

cornerstone of these standards is the conditions in which informed consent is obtained from participants, as well as the scope of this consent. While obtaining consent is not part of the mission of the BRC, it is important that information on consent, as well as on the scope of this consent, is archived within the BRC records. In many instances, new biomarkers of interest may become available for analysis after the inception of the study, making it necessary to assess whether use of this new biomarker is consistent with the scope of the initial informed consent. Another, essential ethical aspect with respect to good BRC practices is the use of secure, anonymous coding systems to protect confidentiality (see section on identifying biospecimens).

The first question to address in implementing a molecular epidemiology study is which biomarker(s) to measure and which material (type of biospecimen) to use as a source. Making decisions on these points depends upon scientific hypotheses and practical constraints, but also on availability and the justification for obtaining biospecimens. Thus, in many instances, it will be preferable or required to use surrogate biospecimens rather than samples of the target tissue in which the biomarker under investigation is supposed to be present. This decision will specifically guide the procedure followed for obtaining biological materials and associated data. The collection procedure is determined by the approval of study design, including biomarker and biospecimen selection, by an appropriate

Table 21.1   Sources for guidelines, procedures and recommendations on BRC

Title	Organization	Link
Common Minimum Technical Standards and Protocols for Biological Resource Centres Dedicated to Cancer Research, Elodie Caboux, Amelie Plymoth and Pierre Hainaut. IARC Working Group Reports Volume 2	IARC, 2007	http://www.iarc.fr/IARCPress/pdfs/ standardsBRC/index.php
Tissue Banking for Biomedical Research	National Cancer Centre	http://www.bioethics-singapore.org/resources/ pdf/AppendixB-Dr%20Kon.pdf
Biorepository Protocols	Australian Biospecimen Network (ABN)	http://www.abrn.net/pdf/ABN_SOPs_Review_ Mar06_final.pdf
European Human Frozen Tumor Tissue Bank (TUBAFROST)	The European Human Tumour Frozen Tissue Bank (TUBAFROST)	http://www.tubafrost.org/research/moreinfo/ deliverables/TUBAFROST%20Deliverable%2 04.1.pdf (Teodorovic *et al.* 2003)
Human Tissue and Biological Samples for Use in Research. Operational and Ethical Guidelines	Medical Research Council (MRC)	www.mrc.ac.uk/consumption/idcplg? IdcService=GET_FILE&dID=9051& dDocName=MRC002420&allowInterrupt=1
Best Practices for Repositories I: Collection, Storage, and Retrieval of Human Biological Materials for Research	International Society for Biological and Environmental Repositories (ISBER)	http://ehs.sph.berkeley.edu/Holland/Biorep/ BestPractices2005.3.5.pdf (International Society for Biological and Environmental Repositories 2005)
National Cancer Institute: Best Practices for Biospecimen Resources	National Cancer Institute (NCI)	http://biospecimens.cancer.gov/NCI_Best_ Practices_040507.pdf
Transport of Infectious Substances	World Health Organization (WHO)	http://www.who.int/csr/resources/publications/ biosafety/WHO_CDS_CSR_LYO_2005_ 22r%20.pdf
UN Recommendations on the Transport of Dangerous Goods. Model Regulations.	United Nations Economic Commission for Europe (UNECE)	http://www.unece.org/trans/danger/publi/unrec/ rev13/13files_e.html
A Cold Greeting: An Introduction to Cryobiology	Biotech	http://www.bioteach.ubc.ca/Bioengineering/ AColdGreeting/
Specimen Collection, Preparation, and Handling	Labcorp	http://www.labcorp.com/datasets/labcorp/html/ frontm_group/frontm/section/speccol.htm
Recommendation Rec(2006)4 of the Committee of Ministers to Member States on Research on Biological Materials of Human Origin	Council of Europe Committee of Ministers	https://wcd.coe.int/ViewDoc.jsp?id=977859& BackColorInternet=9999CC&BackColorIntranet =FFBB55&BackColorLogged=FFAC75
The Standard MM13-A – Collection, Transport, Preparation and Storage of Specimens for Molecular Methods; Approved Guideline'	Clinical and Laboratory	http://www.clsi.org/source/orders/index.cfm?section= Shop&ETask=1&Task=1&SEARCH_TYPE= FIND&FindIn=0&FindSpec=MM13&x=9&y=2
The Human Proteome Organization (HUPO)	Human Proteome	http://www.hupo.org/
Biological Resource Centres: Uunderpinning the Future of Life Sciences and Biotechnology	OECD/International	http://wdcm.nig.ac.jp/brc.pdf
OECD Best Practice Guidelines for Biological Resource Centres	OECD/International	http://www.wfcc.nig.ac.jp/Documents/OECD.pdf
Case Studies of Existing Human Tissue Repositories: 'Best Practices' of a Biospecimen Resource for the Genomic and Proteomic Era	RAND corporation	http://www.rand.org/pubs/monographs/2004/ RAND_MG120.pdf

ethical review board (ERB; Figure 21.1). The prospective participants then receive information and are asked to give informed consent. This consent should include long-term biospecimen storage. It should also provide information to the participant and strictly delineate the use of the biospecimen(s). The biospecimen, collection protocol and participant consent will delineate the type of biomarkers that can be measured in the present and future studies. Handling these aspects is one of the most complex and specific aspects of using biomarkers in molecular epidemiology.

Types of biospecimens

Table 21.2 is a non-exhaustive list of biospecimens that can be collected and the biomarkers that can be measured. Biospecimens can be divided into two sub-types according to the way they are obtained (Holland *et al.* 2005). Type 1 regroups biospecimens acquired through invasive methods (blood and its four different compartments, tissue, etc.). Type 2 are biospecimens acquired through noninvasive methods, such as urine and other body fluids, hair, nail, and easily accessible exfoliated cells. This classification influences the recruitment of participants and the cost of the collection. Biospecimens collected by non-invasive methods will generally not require highly trained clinical staff; in some cases the biospecimens can be collected by the participant themselves and then sent to the collection centre.

Many types of biospecimens are amenable to the detection of different biomarkers, such as proteins and their modifications, DNA or protein adducts of xenobiotics, DNA, RNA, etc. However, exploiting these biomarkers may require different collection, storage and processing protocols. The type and nature of biomarkers to be analysed will thus determine the exact collection, processing and storage protocol.

Collecting biospecimens

In biospecimen collections, three basic aspects are essential with respect to the requirements of running a BRC: biospecimen amount and aliquoting, traceability of collection and processing events, and biospecimen annotation and identification.

Regarding the amount and aliquoting of biospecimens, good practice recommends: (a) that amounts collected are kept to a minimum, taking into account the needs of the specific analytical methods that will be used to detect biomarkers; (b) that biospecimens are distributed in different aliquots to avoid decay that may result from the frequent freezing and thawing of the same aliquot. This is essential, in particular, for biomarkers of limited stability. Whenever possible, the addition of preservative or stabilizing agents may be a suitable option to increase the lifetime of such biomarkers (see Table 21.2).

Keeping records of the temporal sequence of events from collection to storage is paramount for biospecimen traceability and should be seen as one of the basic quality criteria for running a BRC. Undue lag times in this sequence may have a profound impact on subsequent biomarker analysis. The basic good practice recommendation here is that the timing of each step should be kept to a minimum. If waiting periods are unavoidable (e.g. during biospecimen processing, packaging or shipping), a precise record of these periods should be kept. The level of detail of these temporal annotations should be down to the hour. In particular, the time elapsed between collection and final preservation (e.g. deep-freezing) should be recorded. In multi-centre studies, considerations should be given to the fact that the timing of all events from collection to storage should be comparable.

21.3. ANNOTATING, STORING AND PROCESSING BIOSPECIMENS

After the biospecimens have been collected, processed and, if appropriate, diagnosed (e.g. in the case of diseased tissues), they are annotated and enter a BRC to manage their movements and for long-term storage. Four main components can be distinguished in the BRC: (a) the storage facility; (b) the database; (c) the laboratory for biospecimen processing; and (d) the procedures for quality assurance and quality control (Figure 21.1).

Identifying biospecimens

After sampling and before storage the biospecimen will have to be given a unique identifier to protect

Table 21.2 Examples of biospecimens and biomarkers used in molecular epidemiology; special handling protocols are indicated when available

Method of collection	Biospecimens	Examples of biomarkers/ biomaterials	Recommendation and condition for collection and storage	References/ web sites
Invasive	Serum	Micronutrients (vitamins and minerals)	Fasting for a maximum of 12 h	(Chiplonkar et al. 2004)
		Sex hormones	In females recording of day 1 of the ongoing and next menstrual cycle. Follow-up: age at menopause	(Kaaks et al. 2005; Verkasalo et al. 2001)
		Lipids	Fasting for a maximum of 12 h	(Appel et al. 2005)
		Uric acid	Serum separation within the 32 h after venipuncture	(Boyanton and Blick 2002)
		Glucose	Stabilized by sodium fluoride, separation within 3 h after venipuncture	(Zhang et al. 1998)
		Metabonomic	Fasted subjects, clotting for 20–35 min on ice; avoid repeated freezing and thawing	(Teahan et al. 2006)
	Plasma	Vitamin C	Stabilization with metaphosphoric acid, EDTA, perchloric acid or DTT. Short-term storage at −70°C (1 week) and long-term storage at −196°C	(Jenab et al. 2005)
		DNA	Do not use heparin as an anticoagulant (inhibition of PCR); separate plasma within 2 h after venipuncture	(Jen et al. 2000)
	Tissue	Protein	Freezing within 30 min after resection	(Jackson et al. 2006)
		RNA	RNAlater or snap-freezing in liquid nitrogen	(Mutter et al. 2004; Winnepenninckx et al. 2006)
Non-invasive	Urine	Erythropoietin	Immediate addition of complete protease inhibitor cocktail	(Beullens et al. 2006; Khan et al. 2005)
		Fluoride, magnesium, calcium	24 h urine collection	(Anatol et al. 2005; Rylander and Arnaud 2004; Zohouri et al. 2006)
	Buccal cells	DNA	Collection by cytobrush gives the best DNA yield and purity	(Mulot et al. 2005)
	Cervical cells	RNA	Snap-freezing in liquid nitrogen	(Wang et al. 2006)
	Hair	DNA, drugs and metals	Sequential washing with detergent, water and organic solvent	(Lachenmeier et al., 2006; McNevin et al. 2005; Pereira et al. 2004)
	Nail	Metals	Hand-washed with distilled water and medicated soap devoid of metal contamination, nails cut with clean stainless steel scissors, nails washed sequentially with non-ionic detergent, acetone and water	(Mehra and Juneja 2005)
	White blood cells	RNA	Process blood quickly to avoid changes in gene expression	http://www.affymetrix.com/support/technical/technotes/blood_technote.pdf

participant confidentiality. Two levels of confidentiality are possible (Le Roux *et al.* 2003). First, the biospecimen is anonymous; the link between the participant and the biospecimen is completely erased, so that it is impossible to go back to the participant. Second, confidential codes might permit re-tracing of the participants, e.g. for follow-up or return of information to the participant. Whether the identifier is assigned to the participant or the biospecimen has to be decided in the study design. However, care should be taken that each aliquot or different types of biospecimen carries a unique identifier. These identifiers should be compatible with the 'double-blind' standard for laboratory analyses.

Three types of information will be collected at different times and by different actors in the study. All the information collected will not be shared between the actors because of its confidentiality and its relevance. The first type of information relates to the data collected by the medical staff and the specific study questionnaire and the conclusion made by the pathologist after biospecimen analysis. This data are collected and stored by the principal investigator and his/her working group. The second type of information is exclusively about the biospecimen. It will include the data on storage location, condition and history. Data on the biospecimen is usually curated by the BRC. Finally, information on laboratory analysis will be collected when the biomarkers are analysed and fed back to the principal investigator and his/her working group (Figure 21.1). This information represents three specific datasets, each with its own logistics, curation needs and constraints (e.g. confidentiality for subject data; intellectual property for analytical results). The datasets should be interconnected and information should be traceable across each dataset. How this traceability is organized and managed, and the extent of protections between each dataset, should be determined specifically for each particular study. As far as the BRC is concerned, the database information should be sufficient to ensure the proper management, sorting and distribution of the specimens.

Storage facilities

The central activity of a BRC is the storage of the biospecimen in a bio-repository. The storage facility includes dedicated secured environments such as a room or a building, alarm systems that ensure surveillance of the preservation system (e.g. with adequate temperature monitoring and tracking in the case of cryopreserved biospecimens), back-up storage equipment for emergencies, including spare freezers and/or liquid nitrogen (LN_2), a back-up source of electrical power such as a generator, and dedicated, trained staff applying defined standard operating procedures (SOP). Emergency and contingency plans should be developed for the fast rescue of samples in case of system failure. Storage unit breakdowns should be recorded within the BRC database and documented in the specimen history.

Labelling

Before long-term storage, the container of the biospecimen (e.g. microtube, plastic straw, microplate, etc.) has to be clearly labelled. The type of container and of storage conditions directly affect the type of label to be used. Labels have to be resistant to water, temperature between $-20°C$ and $-196°C$ and to freezing and thawing. Appropriate labels can be obtained from commercial sources, e.g. companies such as Brady (Wisconsin) or Modul-Bio (Marseille). Whenever possible, double-labelling is recommended.

The basic information on the label should consist of a unique identifier, according to a syntax that does not allow easy determination of the origin of the biospecimen or the sampling date. Rather, this latter information should be kept in a database. Furthermore, neither the label nor the information in the database should allow direct identification of the person who is the source of the biospecimen. In the case of non-anonymized biospecimens, the key for linkage between storage records and personal identifiers should be kept separate from the storage database. One elegant way to fulfil these requirements, while posting specific information on the label, is the use of linear or two-dimensional bar-code labels. Such bar-codes are helpful to standardize numbers for inter-laboratory exchanges, and prevent human errors and loss of data. They are also useful in automated procedures for biospecimen storage, retrieval or processing.

Conditions of storage

The types of biospecimen and biomarker determine the storage conditions. The range of temperatures used for storage varies between room temperature for archived, paraffin-embedded tissues, for example, to −196°C for viable cells (Table 21.3). The main factors to take into consideration in selecting the storage temperature and method are the physical properties of water and the effect of temperature on the stability of biological molecules. The structure of water and the status of biological molecules in water solutions change at different temperatures and directly affect the stability of the biospecimen and biomarker during freezing and long-term storage. It has been shown that the conformation of water molecules changes between −100°C and −130°C, which might affect cell and tissue structures (Baust 2002). Changes in the structure of water crystals may exert drastic effects that shear biological structures such as macromolecules, cells or organelles.

Three main types of cold storage methods are commonly used, including fridges (0–4°C), mechanical freezers (−20°C to −130°C) and LN₂ tanks (−130°C to −196°C). The term 'cryopreservation' usually applies to storage at very low temperatures.

Fridges are not a recommended system for long-term storage but might be useful as intermediate storage system during specimen processing. It should be noted, however, that purified DNA is relatively stable under alcohol at 4°C. Freezing at temperatures between −20°C and −40°C allows the life span of most bio-molecules to be significantly extended but is not a recommended standard for long-term storage or for the storage of very labile biological structures, such as RNA or post-translational modifications of proteins. Storage at temperatures between −80°C and −130°C provides long-term stability for most bio-molecules but does not preserve cell survival. Furthermore, these temperatures encompass the range of temperatures at which water crystals undergo structural changes, thus exposing the biospecimens to undue structural stress. Finally, liquid nitrogen storage is recommended for the cryopreservation of viable cells. For long-term storage, liquid nitrogen is often preferred but data on differential stability of biospecimen and biomarker at temperatures between −80°C and −196°C are scarce.

Two types of storage in LN₂ can be found, vapour or immersion. While storage under vapour requires much less LN₂ and minimizes hazards for BRC staff, the temperature varies from the top of the tank (−80°C) to the bottom (−170°C) (Rowley and Byrne 1992). Immersion in LN₂ tanks ensures a constant temperature at −196°C but is expensive and represents a source of risks, including burns and asphyxia. Therefore, immersion storage requires dedicated rooms equipped with adequate protections, alarms and ventilation systems.

Table 21.3 Water and biological species status for different storage conditions

Temperature (°C)	Properties of water/liquid nitrogen	Cryopreservation method	Biological relevance
0 to +4	Ice melting	Fridge	
−0.5 to −27	Ice fusion area	Freezer	
−27 to −40	Ice	Freezer	Limit of proteins mobility/ DNA stability
−40 to −80	Limit of water molecules mobility	Freezer	RNA stability
−80 to −130	Ice transition	Freezer/liquid nitrogen	No metabolic activity
			Recommended storage for blood and urine
−130 to −150	Liquid nitrogen vapour	Liquid nitrogen	Recommended storage for tissue
−150 to −196	Liquid nitrogen liquid	Liquid nitrogen	Possible microfractures
			Recommended storage for living cells

Studies have also shown that storage in immersion liquid nitrogen tanks favours cross-contaminations between biospecimens (Fountain *et al.* 1997).

Laboratory processing and shipping

The processing laboratory in the BRC provides a service that handles and prepares the biospecimens before storage, manages the retrieval and aliquoting of the biospecimens before laboratory analysis, oversees their shipping and organizes the re-storing of left-over material after biomarker laboratory analysis (if appropriate). In many instances, the processing, labelling and aliquoting of the biospecimens prior to initial storage may be performed by the collecting laboratory, and the BRC will only get involved at the time of placing the biospecimens into long-term storage. In other cases, the BRC may receive crude biospecimens that will require processing and aliquoting before storage. Biomarker analysis might require the preparation of a by-product, e.g. to study genetic susceptibility markers, DNA has to be extracted from white blood cells or tissues. Another critical activity of the BRC processing laboratory is the shipping of biospecimens or their by-products.

According to current regulations, biospecimens can be transported by any transportation means (air, rail, road, sea) provided that adequate packaging and safety measures are used. A common means of transport for large, multi-centre studies is air freight. Shipping by air has special constraints in relation to toxicological, microbial or biological safety. According to the current regulations of the International Air Transport Association (IATA; http://www.iata.org/index.htm), human biospecimens are considered as 'dangerous goods', i.e. 'articles or substances which are capable of posing a risk to health, safety, property or the environment'. Therefore, the transport of biological materials is regulated by the Recommendations by the Committee of Experts on the Transport of Dangerous Goods (UNCETDG; http://www.unece.org/trans/main/dgdb/dgcomm/ac10age.html), a committee of the United Nations Economic and Social Council. The UN has distinguished nine classes of dangerous goods. The relevant class for human biospecimens is Class 6, division 6.2:

Infectious substances. Division 6.2 is then subdivided into two categories, A and B, and a group of exempt biospecimens:

1. *Category A* comprises substances capable of posing permanent disability, life-threatening, or fatal disease to humans or animals. Category A biospecimens include, but are not restricted to, biospecimens contaminated by highly pathogenic viruses (Ebola, Hantaan, Marburg, Lassa, etc. or cultures of viruses such as Dengue, Human Immunodeficiency Virus (HIV), and Hepatitis B Virus (HBV). The proper shipping name for such substances is UN2814, 'Infectious substances affecting humans', or UN2900, 'Infectious substances affecting animals only'.

2. *Category B* comprises substances which do not meet the above criteria. Most human biospecimens, such as blood samples, tissues, exfoliated cells or urine, not contaminated by highly pathogenic viruses will fall into Category B. The proper shipping name for such substances is UN3373: 'Biological Substance, Category B'.

3. *'Exempt Human (or Animal) Specimens'* includes biospecimens or derived by-products that have been specifically treated to neutralize infectious agents, or for which there is a minimal likelihood that pathogens are present. This exempt group of biospecimens is not subject to categories A and B regulations.

Recommendations and regulations for air transport can be found in two documents, the *Technical Instructions for the Safe Transport of Dangerous Goods by Air* and the *Dangerous Goods Regulations* (DGR), published by the International Civil Aviation Organization (ICAO, http://www.icao.int/) and IATA (International Air Transport Association 2006), respectively.

The packaging of the biospecimens and the attached documentation for air transport depend on the category of the infectious substance (categories A and B or exempt) and its nature (liquid or solid). Packaging instructions and protocols for the different types of biospecimens can be found in the DGR (International Air Transport Association 2006). The common feature for the packaging instructions is the use of a basic triple packaging system. The biospecimen and biomarker to be measured determine the temperature used for transport. Four types of packaging temperature exist, room temperature, $+4°C$ (freezer packs), $-78.5°C$ (dry ice) and liquid nitrogen ($-196°C$).

When preparing to transport biospecimens, it is important to consider shipping time, distance, climate, method of transport and requirements and regulations in the destination country, as well as the type and number of biospecimens to be sent and their intended use. It is essential to consider biospecimen transport during study design, first, because some geographical areas are not easy to reach, but second, because the cost of biospecimen transport is very high. Finally, customs clearance should be taken into consideration. Inappropriate documentation for customs is one of the main factors causing delay in biospecimen shipping.

BRC database

The management of biospecimen storage and history requires the constitution of a database, which manages the geographical position of the biospecimen and its flow in and out of the BRC. The BRC database essentially stores information on the biospecimen and is thus distinct from the study database, which generally stores much more detailed and specific information, as well as data resulting from biospecimen analysis. The BRC database includes the identification code, the nature of the biospecimens, the number of aliquots and by-products, the date of sampling, the title of the study and the contact details of the principal investigator and of the persons responsible for specimen management within each study (if different from the PI). Data on the history of the biospecimen is also desirable. This information should cover the history of the biospecimen from the participant to laboratory analysis, including the sampling protocol, processing and cryopreservation before entry into the BRC, cryopreservation in the BRC and in the laboratory where the biomarkers are analysed. Repeated freezing and thawing, time spent at inadequate temperatures, information on by-product or volume of biospecimen avail-

able (volume of biospecimen sent for laboratory analysis and that left in the BRC) are examples of important data on biospecimen history. Detailed history of biospecimen usage is essential to allow future analysis of biomarkers requiring sensitive technologies and to prevent study bias due to heterogeneous conditions of collection, processing and storage. Use of the biospecimen or by-product by another investigator should also be recorded and documented with amounts of biospecimen used and scope of the study.

The database contains information on the exact geographical position of a biospecimen in the BRC to allow its retrieval. Identification, history and localization allow good traceability of a biospecimen. Finally, it is useful to keep in the database information on the scope of the informed consent and on approval by the ethical review board (ERB) under which biospecimens have been collected, since this information provides the framework for any future biomarker analysis.

Many formats are available that are suitable for maintaining a BRC database. The main constraint in selecting a database format or particular software is that the system should be able to work as a dynamic, specimen and data-flow operating system. Thus, we recommend databases that are structured as components of a Laboratory Information Management System (LIMS).

Quality assurance and quality control

Quality assurance (QA) and quality control (QC) are approaches that provide for an integrated action plan to make sure that the activities meet pre-established standards of quality, and for operational procedures to assess these quality requirements. BRCs have to organize quality assurance and quality control programmes that take into consideration staff and safety, laboratory work, infrastructures and equipment, biospecimens, transport, data management, record keeping, and occupational and environmental risks. While QA programmes specify the nature of these actions, QC programmes provide for the procedures to implement them. QA/QC guarantee the regular training of laboratory staff and their safety. It also makes sure that protocols and procedures used in the

BRC storage and laboratory facilities are defined, written and followed-up as standard operating procedures (SOP) which are reviewed and updated on a regular basis. QA/QC oversee the quality and the security of the biospecimen storage. Thus, they might, for example, set up laboratory experiments to measure specific biospecimen or biomarker stability over time, under different storage conditions and after repeated freezing and thawing. They may also check the setting up, the running and the maintenance of alarm and back-up systems for the storage infrastructure. QA/QC procedures ensure that data entry and management is performed according to SOP. Finally, QA/QC oversee the proper maintenance of infrastructure and equipment, including alarm systems, and implement all appropriate measures to control health hazards for staff, BRC users and the environment.

21.4. ANALYSING BIOMARKERS

The purpose of this chapter is not to give an exhaustive list of the methods and protocols used to measure different types of biospecimens and biomarkers. In molecular epidemiology a biomarker may be used to measure exposure, susceptibility, disease progression or outcome, and the type of biomarker is selected accordingly. However, it is important to stress that biospecimens are a rare and limited resource which should be used with care and with strict principles for priority. This implies using a very pragmatic and specimen-saving approach in selecting the marker, the method to analyse it, and the amount of material required for the analysis. Common sources of biospecimen waste are the use of insufficiently validated methods, inadequate assessment of statistical power, and inadequate consideration of the cost of analysis. Therefore, it is recommended that laboratories conduct detailed feasibility studies before starting systematic biomarker analysis.

After validation of the laboratory analysis, the biospecimens are shipped from the BRC to the laboratory, either from the storage facility or after aliquoting, processing or by-product preparation by the BRC laboratory (Figure 21.1). The BRC and

laboratory staff have to ensure that a person will be present to receive the biospecimens, put them at the proper storage temperature as soon as possible, and provide feedback to the BRC upon reception. The content of the package and the accompanying lists of biospecimens should be thoroughly checked for discrepancies and losses. Any problems during transport or reception must be reported to the BRC to allow the traceability of biospecimen history. The BRC also provides instructions for handling the biospecimens.

When biospecimens are handled in the laboratory, a number of rules have to be respected to allow traceability. Laboratory analyses have to be within the scope of the informed consent. Each biospecimen should be labelled with its unique identifier throughout all laboratory procedures (e.g. on the container, in the laboratory books, in Laboratory Information Management System or database compiling the protocols and the results). At the end of laboratory studies, residual samples or by-products might be available. The handling of such residual material should be discussed during study design between the principal investigator, the BRC and the laboratory using the biospecimens. In principle, left-over biospecimens and their by-products should be returned to the BRC, accompanied by detailed information on storage, freezing and thawing history, volume and characteristics of the by-product (e.g. nature and concentration). Any intentional or accidental freezing and thawing of the biospecimen or cross-contamination must be recorded and reported to the BRC at the end of the study.

In some instances, it may be acceptable for the laboratory to retain residual material. In this case, such products may be further used as fully anonymous biological samples in translational studies or in the development of new methods. This implies: (a) that all analyses planned in the study design have been implemented; and (b) that the products do not meet the conditions needed to be re-entered into the BRC.

Final decisions to destroy biospecimens after use has serious implications. This decision should be taken jointly by the principal investigator and the BRC, and should not be left to the laboratories involved in analysing biomarkers.

21.5. CONCLUSIONS

In molecular epidemiology studies BRCs have an increasingly strategic role. Their design, implementation and management are crucial for present and future studies. National and international regulations for BRCs must be developed to ensure the proper use of human tissues and traceability of biospecimens, and to facilitate networking. These regulations should encompass ethical approval, collection, storage, processing, annotation, retrieval and shipping protocols, as well as accreditation procedures for BRCs as operating units.

Many challenges remain to be addressed. One is the long-term, economic sustainability of BRCs as well as the mechanisms for transmission of responsibilities, biospecimens and data when a BRC closes. Another is the conditions for elimination of surplus biospecimens. International mechanisms have to be set up to facilitate networking and specimen exchange between countries in conditions that meet the highest standards of safety and security. The question of intellectual property and of recognition of the scientific and technical input of the BRCs also needs to be better addressed. However, the most important challenge today is to develop a consensus on the nature, scope and type of informed consent under which the biospecimens may be obtained and analysed. This issue is particularly important in the context of prospective studies, in which biospecimens may be kept in long-term storage before being analysed using methods or techniques that were not known at the time of study design. However, the problem is not limited to prospective studies, as it often happens that interesting opportunities for new analyses may occur during or after the completion of a study. Such new studies may be hampered by the scope of the initial informed consent, which should be very carefully defined. In particular, consent should be sought for specific research objectives, rather than for specific types of analyses. This does not imply requesting broad, general consent: the objectives should be specific and should clearly define the type of information that will be obtained through laboratory testing.

When faced with the possibility of developing analyses that are not within the scope of the initial consent, two main solutions may be envisaged. One is to obtain a new informed consent from each participant for specific analyses, or for broader uses of the biospecimen. A second is for relevant ethical review boards (ERBs) to make surrogate decisions on behalf of the participants (Hansson *et al.* 2006).

Finally, efforts should be made towards the promotion of BRC standards that are acceptable worldwide, and commensurate with different cultural, social and economical statuses. The development of BRCs in low-resource countries is a critical challenge, particularly since there is a huge need for specific molecular epidemiological studies in these countries in order to assess exposures, risks and genetic susceptibility and to identify adequate prevention or early detection strategies.

References

Anatol TI, Pinto PL, Matthew J *et al.* 2005. The relationship of magnesium intake to serum and urinary calcium and magnesium levels in Trinidadian stone formers. *Int J Urol* **12**: 244–249.

Appel LJ, Sacks FM, Carey VJ *et al.* 2005. Effects of protein, mono-unsaturated fat, and carbohydrate intake on blood pressure and serum lipids: results of the OmniHeart randomized trial. *J Am Med Assoc* **294**: 2455–2464.

Baust J. 2002. Molecular mechanisms of cellular demise associated with cryopreservation failure. *Cell Preserv Technol* **1**, 17–31.

Beullens M, Delanghe JR, Bollen M. 2006. False-positive detection of recombinant human erythropoietin in urine following strenuous physical exercise. *Blood* **107**: 4711–4713.

Bingham S, Riboli E. 2004. Diet and cancer – the European Prospective Investigation into Cancer and Nutrition. *Nat Rev Cancer* **4**: 206–215.

Boyanton BL Jr, Blick KE. 2002. Stability studies of twenty-four analytes in human plasma and serum. *Clin Chem* **48**: 2242–2247.

Chiplonkar SA, Agte VV, Tarwadi KV *et al.* 2004. Micronutrient deficiencies as predisposing factors for hypertension in lacto-vegetarian Indian adults. *J Am Coll Nutr* **23**: 239–247.

Fountain D, Ralston M, Higgins N *et al.* 1997. Liquid nitrogen freezers: a potential source of microbial contamination of hematopoietic stem cell components. *Transfusion* **37**: 585–591.

Hansson MG, Dillner J, Bartram CR *et al.* 2006. Should donors be allowed to give broad consent to future biobank research? *Lancet Oncol* **7**: 266–269.

Holland NT, Pfleger L, Berger E *et al.* 2005. Molecular epidemiology biomarkers – sample collection and processing considerations. *Toxicol Appl Pharmacol* **206**: 261–268.

International Air Transport Association. 2006. *Dangerous Goods Regulations*, 47th edn. Montreal, Publisher: International Air Transport Association (IATA).

International Society for Biological and Environmental Repositories. 2005. Best practices for repositories I: collection, storage, and retrieval of human biological materials for research. *Cell Preserv Technol* **3**: 5–48.

Ioannidis JP, Bernstein J, Boffetta P *et al.* 2005. A network of investigator networks in human genome epidemiology. *Am J Epidemiol* **162**: 302–304.

Jackson D, Rowlinson RA, Eaton CK *et al.* 2006. Prostatic tissue protein alterations due to delayed time to freezing. *Proteomics* **6**: 3901–3908.

Jen J, Wu L, Sidransky D. 2000. An overview on the isolation and analysis of circulating tumor DNA in plasma and serum. *Ann N Y Acad Sci* **906**: 8–12.

Jenab M, Bingham S, Ferrari P *et al.* 2005. Long-term cryoconservation and stability of vitamin C in serum samples of the European Prospective Investigation into Cancer and Nutrition. *Cancer Epidemiol Biomarkers Prevent* **14**: 1837–1840.

Kaaks R, Berrino F, Key T *et al.* 2005. Serum sex steroids in premenopausal women and breast cancer risk within the European Prospective Investigation into Cancer and Nutrition (EPIC). *J Natl Cancer Inst* **97**: 755–765.

Khan A, Grinyer J, Truong ST *et al.* 2005. New urinary EPO drug testing method using two-dimensional gel electrophoresis. *Clin Chim Acta* **358**: 119–130.

Lachenmeier K, Musshoff F, Madea B. 2006. Determination of opiates and cocaine in hair using automated enzyme immunoassay screening methodologies followed by gas chromatographic–mass spectrometric (GC–MS) confirmation. *Forensic Sci Int* **159**: 189–199.

Le Roux N, de Montgolfier S, di Donato JH *et al.* 2003. [Collections of human biological resources for research purposes: from regulations to the need of a guide of good collection practices]. *Rev Med Interne* **24**: 803–814.

McNevin D, Wilson-Wilde L, Robertson J *et al.* 2005. Short tandem repeat (STR) genotyping of keratinised hair. Part 2. An optimized genomic DNA extraction procedure reveals donor dependence of STR profiles. *Forens Sci Int* **153**: 247–259.

Mehra R, Juneja M. 2005. Fingernails as biological indices of metal exposure. *J Biosci* **30**: 253–257.

Mulot C, Stucker I, Clavel J *et al.* 2005. Collection of human genomic DNA from buccal cells for genetics studies: comparison between cytobrush, mouthwash, and treated card. *J Biomed Biotechnol* **3**: 291–296.

Mutter GL, Zahrieh D, Liu C *et al.* 2004. Comparison of frozen and RNALater solid tissue storage methods for use in RNA expression microarrays. *BMC Genom* **5**: 88.

Pereira R, Ribeiro R, Goncalves F. 2004. Scalp hair analysis as a tool in assessing human exposure to heavy metals (S. Domingos mine, Portugal). *Sci Total Environ* **327**: 81–92.

Rowley SD, Byrne DV. 1992. Low-temperature storage of bone marrow in nitrogen vapor-phase refrigerators: decreased temperature gradients with an aluminum racking system. *Transfusion* **32**: 750–754.

Rylander R, Arnaud MJ. 2004. Mineral water intake reduces blood pressure among subjects with low urinary magnesium and calcium levels. *BMC Publ Health* **4**: 56.

Schulte PA, Perera FP. 1997. *Transitional Studies*. IARC Scientific Publication. IARC: Lyons, France; 19–29.

Teahan O, Gamble S, Holmes E *et al.* 2006. Impact of analytical bias in metabonomic studies of human blood serum and plasma. *Anal Chem* **78**: 4307–4318.

Teodorovic I, Therasse P, Spatz A *et al.* 2003. Human tissue research: EORTC recommendations on its practical consequences. *Eur J Cancer* **39**: 2256–2263.

Verkasalo PK, Thomas HV, Appleby PN *et al.* 2001. Circulating levels of sex hormones and their relation to risk factors for breast cancer: a cross-sectional study in 1092 pre- and postmenopausal women (United Kingdom). *Cancer Causes Control* **12**: 47–59.

Wang SS, Sherman ME, Rader JS *et al.* 2006. Cervical tissue collection methods for RNA preservation: comparison of snap-frozen, ethanol-fixed, and RNAlater-fixation. *Diagn Mol Pathol* **15**: 144–148.

Winnepenninckx V, Lazar V, Michiels S *et al.* 2006. Gene expression profiling of primary cutaneous melanoma and clinical outcome. *J Natl Cancer Inst* **98**: 472–482.

Zhang DJ, Elswick RK, Miller WG *et al.* 1998. Effect of serum-clot contact time on clinical chemistry laboratory results. *Clin Chem* **44**: 1325–1333.

Zohouri FV, Swinbank CM, Maguire A *et al.* 2006. Is the fluoride:creatinine ratio of a spot urine sample indicative of 24 h urinary fluoride? *Commun Dent Oral Epidemiol* **34**: 130–138.

22

Molecular Epidemiology and Ethics: Biomarkers for Disease Susceptibility

Kirsi Vähäkangas

University of Kuopio, Finland

22.1. INTRODUCTION

While biomarkers for disease or early disease may provide the benefit of early detection and better treatment results, the search for genetic biomarkers for disease susceptibility remains much more controversial (Vineis and Schulte 1995; Vineis 1997; Hainaut and Vähäkangas 1999; Vähäkangas 2001, 2004; Taioli 2005). This is based on the understanding that genetic knowledge is different from other health information (see e.g. www.unesdoc. unesco.org; BMA Ethics Department 2004). The share of benefits and costs may fall very unevenly between society and an individual. On the one hand, biomarkers of susceptibility to an environmental agent can be used to set limits to concentrations of toxic compounds in the environment to protect the whole population, including the most susceptible groups. On the other hand, the same markers can be used to detect sensitive individuals to be discriminated against, e.g. in the insurance business or job market (Barash 2000; Clayton 2003).

The evidence that biomarkers for disease susceptibility are useful in many applications, such as in the field of occupational health (e.g. Soskolne 1997), in drug development (e.g. Roses *et al.* 2007) and in studies pursuing cancer chemoprevention, (e.g. Lieberman 2001), stresses the importance of continuing the research. Also, from the ethical point of view, such research needs to be supported for the benefit of people (Dumez *et al.* 2007). A good biomarker has to be 'valid, reliable and practical' (Schulte 1992). The development of such biomarkers for disease susceptibility is complicated in most diseases by the very complex combined genetic and environmental aetiology, including gene–gene and gene–environment interactions, e.g. in neurodegenerative disorders (DeKosky and Marek 2003; Dyment *et al.* 2004), psychiatric disorders (Bunney *et al.* 2003), cancer (Cavalieri *et al.* 2006; Kyrtopoulos 2006; Stefanovic 2006) and immune system disorders (Akahoshi *et al.* 2006; Martinez 2007).

Predictive testing has different implications, depending on the nature of the condition, the individual to be tested, the developmental stage of the test and whether prevention or treatment possibilities for the disease exist (Table 22.1). If the condition is not preventable or treatable, the only result of the analysis of a susceptibility biomarker may in the end be harm to the tested individual. In addition to practical consequences, there may be psychosocial consequences in the form of anxiety and worry, even depression when people have to face susceptibility to a potentially life-threatening illness (Wexler 1995; Dorval *et al.* 2000; van Oostrom *et al.* 2003; van Korlaar *et al.* 2005). These potential harms to individuals set high requirements for the relevance of the marker and the reliability of the method. The negative consequences may be relevant for family and children if the marker and the condition are heritable. Especially in the case of

Molecular Epidemiology of Chronic Diseases, Edited by C. P. Wild, P. Vineis, and S. Garte
© 2008 John Wiley & Sons, Ltd

Table 22.1. Aspects of biomarkers with different implications on the ethics within studies on biomarkers for disease susceptibility

Aspect	Differences to consider	Comments, examples of ethical aspects	Example of a marker (proposed or in use)	Reference for the example
1. Nature of condition	Time of outbreak	Timing of analysis, disclosure	ApoE and late-onset Alzheimer's disease	Lahiri et al. 2004
	Consequences	Possibility to prevent or treat	Protein C deficiency and susceptibility to venous thrombosis	Van Korlaar et al. 2005
	Heritability	Quality of tests, implications for family	BRCA mutations and breast cancer	Walsh et al. 2006
2. Nature of the marker	Epigenetic (heritable, influenced by environment)	Developmental stage of the field, disclosure	DNA methylation and cancer susceptibility	Davis and Uthus 2004
	Genetic (heritable, unchangeable)	Practicality (high vs. low penetrance), implications for family, privacy	GSTM1 null genotype and lung cancer	Vineis et al. 2005
			Long QT syndrome in children and sudden death	Hendriks et al. 2005
	Phenotypic (influenced by environment)	Validation, sensitivity of health information	LDL and ischaemic heart disease	Benn et al. 2007
3. Stage of development	Possible role in susceptibility	Usefulness, publication bias	Candidate genes in asthma	Martinez 2007
	Well-established association	Disclosure, counselling	Dopamine system genes and ADHD	Li et al. 2006
	Established marker	Disclosure, counselling, psychosocial implications	BRCA mutations and breast cancer	van Oostrom et al. 2003
4. Strength of association	Poor	Usefulness, disclosure	Growth retardation and type II diabetes	Holness and Sugden 2006
	Good	Disclosure, counselling	BRCA mutation and breast cancer	van Oostrom et al. 2003
	Very strong	Disclosure, counselling	>36 CAG repeats in short arm of chromosome 4 and Huntington's disease	Braude et al. 1998
5. Intended use	Prenatal screening	Abortion-related issues, conflicts of interest, autonomy	Sonographic screening for Down's syndrome	Vassy 2006
	Preimplantation screening	Disclosure, security of data	Huntington's disease	Braude et al. 1998
	Occupational screening	Purpose, usefulness of screening, distribution of result	Oxidative stress and DNA-repair gene polymorphisms	Vineis et al. 2005
	Clinical	Validation	UDP-glucuronyl-transferase 1A1*27 and irinotecan toxicity	Ando et al. 2000

children, it is important to remember that the genetic susceptibility remains over a lifetime (Merlo *et al.* 2007). Thus, careful planning is required before the studies are conducted to consider whether, when and how information regarding the results are given to people.

For the development of biomarkers, good epidemiological study design is needed. One of the most critical issues is the size of the population to be studied (Taioli 2005). To achieve enough statistical power, there is a demand for large studies. This has led to the need to set up biobanks of samples with the possibility to link research data with individual health and life style data (Austin *et al.* 2003; Cambon-Thomsen 2004; Godard *et al.* 2003). The logistics of a large biobank have not proved easy (see e.g. Jamrozik *et al.* 2005). However, since the need, possibility and economic prospects exist, the question is not whether they will be set up, but how to overcome the practical and ethical difficulties. In large biobank projects the ethical aspects involve issues related to both the establishment and management of the biobank (Godard *et al.* 2003) as well as to new powerful technologies providing new possibilities to gain and link data (Grody 2003a).

The Declaration of Helsinki (www.wma.net) is one of the most influential documents in biomedical research ethics. Its principles (Table 22.2) continue to be widely accepted and have been adopted in numerous guidelines and laws. To work in the spirit

of this document requires more than a minimum application of the wording of the document. It requires personal involvement and the development of research ethics simultaneously with the development of scientific research (Vähäkangas 2004). Discussion is on-going about the balance between rights and responsibilities of individuals and societies in environmental health studies, including studies on biomarkers for disease susceptibility. Hopefully it leads to practices enabling good and beneficial research as well as autonomy and respect for the studied individuals (Dumez *et al.* 2007).

22.2. ETHICAL ASPECTS IN BIOMARKER DEVELOPMENT FOR DISEASE SUSCEPTIBILITY

Variation of biomarkers

The wide variation in the susceptibility biomarkers in use, under development or being considered and studied is evident from the literature (Table 22.1). Consequently, ethical implications also vary in many respects. The nature of the condition has naturally a major effect on the ethical implications. It is psychologically totally different to find out about susceptibility to a treatable condition (e.g. cardiovascular disease; Benn *et al.* 2007), probably not shortening the lifespan if proper measures are taken, than to find out about a susceptibility to a non-treatable condition that inevitably leads to premature death (e.g. Huntington's disease; Wexler 1995). The feasibility of the development of a genetic biomarker for disease susceptibility has to be considered. According to Grody (2003b), the minimum disease criteria are that the disease is relatively common and serious, there is a manageable number of predominant mutations, high penetrance, defined and consistent natural history and effective interventions. As to the analysis, it should be relatively inexpensive, acceptable to the population, and pre- and post-analysis education should be organized.

With the advancing understanding of cell and molecular biology, ideas for susceptibility biomarkers are increasing. Some fields of research (e.g. epigenetics; Jones 2005), and their possibilities in providing biomarkers are still controversial.

Table 22.2 Basic principles of the Declaration of Helsinki (reprinted from Vähäkangas (2001), with permission from Elsevier)

Item	Principle
Individual	Always comes before research and society
Informed consent	Has to be gained from the research subject after he/she is fully aware of what is going to happen and agrees to it
Quality of research	Is important: scientifically questionable research cannot be ethically good
Ethics review board	Has to approve the research before it is commenced; ethics committees have to be independent

The Human Genome Project (Collins *et al.* 2003) and modern technologies for gene mining have enabled large-scale studies on genetic susceptibilities for diseases, with the possibility to develop genetic biomarkers of susceptibility (Grody 2003a; Cambon-Thomsen 2003). It is now possible to test for susceptibility for many single gene-associated diseases but studies of gene variants for complex diseases have been disappointing (Foster and Sharp 2005; Smith *et al.* 2005; Vineis *et al.* 2005). The reason is the complexity of gene–gene and gene–environment interactions (Foster and Sharp 2005; Vineis *et al.* 2005). High-throughput analysis of quantitative trait loci shows some promise in providing a solution (Grody 2003b; see also Chapter 11). In the cancer field, a genotypic pattern combining polymorphisms and mutations of carcinogen activating and inactivating genes with those of oncogenes, tumour suppressor genes and inflammation-related genes may reveal individual susceptibilities to at least some cancer types (for reviews, see e.g. Hussain *et al.* 2003; Pelkonen *et al.* 2003).

At the early stage of development, unrealistic expectations on the usefulness of a susceptibility biomarker may exist. The condition targeted may be just one symptom of a disease and the developed test not useful for the original purpose. An example of this was an attempt to test for a mutation putatively associated with a susceptibility to hereditary neuropathy in workers who had developed carpal tunnel syndrome at work (see Clayton 2003). However, carpal tunnel syndrome may or may not be a symptom of the very rare hereditary neuropathy. Publication bias may exist in the literature due to the small sample size of published studies, as happened in the association of immunoglobulin binding receptor FcγRIII polymorphism with susceptibility to systemic lupus erythematosus (Karassa *et al.* 2003). There may also be ethnic differences, so that the marker is not generally applicable, e.g. cytotoxic T lymphocyte-associated antigen 4 polymorphism predicting susceptibility to rheumatoid arthritis in Asians but not Europeans (Han *et al.* 2005). Furthermore, conflicts of interest may inhibit publication of results and create a biased impression of the usefulness of a susceptibility biomarker (Healy 2007).

Ethical aspects of genetic biomarkers for susceptibility

Much hope has been vested in the development of genetic biomarkers, but for environmental and occupational field the research so far has not led to any routinely usable biomarkers. Many known genetic traits, such as polymorphisms of drug metabolism, are individually only weakly associated with disease (Vineis and Christiani 2004) and many probably remain unknown, due to the requirement for environmental factors (Foster and Sharp 2005). Especially if the risk of disease is associated more with exposure than with the genotype, limiting exposure is the only feasible approach to prevention and benefits all (Smith et al. 2005). Furthermore, both regulation of gene expression and all the events at the protein level can be modified by the environment, and it has been shown in twin studies that the epigenetic differences between monozygotic twins increase by age, putatively at least partly due to these environmental influences (Fraga *et al.* 2005; Petronis 2006).

Many population-based genetic studies for the development of susceptibility biomarkers are carried out on healthy individuals, with potential benefits in the future for people with the same condition but the possibility of harm for the people in the study. At best, when fully developed, genetic biomarkers for susceptibility may protect individuals by warning about health risks and helping to select the correct prevention strategy. Defining genetic susceptibility to diseases induced by environmental chemicals may also aid in decisions about the level of acceptable daily intake (ADI) value for a chemical compound to protect even the most sensitive individuals in a population. However, the same information can be used to discriminate against an individual by employers (Soskolne 1997; Berry 2003; Clayton 2003). Genetic discrimination on the basis of a gene test or knowledge of a genetic susceptibility through family history has already occurred (Barash 2000; Council for Responsible Genetics: www.genewatch.org). Furthermore, genetic information as the 'vehicle of biological inheritance' (Suzuki and Knudtson 1990) never involves only one individual, but is the common property of a family. A positive predictive test for a genetic aberration, possibly predicting a risk for a genetically inherited disease in future generations, may have a great impact on the family by causing stigmatization and strain in relationships within the family (Barash 2000; Hakimian 2000).

DNA data at this point in time may be very difficult to handle, due to their complexity, but this will probably change with the advancement of bioinformatics. For instance, *BRCA1* and *BRCA2* mutations, which confer a 50–80% lifetime risk for breast cancer, have different consequences, depending on the mutation. While the risk varies with the particular mutation, the reliability of *BRCA* screening depends also on the probability of the condition, the screening technique and the number of mutations screened (Surbone 2001). During genetic analysis, whether within research or routine screening, there is also always a possibility of incidental related findings (Khoury *et al.* 2003). Illes and co-workers (2002), in connection with their NMR-imaging studies of brain involving normal volunteers, have proposed a database to obtain more insight into the effects of such findings. Collection of data on circumstances of detection of false positives in analysis of susceptibility biomarkers would be important as well. According to the proposal by Illes and co-workers (2002) such a database should include information on how the finding was handled, to whom and how it was reported and followed-up, and what the outcome was for the individual. With children, the follow-up would be long enough to get meaningful information of lifetime consequences of incidental findings related to biomarkers.

Quality of research

The relevance and reliability of the data in genetic studies is central to ethical issues, due to the possible persistent nature of genetic information. Such information is also more far-reaching in society than many other types of health information, because close relatives may have the same trait. Similar scrutiny balancing usefulness and harm, and possible registration, as is required in the development of clinical drugs would be a good idea also in the case of genetic biomarkers.

Development of a biomarker from the idea to a usable marker is a long scientific process that proceeds through validation of the method to testing the final value of the marker; a process that truly is 'a bridge over troubled waters'. Important practical considerations are the relevance of the marker used, the specificity and sensitivity of the assay, and the prevalence of the risk factor in the population, which all affect the proportion of true positives (Rawborne 1999; Vineis and Christiani 2004). A careful planning stage of a study requires considerations of ethical aspects including balancing of benefits against possible harm caused by testing (Vineis 2003; Vineis and Christiani 2004).

When the laboratory methods to analyse a biomarker are at a developmental stage, they cannot be reliably used in testing for individual susceptibility to diseases (Koh and Jeyaratnam 1998; Koh *et al.* 1999; Hainaut and Vahakangas 1999; Shields and Harris 2000). Sensitivity and specificity of methods are never 100% (Makni *et al.* 2000; Vineis and Christiani 2004). Publication bias in the scientific literature may lead to wrong conclusions about the consistency of an association between a disease and a biomarker. Pooled data with re-analysis from published and unpublished sources (Taioli and Bonassi 2002) and the increasingly accepted policy to publish well-done studies with negative results (e.g. van der Hel 2003) will hopefully resolve some of these problems.

Many of the difficulties in the development of genetic biomarkers are illustrated by the case of chronic beryllium-induced lung disease. Genetic susceptibility for this condition is linked to human leukocyte antigen (HLA) polymorphism, and this knowledge has been seriously pursued as a basis for the development of a potentially useful biomarker (McCanlies *et al.* 2003). While a specific allele, HLA-DPB1Glu69, has been found to be associated with beryllium disease, the predictive value appeared to be too low for application in an occupational setting. Furthermore, some of the more feasible methods for the analysis of the polymorphism do not detect the at risk allele with a good specificity. The difficulties in interpretation of data as well as the associated ethical problems led to the conclusion that the biomarker is unusable (McCanlies *et al.* 2003; Silver and Sharp 2006).

22.3. ETHICAL ASPECTS OF BIOBANKING

Management of biobanks

In molecular epidemiology for disease susceptibility, the most useful study design would include a possibility to link genetic, genealogical and public health databases. Large, population-based gene or tissue banks to be utilized in such studies are under development (Austin *et al.* 2003; Arnason 2004; Cambon-Thomsen 2004; Smith *et al.* 2005; Maschke 2005). Foster and Sharp (2005) highlight local variations which may have major effects on health, and propose a community-based approach. Scientific, societal and individual benefits and costs (including ethical costs) have to be balanced in biobank management and studies (Table 22.3; Anderlik and Rothstein 2001; WHO 2002; Arnason 2004; Vähäkangas 2004; Andrews 2005; Maschke 2006). Several processes on-going for international harmonization of biobanks to allow powerful analysis on modest associations (Smith *et al.* 2005; Uranga *et al.* 2005) face challenges with respect to,

Table 22.3 Ethical aspects in use of biobanks for genetic studies[*]

Management of the biobank, benefit sharing, solidarity
   Private or public ownership
   Private or public management
   Logistics, e.g. quality of samples, health safety
   standards, security
   Who decides and how
Consent practice, voluntary, autonomy
   Who and how are recruited
   Opt-in or opt-out policy
   Is consent going to be asked
   Specific or open consent
Storage of samples and data, privacy, confidentiality
   Storage place and time
   Labelling of samples
   Access to samples and data
Use of samples and data, possible harm
   Combining of datasets and registries
   Aim of use (research, commercial)
   Quality of studies and methods
   Disclosure of data to volunteers

[*]Vineis 2003; Godard *et al.* 2003; Vähäkangas 2004; Arnason 2004; Clayton 2005; Hansson 2005; for the opinion of European Group on Ethics published in 1998, see http://ec.europa.eu/european_group_ethics/

for example, consent issues, privacy and intellectual property rights (Maschke 2005).

The idea of a biobank includes systematic collection of identifiable DNA data, DNA samples or tissue samples from which DNA can be isolated, and the possibility to connect DNA data with individual data about health and life style (Arnason 2004; Cambon-Thomsen 2004; Petersen 2005; Smith *et al.* 2005). This requires extensive use of information technology, which may endanger privacy (Clayton 2005). Legal and regulatory aspects of long-term control of DNA samples and information are not simple (Knoppers *et al.* 1998; Berry 2003), therefore it is no wonder that no international consensus exists on how to manage biobank efforts in an ethically defensible way (Petersen 2005; Kaye 2006). Even at a national level, most countries have no guidelines or laws on biobanking. National and international organizations have been established to help in dealing with ethical issues (Table 22.4; Smith *et al.* 2005). For instance, in Quebec the IPEG (Institute of Populations, Ethics and Governance) was established to help research ethics committees in the increasingly complex issue of biobanking human tissue.

Auray-Blais and Patenaude (2006) propose an independent medical archivist to manage coded samples from research projects, to protect the data and to supervise a biobank. Licensing of biobanks has been suggested, to ensure good practice in the increasing number of commercial biobanks (Kaiser 2002; Anderlik 2003). An international organization, in the form of a charitable trust as a trustee and fundholder of a biobank has been proposed as an alternative to national initiatives or commercial biobanks (Winickoff and Winickoff 2003). The European Group on Ethics (EGE; http://ec.europa.eu/european_group_ethics/) in its opinion no. 11 on Ethical Aspects on Human Tissue Banking (21 July 1998) just states the difference on opinions about commercialization of tissue banking with justifications, but does not take a stand for or against. Time will show, hopefully after a wide international discussion, which is acceptable to the public(s), and whether different countries can accept similar regulations within Europe or globally (Petersen 2005 and references therein; Maschke 2005). In any case, those who manage biobanks and other

Table 22.4 Some useful sites on medical bioethics including molecular epidemiology and genetic studies

Organization	Website location	Content includes
ASA	http://www.amstat.org	Privacy, confidentiality, and data security
American Society of Human Genetics	www.faseb.org/genetics	DNA banking and analysis
CIOMS	www.cioms.ch/guidelines	International ethical guidelines for biomedical research involving human subjects
Ethical, Legal and Social Implication (ELSI) Research programme of Human Genome project	www.ornl.gov/hgmis/elsi/ elsi.html	Various aspects of genetic information and ethics
GeneWatch UK	www.genewatch.org	Human genetics
Human Genome Project	www.ornl.gov/TechResources/ Human_Genome/	NCHGR-DOE Guidance on Human Subjects Issues in Large-scale DNA Sequencing
IPEG, Quebec	www.cartagene.qc.ca/ partenaires.cfm	Statement of principles of human genome research, statement of principles on the ethical conduct of human research involving populations
NBAC, USA	www.bioethics.gov	Ethical and policy issues in research involving human participants
Pharmacogenetics Research Network	www.nigms.nih.gov/ pharmacogenetics/	Information of the status of art in susceptibility biomarkers for drug toxicity
Nordic Committee on Bioethics	http://www.ncbio.org/news.htm	Legislation on biotechnology in Nordic countries; useful articles on various ethical issues in biotechnology
Nuffield Council on Bioethics	www.nuffieldbioethics.org	Ethics of patenting DNA, ethical issues of genetic screening and pharmacogenetics
UNESCO	www.unesdoc.unesco.org	The Universal Declaration on the Human Genome and Human Rights – from theory to practice
US National Institute of Health (NIH) Office of Biotechnology Activities (OBA)	www.od.nih.gov/oba/	US SACGHS (Secretary's Advisory Committee on Genetics, Health and Society, Health and Human Services): Policy issues associated with undertaking a large US population cohort project on genes, environment and disease. Public Comment Draft, May 2006.
NIH Office of Human Subject Research	http://ohsr.od.nih.gov/guidelines/	Regulations and ethical guidelines, e.g. Nuremberg code, Belmont Report, NIH guidelines
WMA	www.wma.net	Declaration of Helsinki

ASA, American Statistical Association; CIOMS, Council for International Organizations of Medical Sciences; IPEG, Institute of Populations, Ethics and Governance; NCHGR, National Centre for Human Genome Research; DOE, Department of Energy; WMA, World Medical Association

valuable resources in science should not be the ones who also financially benefit from them, to keep the balance of cost and benefit (both financial and ethical) among lay people.

Public consultation about biobanks is on-going in some countries. In the USA, a plan for a large-scale population cohort project to study gene–environment interactions in disease susceptibility has been recently published for public consultation (www.od.nih.gov/oba/sacghs/public–comments.htm). In the UK, a preliminary study with 3800 participants was carried out in 2006 to probe for public acceptance of a biobank and to test the practical procedures. The study obtained favourable backing from international experts and funders of the biobank, and recruitment for the biobank was started at the end of 2006 (www.ukbiobank.ac.uk). The UK Biobank also has an independent Ethics and Governance Council (www.egcukbiobank.org. uk). Public discussion has appeared necessary and has in the past helped in making decisions about controversial ethical issues (see Sade 2003), while a restricted discourse may endanger trust in science (Petersen 2005). Open communication should thus be encouraged in all countries before population-based large biobanks are established.

Consent practice

One of the most difficult questions is that about consent, part of the difficulty being the application of principles from a medical context to population studies (Petersen 2005). The concept of informed consent is based on the Nuremberg Convention and is central to the Declaration of Helsinki, originally formulated in 1964 in the meeting of the World Medical Association in Helsinki (www.wma.net). Its principles appear in numerous international and national documents, e.g. Belmont Report and Council for International Organizations of Medical Sciences (CIOMS) guidelines (see Table 22.4) and have been adopted in many countries in the legislation for research on human subjects. Thus, since the first documents in the 1930s there has been no disagreement about the importance of informed consent in biomedical research on human subjects (Sade 2003). The potential, but from practical (e.g. Arnason 2004) and philosophical (e.g. Kristinsson 2007) points of view arguable, importance of informed consent include social and individual benefits as well as respect for autonomy. As Clayton (2005) points out, in practice the regulations have not been easy to apply.

An informed consent contains two important principles: understanding and autonomy. First, 'informed' means a requirement for a thorough explanation of the background and purpose of the study, methods and procedures, potential benefits and harms (physical, mental and social) as well as rights of the study participants (voluntary, confidentiality and a possibility to withdraw) (Arnason 2004; Vähäkangas 2004). For understanding, this information should be given in clear and under-

standable language fit for the age and education of the target group; this is not an easy task, as shown by Paasche-Orlow and co-workers (2003). To ensure this understanding, the Declaration of Helsinki recommends participation of lay people in the work of ethics committees. Second, autonomy of a person requires a voluntary consent for a specific study. Many authors regard the right of people to decide what their tissue and DNA is used for as an important one (Grady 1999; Andrews 2005; Trouet 2004). Opinions of the importance of a study will differ among lay people. Especially in the case of unsure benefits for people, they should be able to refrain from participation. Furthermore, the CIOMS guidelines (see www.cioms.ch/guide-lines) recommend not only obtaining informed consent, but also periodically renewing it in long-term studies. Contegiacomo and co-workers (2004) launched a concept of 'aware consent' for clinical counselling where information giving is a dynamic process over time, ensuring real understanding. According to the CIOMS guidelines, waiver of the informed consent should come from an ethics committee, and this should be an exception. Also, respectable scientific journals support these guidelines and the requirement for an informed consent in their publication policies (see Sade 2003).

These principles originally formulated for more clinical types of medical research have lately been challenged in population studies, where two possible ways of sample collection have different implications. In a prospective collection, people donating samples are in direct contact with the studies and it is possible to obtain informed consent, while in studies using archived samples, it is impossible to get consent from the donor if he/she has died. Thus, it is understandable that the need for and the form of the consent is being discussed (Vähäkangas 2004; Arnason 2004; Hansson *et al.* 2006; Maschke 2006) and the validity of a specific informed consent in population-based studies has been questioned. Studies targeting polymorphisms associated with a low relative risk pose no danger of discrimination (Hunter and Caporaso 1997; Beskow *et al.* 2001). The impracticality of a specific informed consent in long-term studies is evident (Arnason 2004; Hansson *et al.* 2006; Hansson

2005). Hoeyer and co-workers (2005) found that only a small number of sample donors regarded information about biobank studies as important or were unsatisfied with the information they had been given. Kristinsson (2007) questions the association between autonomy and informed consent on philosophical grounds.

Several alternatives to the traditional specific informed consent are being discussed (Winickoff and Winickoff 2003; Vähäkangas 2004; Maschke 2006). Blanket consent or open consent implies permission to any type of future research. Hansson and co-workers (2006) reject this idea and propose instead a broad consent and a consent for future studies, provided that information is coded, handled safely and harms are minimized. This is close to the proposal by Arnason (2004) of a written authorization for biobank research, based on general knowledge of the principles and practices of the biobank. Public consultation or community consent to ensure the public acceptability of a biobank in the society at large (Kaye 2001; Levitt 2003) may be needed in any case, since some people do not care enough to read the written information (Hoeyer et al. 2005).

When identifiable samples and data are used, Helgesson and Johnsson (2005) stress that, for public trust in research and to ensure participation, it is important to retain the right for study participants to withdraw their consent. People may want to refuse the use of their tissue and samples for certain purposes, e.g. of fear of stigmatization of a group (Trouet 2004). The right to withdraw is opposed by Edwards (2005) and Eriksson and Helgesson (2005), who argue that participants should not have unconditional right to withdraw, due to a general responsibility for health research. Ellis and co-workers (2003) also express concern that present proposals for regulations in the UK risk future important studies by too much sensitivity. After anonymization of samples and data, withdrawal from a study is obviously impossible. New European guidelines (Trouet 2004) and regulations in Canada and USA (Hansson *et al.* 2006) allow free use of anonymized human biological materials for biomedical research. This has been challenged from the human rights perspective: people should be able to determine, or at least know what happens to their tissue, and determine what kind of research it is used for (Trouet 2004; Helgesson and Johnsson 2005; Maschke 2006).

Storage and distribution of samples and data

In biomedical research, the principle of autonomy (Beauchamp and Childress 2001) includes the right to privacy regarding sensitive health information (Fuller *et al.* 1999; Hainaut and Vahakangas 1999). In molecular epidemiology involving healthy volunteers, computerized databases, long-term storage and private ownership of samples may, however, endanger confidentiality. For biobanks health and life style information is required from individual patients and communities. Such data can be very sensitive not only to individuals and their relatives, but also to other members belonging to a certain group or community (Trouet 2004; Foster and Sharp 2005). While it is clear that advanced modern medicine badly needs individualized information to obtain more insight into the aetiology of the diseases, and to develop better individualized treatments and diagnostic means, it is also clear that other institutions in society are also interested in the information (Clayton 2003; BMA 2004). Such interested parties include the insurance industry, banks, criminal investigators, employees, spouses and other relatives.

The most important target group for information distribution are the study participants themselves. The decision of whether to know or not to know about susceptibility is a difficult one. The knowledge of a risk for a disease cannot be erased from the memory and this may have potentially serious psychological consequences (Wexler 1995; Barash 2000; Hakimian 2000; Dorwal *et al.* 2000). Andrews (2005) reminds that the samples are 'pieces of my body, your body, and all of our bodies'. It would be important to study how willing scientists themselves, with the best possible understanding of the implications, are to donate their tissues for scientific studies and biobanks. The few hints from the literature tell of variation, with some scientists wanting to 'die ignorant of their weaknesses' (Butcher 2007).

Populations, individuals and autonomy

There are several special issues when considering groups of participants. In big multinational projects on genetic biomarkers, different ethnic groups will be included. The frequency of genetic variants may be very different in different ethnic groups, making some populations more susceptible as a group (Solskone 1997; Olden and White 2005). At worst, this may lead to stigmatization or discrimination, e.g. in the job market of the whole ethnic group. A variable level of education may be an obstacle in some countries with regard to the principle of informed consent and public consultation.

Children form a special group among study participants because they are largely dependent on the decisions made by their parents. Depending on age, they have less capacity to handle difficult issues and thus less autonomy (Grodin and Glantz 1994). Formulation of information understandable for different age groups has also been challenging (Paasche-Orlow *et al.* 2003). The special issues regarding children in biomedical research include at least the following aspects (Neri *et al.* 2006; Gilbert 2005; Merlo *et al.* 2007): (a) children and unborn children are more vulnerable physically and mentally than adults; (b) children have less capacity to understand complex issues and long-term consequences; (c) children have less knowledge of the world, usually trust adults and may be easily persuaded to change their opinions; (d) small and unborn children do not have autonomy; and (e) children have to live in the future we create. Especially, genetic studies may affect the whole life-time of children. The use of identifiable genetic data can be permanently attached to the child and have long-term consequences years after the data were created (WHO 2002; Clayton 2003; Vähäkangas 2001, 2004). Even when consent to a study is voluntarily given by a parent and the child, it should always be considered just a temporary one until the child reaches adulthood.

Interestingly, gender is probably also an important variable to take into account when considering ethics of recruiting participants. Females seem to be more prone to participate in scientific studies, and for this reason may also be more vulnerable to risks of the studies (see Arar *et al.* 2005).

22.4. MOLECULAR EPIDEMIOLOGY AND SOCIETY

Science, money and public trust

Because of recent widely publicized scientific fraud cases (Neill 2006; Horton 2006), and catastrophes in drug development and clinical trials (Halpern 2005), the public confidence in biomedical science, scientists and the ability of governments to regulate complex biomedical and ethical issues is declining (Petersen 2005). Unfortunate cases of a 'black market' of human samples to the drug industry, with undisclosed conflict of interest, have also surfaced (Lenzer 2006). In all of these cases twisted career or financial interests have played a major role. Governments and universities put great pressure on scientists to translate scientific results and discoveries into marketable products and have interest in being part of commercial efforts making it possible. Scientists have the responsibility to regain trust by showing more enthusiasm for science than for money, and personal responsibility for good ethics in science (Vähäkangas 2004; Nylenna and Simonsen 2006). Because sponsorship of research by industry affects conclusions, publication and data sharing (Bekelman *et al.* 2003), it is currently self-evident that any financial conflicts of interest should be disclosed by scientists. Study participants have also expressed their interest in such ties (Weinfurt *et al.* 2006; Kim *et al.* 2004).

Over the past 10 years, new high-throughput techniques (genomics, proteomics, metabonomics) have revolutionized the speed of new information but also created new ethical problems (Grody 2003a; Cambon-Thomsen 2003; Knudsen 2005). For genomics, biobanks providing large numbers of samples are essential (Cambon-Thomsen 2003). So far, '-omics' and bioinformatics techniques have largely been developed by companies which, no doubt, are looking forward to applications which bring money back to the company. This may end up with premature marketing with unjustified claims of benefit, eroding trust not only in the commercial companies but also on biomarker research (Vineis and Chistiani 2004). Another important ethical aspect is the fact that in a society resources are limited and large investments in one sector usually mean lack of resources somewhere else. Weaker regulation of privacy and the effect of a possibly wrong result of a genetic test are among the recognized ethical problems in marketing tests based on genetic biomarkers (Vineis and Christiani 2004).

Not only the developed methods but also the human samples and tissues themselves, or cell-lines derived from them, have significant monetary value, which is already being realized by hospitals, institutions and companies (Andrews 2005). The fact that such trade occurs on some occasions without the consent of the donor does not increase public trust (Lenzer 2006). A lot of samples exist in pathological archives and clinical laboratories. How these can be used and combined with health and life style information, and for which purpose, are important questions (Marshall 1998; Pelias and Markward 2001). Information relating to mental illness or (mis)use of alcohol, for example, may be very sensitive and putatively prone to uses against the benefit of a person. No consensus exists on under what terms it should be possible, for example, to sell sample sets with attached health and life style information to private industry. Andrews (2005) also points out the possible unwillingness of scientists to collaborate and consequent hindrance of research by wanting to patent a gene associated to a disease. The Nuffield Council of Bioethics (2002) has warned about such difficulties and recommends that it should not be possible to patent genes, only the applications based on genetic knowledge. This is based on the argument that genes are the common property of mankind which cannot be invented, only found, and only inventions should be patentable.

Communication

Communication issues within as well as outside the scientific community are critical for successful biomarker studies and the application of knowledge (Beskow et al. 2001; Barbour 2003; Petersen 2005). With research participants, issues of communication arise at different stages of research, both when recruiting people into studies, while the study is on-going (especially in long-term prospective studies) and after the results have been gained. What to communicate to research participants is debated in the literature. In most recommendations the following points are regarded as important (e.g. CIOMS international ethical guidelines for biomedical research involving human subjects: www.cioms.ch/guidelines): the purpose of the study, the samples and information collected, the storage and use of samples and information, and voluntary participation.

Lately, with increasing commercial interest in biomedical research, including biomarkers for disease susceptibility, conflict of interest issues have gained more importance. Bekelman and co-workers (2003), in their review of eight articles including altogether 1140 original studies, showed a statistically significant association of financial ties with conclusions, and restrictions on publication and data sharing. Recent large studies on the views of various groups of potential research participants imply that people want to know about conflicts of interest (Weinfurt et al. 2006; Kim et al. 2004).

Informing people in an understandable way is a challenge. According to the polls, the Icelandic population supports, but does not understand very well, the deCODE project, banking DNA and combining the DNA data with genealogical and health data (Arnason 2004). Involvement of scientists in public consultation will help if they present their arguments for and against in the media in clear language. Confusing terms should be avoided altogether (Caulfield and Brownsword 2006). Language may be a major obstacle even for professionals in cases where the terminology has not been harmonized. One example is the labelling of samples, the terminology of which continues to confuse research participants as well as other practical work, for instance in ethics committees (Vähäkangas et al. 2004; Table 22.5).

Whether, to whom and how to distribute knowledge of the potential detected risk has to be resolved in biomarker studies on a case-by-case basis, with a special emphasis on the effects for sensitive groups (children, ethnic and other minorities). In any case, development of a strategy to deal with positive test results is needed (Schulte 1992). Long-term psychological stress after genetic testing for cancer susceptibility is related, in addition to the seriousness of the condition, to the communication with the family as well

Table 22.5   Definitions for the terminology in the identification of tissue and DNA samples: a proposal to simplify terminology for better communication with research participants and within ethics committees (Vähäkangas 2004)

Proposed term	Practical synonyms*	Possibility to identify person	Explanation
Identified		Possible	Name and/or social security number appear on the sample
Coded	Identifiable Linked Single-coded	Possible	The sample does not have identification, only a code, but the person is identifiable with relative ease
Encrypted	Double-coded	Possible	The sample does not have identification, only a code, and it requires extra effort to identify the person. However, identification is still possible
Anonymous	Anonymized Unidentified Unlinked	Not possible	The sample lacks identifiers or codes; no possibility to link data from the sample to a person

*Terms with equal meaning from the perspective of research participants (Godard *et al.* 2003; Knudsen 2005).

as during recruitment, disclosure and counselling (van Oostrom *et al.* 2003; Contegiacomo *et al.* 2004; van Korlaar *et al.* 2005).

Education in ethics

Scientists themselves recognize the need for training in ethics as part of scientific education (e.g. Rossignol and Goodmonson 1995). They have to be able to reflect on the existing laws and literature, to write about ethics based on their considerations, for instance in applications for grants and ethics committees, and to make decisions that last beyond the current trends in science. This is a major undertaking for a laboratory scientist. Furthermore, there is constant evolution, not only in science but also in the ethics of science (Pellegrino 1993; van Leeuwen and Bernat 2006).

It would be important to realize that appropriate knowledge of legislation and ethics is essential for the protection of both the public and the scientists (Bravo *et al.* 2003; Vähäkangas 2004; Arnason 2004). It would also enable scientists to take part in public discussion about ethics and to better understand the work of ethics committees. Rather than regarding the ethics committees as a nuisance, scientists should work together with the committee members, most of them their peers, for prevention of future difficulties. This is important also considering that the attitudes and practices of senior scientists are inherited by younger colleagues as a research tradition. Thus, the education

of professional ethics should be targeted both for doctoral students, staff scientists and group leaders alike. Since good science is inseparable from good scientific ethics, the teaching of ethics should be integrated in all teaching of scientific practise.

Multidisciplinary dialogue is necessary in bioethics (deVries *et al.* 2006). Pellegrino (1993) regards a continuing dialogue with moral philosophers as the prerequisite of the critical analysis of ethical decisions in medicine. A thorough public discussion about the implications of modern molecular epidemiology requires interaction between professionals, lay organizations and other fields of society (Vähäkangas 2004; van Leeuwen and Bernat 2006). The rights of study participants are ineffective if the population does not know about, or understand, their rights: it is probable that in many countries the 'population sorely stands in need of such education' (Arnason 2004). It is the task of scientists, the school system and the mass media to keep the level of basic knowledge in science high enough to enable people to collaborate in scientific studies and not just act as 'guinea pigs'.

22.5. CONCLUSIONS

Biomarkers for disease susceptibility have potential among other tools in improving public health. For real benefit, biomarkers and methods used to analyse them should be carefully validated for their purpose, in analogous fashion to new drugs. It might be worthwhile to have a registration process for their use in clinical and occupational medicine. The results gained using non-validated markers or methods during the development phase should never be revealed to people, due to putative unnecessary anxiety and harm. Genetic counselling by a specialized doctor or nurse should be included whenever genetic data are given to individuals.

The development of biomarkers would greatly benefit from large biobanks. However, logistic difficulties and related ethical considerations are being faced by initiators of the biobank projects. Open national and international discussion among scientists, as well as national public consultations, are needed for the public to retain its trust in science and scientists. Financial gain should not be allowed to corrupt the process. Retaining public ownership of biobanks would probably best assure the open function and benefit for society as well as for individuals.

The concept of informed consent has not lost its importance. According to the Declaration of Helsinki, informed consent is important to ensure autonomy. Children should be able to give their own consent and reconsider participation in molecular epidemiology studies or biobank projects when they have reached legal age of adulthood. As to old samples, in most cases it is possible to consult the owners or their relatives. The fact that this may require time and effort should not overrule its importance. Independent ethics committees could be involved in cases where an individual informed consent is absolutely impossible. In any case, public consultation, followed by a community consent should be a prerequisite for nationwide biobank projects.

Acknowledgements

Stimulating discussions with Professor Emeritus Esko Länsimies and Professor Emeritus Reijo E. Heinonen are gratefully acknowledged. I am also grateful for the important constructive criticism from Dr. Paolo Vineis and from an anonymous referee.

References

Akahoshi M, Nakashima H, Shirakawa T. 2006. Roles of genetic variations in signalling/immunoregulatory molecules in susceptibility to systemic lupus erythematosus. *Semin Immunol* **18**: 224–229.

Anderlik M. 2003. Commercial biobanks and genetic research: ethical and legal issues. *Am J Pharmacogenom* **3**: 203–215.

Anderlik MR, Rothstein MA. 2001. Privacy and confidentiality of genetic information: what rules for the new science? *Ann Rev Genom Hum Gene* **2**: 401–433.

Ando Y, Saka H, Ando M *et al.* 2000. Polymorphisms of UDP-glucuronosyltransferase gene and irinotecan toxicity: a pharmacogenetic analysis. *Cancer Res* **60**: 6921–6926.

Andrews LB. 2005. Harnessing the benefits of biobanks. *J Law Med Ethics* **33**: 22–30.

Arar NH, Hazuda H, Steinbach R *et al.* 2005. Ethical issues associated with conducting genetic family studies of complex disease. *Ann Epidemiol* **15**: 712–719.

Arnason V. 2004. Coding and consent: moral challenges of the database project in Iceland. *Bioethics* **18**: 27–49.

Auray-Blais C, Patenaude J. 2006. A biobank management model applicable to biomedical research. *BMC Med Ethics* **7**: E4: www.biomedcentral.com/1472–6939/7/4

Austin MA, Harding S, McElroy C. 2003. Genebanks: a comparison of eight proposed international genetic databases. *Commun Genet* **6**: 37–45.

Barash CI. 2000. Genetic discrimination and screening for hemochromatosis: then and now. *Genet Testing* **4**: 213–218.

Barbour V. 2003. UK biobank: a project in search of a protocol. *Lancet* **361**: 1734–1738.

Beauchamp TL, Childress JF. 2001. *Principles of Biomedical Ethics*, 5th edn. Oxford University Press: New York.

Bekelman JE, Li Y, Gross CP. 2003. Scope and impact of financial conflicts of interest in biomedical research: a systematic review. *J Am Med Assoc* **289**: 454–465.

Benn M, Nordestgaard BG, Jensen GB *et al.* 2007. Improving prediction of ischemic cardiovascular disease in the general population using apolipoprotein B: the Copenhagen City Heart Study. *Arterioscler Thromb Vasc Biol* **27**: 661–670.

Berry RM. 2003. Genetic information and research: emerging legal issues. *HEC Forum* **15**: 70–99.

Beskow LM, Burke W, Merz JF *et al.* 2001. Informed consent for population-based research involving genetics. *J Am Med Assoc* **286**: 2315–2321.

Braude PR, De Wert GM, Evers-Kiebooms G *et al.* 1988. Non-disclosure preimplantation genetic diagnosis for Huntington's disease: practical and ethical dilemmas. *Prenat Diagn* **18**: 1422–1426.

Bravo G, Paquet M, Dubois M-F. 2003. Knowledge of the legislation governing proxy consent to treatment and research. *J Med Ethics* **29**: 44–50.

BMA (British Medical Association, Ethics Department). 2004. *Medical Ethics Today. The BMA's Handbook of Ethics and Law*. 2nd edn. BMJ Books: BMA House, London.

Bunney WE, Bunney BG, Vawter MP *et al.* 2003. Microarray technology: a review of new strategies to discover candidate vulnerability genes in psychiatric disorders. *Am J Psychiat* **160**: 657–666.

Butcher J. 2007. Kari Stefansson: a general of genetics. *Lancet* **369**: 267.

Cambon-Thomsen A. 2003. Biobanks for genomics and genomics for biobanks. *Comp Funct Genom* **4**: 628–634.

Cambon-Thomsen A. 2004. The social and ethical issues of post-genomic human biobanks. *Nat Rev Genet* **5**: 866–873.

Caulfield T, Brownsword R. 2006. Human dignity: a guide to policy making in the biotechnology era? *Nat Rev Genet* **7**: 72–76.

Cavalieri E, Chakravarti D, Guttenplan J *et al.* 2006. Catechol estrogen quinones as initiators of breast and other human cancers: implications for biomarkers of susceptibility and cancer prevention. *Biochim Biophys Acta* **1766**: 63–78.

Clayton EW. 2003. Ethical, legal and social implications of genomic medicine. *N Engl J Med* **349**: 562–569.

Clayton EW. 2005. Informed consent and biobanks. *J Law Med Ethics* **33**: 15–21.

Collins FS, Morgan M, Patrinos A. 2003. The Human Genome Project: lessons from large-scale biology. *Science* **300**: 286–290.

Contegiacomo A, Pensabene M, Capuano I *et al.* 2004. An oncologist-based model of cancer genetic counselling for hereditary breast and ovarian cancer. *Ann Oncol* **15**: 726–732.

Davis CD, Uthus EO. 2004. DNA methylation, cancer susceptibility, and nutrient interactions. *Exp Biol Med Maywood* **229**: 988–995.

DeKosky ST, Marek K. 2003. Looking backward to move forward: early detection of neurodegenerative disorders. *Science* **302**: 830–834.

de Vries R, Turner L, Orfali K *et al.* 2006. Social science and bioethics: the way forward. *Sociol Health Illn* **28**: 665–677.

Dorval M, Patenaude AF, Schneider KA *et al.* 2000. Anticipated vs. actual emotional reactions to disclosure of results of genetic tests for cancer susceptibility: findings from *p53* and *BRCA1* testing programs. *J Clin Oncol* **18**: 2135–2142.

Dumez B, Van Damme K, Casteleyn L. 2007. Research on the socio-ethical impact of biomarker use and the communication processes in ECNIS NoE and NewGeneris IP. *Int J Hyg Environ Health* 210: 263–265.

Dyment DA, Ebers GC, Sadovnick AD. 2004. Genetics of multiple sclerosis. *Lancet Neurol* **3**: 104–110.

Edwards SJ. 2005. Research participation and the right to withdraw. *Bioethics* **19**: 112–130.

Ellis I, Mannion G, Warren-Jones A. 2003. Retained human tissues: a molecular genetics goldmine or modern grave robbing? A legal approach to obtaining and using stored human samples. *Med Law* **22**: 357–372.

Eriksson S, Helgesson G. 2005. Potential harms, anonymization, and the right to withdraw consent to biobank research. *Eur J Hum Genet* **13**: 1071–1076.

Foster MW, Sharp RR. 2005. Will investments in large-scale prospective cohorts and biobanks limit our ability to discover weaker, less common genetic and environmental contributors to complex diseases? *Environ Health Perspect* **113**: 119–122.

Fraga MF, Ballestar E, Paz MF *et al.* 2005. Epigenetic differences arise during the lifetime of monozygotic twins. *Proc Natl Acad Sci USA* **102**: 10604–10609.

Fuller BP, Kahn MJ, Barr PA *et al.* 1999. Privacy in genetics research. *Science* **285**: 1359–1361.

Gilbert SG. 2005. Ethical, legal, and social issues: our children's future. *NeuroToxicology* **26**: 521–530.

Godard B, Schmidtke J, Cassiman JJ *et al.* 2003. Data storage and DNA banking for biomedical research: informed consent, confidentiality, quality issues, ownership, return of benefits. A professional perspective. *Eur J Hum Genet* **11**(suppl 2): S88-S122.

Grady C. 1999. Ethics and genetic testing. *Adv Int Med* **44**: 389–411.

Grodin MA, Glantz LH (eds). 1994. *Children as Research Subjects. Science, Ethics and Law*. Oxford University Press: New York.

Grody WW. 2003a. Ethical issues raised by genetic testing with oligonucleotide microarrays. *Mol Biotechnol* **23**: 127–138.

Grody WW. 2003b. Molecular genetic screening. *Annu Rev Med* **54**: 473–490.

Hainaut P, Vähäkangas K. 1999. Genetic analysis of metabolic polymorphisms in molecular epidemiological studies: social and ethical implications. In *Metabolic Polymorphisms and Susceptibility to Cancer*, Vineis P *et al.* (eds). IARC Scientific Publication No. 148. IARC: Lyon, France; 395–402.

Hakimian R. 2000. Disclosure of Huntington's disease to family members: the dilemma of known but unknowing parties. *Genet Testing* **4**: 359–364.

Halpern GM. 2005. COX-2 inhibitors: a story of greed, deception and death. *Inflammopharmacology* **13**: 419–425.

Han S, Li Y, Mao Y *et al.* 2005. Meta-analysis of the association of *CTLA-4* exon-1 +49A/G polymorphism with rheumatoid arthritis. *Hum Genet* **118**: 123–132.

Hansson MG. 2005. Building on relationship of trust in biobank research. *J Med Ethics* **31**: 415–418.

Hansson MG, Dillner J, Bartram CR *et al.* 2006. Should donors be allowed to give broad consent to future biobank research? *Lancet Oncol* **7**: 266–269.

Healy D. 2007. One flew over the conflict of interest nest. *World Psychiat* **6**: 26–27.

Helgesson G, Johnsson L. 2005. The right to withdraw consent to research on biobank samples. *Med Health Care Philos* **8**: 315–321.

Hendriks KS, Grosfeld FJ, van Tintelen JP *et al.* 2005. Can parents adjust to the idea that their child is at risk for a sudden death? Psychological impact of risk for long QT syndrome. *Am J Med Genet* **138A**: 107–112.

Holness MJ, Sugden MC. 2006. Epigenetic regulation of metabolism in children born small for gestational age. *Curr Opin Clin Nutr Metab Care* **9**: 482–488.

Horton R. 2006. Retraction – Non-steroidal anti-inflammatory drugs and the risk of oral cancer: a nested case-control study. *Lancet* **367**: 382.

Hoeyer K, Olofsson B-O, Mjörndal T *et al.* 2005. The ethics of research using biobanks. Reason to question the importance attributed to informed consent. *Arch Intern Med* **165**: 97–100.

Hunter D, Caporaso N. 1997. Informed consent in epidemiological studies involving genetic markers. *Epidemiology* **8**: 596–599.

Hussain SP, Hofseth LJ, Harris CC. 2003. Radical causes of cancer. *Nature Rev* **3**: 276–285.

Illes J, Desmond JE, Huang LF *et al.* 2002. Ethical and practical considerations in managing incidental findings in functional magnetic resonance imaging. *Brain Cogn* 50: 358–365.

Jamrozik K, Weller DP, Heller RF. 2005. Biobank: who'd bank on it? *Med J Aust* **182**: 56–57.

Jones PA. 2005. Overview of cancer epigenetics. *Semin Hematol* **42**(suppl 2): S3–8.

Kaiser J. 2002. Population databases boom, from Iceland to the US. *Science* **298**: 1158–1161.

Karassa FB, Trikalinos TA, Ioannidis JP. 2003. Fcγ RIIIA-SLE meta-analysis investigators. The Fcγ RIIIA-F158 allele is a risk factor for the development of lupus nephritis: a meta-analysis. *Kidney Int* **63**: 1475–1482.

Kaye J. 2001. Genetic research on the UK population – do new principles need to be developed? *Trends Mol Med* 7: 528–530.

Kaye J. 2006. Do we need a uniform regulatory system for biobanks across Europe? *Eur J Hum Genet* **14**: 245–248.

Khoury MJ, McCabe LL, McCabe ERB. 2003. Population screening in the age of genomic medicine. *N Engl J Med* **348**: 50–58.

Kim SYH, Millard RW, Nisbet P *et al.* 2004. Potential research participants' views regarding researcher and institutional financial conflicts of interest. *J Med Ethics* **30**: 73–79.

Knoppers BM, Hirtle M, Lormeau S *et al.* 1998. Control of DNA samples and information. *Genomics* **50**: 385–401.

Knudsen LE. 2005. Global gene mining and the pharmaceutical industry. *Toxicol Appl Pharmacol* **207**(2 suppl): 679–683.

Koh D, Jeyaratnam J. 1998. Biomarkers, screening and ethics. *Occup Med Lond* **48**: 27–30.

Koh D, Seow A, Ong CN. 1999. Applications of new technology in molecular epidemiology and their relevance to occupational medicine. *Occup Environ Med* **56**: 725–729.

Kristinsson S. 2007. Autonomy and informed consent: a mistaken association. *Med Health Care Philos* **10**: 253–264.

Kyrtopoulos SA. 2006. Biomarkers in environmental carcinogenesis research: striving for a new momentum. *Toxicol Lett* **162**: 3–15.

Lahiri DK, Sambamurti K, Bennett DA. 2004. Apolipoprotein gene and its interaction with the environmentally driven risk factors: molecular, genetic and epidemiological studies of Alzheimer's disease. *Neurobiol Aging* **25**: 651–660.

Lenzer J. 2006. Researcher received undisclosed payments of 300 000 dollars from Pfizer. *Br Med J* **333**: 1237.

Levitt M. 2003. Public consultation in bioethics. What's the point of asking the public when they have neither scientific nor ethical expertise? *Health Care Anal* **11**: 15–25.

Li D, Sham PC, Owen MJ, He L. 2006. Meta-analysis shows significant association between dopamine system genes and attention deficit hyperactivity disorder (ADHD). *Hum Mol Genet* **15**: 2276–2284.

Lieberman R. 2001. Prostate cancer chemoprevention: Strategies for designing efficient clinical trials. *Urology* **57**(4 suppl 1): 224–229.

Makni H, Franco EL, Kaiano J *et al.* 2000. *p53* polymorphism in codon 72 and risk of human papillomavirus-induced cervical cancer: effect of inter-laboratory variation. *Int J Cancer* **87**: 528–533.

Marshall E. 1998. Panel proposes tighter rules for tissue studies. *Science* **282**: 2165–2166.

Martinez FD. 2007. Gene–environment interactions in asthma: with apologies to William of Ockham. *Proc Am Thorac Soc* **4**: 26–31.

Maschke KJ. 2005. Navigating an ethical patchwork – human gene banks. *Nat Biotechnol* **23**: 539–545.

Maschke KJ. 2006. Alternative consent approaches for biobank research. *Lancet Oncol* **7**: 193–194.

McCanlies EC, Kreiss K, Andrew M *et al.* 2003. HLA-DPB1 and chronic beryllium disease: a HuGE review. *Am J Epidemiol* **157**: 388–398.

Neri M, Bonassi S, Knudsen LE *et al.* 2006. Children's exposure to environmental pollutants and biomarkers of genetic damage. I. Overview and critical issues. *Mutat Res* **612**: 1–13.

Merlo D, Knudsen L, Matusiewicz K *et al.* 2007. Ethics in studies on children and environmental health. *J Med Ethics* **33**: 408–413.

Neill US. 2006. Stop misbehaving! *J Clin Invest* **116**: 1740–1741.

Nuffield Council on Bioethics. 2002. *The Ethics of Patenting DNA – A Discussion Paper*: www.nuffield-bioethics.org

Nylenna M, Simonsen S. 2006. Scientific misconduct: a new approach to prevention. *Lancet* **367**: 1882–1884.

Olden K, White SL. 2005. Health-related disparities: influence of environmental factors. *Med Clin N Am* **89**: 721–738.

Paasche-Orlow MK, Taylor HA, Brancati FL. 2003. Readability standards or informed consent forms as compared with actual readability. *N Engl J Med* **348**: 721–726.

Pelias MK, Markward NJ. 2001. Newborn screening, informed consent, and future use of archived tissue samples. *Genet Testing* **5**: 179–185.

Pellegrino ED. 1993. The metamorphosis of medical ethics. A 30-year retrospective. *J Am Med Assoc* **269**: 1158–1162.

Pelkonen O, Vähäkangas K, Raunio H. 2003. Xenobiotic metabolism and cancer susceptibility. In Vainio H, Hietanen E (eds). *Handbook of Experimental Pharmacology, Vol 156: Mechanisms in Carcinogenesis and Cancer Prevention* ed. Vainio H, Hietanen E, Lavoisier, Paris, pp 253–269.

Petersen A. 2005. Securing our genetic health: engendering trust in UK biobank. *Sociol Health Illness* **27**: 271–292.

Petronis A. 2006. Epigenetics and twins: three variations on the theme. *Trends Genet* **22**: 347–350

Rawbone RG. 1999. Future impact of genetic screening in occupational and environmental medicine. *Occup Environ Med* **56**: 721–724.

Roses AD, Saunders AM, Huang Y *et al.* 2007. Complex disease-associated pharmacogenetics: drug efficacy, drug safety, and confirmation of a pathogenetic hypothesis (Alzheimer's disease). *Pharmacogenom J* **7**: 10–28.

Rossignol AM, Goodmonson S. 1995. Are ethical topics in epidemiology included in the graduate epidemiology curricula? *Am J Epidemiol* **142**: 1265–1268.

Sade RM. 2003. Publication of unethical research studies. The importance of informed consent *Ann Thorac Surg* **75**: 325–328.

Schulte PA. 1992. Biomarkers in epidemiology: scientific issues and ethical implications. *Environ Health Perspect* **98**: 143–147.

Shields PG, Harris CC. 2000. Cancer risk and low-penetrance susceptibility genes in gene–environment interactions. *J Clin Oncol* **18**: 2309–2315.

Silver K, Sharp RR. 2006. Ethical considerations in testing workers for the Glu69 marker of genetic susceptibility to chronic beryllium disease. *J Occup Environ Med* **48**: 434–443.

Smith GD, Ebrahim S, Lewis S. 2005. Genetic epidemiology and public health: hope, hype and future prospects. *Lancet* **366**: 1484–1498.

Soskolne CL. 1997. Ethical, social, and legal issues surrounding studies of susceptible populations and individuals. *Environ Health Perspect* **105**(suppl 4): 837–841.

Stefanovic V, Toncheva D, Atanasova S *et al.* 2006. Etiology of Balkan endemic nephropathy and associated urothelial cancer. *Am J Nephrol* **26**: 1–11.

Surbone A. 2001. Ethical implications of genetic testing for breast cancer susceptibility. *Crit Rev Oncol Hematol* **40**: 149–157.

Suzuki D, Knudtson P. 1990. *Genethics. The Clash Between the New Genetics and Human Values*, revised edn. Harward University Press: Cambridge, MA.

Taioli E. 2005. Biomarkers of genetic susceptibility to cancer: applications to epidemiological studies. *Future Oncol* **1**: 51–56.

Taioli E, Bonassi S. 2002. Methodological issues in pooled analysis of biomarker studies. *Mutat Res* **512**: 85–92.

Trouet C. 2004. New European guidelines for the use of stored human biological materials in biomedical research. *J Med Ethics* **30**: 99–103.

Uranga AM, Arribas MCM, Jaeger C *et al.* 2005. Outstanding ethical–legal issues on biobanks. An overview on the regulations of the member states of the EuroBiobank project. *Rev Der Gen* **H22**: 103–114.

Vähäkangas KH. 2001. Ethical implications of genetic analysis of individual susceptibility to diseases. *Mutat Res* **482**: 105–110.

Vähäkangas K. 2004. Ethical aspects of molecular epidemiology of cancer. *Carcinogenesis* **25**: 465–471.

van Leeuwen DJ, Bernat JL. 2006. Ethical, social and legal implications of genetic testing in liver disease. *Hepatology* **43**: 1195–1201.

van der Hel OL, Bueno de Mesquita HB, Roest M *et al.* 2003. No modifying effect of *NAT1*, *GSTM1* and *GSTT1* on the relation between smoking and colorectal cancer risk. *Cancer Epidemiol Biomarkers Prevent* **12**: 681–682.

van Korlaar IM, Vossen CY, Rosendaal FR *et al.* 2005. Attitudes toward genetic testing for thrombophilia in asymptomatic members of a large family with heritable protein C deficiency. *J Thromb Haemost* **3**: 2437–2444.

van Oostrom I, Meijers-Heijboer H, Lodder LN *et al.* 2003. Long-term psychological impact of carrying a *BRCA1/2* mutation and prophylactic surgery: a 5-year follow-up study. *J Clin Oncol* **21**: 3867–3874.

Vassy C. 2006. From a genetic innovation to mass health programmes: the diffusion of Down's syndrome prenatal screening and diagnostic techniques in France. *Soc Sci Med* **63**: 2041–2051.

Vineis P. 2003. The randomized controlled trial in studies using biomarkers. *Biomarkers* **8**: 13–32.

Vineis P. 1997. Ethical issues in genetic screening for cancer. *Ann Oncol* **8**: 945–949.

Vineis P, Ahsan H, Parker M. 2005. Genetic screening and occupational and environmental exposures. *Occup Environ Med* **62**: 657–662.

Vineis P, Christiani DC. 2004. Genetic testing for sale. *Epidemiology* **15**: 3–5.

Vineis P, Schulte PA. 1995. Scientific and ethical aspects of genetic screening of workers for cancer risk: the case of the *N*-acetyltransferase phenotype. *J Clin Epidemiol* **48**: 189–197.

Walsh T, Casadei S, Coats KH *et al.* 2006. Spectrum of mutations in *BRCA1*, *BRCA2*, *CHEK2*, and *TP53* in families at high risk of breast cancer. *J Am Med Assoc* **295**: 1379–1388.

Weinfurt KP, Friedman JY, Allsbrook JS *et al.* 2006. Views of potential research participants on financial conflicts of interest: barriers and opportunities for effective disclosure. *J Gen Intern Med* **21**: 901–906.

Wexler A. 1995. *Mapping Fate*. A memory of family, Risk and Genetic Research, Times Books, New York.

Winickoff DE, Winickoff RN. 2003. The charitable trust as a model for genomic biobanks. *N Engl J Med* **349**: 1180–1184.

WHO (World Health Organization). 2002. *Genomics and World Health. Report of The Advisory Committee on Health Research*. WHO: Geneva.

23

Biomarkers for Dietary Carcinogens: The Example of Heterocyclic Amines in Epidemiological Studies

Rashmi Sinha[1], Amanda Cross[1] and Robert J. Turesky[2]

[1]*National Institutes of Health, Rockville, MD, and* [2]*Department of Health, Empire State Plaza, Albany, New York, USA*

23.1. INTRODUCTION

Meats cooked by high temperature cooking techniques, such as pan-frying or grilling/barbecuing, contain heterocyclic amines (HCAs). This group of compounds are formed from the reaction, at high temperatures, between creatine or creatinine (found in muscle meats), amino acids and sugars (Nagao *et al.* 1983; Sugimura *et al.* 2004; Sugimura and Wakabayashi 1991; Wakabayashi *et al.* 1992). In 1993, the International Agency for Research on Cancer concluded that the following HCAs were carcinogenic: 2-amino-3-methylimidazo [4,5-*f*]quinoline (IQ); 2-amino-3,4-dimethylimidazo [4,5-*f*]quinoline(MeIQ); 2-amino-3,8-dimethylim-idazo[4,5-*f*]quinoxaline (MeIQx); and 2-amino-1-methyl-6-phenylimidazo(4,5-*b*)pyridine (PhIP) were carcinogenic (IARC 1993). Over 20 individual HCAs have been identified, and most are potent bacterial mutagens. Furthermore, HCAs produce tumours in a variety of organs, such as the liver, lung, forestomach, colon, prostate, mammary gland and lymphomas, in rodent and non-human primate models. The most abundant HCAs in cooked meat are PhIP and MeIQx; in rodents, PhIP has been associated with an increased risk of intestinal and mammary adenocarcinomas (Ghoshal *et al.* 1994; Ito *et al.* 1991), as well as prostate tumours (Shirai *et al.* 1999, 2002) and MeIQx with liver and lung tumours as well as lymphomas and leukae-mias (Kato *et al.* 1988; Ohgaki *et al.* 1987). Using crude surrogates for HCA exposure from meat (e.g. 'doneness' level, surface browning, cooking method, and intake of gravy), epidemiological studies of colon and breast cancer have produced suggestive, but inconsistent, results (Augustsson *et al.* 1999; Delfino *et al.* 2000; Ferguson 2002; Gerhardsson *et al.* 1991; Gunter *et al.* 2006; Lang *et al.* 1994; Le Marchand *et al.* 2002; Muscat and Wynder 1994; Nowell *et al.* 2002; Sinha *et al.* 2001, 2005; Steineck *et al.* 1990).

23.2. INTAKE ASSESSMENT OF HCAS

The main method in epidemiological studies for estimating intake of meat or cooked-meat carcinogens has been food frequency questionnaires (FFQs); Anderson *et al.* 2002; Augustsson *et al.* 1999; Butler *et al.* 2003; Cross *et al.* 2005, 2006; De Stefani *et al.* 1997; Ferguson 2002; Gerhardsson *et al.* 1991; Kampman *et al.* 1999; Probst-Hensch *et al.* 1997; Sinha *et al.* 2000; Zheng *et al.* 1998). Recently, there has been controversy about the use of FFQs as compared to food records/diaries for estimating different dietary components in epidemiological studies (Bingham 1997; Bingham

Molecular Epidemiology of Chronic Diseases, Edited by C. P. Wild, P. Vineis, and S. Garte
© 2008 John Wiley & Sons, Ltd

et al. 1997, 2001; Kipnis *et al.* 2002; Kristal *et al.* 2005; Willett and Hu 2006, 2007). The main nutritional components that have been investigated for misclassification by FFQs are calories and fat in relation to breast cancer. At present, there is no research available from food records/diaries or 24-hour recalls, except from validation studies (Cantwell *et al.* 2004), pertinent to the question of misclassification by FFQs of meat and HCAs. There is no reason to believe that every dietary component will be subject to a similar level of mis-classification by FFQs as calories and fat. There may be some dietary components, such as meat, which are consumed relatively consistently and can be remembered with less bias. Calories and fat may be more difficult to estimate, as they are often hidden in many foods and so are more susceptible to misclassification.

There are other issues that are relevant to the type of dietary instrument that is being used. To estimate HCAs, one needs detailed information on cooking methods and doneness levels. Most FFQs, food records/diaries or 24-hour recalls do not elicit the level of detail that is required. Quite often the subject burden and competing scientific questions do not allow us to obtain the level of detail required, especially in prospective cohort studies. However, FFQ meat modules have been incorporated into several cohorts (Alavanja *et al.* 1996; Byrne *et al.* 1998). Currently, the existing literature on HCA exposure from meats is mainly from case–control studies, which are retrospective in nature and more likely suffer from recall bias.

Another major issue in relation to assessing exposure to mutagenic compounds in foods, such as HCAs, is the lack of databases that can be linked to the information collected from the dietary questionnaires (Sinha *et al.* 2005). Developing such databases is a major undertaking in terms of time and expense; it is somewhat possible for FFQs, as these questionnaires contain limited numbers of food items. For open-ended methods of collecting dietary information, such as food records/diaries or 24-hour recalls, there can be so many combinations of meat items and cooking methods as to render development of a HCA database virtually impossible for any individual group of researchers. Therefore, the existing research for HCA database

development has used a targeted FFQ approach (Alavanja *et al.* 1996; Augustsson *et al.* 1997, 1999; Sinha *et al.* 1995, 1998a, 1998b). These limitations in estimating HCAs using questionnaires have led researchers towards evaluating reliable biomarkers of HCAs. 1–2 line of verbiage about global biomarkers used for HAAs in concert with the FFQ and allude to Figure 23.1.

23.3. HCA METABOLISM

HCAs undergo extensive metabolism by phase I and II enzymes in experimental animals and humans (Alexander *et al.* 1995; King, Kadlubar and Turesky 2000; Sugimura 1997; Sugimura and Wakabayashi 1991). The major pathways of HCA metabolism are depicted in Figure 23.2. Oxidation reactions can occur at the heterocyclic ring, methyl and the exocyclic amine groups. These reactions are catalysed by CYP1A2, which is mainly expressed in liver (Butler *et al.* 1989), and also by CYP1A1 and CYP1B1 in extrahepatic tissues (Crofts, Sutter and Strickland 1998; Shimada and Guengerich 1991). Direct sulphamation and glucuronidation of the exocyclic amino groups of the HCAs are cata-lysed by sulphotransferases (SULTs) and glucuron-osyl transferases (UGTs) (Alexander *et al.* 1995; King *et al.* 2000) and form detoxicated products. *N*-acetylation, which is an important mechanism of detoxication of primary arylamines (Hein 2002), is not a prominent pathway of detoxication of HCAs containing the *N*-methyl-2-aminoimidazole moiety; however, NATs do catalyse the detoxication of 2-amino-9*H*-pyridole(2,3-*b*)indole (AαC) and other pyrolysate HCA mutagens (King *et al.* 2000).

The *N*-hydroxy-HCA metabolites are the genotoxic species that modify DNA. These metabolites can directly react with DNA, but the ulitmate carcinogenic species are thought to be acetate or sulphate esters of the *N*-hydroxy-HCAs, the formation of which is catalysed by NATs or SULTs expressed in liver or extrahepatic tissues (Schut and Snyderwine 1999; Turesky and Vouros 2004). For many *N*-hydroxy-HCA substrates, NAT2 is catalytically superior to NAT1 in bio-activation, although the *N*-hydroxy metabolite of PhIP is activated by both isoforms (King *et al.* 2000). SULT1A1 is the most active SULT in

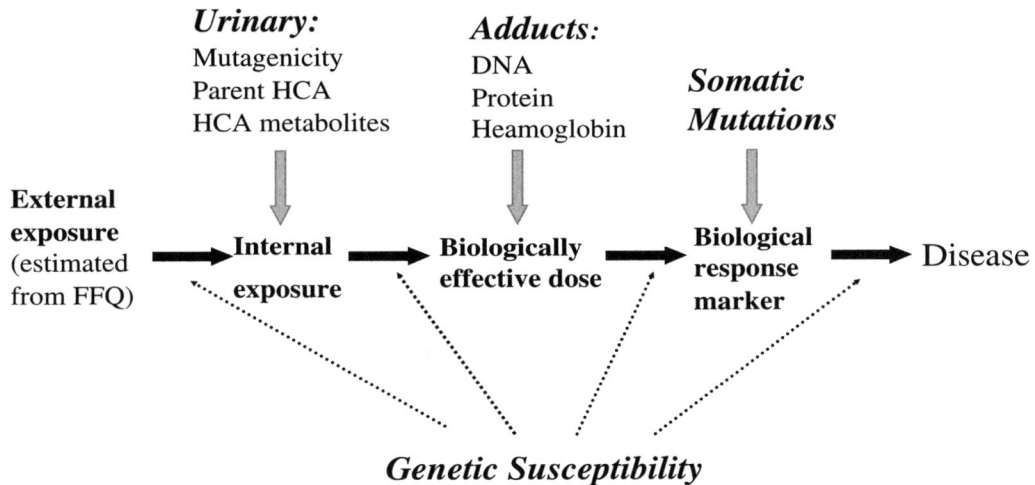

Figure 23.1

bioactivation of *N*-hyroxylated metabolites of PhIP and 2-amino-3-methyl-9*H*-pyridole(2,3-*b*)indole (MeAαC) (Glatt *et al.* 2004; Muckel, Frandsen and Glatt 2002; Wu *et al.*2000); these esters are unstable and undergo heterolytic cleavage to

generate the reactive nitrenium ions which adduct to DNA (Figure 23.2). The principal reaction of the *N*-hydroxy HCA derivatives with DNA occurs at deoxyguanosine (dG) to produce dG-C8–HCA adducts, where bond formation occurs

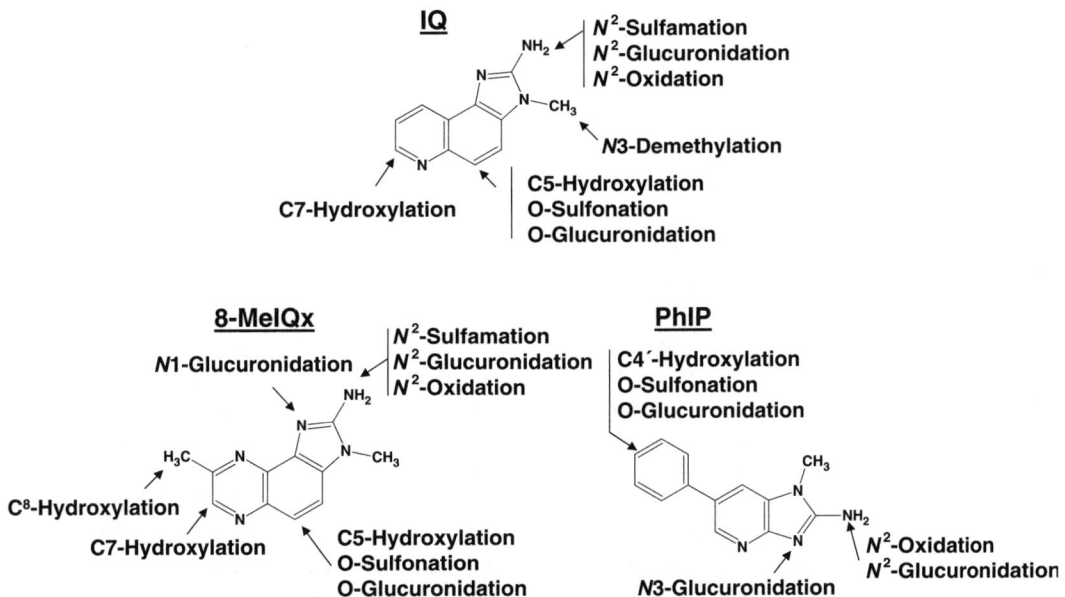

Figure 23.2   Major pathways of HCA metabolism in experimental laboratory animals and humans

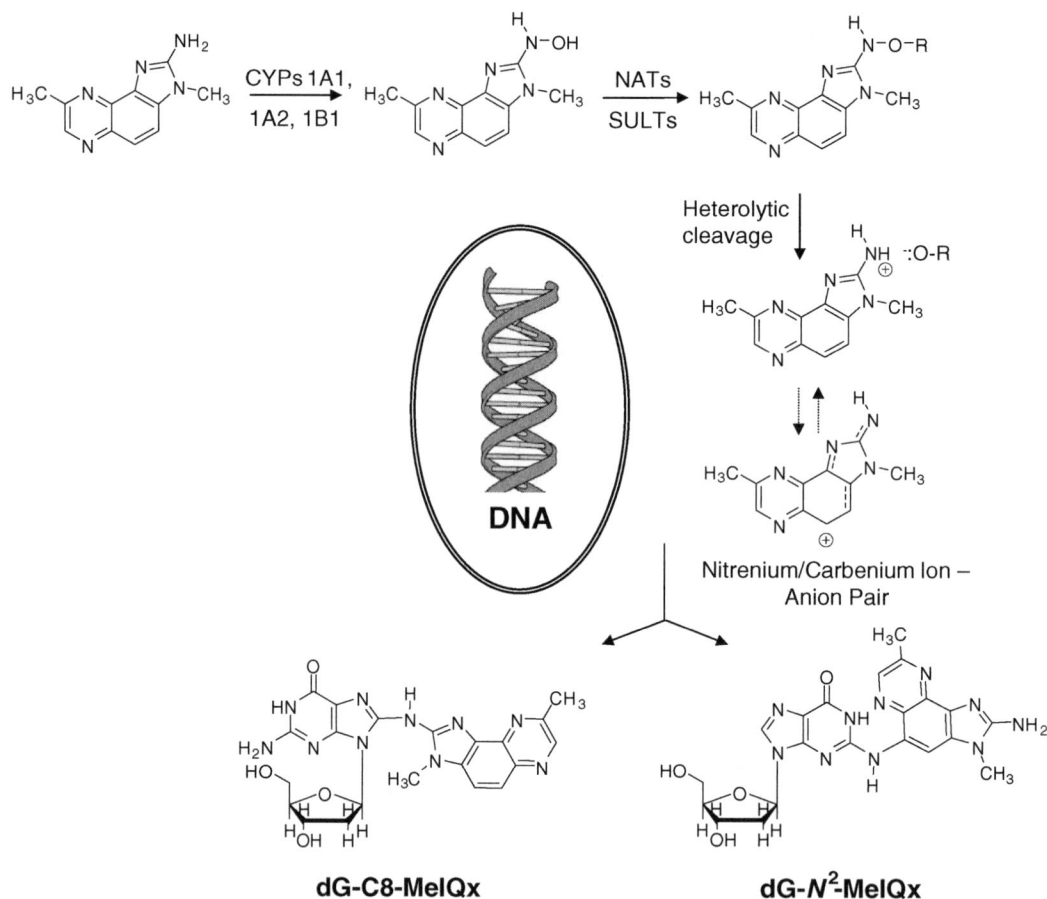

Figure 23.3   Metabolic activation of MeIQx and formation of dG-C8 and dG-N^2 adducts

between the C8 atom of dG and the activated exocyclic amine group of the HCA (Schut and Snyderwine 1999; Turesky and Vouros 2004). In the cases of IQ and 2-amino-3,8-dimethylimidazo (4,5-*f*)quinoxaline (8-MeIQx), DNA adducts also form at the N^2 group of dG and the C-5 atom of the heterocyclic ring structures, indicating charge delocalization of the nitrenium ion over the heteronuclei of these respective HCAs (Figure 23.3) (Turesky and Vouros 2004). Using synthetic biomimetic methods, the amount of dG-N^2 adducts formed is small relative to the dG-C8 isomers; however, the dG-N^2 adducts persist *in vivo* to become the prominent lesions in slowly dividing tissues of rats and non-human primates during chronic exposure to IQ (Turesky and Vouros 2004), or following different dose regimens with MeIQx (Paehler *et al.* 2002).

Human CYP1A2 does not efficiently catalyse the detoxification of PhIP through 4′ hydroxylation, but this enzyme does catalyse the oxidation of the C^8-methyl group of 8-MeIQx to form the carboxylic acid, 2-amino-3-methylimidazo (4,5-*f*)quinoxaline-8-carboxylic acid (IQx-8-COOH), the major pathway of metabolism and detoxification of 8-MeIQx. In human liver samples, the amount of CYP1A2 varies across a > 50-fold range (5–250 pmol CYP11A2/mg microsomal protein; King *et al.* 2000; Turesky *et al.* 2001) and several genetic polymorphisms have been reported in the promoter region of the *CYP1A2* gene (Sachse *et al.* 2003), which lead to modest differences in the inducibility of CYP1A2 protein expression. However, much of the large interindividual differences observed in CYP1A2 expression are likely attributed to varying exposures to environmental and dietary constituents, which serve as inducers of this enzyme (Mori *et al.* 2003; Sinha *et al.* 1994).

Urinary biomarkers as a measure of internal exposure

Urine is a useful biological fluid for the measurement of exposure to various classes of carcinogens, since large quantities may be obtained non-invasively (Hecht 2002). The measurement of HCAs and their metabolites in urine can be used to assess the capacity of an individual to bioactivate and detoxify these carcinogens and to assess the impact of xenobiotic enzyme polymorphisms on health risk (Kelada *et al.* 2003). HCAs are rapidly absorbed from the gastrointestinal tract and eliminated in urine as multiple metabolites with several percent of dose present as the unmetabolized parent compounds within 24 hours of consuming grilled meats. Based upon pharmacokinetic studies with furafylline, a mechanism-based inhibitor of CYP1A2, up to 91% of the elimination of 8-MeIQx and 70% of the elimination of PhIP could be accounted for by CYP1A2-catalysed metabolism (Boobis *et al.* 1994). Since these investigations were reported, the major metabolite of 8-MeIQx in human urine, accounting for about 50–70% of the dose in urine, was identified as the 8-carboxylic acid derivative, and its formation is catalysed by CYP1A2 (Turesky *et al.* 1998b, 2001). Recent developments in tandem solvent/solid phase extraction procedures have been established to isolate a variety of HCAs and their metabolites from human urine by liquid chromatography–electrospray ionization–mass spectrometry (LC–ESI–MS) techniques (Holland *et al.* 2004; Kulp *et al.* 2004; Turesky *et al.* 1998b; Walters *et al.* 2004). These urinary biomarkers may be used to investigate the relationship between HCA exposure, rapid CYP1A2 activity, metabolism and cancer risk.

Urine samples in epidemiological studies may be useful for a measure of internal dose. There are several measures, such as mutagenic activity (free and acid-hydrolysed), HCAs (free and acid-hydrolysed) and metabolites of HCAs, such as N^2-glucuronide, and sulphamate metabolites. In a controlled metabolic study in which subjects consumed well-done meat with known amount of HCAs,

there was a correlation of 0.4–0.6 between intake and urinary output (Augustsson *et al.* 1999; Stillwell *et al.* 1997, 1999a, 1999b; 2002). As HCAs are known to be metabolized within 12–24 h, indicative of a short half-life, they are not an ideal measure of 'usual' intake in aetiological studies, especially if there is substantial day-to-day variability. Measurement of HCAs in multiple urine samples over a long period may circumvent this short-coming. Furthermore, urine analysis could still be used to validate intake of HCAs as estimated by questionnaires.

Adducts as a measure of biologically effective dose

Ideally, long-term biomarkers of HCA exposure and consequent genetic damage need to be incorporated into population-based studies, to reliably assess exposure and health risk. The chemical modification of DNA by a genotoxicant is believed to be the initiating event that may ultimately lead to cancer (Miller 1978). However, it is generally not feasible to procure and measure DNA adducts and genetic damage in target tissues of tumourigenesis in population-based studies of healthy subjects. Consequently, surrogate markers of the biologically effective dose and genetic damage are developed in tissues and fluids that can be obtained non-invasively. Potential surrogate markers include the HCA–DNA adducts in lymphocytes and HCA metabolites bound to circulating blood proteins, such as haemoglobin (Hb) or serum albumin (SA). The measurements of these biomarkers can provide an estimate of exposure and the biologically effective dose (Skipper and Tannenbaum 1990), but they do not provide a measure of genetic damage directly in the target tissue.

DNA and protein adducts of HCAs have been detected in experimental animal models by ^{32}P-postlabelling, accelerator mass spectrometry (AMS), and LC–ESI–MS techniques (Paehler *et al.* 2002; Pfau *et al.* 1997; Schut and Snyderwine 1999; Skipper and Tannenbaum 1990; Totsuka *et al.* 1996; Turesky and Vouros 2004; Turteltaub and Dingley 1998). However, there is a paucity of data reporting on HCA biomarkers in humans because their detection and quantification remains a challenging analytical problem: the concentration of HCAs in the diet is at the parts-per-billion level and the quantity of these HCA biomarkers formed in humans occurs at very low levels.

AMS studies have shown that PhIP and MeIQx form adducts in human breast and colorectal tissues, when radiolabelled compounds were given to subjects at dose levels comparable to those found in the daily human diet (Dingley *et al.* 1999; Lightfoot *et al.* 2000; Turteltaub *et al.* 1997). One ^{32}P-postlabelling study revealed the presence of the dG-C8–MeIQx adduct in several colon and kidney tissues at levels of several adducts per 10^9 DNA bases (Totsuka *et al.* 1996). Alkaline hydrolysis of DNA, followed by derivatization and gas chromatography–electron capture detection to measure PhIP, has been used as an indirect measure of the base-labile dG-C8–PhIP adduct in colorectal mucosa DNA, where the levels of putative dG-C8–PhIP adduct were detected at up to several adducts per 10^7 DNA bases (Friesen *et al.* 1994). Another study using alkaline treated DNA reported the presence of the putative of dG-C8–PhIP adduct in white blood cells of subjects at levels approaching ~3 adducts per 10^8 DNA bases by LC–ESI–MS (Magagnotti *et al.* 2003), although a second study failed to identify PhIP adducts in white blood cells using this approach (Murray *et al.* 2001). DNA adducts of PhIP, presumably the dG-C8–PhIP lesion, were also detected in human breast tissue at levels of ~1 adduct/10^7 bases upon immunohistochemistry (Zhu *et al.* 2003). Thus, some of the data reported in the literature reveals that even low concentrations of HCAs formed in grilled meats are capable of inducing damage to DNA in humans. However, the proof of adduct structures in all of these reported studies are equivocal. With the recent improvements in sensitivity of LC–ESI–MS instrumentation, the characterization and quantification of DNA adducts at levels approaching 1 adduct/10^9 DNA bases should be feasible by LC–ESI–MS techniques. Indeed, DNA adducts of the aromatic amine 4-aminobiphenyl have been quantitated, by LC–MS, in pancreatic tissues of human subjects at levels in the range 1–60 adducts/10^8 DNA bases (Ricicki *et al.* 2005).

HCA metabolites bind more avidly to SA than to Hb in rodents and humans, although the percentage of the dose bound to either protein is very low (Dingley *et al.* 1998, 1999; Garner *et al.* 1999; Lynch *et al.* 1991; Turesky *et al.* 1987). The cysteine[34] residue of SA of the rodents binds

to *N*-oxidized metabolites of IQ (Turesky *et al.* 1987), 8-MeIQx (Lynch *et al.* 1991) and PhIP (Chepanoske *et al.* 2004) to form sulphinamide or sulphenamide derivatives; however, only about 0.01% of the dose of HCAs forms this adduct. These adducts undergo hydrolysis to regenerate the parent amine under mildly acidic conditions and can be readily isolated from SA (Skipper and Tannenbaum 1990). A pilot study in human cancer patients found that the level of PhIP bound to SA in humans was up to 40-fold higher than that bound to SA of rats given the same dose of PhIP, based upon AMS measurements (Dingley *et al.* 1999; Garner *et al.* 1999). The higher PhIP dose bound to human SA may be due to the superior catalytic activity of human CYP1A2 in the bioactivation of PhIP (Turesky *et al.* 1998a). Unfortunately, the percentage of the dose bound as sulphinamide or sulphenamide derivatives was not determined. Subsequently, one investigation conducted on healthy human subjects reported that acid-labile, putative sulphinamide/sulphenamide SA adducts of PhIP were 10-fold higher in meat-eaters than in vegetarians(Magagnotti *et al.* 2000). The structure of sulphinamide/sulphenamide SA adducts of PhIP adduct attributed to the acid-labile lesion remains to be determined. It seems likely that a portion of the acid-labile PhIP adduction products was formed at the cysteine[34] residue in human SA. Moreover, the chemical stability of adduct is unknown and additional studies are required to validate the acid-labile PhIP adduction products as a biomarker prior to its use in population-based studies.

23.4. CONCLUSIONS AND FUTURE RESEARCH

An interdisciplinary approach needs to be taken when examining the role of dietary HCAs in cancer aetiology. Although dietary questionnaires can be further improved, we will always be confronted with limitations in recall when assessing diet in large epidemiological studies. However, recent advances in the sensitivity of MS instrumentation have allowed for the detection of HCAs and their metabolites at trace levels in human urine (Alexander *et al.* 2002; Holland *et al.* 2004; Kulp *et al.* 2004; Lynch *et al.* 1992; Turesky *et al.* 1998b). Preliminary analytical data suggest that protein adducts and white blood cell DNA adducts of PhIP may be useful biomarkers (Dingley *et al.* 1999; Magagnotti *et al.* 2000, 2003) to assess the interactive effects of HCA exposure, DNA damage, genetic polymorphisms in metabolizing enzymes and cancer risk. Two recent studies have also reported the accumulation of HCAs in human hair, which may serve as a potential long-term biomarker to assess chronic exposure of HCAs (Alexander *et al.* 2002; Kobayashi *et al.* 2005). Futher studies on the development and validation of biomarkers of PhIP and other HCAs, which can be incorporated into large, population-based studies, may refine the human risk assessment of these genotoxicants and clarify their role in the aetiology of dietary-related cancers.

References

Alavanja MC, Sandler DP, McMaster SB *et al.* 1996. The Agricultural Health Study. *Environ. Health Perspect* **104**: 362–369.

Adamson RH, Gustafsson JA, Ito N *et al.* 1995. Heterocyclic amines in cooked foods: Possible human carcinogens. 23rd Proceedings of the Princess Takamatsu Cancer Society. Princeton Scientific: Princeton, NJ; 59–68.

Alexander J, Heidenreich B, Reistad R *et al.* 1995. Metabolism of the food carcinogen 2-amino-1-methyl-6-phenylimidazo[4,5-*b*]pyridine (PhIP) in the rat and other rodents. In *Heterocyclic amines in cooked foods: Possible human carcinogens. 23rd Proceedings of the Princess Takamatsu Cancer Society*, RH Adamson *et al.* eds., Princeton Scientific Publishing Co., Inc., New Jersey, pp. 59–68.

Alexander J, Reistad R, Hegstad S *et al.* 2002. Biomarkers of exposure to heterocyclic amines: approaches to improve the exposure assessment. *Food Chem Toxicol* **40**: 1131–1137.

Anderson KE, Sinha R, Kulldorff M *et al.* 2002. Meat intake and cooking techniques: associations with pancreatic cancer. *Mutat Res* **506–507**: 225–231.

Augustsson K, Skog K, Jagerstad M *et al.* 1999. Dietary heterocyclic amines and cancer of the colon, rectum, bladder, and kidney: a population-based study. *Lancet* **353**: 703–707.

Augustsson K, Skog K, Jagerstad M *et al.* 1997. Assessment of the human exposure to heterocyclic amines. *Carcinogenesis* **18**: 1931–1935.

Bingham SA. 1997. Dietary assessments in the European prospective study of diet and cancer (EPIC). *Eur J Cancer Prevent* **6**: 118–124.

Bingham SA, Gill C, Welch A *et al.* 1997. Validation of dietary assessment methods in the UK arm of EPIC using weighed records, and 24-hour urinary nitrogen and potassium and serum vitamin C and carotenoids as biomarkers. *Int J Epidemiol* **26**(suppl 1): S137–151.

Bingham SA, Welch AA, McTaggart A *et al.* 2001. Nutritional methods in the European Prospective Investigation of Cancer in Norfolk. *Public Health Nutr* **4**: 847–858.

Boobis AR, Lynch AM, Murray S *et al.* 1994. CYP1A2-catalysed conversion of dietary heterocyclic amines to their proximate carcinogens is their major route of metabolism in humans. *Cancer Res* **54**: 89–94.

Butler LM, Sinha R, Millikan RC *et al.* 2003. Heterocyclic amines, meat intake, and association with colon cancer in a population-based study. *Am J Epidemiol* **157**: 434–445.

Butler MA, Iwasaki M, Guengerich FP *et al.* 1989. Human cytochrome P-450PA (P-450IA2), the phenacetin O-deethylase, is primarily responsible for the hepatic 3-demethylation of caffeine and *N*-oxidation of carcinogenic arylamines. *Proc Natl Acad Sci USA* **86**: 7696–7700.

Byrne C, Sinha R, Platz EA *et al.* 1998. Predictors of dietary heterocyclic amine intake in three prospective cohorts. *Cancer Epidemiol Biomarkers Prevent* **7**: 523–529.

Cantwell M, Mittl B, Curtin J *et al.* 2004. Relative validity of a food frequency questionnaire with a meat-cooking and heterocyclic amine module. *Cancer Epidemiol Biomarkers Prevent* **13**: 293–298.

Chepanoske CL, Brown K, Turteltaub KW *et al.* 2004. Characterization of a peptide adduct formed by *N*-acetoxy-2-amino-1-methyl-6-phenylimidazo[4,5-*b*]pyridine (PhIP), a reactive intermediate of the food carcinogen PhIP. *Food Chem Toxicol* **42**: 1367–1372.

Crofts FG, Sutter TR, Strickland PT. 1998. Metabolism of 2-amino-1-methyl-6-phenylimidazo[4,5-*b*]pyridine by human cytochrome P4501A1, P4501A2 and P4501B1. *Carcinogenesis* **19**: 1969–1973.

Cross AJ, Peters U, Kirsh VA *et al.* 2005. A prospective study of meat and meat mutagens and prostate cancer risk. *Cancer Res* **65**: 11779–11784.

Cross AJ, Ward MH, Schenk M *et al.* 2006. Meat and meat-mutagen intake and risk of non-Hodgkin lymphoma: results from a NCI–SEER case–control study 1. *Carcinogenesis* **27**: 293–297.

De Stefani E, Ronco A, Mendilaharsu M *et al.* 1997. Meat intake, heterocyclic amines, and risk of breast cancer: a case–control study in Uruguay. *Cancer Epidemiol Biomarkers Prevent* **6**: 573–581.

Delfino RJ, Sinha R, Smith C *et al.* 2000. Breast cancer, heterocyclic aromatic amines from meat and *N*-acetyltransferase 2 genotype. *Carcinogenesis* **21**: 607–615.

Dingley KH, Curtis KD, Nowell S *et al.* 1999. DNA and protein adduct formation in the colon and blood of humans after exposure to a dietary-relevant dose of 2-amino-1-methyl-6-phenylimidazo[4,5-*b*]pyridine. *Cancer Epidemiol Biomarkers Prevent* **8**: 507–512.

Dingley KH, Freeman SP, Nelson DO *et al.* 1998. Covalent binding of 2-amino-3,8-dimethylimidazo [4,5-f]quinoxaline to albumin and hemoglobin at environmentally relevant doses. Comparison of human subjects and F344 rats. *Drug Metab Dispos* **26**: 825–828.

Ferguson LR. 2002. Meat consumption, cancer risk and population groups within New Zealand. *Mutat Res* **506–507**: 215–224.

Friesen MD, Kaderlik K, Lin D *et al.* 1994. Analysis of DNA adducts of 2-amino-1-methyl-6-phenylimidazo[4,5-*b*]pyridine in rat and human tissues by alkaline hydrolysis and gas chromatography/electron capture mass spectrometry: validation by comparison with 32P-postlabeling. *Chem Res Toxicol* **7**: 733–739.

Garner RC, Lightfoot TJ, Cupid BC *et al.* 1999. Comparative biotransformation studies of MeIQx and PhIP in animal models and humans. *Cancer Lett* **143**: 161–165.

Gerhardsson DV, Hagman U, Peters RK *et al.* 1991. Meat, cooking methods and colorectal cancer: a case–referent study in Stockholm. *Int J Cancer* **49**: 520–525.

Ghoshal A, Preisegger KH, Takayama S *et al.* 1994. Induction of mammary tumors in female Sprague–Dawley rats by the food-derived carcinogen 2-amino-1-methyl-6-phenylimidazo[4,5-*b*]pyridine and effect of dietary fat. *Carcinogenesis* **15**: 2429–2433.

Glatt H, Pabel U, Meinl W *et al.* 2004. Bioactivation of the heterocyclic aromatic amine 2-amino-3-methyl-9*H*-pyrido [2,3-*b*]indole (MeAαC) in recombinant test systems expressing human xenobiotic-metabolizing enzymes. *Carcinogenesis* **25**: 801–807.

Gunter MJ, Stolzenberg-Solomon R, Cross AJ *et al.* 2006. A prospective study of serum C-reactive protein and colorectal cancer risk in men. *Cancer Res* **66**: 2483–2487.

Hecht SS. 2002. Human urinary carcinogen metabolites: biomarkers for investigating tobacco and cancer. *Carcinogenesis* **23**: 907–922.

Hein DW. 2002. Molecular genetics and function of *NAT1* and *NAT2*: role in aromatic amine metabolism and carcinogenesis. *Mutat Res* **506–507**: 65–77.

Holland RD, Taylor J, Schoenbachler L *et al.* 2004. Rapid biomonitoring of heterocyclic aromatic amines in human urine by tandem solvent solid-phase extraction–liquid chromatography–electrospray ionization mass spectrometry. *Chem Res Toxicol* **17**: 1121–1136.

IARC. 1993. *Some Naturally Occurring Substances: Food Items and Constituents, Heterocyclic Aromatic Amines and Mycotoxins*. IARC Monographs on the

Evaluation of Carcinogenic Risks to Humans No. 56. IARC: Lyon, France.

Ito N, Hasegawa R, Sano M *et al.* 1991. A new colon and mammary carcinogen in cooked food, 2-amino-1-methyl-6-phenylimidazo[4,5-*b*]pyridine (PhIP). *Carcinogenesis* **12**: 1503–1506.

Kampman E, Slattery ML, Bigler J *et al.* 1999. Meat consumption, genetic susceptibility, and colon cancer risk: a United States multicenter case–control study. *Cancer Epidemiol Biomarkers Prevent* **8**: 15–24.

Kato T, Ohgaki H, Hasegawa H *et al.* 1988. Carcinogenicity in rats of a mutagenic compound, 2-amino-3,8-dimethylimidazo[4,5-f]quinoxaline. *Carcinogenesis* **9**: 71–73.

Kelada SN, Eaton DL, Wang SS *et al.* 2003. The role of genetic polymorphisms in environmental health. *Environ Health Perspect* **111**: 1055–1064.

King RS, Kadlubar FF, Turesky RJ. 2000. *In vivo* metabolism of heterocyclic amines. In *Food-borne Carcinogens: Heterocyclic Amines*, Nagao M, Sugimura T (eds). Wiley: Chichester, UK; 90–111.

Kipnis V, Midthune D, Freedman L *et al.* 2002. Bias in dietary-report instruments and its implications for nutritional epidemiology. *Public Health Nutr* **5**: 915–923.

Kobayashi M, Hanaoka T, Hashimoto H *et al.* 2005. 2-Amino-1-methyl-6-phenylimidazo[4,5-*b*]pyridine (PhIP) level in human hair as biomarkers for dietary grilled/stir-fried meat and fish intake. *Mutat Res* **588**: 136–142.

Kristal AR, Peters U, Potter JD. 2005. Is it time to abandon the food frequency questionnaire? *Cancer Epidemiol Biomarkers Prevent* **14**: 2826–2828.

Kulp KS, Knize MG, Fowler ND *et al.* 2004. PhIP metabolites in human urine after consumption of well-cooked chicken. *J Chromatogr B Analyt Technol Biomed Life Sci* **802**: 143–153.

Lang NP, Butler MA, Massengill J *et al.* 1994. Rapid metabolic phenotypes for acetyltransferase and cytochrome P4501A2 and putative exposure to food-borne heterocyclic amines increase the risk for colorectal cancer or polyps. *Cancer Epidemiol Biomarkers Prevent* **3**: 675–682.

Le Marchand L, Hankin JH, Pierce LM *et al.* 2002. Well-done red meat, metabolic phenotypes and colorectal cancer in Hawaii. *Mutat. Res* **506–507**: 205–214.

Lightfoot TJ, Coxhead JM, Cupid BC *et al.* 2000. Analysis of DNA adducts by accelerator mass spectrometry in human breast tissue after administration of 2-amino-1-methyl-6-phenylimidazo[4,5-*b*]pyridine and benzo[α]pyrene. *Mutat Res* **472**: 119–127.

Lynch AM, Knize MG, Boobis AR *et al.* 1992. Intra- and interindividual variability in systemic exposure in humans to 2-amino-3,8-dimethylimidazo[4,5-f]quinoxaline and 2-amino-1-methyl- 6-phenylimidazo[4,5-*b*]pyridine, carcinogens present in cooked beef. *Cancer Res* **52**: 6216–6223.

Lynch AM, Murray S, Boobis AR *et al.* 1991. The measurement of MeIQx adducts with mouse haemoglobin *in vitro* and *in vivo*: implications for human dosimetry. *Carcinogenesis* **12**, 1067–1072.

Magagnotti C, Orsi F, Bagnati R *et al.* 2000. Effect of diet on serum albumin and hemoglobin adducts of 2-amino-1-methyl-6-phenylimidazo[4,5-*b*]pyridine (PhIP) in humans. *Int J Cancer* **88**: 1–6.

Magagnotti C, Pastorelli R, Pozzi S *et al.* 2003. Genetic polymorphisms and modulation of 2-amino-1-methyl-6-phenylimidazo[4,5-*b*]pyridine (PhIP)–DNA adducts in human lymphocytes. *Int J Cancer* **107**: 878–884.

Miller EC. 1978. Some current perspectives on chemical carcinogenesis in humans and experimental animals: Presidential Address. *Cancer Res* **38**: 1479–1496.

Mori Y, Koide A, Kobayashi Y *et al.* 2003. Effects of cigarette smoke and a heterocyclic amine, MeIQx on cytochrome P-450, mutagenic activation of various carcinogens and glucuronidation in rat liver. *Mutagenesis* **18**: 87–93.

Muckel E, Frandsen H, Glatt HR. 2002. Heterologous expression of human *N*-acetyltransferases 1 and 2 and sulfotransferase 1A1 in *Salmonella typhimurium* for mutagenicity testing of heterocyclic amines. *Food Chem Toxicol* **40**: 1063–1068.

Murray S, Lake BG, Gray S *et al.* 2001. Effect of cruciferous vegetable consumption on heterocyclic aromatic amine metabolism in man. *Carcinogenesis* **22**: 1413–1420.

Muscat JE, Wynder EL. 1994. The consumption of well-done red meat and the risk of colorectal cancer. *Am J Publ Health* **84**: 856–858.

Nagao M, Fujita Y, Wakabayashi K *et al.* 1983. Ultimate forms of mutagenic and carcinogenic heterocyclic amines produced by pyrolysis. *Biochem Biophys Res Commun* **114**: 626–631.

Nowell S, Coles B, Sinha R *et al.* 2002. Analysis of total meat intake and exposure to individual heterocyclic amines in a case–control study of colorectal cancer: contribution of metabolic variation to risk. *Mutat Res* **506–507**: 175–185.

Ohgaki H, Hasegawa H, Suenaga M *et al.* 1987. Carcinogenicity in mice of a mutagenic compound, 2-amino-3,8-dimethylimidazo[4,5-f]quinoxaline (MeIQx) from cooked foods. *Carcinogenesis* **8**: 665–668.

Paehler A, Richoz J, Soglia J *et al.* 2002. Analysis and quantification of DNA adducts of 2-amino-3,8-dimethylimidazo[4,5-f]quinoxaline in liver of rats by liquid chromatography/electrospray tandem mass spectrometry. *Chem Res Toxicol* **15**: 551–561.

Pfau W, Schulze C, Shirai T *et al.* 1997. Identification of the major hepatic DNA adduct formed by the food mutagen 2-amino-9*H*-pyrido[2,3-β]indole (AαC). *Chem Res Toxicol* **10**: 1192–1197.

Probst-Hensch NM, Sinha R, Longnecker MP *et al.* 1997. Meat preparation and colorectal adenomas in a large sigmoidoscopy-based case–control study in California (United States). *Cancer Causes Control* **8**: 175–183.

Ricicki EM, Soglia JR, Teitel C *et al.* 2005. Detection and quantification of *N*-(deoxyguanosin-8-yl)-4-aminobiphenyl adducts in human pancreas tissue using capillary liquid chromatography–microelectrospray mass spectrometry. *Chem Res Toxicol* **18**: 692–699.

Sachse C, Bhambra U, Smith G *et al.* 2003. Polymorphisms in the cytochrome P450 *CYP1A2* gene (*CYP1A2*) in colorectal cancer patients and controls: allele frequencies, linkage disequilibrium and influence on caffeine metabolism. *Br J Clin Pharmacol* **55**: 68–76.

Schut HA, Snyderwine EG. 1999. DNA adducts of heterocyclic amine food mutagens: implications for mutagenesis and carcinogenesis. *Carcinogenesis* **20**: 353–368.

Shimada T, Guengerich FP. 1991. Activation of amino-α-carboline, 2-amino-1-methyl-6-phenylimidazo[4,5-β]pyridine and a copper phthalocyanine cellulose extract of cigarette smoke condensate by cytochrome P-450 enzymes in rat and human liver microsomes. *Cancer Res* **51**: 5284–5291.

Shirai T, Cui L, Takahashi S *et al.* 1999. Carcinogenicity of 2-amino-1-methyl-6-phenylimidazo [4,5-*b*]pyridine (PhIP) in the rat prostate and induction of invasive carcinomas by subsequent treatment with testosterone propionate. *Cancer Lett* **143**: 217–221.

Shirai T, Kato K, Futakuchi M *et al.* 2002. Organ differences in the enhancing potential of 2-amino-1-methyl-6-phenylimidazo[4,5-*b*]pyridine on carcinogenicity in the prostate, colon and pancreas. *Mutat Res* **506–507**: 129–136.

Sinha R, Gustafson DR, Kulldorff M *et al.* 2000. 2-amino-1-methyl-6-phenylimidazo[4,5-*b*]pyridine, a carcinogen in high-temperature-cooked meat, and breast cancer risk. *J Natl Cancer Inst* **92**: 1352–1354.

Sinha R, Knize MG, Salmon CP *et al.* 1998a. Heterocyclic amine content of pork products cooked by different methods and to varying degrees of doneness. *Food Chem Toxicol* **36**: 289–297.

Sinha R, Kulldorff M, Chow WH *et al.* 2001. Dietary intake of heterocyclic amines, meat-derived mutagenic activity, and risk of colorectal adenomas. *Cancer Epidemiol Biomarkers Prevent* **10**: 559–562.

Sinha R, Peters U, Cross AJ *et al.* 2005. Meat, meat cooking methods and preservation, and risk for colorectal adenoma. *Cancer Res* **65**: 8034–8041.

Sinha R, Rothman N, Brown ED *et al.* 1994. Pan-fried meat containing high levels of heterocyclic aromatic amines but low levels of polycyclic aromatic hydrocarbons induces cytochrome P4501A2 activity in humans. *Cancer Res* **54**: 6154–6159.

Sinha R, Rothman N, Brown ED *et al.* 1995. High concentrations of the carcinogen 2-amino-1-methyl-6-phenylimidazo-[4,5-*b*]pyridine (PhIP) occur in chicken but are dependent on the cooking method. *Cancer Res* **55**: 4516–4519.

Sinha R, Rothman N, Salmon CP *et al.* 1998b. Heterocyclic amine content in beef cooked by different methods to varying degrees of doneness and gravy made from meat drippings. *Food Chem Toxicol* **36**: 279–287.

Skipper PL, Tannenbaum SR. 1990. Protein adducts in the molecular dosimetry of chemical carcinogens. *Carcinogenesis* **11**: 507–518.

Steineck G, Hagman U, Gerhardsson M *et al.* 1990. Vitamin A supplements, fried foods, fat and urothelial cancer. A case–referent study in Stockholm in 1985–1987. *Int J Cancer* **45**: 1006–1011.

Stillwell WG, Kidd LC, Wishnok JS *et al.* 1997. Urinary excretion of unmetabolized and phase II conjugates of 2-amino-1-methyl-6-phenylimidazo[4,5-*b*]pyridine and 2-amino-3,8-dimethylimidazo[4,5-f]quinoxaline in humans: relationship to cytochrome P4501A2 and N-acetyltransferase activity. *Cancer Res* **57**: 3457–3464.

Stillwell WG, Sinha R, Tannenbaum SR. 2002. Excretion of the N(2)-glucuronide conjugate of 2-hydroxyamino-1-methyl-6-phenylimidazo[4,5-*b*]pyridine in urine and its relationship to *CYP1A2* and *NAT2* activity levels in humans. *Carcinogenesis* **23**: 831–838.

Stillwell WG, Turesky RJ, Sinha R *et al.* 1999a. Biomonitoring of heterocyclic aromatic amine metabolites in human urine. *Cancer Lett* **143**: 145–148.

Stillwell WG, Turesky RJ, Sinha R *et al.* 1999b. N-oxidative metabolism of 2-amino-3,8-dimethylimidazo[4,5-f]quinoxaline (MeIQx) in humans: excretion of the N2-glucuronide conjugate of 2-hydroxyamino-MeIQx in urine. *Cancer Res* **59**: 5154–5159.

Sugimura T. 1997. Overview of carcinogenic heterocyclic amines. *Mutat Res* **376**: 211–219.

Sugimura T, Wakabayashi K. 1991. Heterocyclic amines: new mutagens and carcinogens in cooked foods. *Adv Exp Med Biol* **283**: 569–578.

Sugimura T, Wakabayashi K, Nakagama H *et al.* 2004. Heterocyclic amines: mutagens/carcinogens produced during cooking of meat and fish. *Cancer Sci* **95**: 290–299.

Totsuka Y, Fukutome K, Takahashi M *et al.* 1996. Presence of N2-(deoxyguanosin-8-yl)-2-amino-3,8-dimethylimidazo[4,5-f]quinoxaline (dG-C8-MeIQx) in human tissues. *Carcinogenesis* **17**: 1029–1034.

Turesky RJ, Constable A, Richoz J *et al.* 1998a. Activation of heterocyclic aromatic amines by rat and human liver microsomes and by purified rat and human cytochrome P450 1A2. *Chem Res Toxicol* **11**: 925–936.

Turesky RJ, Garner RC, Welti DH *et al.* 1998b. Metabolism of the food-borne mutagen 2-amino-3,8-dimethylimidazo[4,5-f]quinoxaline in humans. *Chem Res Toxicol* **11**: 217–225.

Turesky RJ, Parisod V, Huynh-Ba T *et al.* 2001. Regioselective differences in C(8)- and N-oxidation of 2-amino-3,8-dimethylimidazo[4,5-f]quinoxaline by human and rat liver microsomes and cytochromes P450 1A2. *Chem Res Toxicol* **14**: 901–911.

Turesky RJ, Skipper PL, Tannenbaum SR. 1987. Binding of 2-amino-3-methylimidazo[4,5-f]quinoline to hemoglobin and albumin *in vivo* in the rat. Identification of an adduct suitable for dosimetry. *Carcinogenesis* **8**: 1537–1542.

Turesky RJ, Vouros P. 2004. Formation and analysis of heterocyclic aromatic amine–DNA adducts *in vitro* and *in vivo*. *J Chromatogr B Analyt Technol Biomed Life Sci* **802**: 155–166.

Turteltaub KW, Dingley KH. 1998. Application of accelerated mass spectrometry (AMS) in DNA adduct quantification and identification. *Toxicol Lett* **102–103**: 435–439.

Turteltaub KW, Mauthe RJ, Dingley KH *et al.* 1997. MeIQx–DNA adduct formation in rodent and human tissues at low doses. *Mutat Res* **376**: 243–252.

Wakabayashi K, Nagao M, Esumi H *et al.* 1992. Food-derived mutagens and carcinogens. *Cancer Res* **52**: 2092–2098s.

Walters DG, Young PJ, Agus C *et al.* 2004. Cruciferous vegetable consumption alters the metabolism of the dietary carcinogen 2-amino-1-methyl-6-phenylimidazo[4,5-*b*]pyridine (PhIP) in humans. *Carcinogenesis* **25**: 1659–1669.

Willett WC, Hu FB. 2006. Not the time to abandon the food frequency questionnaire: point. *Cancer Epidemiol Biomarkers Prevent* **15**: 1757–1758.

Willett WC, Hu FB. 2007. The food frequency questionnaire. *Cancer Epidemiol Biomarkers Prevent* **16**, 182–183.

Wu RW, Panteleakos FN, Kadkhodayan S *et al.* 2000. Genetically modified Chinese hamster ovary cells for investigating sulfotransferase-mediated cytotoxicity and mutation by 2-amino-1-methyl-6- phenylimidazo[4,5-*b*]pyridine. *Environ Mol Mutagen* **35**: 57–65.

Zheng W, Gustafson DR, Sinha R *et al.* 1998. Well-done meat intake and the risk of breast cancer. *J Natl Cancer Inst* **90**: 1724–1729.

Zhu J, Chang P, Bondy ML *et al.* 2003. Detection of 2-amino-1-methyl-6-phenylimidazo[4,5-*b*]-pyridine–DNA adducts in normal breast tissues and risk of breast cancer. *Cancer Epidemiol Biomarkers Prevent* **12**: 830–837.

24

Practical Examples: Hormones

Sabina Rinaldi[1] and Rudolf Kaaks[2]

[1]IARC, Lyon, France and [2]German Cancer Research Centre, Heidelberg, Germany

24.1. INTRODUCTION

In an increasing number of well-designed epidemiological studies, levels of endogenous hormones have been related to the development/progression of several cancers, such as breast, endometrial and prostate cancers (see Box 24.1) (Kaaks *et al.* 2005a, 2005b; Rinaldi S *et al.* 2006; Eliassen *et al.* 2006; Missmer *et al.* 2004; Lukanova *et al.* 2004a, 2004b; Chan *et al.* 2002; Stattin *et al.* 2004), as well as to other chronic disease, such as osteoporosis (Lambrinoudaki *et al.* 2006; Rapuri 2004) and cardiovascular disease (Arnlov 2006; de Lecinana *et al.* 2007). In such studies, an essential requirement for an accurate estimation of hormone–disease associations is that relative hormone concentrations be measured with a maximum level of reliability, because for a correct estimation of relative risks, subjects should be ranked accurately by their hormone concentrations. Random errors in the measurements attenuate relative risk estimates and decrease the power of statistical tests for hormone–disease associations. Some important aspects of an accurate estimation of the associations between hormone levels and disease risk are the accuracy of laboratory methods used, the ways in which biological samples are collected and stored, the difficulty of a single hormone measurement (i.e. at one single point in time) to represent longer-term exposures, and the physiological complexity of relationships between circulating hormones vs. the concentrations of hormones locally within specific tissues. Since most of the specimens collected in existing biological banks for prospective cohort studies are blood samples, we will focus our attention on hormone measurements in blood plasma or serum.

24.2. HORMONE MEASUREMENTS FOR LARGE-SCALE EPIDEMIOLOGICAL STUDIES

The requirements for the measurements of hormones for large-scale epidemiological studies are different from the requirements in clinical settings.

First, in clinical settings, the major concern is to reliably measure a marker that, when present in very high or very low concentrations, can give an indication of a certain disease, and can help in making a clinical diagnosis. Methods applied in clinical settings, therefore, should as a minimum be able to distinguish between pathological vs. non-pathological (physiological) conditions. In contrast, in epidemiological studies, the focus is generally on the comparison of hormone concentrations in blood samples from subjects that do not necessarily suffer from extreme endocrine disorders, and whose plasma hormone levels are largely within the normal physiological range.

A second difference between clinical and epidemiological settings is that, in the clinical setting, the conditions under which blood samples are collected, e.g. fasting conditions, can be controlled, and sample volumes are generally sufficient to carry out the assays even with comparatively less sensitive methods. In epidemiology, by contrast,

Box 24.1   Endogenous sex steroids and breast cancer in post-menopausal women

Most of the established risk factors for breast cancer (age at first full-term pregnancy, parity, breast feeding, early menarche, late menopause) are related to alterations in endogenous hormone metabolism. Increased total and bioavailable endogenous oestrogens and androgens have for a long time been implicated as potential risk factors for breast cancer. Recently, results from case–control studies nested within the European Prospective Investigation into Cancer and nutrition (EPIC) cohort, including more than 600 breast cancer cases and twice as many controls, showed a strong increase in breast cancer risk among post-menopausal women who had elevated serum concentrations of androgens and oestrogens (Kaaks *et al.* 2005).

Relative risk (RR) of breast cancer among postmenopausal women by quintiles of serum steroid concentrations, EPIC study. Odds ratios, estimated by conditional logistic regression with study centre, age at blood donation, time of the day for blood donation and fasting status at blood donation as matching factors for breast cancer cases and control subjects

hormone	quintile	Cases/Controls	OR	95%CI	Ptrend
DHEAS					
	1	101/254	1.00		
	2	128/252	1.28	0.94-1.75	0.0002
	3	106/255	1.06	0.76-1.48	
	4	164/252	1.68	1.23-2.30	
	5	162/254	1.69	1.23-2.33	
Androstenedione					
	1	90/254	1.00		
	2	129/253	1.49	1.08-2.07	<0.0001
	3	126/253	1.43	1.03-1.99	
	4	156/253	1.82	1.32-2.50	
	5	162/254	1.94	1.40-2.69	
Testosterone					
	1	107/259	1.00		
	2	112/254	1.14	0.82-1.58	<0.0001
	3	129/255	1.33	0.96-1.84	
	4	149/257	1.56	1.12-2.16	
	5	171/255	1.85	1.33-2.57	
Oestrone					
	1	85/237	1.00		
	2	120/238	1.59	1.12-2.27	0.0001
	3	136/237	1.89	1.32-2.70	
	4	148/239	2.09	1.46-3.01	
	5	141/237	2.07	1.42-3.02	
Oestradiol					
	1	102/259	1.00		
	2	106/259	1.09	0.78-1.53	<0.0001
	3	136/261	1.44	1.04-2.00	
	4	146/257	1.71	1.22-2.41	
	5	182/261	2.28	1.61-3.23	
SHBG					
	1	155/260	1.00		
	2	156/260	0.98	0.75-1.30	
	3	117/261	0.72	0.54-0.98	0.004
	4	142/260	0.87	0.65-1.17	
	5	103/260	0.61	0.44-0.84	
Free testosterone					
	1	82/255	1.00		
	2	138/257	1.83	1.30-2.58	<0.0001
	3	141/254	1.92	1.37-2.70	
	4	134/257	1.86	1.31-2.66	
	5	172/255	2.50	1.76-3.55	
Free oestradiol					
	1	97/259	1.00		
	2	122/259	1.31	0.94-1.82	<0.0001
	3	120/259	1.30	0.93-1.82	
	4	156/259	1.80	1.30-2.50	
	5	176/259	2.13	1.52-2.98	

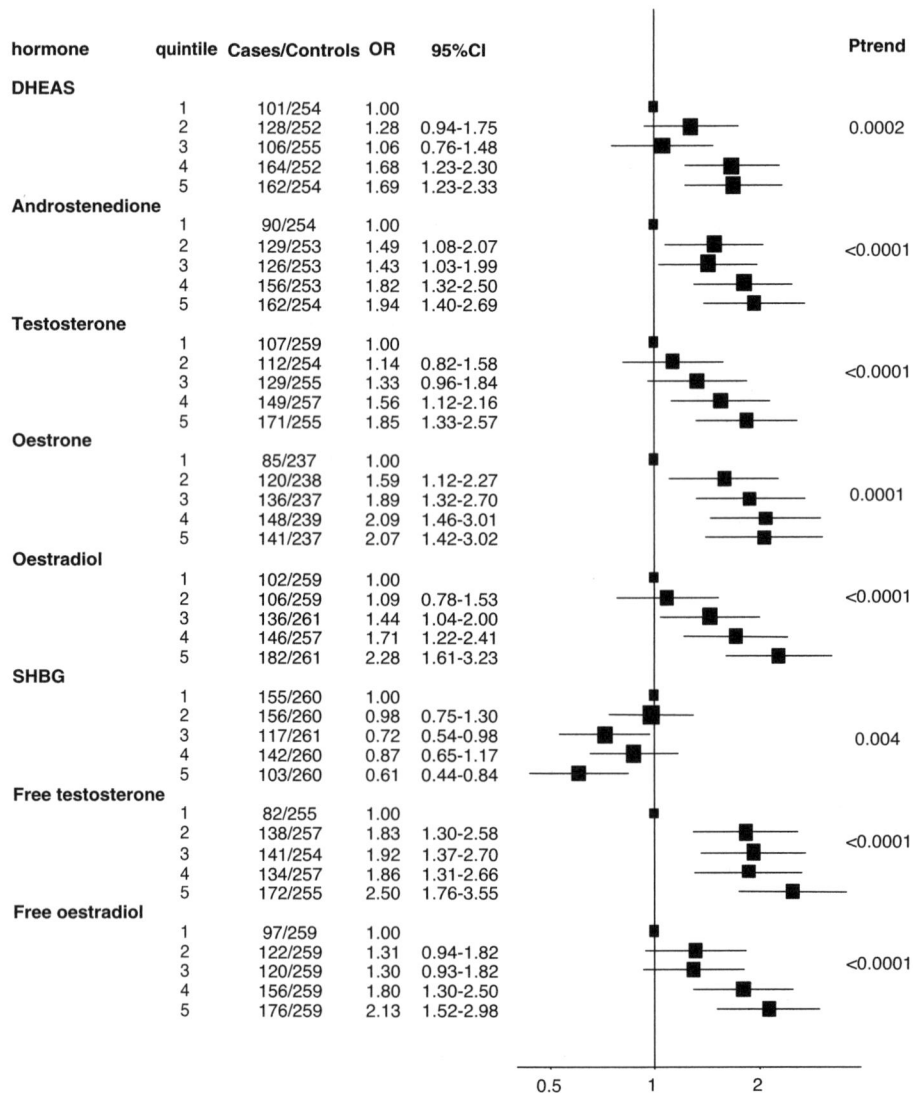

0.5   1   2

the conditions under which samples are collected cannot always be fully controlled, and plasma or serum volumes available for a given study are often more limited, as the same blood samples generally have to be used for a large number of different investigations. Therefore, assays for epidemiological studies should require only small volumes of sample. In addition, since epidemiological studies tend to include large numbers of study subjects, the assays should also be relatively fast and inexpensive, and preferably it should be possible to use automated equipment. In general, immunoassay is one of the techniques that best fulfil these criteria.

A further practical advantage of immunoassays is that to a high degree these can often be automated. In very large studies, automation of analyses generally helps avoid technical mistakes, such as errors in pipetting or erroneous interchanges of samples from different study subjects, and computer network connections of automated laboratory equipment (e.g. pipetting robots, ELISA readers) facilitates the fast and error-free transfer and treatment of data generated. A higher speed of assays also has an advantage in terms of costs, and will more easily allow the measurement of several hormones on the same day, avoiding repeated thawing–freezing cycles or sub-aliquoting of serum samples. Finally, it will generally also allow large series of serum samples from a single study to be measured all by reagents that were produced in a single lot by the manufacturer, which will lead to better standardization of measurements across assay batches.

Besides the technical considerations above, the most important requirement for prospective epidemiological studies is the accuracy (i.e. validity and reproducibility) of the measurements. Misclassification of subjects due to the low reproducibility and accuracy of an analytical method would lead to a misclassification of the subjects according to their hormone levels, and to an attenuation of the relative risk estimates of the disease. Therefore, before using an assay, one should examine whether it provides reproducible results within the range of the concentrations of normal physiological range. In addition, it should preferably be tested whether the measurements are valid. The latter, however, will require comparisons with

reference measurements known to be highly accurate, or otherwise at least measurements that do not have the same sources of error as the assay method to be tested. For example, measurements of serum oestradiol concentrations in the blood of postmenopausal women, obtained by commercially available direct immunoassays, may be tested against immunoassays done after organic extraction and prepurification of sex steroid fractions, and using a different immunoassay based on other antibodies (see Figure 24.1; Rinaldi *et al.* 2001).

While accurate ranking of individuals by relative hormone levels is the primary requisite for the usefulness of hormone assays for epidemiology, a second desirable characteristic of measurements is that they should also be obtained on a valid absolute scale. Correct scaling of measurements enhances comparability of results between studies, and the pooling of study data for combined re-analyses. In practice, it has been observed that absolute levels for a given hormone can vary substantially, depending on the assay method used, and sometimes vary even when using the same method in different laboratory contexts (Gail *et al.* 1996; Hankinson *et al.* 1994; McShane *et al.* 1996; Potischman *et al.* 1994). For example, immunoassays from different manufacturers have shown up to 20-fold differences for mean levels of oestradiol in blood serum from post-menopausal women, and up to almost three-fold differences for mean levels of androstenedione (Rinaldi *et al.* 2001; Schioler V, Thode J 1988). The possible causes for such differences in between-assay absolute levels are many. One of the major sources of errors for direct immunoassays is what is often referred to as 'matrix effects', i.e. specific interferences by other substances in the biological material (matrix) that can interfere in the antigen–antibody reaction of the immunoassay. Bias in mean group-level measurements may thus occur, because of systematic differences between the matrices of natural samples (serum or plasma), and the matrix of the standards used to quantify the absolute concentration, which are generally reconstituted, artificial sera. Extraction steps (by liquid organic solvents or by solid-phase extractions) before measurements by immunoassays, as those applied when using indirect methods, can be used to eliminate such

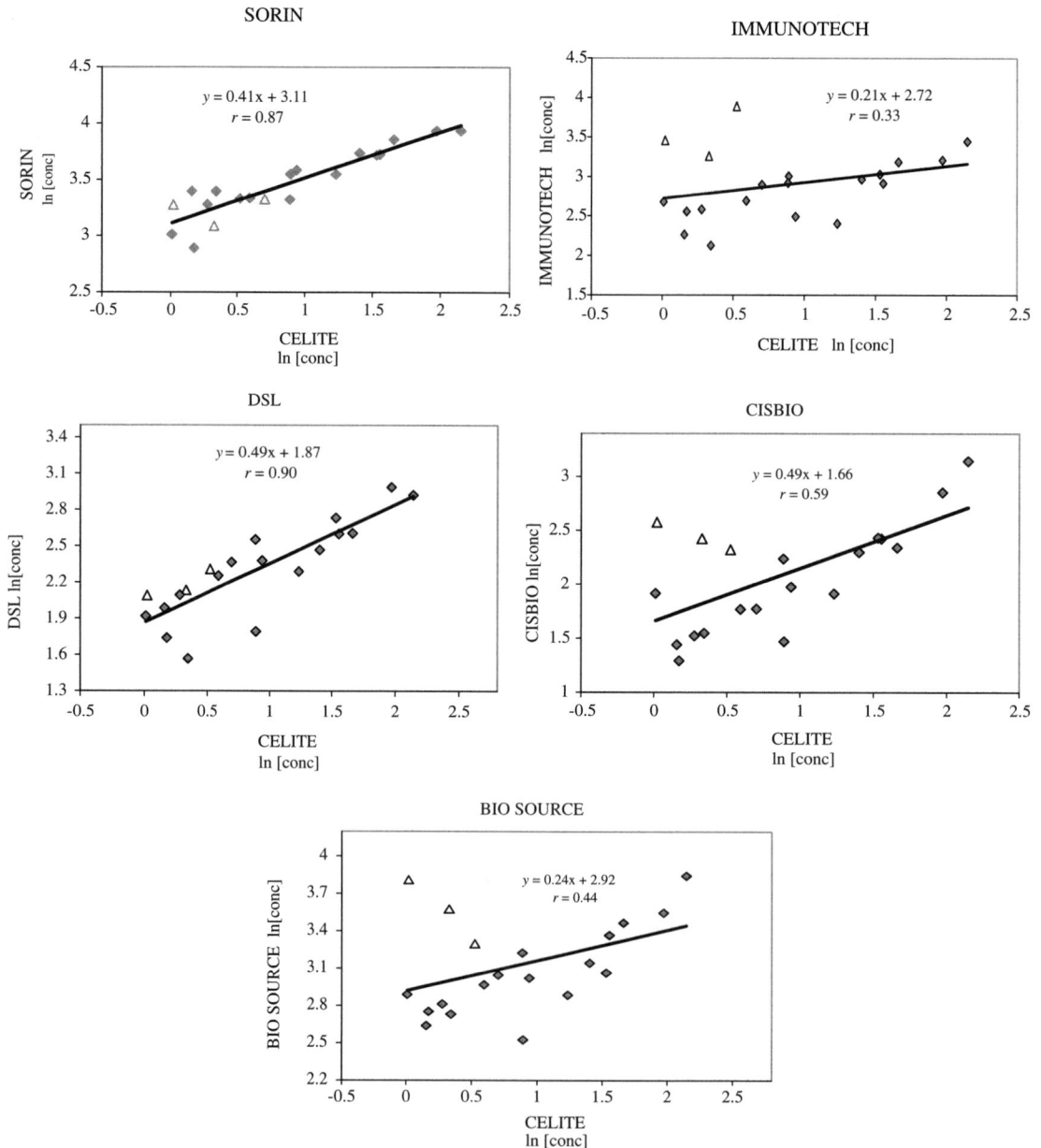

Figure 24.1 Scatter plots for oestradiol measurements (pg/ml) in 20 post-menopausal women who were part of a large-scale epidemiological study. Measurements plotted were obtained by direct assays from different brands and by a reference method (celite method, immunoassay after chromatographic purification). Endogenous oestradiol concentrations were measured in 20 serum samples from post-menopausal women who were part of a large-scale epidemiological study. Measurements were made using commercially available radioimmunoassays (from five different brands) and by an in-house radioimmunoassay after extraction and purification of the serum samples on celite columns (indirect assay, considered here as the reference method). When comparing results from direct and indirect assays, we observed good correlations for two brands of the five tested (DSL and SORIN), and low correlations for the other three brands. These low correlations were mainly due to the presence of three serum samples (indicated as triangles in the graphs), suggesting strong matrix effects influencing some of the direct assay. The results of this validation study are presented in Rinaldi *et al.* (2001)

matrix effects. Unfortunately, such extraction and prepurification steps are still quite cumbersome to perform, and generally also require much larger sample volumes compared to direct immunoassays. However, indirect assays after extraction, and further prepurification of analytes, e.g. by chromatography, can be good tools for the validation of simpler and faster assays that would be more easily applied to large-scale epidemiological studies.

24.3. LABORATORY METHODS

Reference methods

Mass-spectrometric identification and quantification after organic extraction and chromatographic pre-purification (see Box 24.2) [gas chromatography–mass spectrometry (GC–MS) or liquid chromatography (LC)–MS–MS methods] is generally considered the gold standard method for measurement of steroid hormones, because of the high specificity of the detection method used (Shimada *et al.* 2001; Stanczyk 2006; Giese 2003). However, in spite of improvements during recent years, such methods do not always reach the level

of sensitivity that is required to measure androgens or oestrogens in relatively small quantities of blood (a few ml, i.e. the quantities that are available in biological repositories of large-scale epidemiological studies), especially when hormone concentrations are low. A typical example of the difficulty of hormone measurements is the measurement of endogenous oestradiol (E_2) or its metabolites in post-menopausal women, since endogenous E_2 in this population is present only at a concentration of a few pg/ml (see Box 24.3). In addition, GC–MS or LC–MS–MS methods are of long duration and have remained relatively cumbersome to perform, so that their application to large-scale epidemiological studies remains difficult.

For sex steroids, other accurate methods exist that can also be used for reference measurements. This includes 'indirect' radio-immunological assays after extraction and chromatographic separation (Wheeler 2006). This purification step strongly reduces the eventual matrix effects of the samples. Those measurements were once the standard technique in many clinical laboratories, and require smaller volumes compared to the methods using

---

Box 24.2 Principles of gas chromatography–mass spectrometry detection

Before analyses through a mass spectrometer, the mixture of compounds to analyse (e.g. an organic extract from serum samples of sex steroids, and other liposoluble substances) is injected into a gas chromatograph (a), where the samples are vapourized into a chamber and sent to an analytical capillary column through the injection of a gas. The different compounds are separated by the physicochemical interactions with the analytical column. After this first separation step, the compounds go into the mass spectrometer (b), where they are electrically charged ('ionized') and broken into fragments. The fragmented ions are accelerated by manipulation of the charged particles through the mass spectrometer, while uncharged particles are eliminated. Ions go down different paths based on their mass:charge ratio (*m/z*) and their different trajectories are read by a detector. This allows the development of a mass spectrum (c), which allows a highly specific identification of amounts of each of the ion fragments.

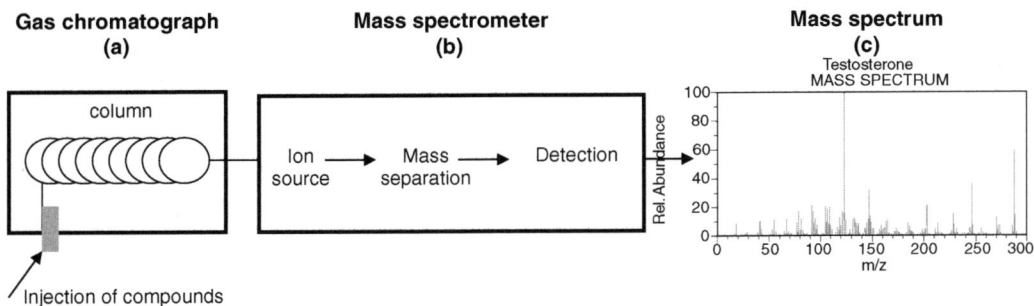

Box 24.3 Measurements of oestrogen metabolites in biological fluids, and their application to epidemiology

Although the mechanisms by which oestrogens increase carcinogenesis are not entirely clear, substantial evidence indicates that, in addition to the mitogenic properties of oestrogens, the hydroxylation of oestrogens may lead to genotoxic metabolites (Cavalieri *et al.* 2006). Preliminary epidemiological evidence suggests that oestrogen metabolism via the 16α-hydroxy pathway is increased in breast cancer patients compared to controls, and an inverse relationship has been found between breast cancer risk and the ratio of urinary concentrations of 2-hydroxy/4-hydroxy or 2-methoxy/4-methoxy oestrogens. However, only a few epidemiological studies have looked at the relationship between oestrogen metabolite concentrations in biological fluids and cancer risks. A major reason for this is the lack of an analytical method with sufficient sensitivity and specificity to measure these metabolites accurately in small quantities of plasma, the biological sample most often collected in prospective cohort studies.

Several analytical methods have been proposed for the measurements of oestrogen metabolites in biological fluids, including radioimmunoassays (RIAs) (McGuinness *et al.* 1994; Emons *et al.* 1982; Kono *et al.* 1982), enzyme-linked immunoassays (ELISAs) (Falk *et al.* 2000), high-performance liquid chromatography (HPLC) with electrochemical (EC) detection (Castagnetta *et al.*1992), LC–mass spectrometry (LC–MS–MS; Beck *et al.* 2005; Xu *et al.* 2002, 2005) and gas chromatography–mass spectrometry (GC–MS; Lee and Peart 1998; Castagnetta *et al.* 1992; Gerhardt *et al.* 1989; Xiao 2001; Zacharia *et al.* 2004). Immunoassay would be the method that best meets the requirement of epidemiological studies, since it requires small volumes of sample for analyses. However, RIAs have been developed for only a small selection of existing sex steroids and their metabolites, and for certain sex steroid metabolites they have may have only limited specificity, accuracy or reproducibility (Kono *et al.* 1982). Valid ELISAs for the measurements of 2- and 16-hydroxy forms are commercially available, but they have been developed mainly for measurements in urine, and less so in serum/plasma samples (Falk *et al.* 2000). HPLC methods with EC detection, or LC–MS–MS and GC–MS methods, have been applied mainly to urine samples (Castagnetta *et al.* 1992; Xu *et al.* 2005; Xiao *et al.* 2001) or to rat plasma (Zacharia *et al.* 2004), or in humans with very high metabolite concentrations (Lakhani *et al.* 2005). Their applicability for measurements in the normal, non-pathological range on a large scale still needs to be investigated.

GC–MS or LC–MS–MS. These methods too, however, remain quite labour-intensive and slow, and therefore cannot be easily applied for routine measurements in large-scale epidemiological studies.

Other hormones (e.g. IGF-I, its binding proteins or SHBG), proteins with much higher molecular weight compared to steroids, cannot be measured by GC–MS. Reference methods for the measurements of these proteins include measurements by binding capacity (Nisula *et al.* 1978), Western ligand blotting and Western immunoblotting (Rajaram 1997). These methods are quite laborious, require experienced technicians, and cannot always be used for quantitative analyses.

Direct immunoassays

Direct immunoassays are assays that are performed without prior extraction and/or prepurification steps (see Box 24.4). Generally, such direct assays require very small sample volumes (10–200 μl) of serum, are easy to perform and to automate, and therefore can be easily applied to large series of measurements (Rinaldi *et al.* 2001). However, since the assays are applied to biological specimens without any pre-purification step, this type of assay does require some further attention and careful evaluation before use. For example, cross-reactivity tests should be used to verify that the antibodies used in the immunoassay are highly specific for the hormone of interest, and preliminary tests may

---

Box 24.4   Principles of a radioimmunoassay for the measurement of serum sex steroid concentrations (e.g. testosterone; 'competitive' assay)

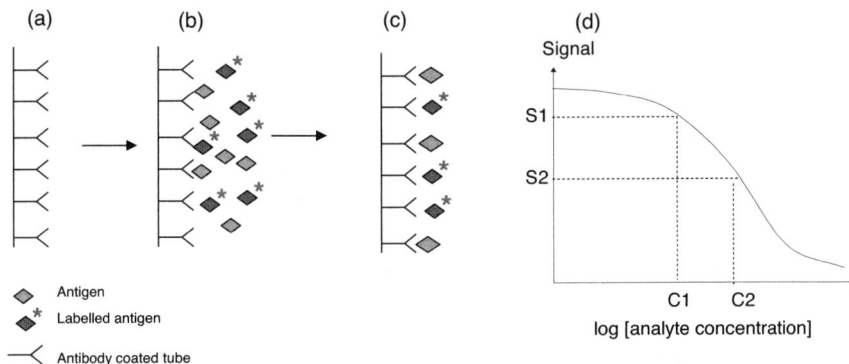

In antibody-coated tubes (a), serum samples or calibrators (immunoassay standards) are incubated with ^{125}I-labelled testosterone (b). After an incubation time, the liquid content of the tubes is washed away (c), and the amount of radioactivity bound to the antibodies is determined using a gamma counter. The unknown testosterone concentrations are obtained by interpolation from the standard curved based on the concentrations of the calibrators (d). The higher the concentration of serum testosterone, the lower the radioactivity signal read.

---

also be needed to examine whether there are any non-specific interferences by other substances (e.g. lipids) that can be present in blood plasma (matrix effects). These preliminary tests are of particular importance for hormones that are present in very low concentrations, such as E_2 in post-menopausal women. For most hormones, direct immunoassays have been produced and commercialized by several companies, and therefore are easily available. However, these assays are mostly produced for use in clinical settings, so their application to epidemiology must be carefully evaluated.

24.4. VALIDATION AND REPRODUCIBILITY OF HORMONE MEASUREMENTS

Validation of hormone assays ideally is based on comparisons with reference measurements that are (almost) perfectly accurate (e.g. mass-spectrometric measurements). In the absence of such a perfect reference method, one may also test against alternative measurements obtained by techniques that may be subject to some error, but where the errors are known to be entirely independent from those in the assay

to be tested, e.g. direct immunoassay vs. a well-characterized indirect assay, i.e. after extraction and pre-purification, to avoid matrix effects, based on different antibodies.

Validation studies should be performed on a sufficient number of samples to give reliable results from the statistical analyses. Ideally, validation studies should be performed on a representative subsample of subjects that are part of the main epidemiological study; this will allow a better evaluation not only of variances of errors in the measurements, but also the between-subject variation that is representative of the variation within the cohort (Rinaldi *et al.* 2001). As accurate ranking of subjects by relative hormone levels is a primary requisite for the usefulness of hormone measurements in epidemiological studies, a key statistic to look at is coefficient of correlation between measurements (Pearson, Spearman or intra-class correlation coefficients). A high correlation coefficient between measurements obtained by the method to be validated and those obtained by the reference method means that subjects will be ranked very similarly from low to high hormone levels by either method. Estimates of the correlation coefficients can also be used to correct for attenuation biases in relative risk estimates (de Klerk *et al.* 1989).

In the absence of any reasonable reference method with which to compare an assay, one should at the very least test the reproducibility of the assays, within and between laboratories. A number of studies on the inter-laboratory concordance of hormone assays have been published (Gail *et al.* 1996; Hankinson *et al.* 1994). These generally show a high degree of variability in the reproducibility of assays and a large variation in absolute levels when comparing different measurements from different assays, or measurements performed using the same assay in different laboratories. Besides the reproducibility, it may also be possible to compare the results from different direct assays; although none would be considered a perfect reference, concordance of results in terms of high correlations will at least give some confidence in the validity of the relative ranking by the methods used. Additional practical criteria for selection of the preferred assay may then further depend, for example, on robustness of assay protocols and standard curves, or on the perceived validity of the absolute scale on which measurements are made (e.g. with respect to a known physiological reference range).

24.5. SAMPLE COLLECTION AND LONG-TIME STORAGE

Differences in the protocols and in the handling of the samples during their collection can add substantial sources of variation in hormone measurements. Fasting vs. non-fasting status of the subjects, for example, can influence the matrix of the blood sample, and can influence the results of hormones (e.g. insulin) that depend on fasting status. Additional sources of variation are time at blood donation (some hormones have circadian or seasonal rhythms, i.e. their concentrations vary through the day or through a whole year), phase of the menstrual cycle in pre-menopausal women (above all for the measurement of oestrogens), and the time between blood withdrawal and final storage (hormones can still be metabolized by enzymes present in whole blood). The temperature

of storage is also an important issue. Storage facilities exist that range from −20°C (freezers) to −196°C degrees (liquid nitrogen containers). Several papers have been published on the stability of hormones (sex steroids and growth factors) in frozen samples (Garcia-Closas *et al.* 2000), and they agreed on the fact that most of these hormones are stable for several years when stored in freezers at −70°C. To be sure to minimize all these sources of errors, selection of cases and matched controls should take all these variables into account in the matching criteria.

24.6. DOES A SINGLE HORMONE MEASUREMENT REPRESENT LONG-TERM EXPOSURE?

Another important issue in epidemiological studies relying on laboratory biomarkers is the estimation of long-term exposure by a single measurement. At the time of recruitment into the study, usually only one blood withdrawal is made per subject. As mentioned above, however, hormone levels may vary for a given subject, either systematically (e.g. during the menstrual cycle) or in a less predictable ('random') manner. It is therefore very important that the measurement of the endogenous hormones present in this unique blood sample be representative of the average level over a longer time period.

The more representative the measurement of a biomarker in a single blood sample is for a longer time period, the larger the possibility of detecting true differences in exposure among the study subjects. Studies relying on measurements of pulsatile hormones, such as GH or LH, would be difficult to interpret, because of the difficulty of classifying subjects with respect to their long-term exposure to these hormones. By contrast, androgen concentrations in blood have been shown to be quite reproducible over short periods of time (a few years) in women. Even though hormone levels tend to decrease with age, correlations of repeated measurements over time are reasonably high, with intra-class correlations (ICC) of up to 0.89–0.93 for dehydroepiandrosterone sulphate (DHEAS) and about 0.57–0.88 for androstenedione and testosterone (Lukanova *et al.* 2003; Muti *et al.* 1996). Concentrations of sex hormone-binding globulin (SHBG) in blood also show a high degree of reproducibility over time, with ICCs of 0.85–0.90 (Eliassen *et al.* 2006; Lukanova *et al.* 2003, 2004a). Oestrogen concentrations have been shown a good reproducibility over time in post-menopausal women, with ICCs of 0.58–0.80 and 0.66-0.83 for oestradiol and oestrone, respectively, but have a much lower reproducibility over time in pre-menopausal women, because of the large intra-individual variations in oestrogen levels during the menstrual cycle. Nevertheless, when considering oestradiol concentrations within a single phase of menstrual cycle, the reproducibility over time improves, with ICCs of 0.45–0.62 within the luteal phase and 0.38–0.53 within the follicular phase (Muti *et al.* 1996; Missmer *et al.* 2006). Contrary to its total absolute levels, the percentage of oestradiol free or unbound to SHBG has been shown to be highly reproducible over time, with ICCs between 0.72 and 0.94 in both pre- menopausal (even when the phase of the menstrual cycle is disregarded) and post-menopausal women (Toniolo P *et al.* 1994). Like androgens and oestrogens, IGF-I and IGFBP-3 concentrations also decrease with increasing age (Rajaram 1997). However, ICCs for reproducibility over time for periods of about 1–5 years are quite high, of at least 0.66 for IGF-I and about 0.86 for IGFBP-3 (Lukanova *et al.* 2004b; Missmer *et al.* 2006).

24.7. INTERPRETATION OF MEASUREMENTS OF CIRCULATING HORMONES

Most of the biobanks related to large-scale prospective studies do not have tissue repositories and mainly collect serum or plasma. It would be logical to think that cancer development or progression would be more related to the concentrations of hormones at the level of the tissues in which tumours actually develop, rather than to circulating levels. However, for obvious reasons, it is not possible to collect samples of breast, prostate or other tissues from healthy control subjects. Thus, most epidemiological studies have focused on levels of hormones in the circulation that are likely to be at least an important determinant of tissue levels. The quantitative relationships between circulating

hormone concentrations and tissue concentrations, however, are still quite unknown for many hormones.

In pre-menopausal women, androgens are produced by both the adrenal glands and the ovaries, while oestradiol is mainly produced by the ovaries, and it is then released into the circulation. Partially, however, active androgens (testosterone, dihydrotestosterone [DHT]) and oestrogens (oestrone, oestradiol) can also be synthesized locally within tissues, from precursors molecules (e.g. dehydroepiandrosterone (DHEA), DHEAS), distributed into the circulation by the adrenal and ovarian glands (Labrie *et al.* 2003). Likewise, in adipose tissue, androgens (androstenedione, testosterone) are converted into oestrogens (oestrone, oestradiol) and in post-menopausal women most of the circulating oestrogens are no longer produced by the ovaries, but mainly by the peripheral aromatization of androgens in adipose tissue. In postmenopausal women, the concentrations of oestrogens in breast tumours is at least 20 times higher than the concentration present in plasma (Simpson *et al.* 2005), and no significant difference between oestradiol concentrations in breast cancer tissues between pre- and post-menopausal women is observed (Labrie *et al.* 2003).

Metabolic conversions in peripheral tissues (such as liver, kidney, prostate, skin) are very important: locally, testosterone can be converted into the more potent androgen dehydrotestosterone (DHT) through the action of the 5α-reductase enzyme, which can be in turn be converted into 5α-androstane, 3α,17β-diol and their glucuronide and sulphate forms (Labrie *et al.* 2003) (Figure 24.2). Testosterone can also be converted into oestradiol, which in its turn can be further metabolized into hydroxy and methoxy oestrogens (Figure 24.3). The hydroxylation of 17β-oestradiol (E_2) and oestrone (E_1) occurs via two major pathways, one involving 16α-hydroxylation to 16α-(OH)E_1 and oestriol (E_3), and a second leading to catechol oestrogens, such as 2-(OH) and 4-(OH) E_1 and E_2 (Ball and Knuppen 1980). The 2-(OH) and 4-(OH) oestrogens are further metabolized to methoxy oestrogens (Pribluda and Green 1998).

The important conversion of steroids at tissue levels shows that the measurements of circulating androgens and oestrogens in pre-menopausal women may not be representative of total androgenic or oestrogenic activity, or of activity of androgens and

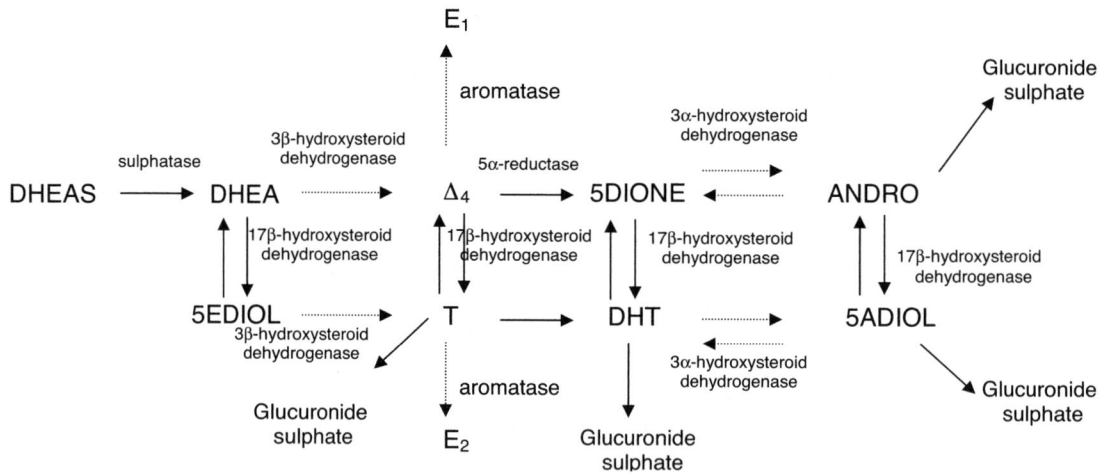

Figure 24.2 Androgen metabolism in tissues. DHEAS, dehydroepiandrosterone sulphate; DHEA, dehydroepiandrosterone; Δ4, androstendione; 5EDIOL: androst-5-ene-3β,17β-diol; 5DIONE: 5α-androstanedione; ANDRO: androsterone; 5ADIOL, 5α-androstanediol; T, testosterone; DHT, dehydrotestosterone; E₁, oestrone; E₂, oestradiol

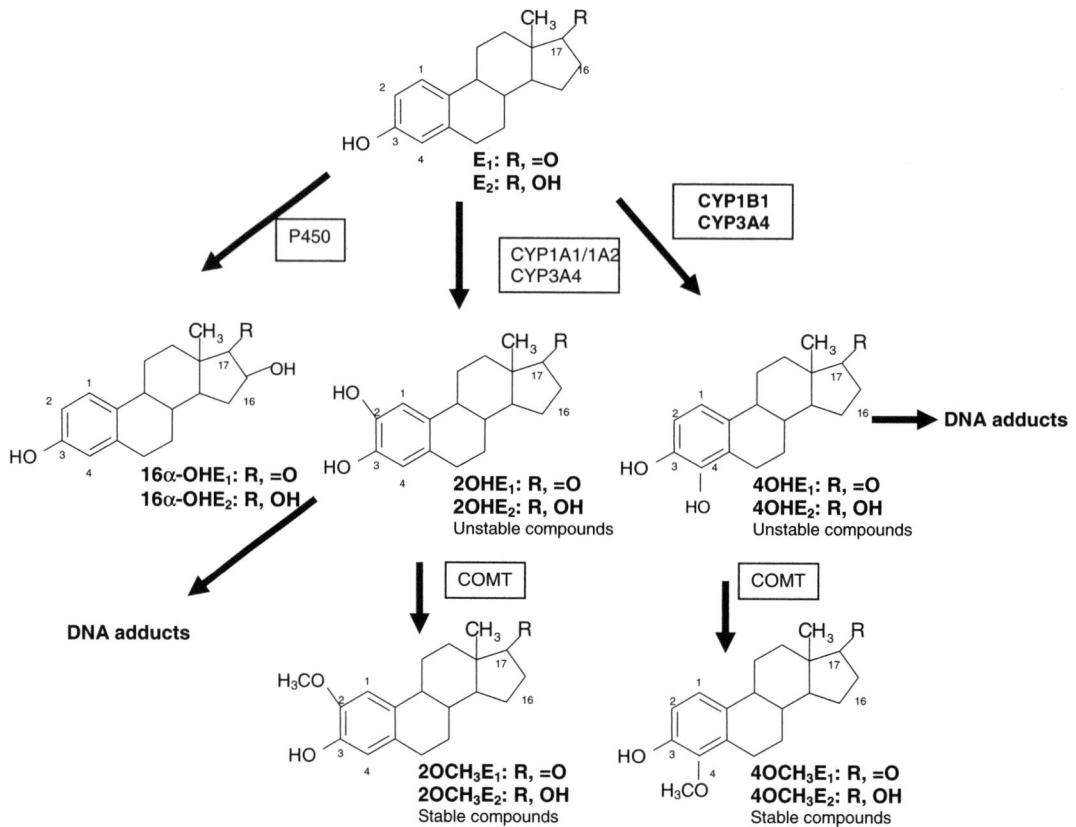

Figure 24.3 Oestrogen metabolism in organ tissues. E₁, oestrone; E₂, oestradiol; COMT, cathecol-*O*-methyltransferase; CYP, cytochrome P450.

oestrogens in tissues, but that they may represent only the part of steroids that are synthesized in the adrenal and the ovary (Labrie *et al*. 2003).

IGF-I, IGFBP-3 and other IGF-binding proteins are hormones that are produced in the liver, which contributes to more than 80% of these peptides in the circulation, but also in most other tissues. This, again, complicates the interpretation of findings relating cancer risk to levels of IGF-I and/or IGF-I in the circulation. It has been estimated that blood concentrations of IGF-I are about 50% higher than those found in tissues (Holly 2004) and there appears to be a large extracellular reservoir of IGF-I that is maintained in complex with IGFBP-3, IGFBP-5 and other binding proteins. The regulation of IGF-I bioavailability to IGF-receptors is very complex and depends, amongst other things, on the concentrations of IGF-I, as well as of IGFBP-3 and at least five other IGF-binding proteins. It is assumed that blood levels of IGF-I and its major binding protein, IGFBP-3, are correlated to the levels found in tissues, but the magnitude of this correlation is still quite unknown. For example, we know that oestrogen replacement therapy reduces the hepatic synthesis and circulating levels of IGF-I, but it may not reduce the concentration of IGF-I tissue levels. Higher IGFBP-3 levels can increase IGF-I levels by increasing its half-life in the circulation, but this increase is not necessarily reflected by higher IGF-I bioactivity within tissues.

An example of the relationship between tissue hormone levels and circulating hormone levels is given by a liver-specific specific IGF-1 deleted mouse model, LID (Yakar *et al*. 2004). This mouse model has 75% reduced circulating IGF-I. However, despite this reduction, the growth and the development of the mice had been found to be normal, even though a further reduction of IGF-I concentrations did lead to growth retardation. One explanation for the normal growth could probably be the 'normal' free IGF-I concentrations that have been found in the mouse. When adenocarcinomas were implanted in these mice, reduced IGF-I circulating levels resulted in reduced tumour development, tumour growth and metastasis compared to mice with normal IGF-I levels, indicating that circulating levels are somehow related to tissue levels, and that

IGF-I circulating in blood plays a role in cancer progression. More studies are, however, needed to clarify this relationship.

24.8. CONCLUSIONS

Epidemiological studies have shown important relationships between levels of circulating hormones and cancer risk. For example, breast cancer risk has been found to be increased among postmenopausal women who have elevated blood concentrations of oestrogens as well as androgens (Kaaks *et al*. 2005a; Missmer *et al*. 2004), and prostate cancer risk has been found to be increased among men who have elevated blood concentrations of IGF-I. These successes have been the result of careful epidemiological study design (e.g. using case–controls studies nested within prospective cohorts, with careful matching of the case and control subjects), and the careful selection of hormone assays (Rinaldi *et al*. 2001, 2005). Other long presumed relationships, such as that between circulating levels of total testosterone and prostate cancer risk, have remained elusive (Eaton *et al*. 1999; Severi *et al*. 2006). These results provide some further pieces of evidence about possible aetiological pathways involved in the development of these diseases. However, further studies are needed to understand in greater detail the complex physiological relationships between levels of hormones in the circulation and local tissue levels.

References

Arnlov J, Pencina MJ, Amin S *et al*. 2006. Endogenous sex hormones and cardiovascular disease incidence in men. *Ann Intern Med* **145**: 176–184.

Ball P, Knuppen R. 1980. Catecholoestrogens (2- and 4-hydroxyoestrogens): chemistry, biogenesis, metabolism, occurrence and physiological significance. *Acta Endocrinol Suppl (Copenh)* **232**: 1–127.

Beck IC, Bruhn R, Gandrass J *et al*. 2005. Liquid chromatography–tandem mass spectrometry analysis of estrogenic compounds in coastal surface water of the Baltic Sea. *J Chromatogr A* **1090**: 98–106.

Castagnetta LA, Granata OM, Arcuri FP *et al*. 1992. Gas chromatography/mass spectrometry of catechol estrogens. *Steroids* **57**: 437–443.

Cavalieri E, Frenkel K, Liehr JG *et al.* 2000. Estrogens as endogenous genotoxic agents – DNA adducts and mutations. *J Natl Cancer Inst Monogr* 75–93.

Chan JM, Stampfer MJ, Ma J *et al.* 2002. Insulin-like growth factor-I (IGF-I) and IGF binding protein-3 as predictors of advanced-stage prostate cancer. *J Natl Cancer Inst* **94**: 1099–1106.

de Klerk NH, English DR, Armstrong BK. 1989. A review of the effects of random measurement error on relative risk estimates in epidemiological studies. *Int J Epidemiol* **18**: 705–712.

de Lecinana MA, Egido JA, Fernandez C *et al.* 2007. Risk of ischemic stroke and lifetime estrogen exposure. *Neurology* **68**: 33–38.

Eaton NE, Reeves GK, Appleby PN *et al.* 1999. Endogenous sex hormones and prostate cancer: a quantitative review of prospective studies. *Br J Cancer* **80**: 930–934.

Eliassen AH, Missmer SA, Tworoger SS *et al.* 2006. Endogenous steroid hormone concentrations and risk of breast cancer among premenopausal women. *J Natl Cancer Inst* **98**: 1406–1415.

Emons G, Klinger B, Haupt O, Ball P. 1982. Radioimmunoassay for 4-hydroxyestrone 4-methyl ether in human urine. *Horm Metab Res* **14**: 376–379.

Falk RT, Rossi SC, Fears TR *et al.* 2000. A new ELISA kit for measuring urinary 2-hydroxyestrone, 16α-hydroxyestrone, and their ratio: reproducibility, validity, and assay performance after freeze–thaw cycling and preservation by boric acid. *Cancer Epidemiol Biomarkers Prevent* **9**: 81–87.

Gail MH, Fears TR, Hoover RN *et al.* 1996. Reproducibility studies and interlaboratory concordance for assays of serum hormone levels: estrone, estradiol, estrone sulfate, and progesterone. *Cancer Epidemiol Biomarkers Prevent* **5**: 835–844.

Garcia-Closas M, Hankinson SE, Ho S *et al.* 2000. Factors critical to the design and execution of epidemiologic studies and description of an innovative technology to follow the progression from normal to cancer tissue. *J Natl Cancer Inst Monogr* 147–156.

Gerhardt K, Ludwig-Kohn H, Henning HV *et al.* 1989. Identification of oestrogen metabolites in human urine by capillary gas chromatography and mass spectrometry. *Biomed Environ Mass Spectrom* **18**: 87–95.

Giese RW. 2003. Measurement of endogenous estrogens: analytical challenges and recent advances. *J Chromatogr A* **1000**: 401–412.

Hankinson SE, Manson JE, London SJ *et al.* 1994. Laboratory reproducibility of endogenous hormone levels in postmenopausal women. *Cancer Epidemiol Biomarkers Prevent* **3**: 51–56.

Holly J. 2004. Physiology of the IGF system. *Novartis Found Symp* **262**: 19–26.

Kaaks R, Berrino F, Key T *et al.* 2005b. Serum sex steroids in premenopausal women and breast cancer risk within the European Prospective Investigation into Cancer and Nutrition (EPIC). *J Natl Cancer Inst* **97**: 755–765.

Kaaks R, Rinaldi S, Key TJ *et al.* 2005a. Postmenopausal serum androgens, oestrogens and breast cancer risk: the European prospective investigation into cancer and nutrition. *Endocr Relat Cancer* **12**: 1071–1082.

Kono S, Merriam GR, Brandon DD *et al.* 1982. Radioimmunoassay and metabolism of the catechol estrogen 2-hydroxyestradiol. *J Clin Endocrinol Metab* **54**: 150–154.

Labrie F, Luu-The V, Labrie C *et al.* 2003. Endocrine and intracrine sources of androgens in women: inhibition of breast cancer and other roles of androgens and their precursor dehydroepiandrosterone. *Endocr Rev* **24**: 152–182.

Lakhani NJ, Lepper ER, Sparreboom A *et al.* 2005. Determination of 2-methoxyestradiol in human plasma, using liquid chromatography/tandem mass spectrometry. *Rapid Commun Mass Spectrom* **19**: 1176–1182.

Lambrinoudaki I, Christodoulakos G, Aravantinos L *et al.* 2006. Endogenous sex steroids and bone mineral density in healthy Greek postmenopausal women. *J Bone Miner Metab* **24**: 65–71.

Lee HB, Peart TE. 1998. Determination of 17 β-estradiol and its metabolites in sewage effluent by solid-phase extraction and gas chromatography/mass spectrometry. *J AOAC Int* **81**: 1209–1216.

Lukanova A, Lundin E, Akhmedkhanov A *et al.* 2003. Circulating levels of sex steroid hormones and risk of ovarian cancer. *Int J Cancer* **104**: 636–642.

Lukanova A, Lundin E, Micheli A *et al.* 2004a. Circulating levels of sex steroid hormones and risk of endometrial cancer in postmenopausal women. *Int J Cancer* **108**: 425–432.

Lukanova A, Zeleniuch-Jacquotte A, Lundin E *et al.* 2004b. Prediagnostic levels of C-peptide, IGF-I, IGFBP -1, -2 and -3 and risk of endometrial cancer. *Int J Cancer* **108**: 262–268.

McGuinness BJ, Power MJ, Fottrell PF. 1994. Radioimmunoassay of 2-hydroxyestrone in urine. *Clin Chem* **40**: 80–85.

McShane LM, Dorgan JF, Greenhut S *et al.* 1996. Reliability and validity of serum sex hormone measurements. *Cancer Epidemiol Biomarkers Prevent* **5**: 923–928.

Missmer SA, Eliassen AH, Barbieri RL *et al.* 2004. Endogenous estrogen, androgen, and progesterone concentrations and breast cancer risk among postmenopausal women. *J Natl Cancer Inst* **96**: 1856–1865.

Missmer SA, Spiegelman D, Bertone-Johnson ER *et al.* 2006. Reproducibility of plasma steroid hormones, prolactin, and insulin-like growth factor levels among premenopausal women over a 2- to 3-year period. *Cancer Epidemiol Biomarkers Prevent* **15**: 972–978.

Muti P, Trevisan M, Micheli A *et al.* 1996. Reliability of serum hormones in premenopausal and postmenopausal women over a one-year period. *Cancer Epidemiol Biomarkers Prevent* **5**: 917–922.

Nisula BC, Loriaux DL, Wilson YA. 1978. Solid phase method for measurement of the binding capacity of testosterone-estradiol binding globulin in human serum. *Steroids* **31**: 681–690.

Potischman N, Falk RT, Laiming VA *et al.* 1994. Reproducibility of laboratory assays for steroid hormones and sex hormone-binding globulin. *Cancer Res* **54**: 5363–5367.

Pribluda VS, Green SJ. 1998. A good estrogen. *Science* **280**: 987–988.

Rajaram S, Baylink DJ, Mohan S. 1997. Insulin-like growth factor-binding proteins in serum and other biological fluids: regulation and functions. *Endocr Rev* **18**: 801–831.

Rapuri PB, Gallagher JC, Haynatzki G. 2004. Endogenous levels of serum estradiol and sex hormone binding globulin determine bone mineral density, bone remodeling, the rate of bone loss, and response to treatment with estrogen in elderly women. *J Clin Endocrinol Metab* **89**: 4954–4962.

Rinaldi S, Dechaud H, Biessy C *et al.* 2001. Reliability and validity of commercially available, direct radioimmunoassays for measurement of blood androgens and estrogens in postmenopausal women. *Cancer Epidemiol Biomarkers Prevent* **10**: 757–765.

Rinaldi S, Kaaks R, Zeleniuch-Jacquotte A *et al.* 2005. Insulin-like growth factor-I, IGF binding protein-3, and breast cancer in young women: a comparison of risk estimates using different peptide assays. *Cancer Epidemiol Biomarkers Prevent* **14**(1): 48–52.

Rinaldi S, Peeters PH, Berrino F *et al.* 2006. IGF-I, IGFBP-3 and breast cancer risk in women: the European Prospective Investigation into Cancer and Nutrition (EPIC). *Endocr Relat Cancer* **13**: 593–605.

Schioler V, Thode J. 1988. Six direct radioimmunoassays of estradiol evaluated. *Clin Chem* **34**: 949–952.

Severi G, Morris HA, MacInnis RJ *et al.* 2006. Circulating steroid hormones and the risk of prostate cancer. *Cancer Epidemiol Biomarkers Prevent* **15**(1): 86–91.

Shimada K, Mitamura K, Higashi T. 2001. Gas chromatography and high-performance liquid chromatography of natural steroids. *J Chromatogr A* **935**: 141–172.

Simpson ER, Misso M, Hewitt KN *et al.* 2005. Estrogen – the good, the bad, and the unexpected. *Endocr Rev* **26**: 322–330.

Stanczyk FZ. 2006. Measurement of androgens in women. *Semin Reprod Med* **24**: 78–85.

Stattin P, Rinaldi S, Biessy C *et al.* 2004. High levels of circulating insulin-like growth factor-I increase prostate cancer risk: a prospective study in a population-based nonscreened cohort. *J Clin Oncol* **22**: 3104–3112.

Toniolo P, Koenig KL, Pasternack BS *et al.* 1994. Reliability of measurements of total, protein-bound, and unbound estradiol in serum. *Cancer Epidemiol Biomarkers Prevent* **3**: 47–50.

Wheeler MJ. 2006. Measurement of androgens. *Methods Mol Biol* **324**: 197–211.

Xiao XY, McCalley DV, McEvoy J. 2001. Analysis of estrogens in river water and effluents using solid-phase extraction and gas chromatography–negative chemical ionization mass spectrometry of the pentafluorobenzoyl derivatives. *J Chromatogr A* **923**: 195–204.

Xu X, Veenstra TD, Fox SD *et al.* 2005. Measuring fifteen endogenous estrogens simultaneously in human urine by high-performance liquid chromatography–mass spectrometry. *Anal Chem* **77**: 6646–6654.

Xu X, Ziegler RG, Waterhouse DJ *et al.* 2002. Stable isotope dilution high-performance liquid chromatography–electrospray ionization msass spectrometry method for endogenous 2- and 4-hydroxyestrones in human urine. *J Chromatogr B Analyt Technol Biomed Life Sci* **780**: 315–330.

Yakar S, Pennisi P, Zhao H *et al.* 2004. Circulating IGF-1 and its role in cancer: lessons from the *IGF-1* gene deletion (LID) mouse. *Novartis Found Symp* **262**: 3–9.

Zacharia LC, Dubey RK, Jackson EK. 2004. A gas chromatography/mass spectrometry assay to measure estradiol, catecholestradiols, and methoxyestradiols in plasma. *Steroids* **69**: 255–261.

25

Aflatoxin, Hepatitis B Virus and Liver Cancer: A Paradigm for Molecular Epidemiology

J. D. Groopman,[1] T. W. Kensler[1] and Chris Wild[2]

[1]*Johns Hopkins University, Baltimore, MD, USA, and* [2]*University of Leeds, Leeds, UK*

25.1. INTRODUCTION

Hepatocellular carcinoma (HCC) is a major cause of cancer morbidity and mortality in many parts of the world, including Asia and sub-Saharan Africa, with upwards of 600 000 new cases each year and over 200 000 deaths annually in the People's Republic of China (PRC) alone (Arbuthnot and Kew 2001; Block *et al.* 2003; Wang *et al.* 2002). The major aetiological factors associated with development of HCC in these regions are infection in early life with hepatitis B virus (HBV) and life-time exposure to high dietary levels of aflatoxins, including the most potent member of this group of toxins, aflatoxin B_1 (AFB$_1$) (Aguilar *et al.* 1993; Block *et al.* 2003; Kensler *et al.* 2003). Over the past 20 years, the role of the hepatitis C virus (HCV) has been recognized as contributing to rising HCC rates in the USA and Japan (Tanaka *et al.* 2002). Detailed knowledge of the aetiology of HCC has spurred many mechanistic studies to understand the aetiology and pathogenesis of this nearly always fatal disease, and this knowledge is beginning to be translated to preventive interventions in high-risk populations (Kensler *et al.* 2003, 2004; Wild and Hall 2000).

The public health significance of HBV as a risk factor for HCC is staggering, with over 400 million chronic carriers, of which 10–25% will develop HCC (Arbuthnot and Kew 2001; Block *et al.* 2003). The biology, mode of transmission and epidemiology of this viral infection continues to be actively investigated and has been recently reviewed (Kirk *et al.* 2006; Lok *et al.* 2001; Lok and McMahon 2001). AFB$_1$ has also been suspected to contribute to human HCC since the 1960s, when its potent activity as a carcinogen in many species of animals, including rodents, non-human primates and fish, were extensively described (Busby 1984). Thus, the wide cross-species potency and the demonstrated contamination of the human diet provided the justification for suspecting that AFB$_1$ could contribute to human cancer.

Elucidation of the roles of HBV and aflatoxins in the initiation and progression to HCC was a complex challenge, which has been significantly helped by the development and validation of biomarkers subsequently applied in studies of aetiology and to optimize strategies for interventions. This chapter outlines how a strategy to apply biomarkers in this way was pursued; this area of research serves as a possible template for adaptation to other exposure scenarios and cancer endpoints in the field of molecular epidemiology.

Molecular Epidemiology of Chronic Diseases, Edited by C. P. Wild, P. Vineis, and S. Garte
© 2008 John Wiley & Sons, Ltd

25.2. DEFINING MOLECULAR BIOMARKERS

Molecular biomarkers are typically used as indicators of exposure, effect or susceptibility within the continuum of a paradigm that has evolved over the past 20 years. A biomarker of exposure refers to measurement of the specific agent of interest, its metabolite(s), or its specific interactive products in a body compartment or fluid, which indicates the presence and magnitude of current and past exposure. A biomarker of effect indicates the presence and magnitude of a biological response to exposure to an environmental agent. Such a biomarker may be an endogenous component, a measure of the functional capacity of the system, or an altered state recognized as impairment or disease. A biomarker of susceptibility is an indicator or a measure of an inherent or acquired ability of an organism to respond to the challenge of exposure to a specific xenobiotic substance or other toxicant. Such a biomarker may be the unusual presence or absence of an endogenous component, including specific genetic variants, or an abnormal functional response to an administered challenge (Wang *et al.* 2001). Measures of these biomarkers through molecular epidemiology studies thus have great utility in addressing the relationships between exposure to environmental agents and development of clinical diseases, and in identifying those individuals at high risk for the disease (Hulka 1991; Wogan 1989). Such biomarkers also allow investigation of the underlying mechanisms of disease and therefore may contribute to establishing the biological plausibility of an exposure–disease association. Collectively these data therefore help to inform the risk assessment process, where the effectiveness of regulations can be tested against biological measurements of exposure and effect.

25.3. VALIDATION STRATEGY FOR MOLECULAR BIOMARKERS

There is a marked difference between the ability to measure a particular biomarker in a human biological sample employing high quality analytical approaches and the ability to interpret that information on the basis of thorough validation of the biomarker. The validation step involves the careful characterization of the relationship between the biomarker and, for example, environmental exposure to the agent of interest or to the consequent progression of disease. This process of biomarker validation is well-served by parallel experimental and human studies (Groopman and Kensler 1999). This is not to negate the importance of the appropriate analytical technology, but rather to emphasize the need for a full characterization of the properties of the biomarker in the context of the exposure–disease continuum in order for the measurements to be informative.

Conceptually, as shown in Figure 25.1, an appropriate animal model can be used to determine the associative or causal role of the biomarker on the disease pathway, and to establish relations between dose (exposure) and response. The putative biomarker can then be validated in pilot human studies, where sensitivity, specificity, accuracy and reliability parameters can be established. Data obtained in these studies can then be used to assess intra- or inter-individual variability, background levels, relationship of the biomarker to external dose or to disease status, as well as feasibility for use in larger population-based studies. For a full interpretation of the information that the biomarker provides, prospective epidemiological studies may be necessary to demonstrate the role of the biomarker in the overall pathogenesis of the disease. Finally, these biomarkers can be translated as efficacy endpoints in interventions in both experimental models and high-risk human populations.

25.4. DEVELOPMENT AND VALIDATION OF BIOMARKERS FOR HUMAN HEPATOCELLULAR CARCINOMA

Early aetiological studies of aflatoxin, HBV and HCC

As described earlier, HCC is among the leading causes of cancer death in many parts of the economically developing world. The map in Figure 25.2, based upon the IARC cancer database, illustrates the unequal distribution of this disease (http://www-dep.iarc.fr/). Since the level of HCC is

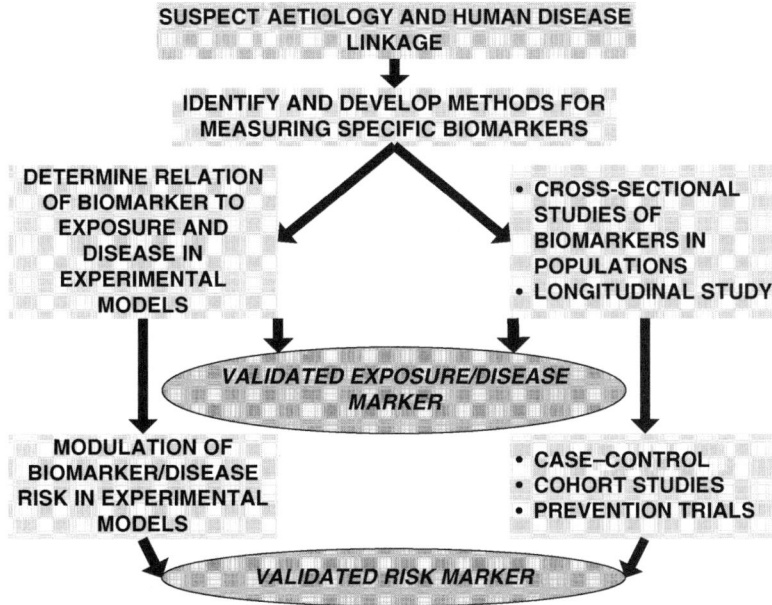

Figure 25.1   Validation scheme for molecular biomarker research

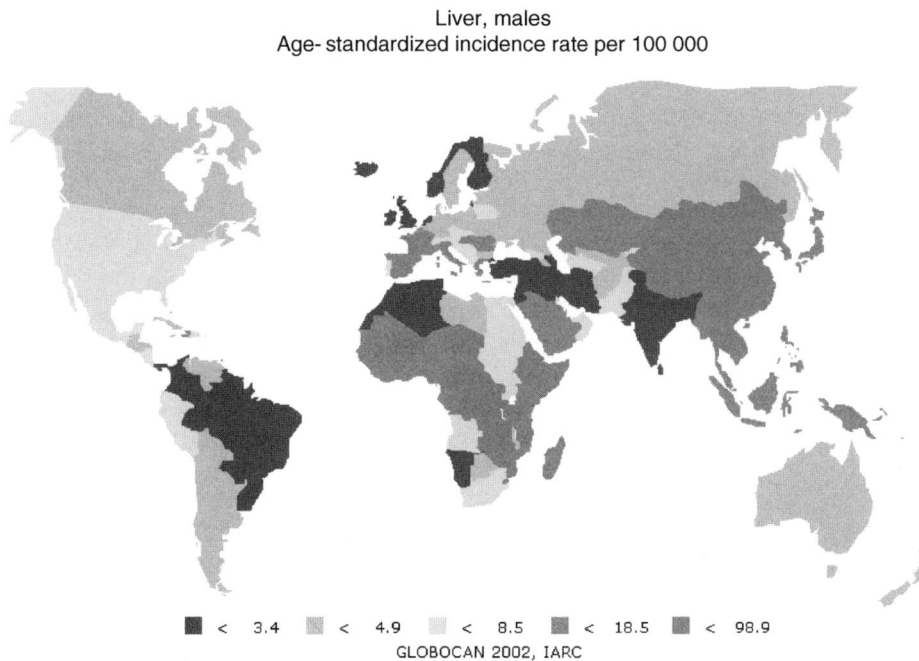

Figure 25.2   Age-standardized incidence of liver cancer in men world-wide

coincident with regions where aflatoxin exposure is high, efforts started in the 1960s to investigate this possible association. As in all ecological investigations, the work was hindered by the lack of adequate data on aflatoxin intake, excretion and metabolism in people, the underlying susceptibility factors such as diet and viral exposure, as well as by the incomplete statistics on world-wide cancer morbidity and mortality. Despite these deficiencies, early studies did provide data illustrating that increasing HCC rates corresponded to increasing levels of dietary aflatoxin exposure (Bosch and Munoz 1988). The commodities most often found to be contaminated by aflatoxins were peanuts (groundnuts), various other nuts, cottonseed, corn (maize) and rice (Eaton and Groopman 1994). The requirements for aflatoxin production are relatively non-specific, since moulds can produce them on almost any foodstuff, with therefore a wide range of commodities contaminated at final concentrations which can vary from < 1 µg/kg (1 p.p.b.) to > 12 000 µg/kg (12 p.p.m.) (Ellis *et al.* 1991). In addition, contamination of, for example, maize kernels or a batch of peanuts is highly heterogeneous and hence representative sampling is problematic. Consequently, measurement of human exposure to aflatoxin by sampling foodstuffs or by dietary questionnaires is extremely imprecise. For this reason aflatoxin exposure biomarkers were considered to have great potential for accurate assessment of exposure (Wild and Turner 2002).

Concurrent with the aflatoxin research described above were a series of studies describing a role for HBV in HCC pathogenesis. Interestingly the advances in understanding the role of HBV were facilitated by sensitive and specific biomarkers of HBV infection, notably the presence of HBV surface antigen (HBsAg) in the serum. Other biomarkers allowed discrimination between past infection, chronic infection and active viral replication. Thus, the surface antigen and core antigens and antibodies reflect the temporality of the infection (Lok and McMahon 2001; Yim and Lok 2006). This ability to characterize the timing and nature of exposure revealed that the age of initial infection was directly related to development of the chronic carrier state and subsequent risk of HCC. Approximately 90% of HBV infections acquired in infancy or early childhood become chronic compared to only 10% acquired in adulthood (Lok *et al.* 2001). Finally, the global burden of HBV infection was shown to vary geographically; China, south-east Asia and sub-Saharan Africa have some of the highest rates of chronic HBV infection in the world, with prevalences of at least 8% (Figure 25.3)

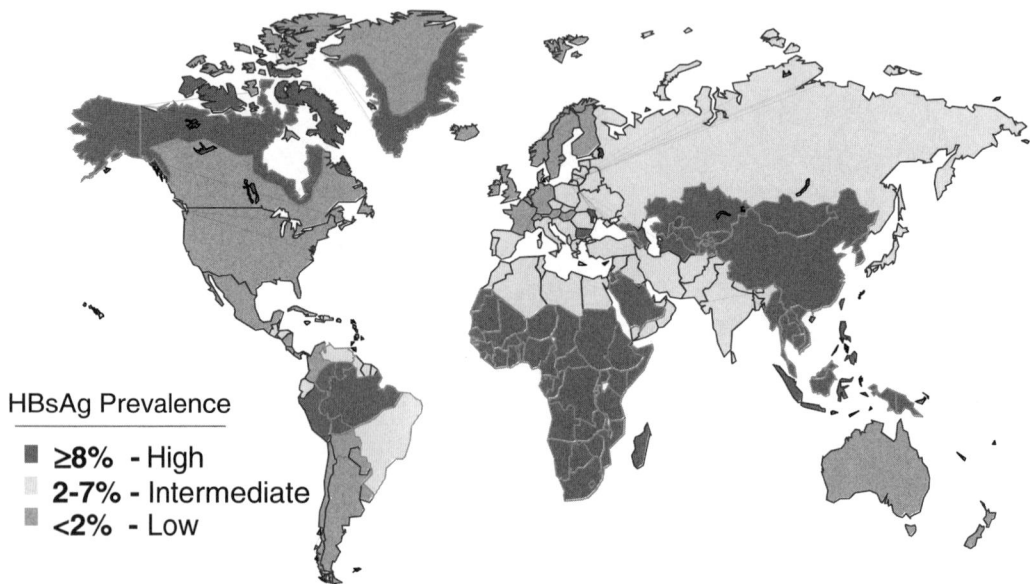

Figure 25.3   Prevalence of hepatitis B surface antigen positivity across the world

(Arbuthnot and Kew 2001). Overall, the advances in understanding the role of HBV in human health have clearly been facilitated by appropriate biomarkers. In contrast, the lack of parallel advances in aflatoxin biomarkers initially hampered progress.

Development of methodologies for measuring biomarkers

In the case of AFB_1 the measurement of the DNA and protein adducts were of interest in the study of HCC because they are direct products of (or surrogate markers for) damage to a critical cellular macromolecular target. The chemical structures of the major aflatoxin DNA and protein adducts were described (Essigmann *et al.* 1977; Sabbioni *et al.* 1987). The finding that the major aflatoxin–nucleic acid adduct AFB_1–N^7-Gua was excreted exclusively in urine of exposed rats spurred interest in

using this metabolite as a biomarker (Bennett *et al.* 1981). The serum aflatoxin–albumin adduct was also examined as a biomarker of exposure, since the longer half-life of albumin would be expected to integrate exposures over longer time periods. Studies in experimental models found that the formation of aflatoxin–DNA adducts in liver, urinary excretion of the aflatoxin–nucleic acid adduct and formation of the serum albumin adduct were highly correlated (Groopman *et al.* 1992a).

A number of different methodologies may be used to measure aflatoxin exposure biomarkers, each with different advantages and limitations (see Box 25.1). An immunoaffinity clean-up/HPLC procedure was developed to isolate and measure aflatoxin metabolites in biological samples (Egner *et al.* 2001; Groopman *et al.* 1984, 1985). With this approach, initial validation studies were perfomed to investigate the dose-dependent excretion

---

Box 25.1   Analytical methodologies for exposure biomarkers

Many different analytical methods are available for quantitation of chemical adducts in biological samples (Poirier *et al.* 2000; Santella 1999; Wang and Groopman 1998). Each methodology has unique specificity and sensitivity and, depending on the application, the user can choose which is most appropriate. For example, to measure a single aflatoxin metabolite, a chromatographic method can resolve mixtures of aflatoxins into individual compounds, providing that the extraction procedure does not introduce large amounts of interfering chemicals. Antibody-based methods are often more sensitive than chromatography, but immunoassays are less selective because the antibody may cross-react with multiple metabolites. An area of considerable importance that has received far less attention than it should has been in the area of internal standard development. All quantitative measurements require the use of an internal standard to account for sample-to-sample variations in the analyte recoveries. In the case of mass spectrometry, internal standards generally employ an isotopically labelled material that is identical to the chemical being measured, albeit of different molecular mass. Obtaining such isotopically labelled materials requires chemical synthesis, if they are not commercially available, and has impeded the application of internal standards in many studies. In the case of immunoassays, internal standards pose a different challenge, since the addition of an internal standard that is recognized by an antibody results in an incremental contribution to the positive value. The dynamic range is usually , <100 in immunoassays, and therefore great care must be taken to spike a sample with an internal standard so that one can obtain a valid result. In contrast, most chromatographic methods result in dynamic ranges of analyses that can be over a 10 000-fold range. The mass spectrometry methods are not only applicable for the quantitation of small molecules such as aflatoxin, but have been extended to measure mutations in DNA fragments that are mechanistically linked to the aetiopathogenesis of HCC (Jackson *et al.* 2001; Laken *et al.* 1998; Lleonart *et al.* 2005).

Many of the aflatoxin studies used different analytical methods and therefore the quantitative comparison of different datasets has been extremely problematic. However, a recent study compared methods of ELISA and mass spectrometry (MS) for aflatoxin–albumin adducts and found high correlation between these two ($r = 0.856, p < 0.0001$) (Scholl *et al.* 2006).

of urinary aflatoxin biomarkers in rats after a single exposure to AFB_1 (Groopman et al. 1992c). A linear relationship was found between AFB_1 dose and excretion of the $AFB–N^7$-Gua adduct in urine over the initial 24 h period of exposure. In contrast, excretion of other oxidative metabolites, such as AFP_1, showed no linear association with dose. Subsequent studies in rodents that assessed the formation of aflatoxin macromolecular adducts after chronic administration also supported the use of DNA and protein adducts as molecular measures of exposure (Egner et al. 1995; Kensler et al. 1986; Wild et al. 1986). For example, in rats treated with relatively low doses of AFB_1 (3.5µg) twice daily for 24 days, there was an accumulation of aflatoxin binding to peripheral blood albumin, followed by steady-state levels illustrating the potential for this biomarker (aflatoxin–albumin adduct) to integrate exposure over time (Wild et al. 1986).

Relationship of aflatoxin biomarkers to exposure and disease in experimental animals

In the early 1980s studies to identify effective chemoprevention strategies for aflatoxin carcinogenesis were initiated. The hypothesis was that reduction of aflatoxin–DNA adduct levels by chemopreventive agents would be mechanistically related to, and therefore predictive of, cancer preventive efficacy. Preliminary data with a variety of established chemopreventive agents demonstrated that after a single dose of aflatoxin, levels of DNA adducts were reduced (Kensler et al. 1985). A more comprehensive study using multiple doses of aflatoxin and the chemopreventive agent ethoxyquin reduced the area and volume of liver occupied by presumptive preneoplastic foci by > 95% and dramatically reduced binding of AFB_1 to hepatic DNA, from 90% initially to 70% at the end of a 2 week dosing period. Intriguingly, no differences in residual DNA adduct burden were, however, discernible several months after dosing, despite the profound reduction in tumour burden (Kensler et al. 1986). The experiment was then repeated with several different chemopreventive agents and in all cases aflatoxin-derived DNA and protein adducts were reduced; however, even under optimal conditions, the reduction

in the macromolecular adducts always underrepresented the effect on tumour burden (Bolton et al. 1993; Roebuck et al. 1991). Therefore these macromolecular adducts can track with disease outcome on a population level, but in the multistage process of cancer the absolute level of adduct provides only a necessary but insufficient measure of tumour risk.

Modulation of biomarkers and disease in animal chemoprevention studies

Using the chemopreventive agent oltipraz, Roebuck et al. (1991) established correlations between reductions in levels of AFB_1–N^7-Gua excreted in urine and incidence of HCC in aflatoxin-exposed rats. Overall, reduction in biomarker levels reflected protection against carcinogenesis, but these studies did not address the quantitative relationship between biomarker levels and individual risk. Thus, in a follow-up study, rats dosed with AFB_1 daily for 5 weeks were randomized into three groups: no intervention; delayed-transient intervention with Oltipraz during weeks 2 and 3 of exposure; and persistent intervention with Oltipraz for all 5 weeks of dosing (Kensler et al. 1997). Serial blood samples were collected from each animal at weekly intervals throughout aflatoxin exposure for measurement of aflatoxin–albumin adducts. The integrated level of aflatoxin–albumin adducts over the exposure period decreased 20% and 39% in the delayed-transient and persistent Oltipraz intervention groups, respectively, as compared with no intervention. Similarly, the total incidence of HCC dropped significantly from 83% to 60% and 48% in these groups. Overall, there was a significant association between integrated biomarker level and risk of HCC ($p = 0.01$). When the predictive value of aflatoxin–serum albumin adducts was assessed within treatment groups, however, there was no association between integrated biomarker levels and risk of HCC ($p = 0.56$). These data clearly demonstrated that levels of the aflatoxin–albumin adducts could predict population-based changes in disease risk, but had no power to identify individual rats destined to develop HCC. Because of the multistage process of carcinogenesis, in order to determine individual risk of disease, a panel of biomarkers reflecting different stages will be required.

Validation of aflatoxin biomarkers in cross-sectional studies in human populations

Initial studies of aflatoxin biomarkers in human populations began in The Philippines (Campbell *et al.* 1970), where it was demonstrated that an oxidative metabolite of aflatoxin, AFM_1, could be measured in urine as an internal dose marker (Campbell *et al.* 1970). Subsequent work conducted in the People's Republic of China and The Gambia, West Africa, areas with high incidences of HCC, determined that the levels of urinary aflatoxin biomarkers showed a dose-dependent relationship with aflatoxin intake (Groopman *et al.* 1992b, 1992d). However, as in the earlier experimental studies this relationship was dependent on the specific urinary marker under study, e.g. with AFB_1–N^7-Gua and AFM1 showing strong correlations with intake whereas urinary AFP1 showed no such link, Gan *et al.* (1988) and Wild *et al.* (1992) similarly monitored levels of aflatoxin–albumin adducts and observed a highly significant association between intake of aflatoxin and level of adduct. This type of study, to measure dietary aflatoxin intake and biomarkers at the individual level, is crucial in validating a biomarker for exposure assessment and is often overlooked in molecular epidemiology. Of particular interest in The Gambia was the observation that whilst urinary aflatoxin metabolites reflected day-to-day variations in aflatoxin intake, the aflatoxin–albumin adducts integrated exposure over the week-long study (Wild *et al.* 1992; see Figure 25.4, Box 25.2).

A further advance in aflatoxin biomarker development resulted from studies on the *p53* tumour suppressor gene, the most commonly mutated gene detected in human cancer (Greenblatt *et al.* 1994). Ecological studies of *p53* mutations in HCC occurring in populations exposed to high levels of dietary aflatoxin found high frequencies of guanine-to-thymine transversions, with clustering at codon 249 (Bressac *et al.* 1991; Hsu *et al.* 1991). In contrast, no mutations at codon 249 were found in *p53* in HCC from Japan and other areas where there was less exposure to aflatoxin (Aguilar *et al.* 1994;

Ozturk 1991). The occurrence of this specific mutation has been mechanistically associated with AFB_1 exposure in experimental models, including bacteria (Foster *et al.* 1983), and through demonstration that aflatoxin-8,9-epoxide could bind to codon 249 of *p53* in a plasmid *in vitro* (Puisieux *et al.* 1991). Mutational analysis of the *p53* gene in human HepG2 cells and hepatocytes exposed to AFB_1 found preferential induction of the transversion of guanine to thymine in the third position of codon 249 (Aguilar *et al.* 1993; Denissenko *et al.* 1997, 1998). In summary, studies of the prevalence of codon 249 mutations in HCC cases from patients in areas of high or low exposure to aflatoxin suggest that a G→T transversion at the third base is associated with aflatoxin exposure, and *in vitro* data would seem to support this hypothesis.

Data from these initial cross-sectional biomarker studies demonstrated short-term dose–response relationships for a number of the aflatoxin metabolites, including the major nucleic acid adduct, serum albumin adduct and AFM_1. This supported the validity of these exposure biomarkers for use in epidemiological studies, including investigations of intervention strategies and studies of the mechanisms underlying susceptibility. The *p53* codon 249 mutation data also encouraged the use of this biomarker in further studies of disease aetiology and progression.

Longitudinal study of biomarkers in humans

Longitudinal studies are extremely important in the further development and validation process for biomarkers. These investigations permit an understanding of the stability of the biomarker in storage as well as the tracking potential of each biomarker, both of which are essential for the evaluation of the predictive power of the biomarker. One approach to establishing the stability of aflatoxin biomarkers was by supplementing urine samples with aflatoxins at the time of collection and then analysing repeated samples over the course of 8 years (unpublished data). Similarly, aflatoxin–albumin adducts in human sera were found to be detectable for at least 15 years after collection when stored at −20°C. (Wild *et al.* 1990). Therefore, at least for some of the aflatoxin biomarkers, degradation over

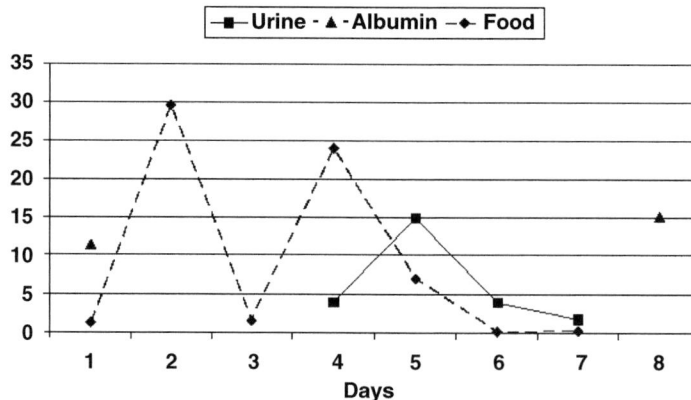

Figure 25.4 Intra-individual variation in aflatoxin biomarkers. Daily variations in aflatoxin dietary intake (μg/day), urinary aflatoxin metabolites (ng/mg creatinine) and aflatoxin–albumin adducts (pg AFB_1–lysine equivalent/mg albumin 3 10^{-1}) (for further details, see (Groopman *et al.* 1992b; Wild *et al.* 1992)

time has not been a major problem; however, similar studies are required for all chemical-specific biomarkers.

An objective in the development of any biomarker is to use it as a predictor of past and future exposure status in people. This concept is embodied in the principle of tracking, which is an index of how well an individual's biomarker remains positioned in a rank order relative to other individuals in a group over time. Tracking within a group of individuals is

---

Box 25.2   Intraindividual fluctuation in biomarker levels

Exposure biomarkers will reflect different periods of dietary intake of aflatoxin, depending on the chemical stability of the biomarker and the rate of turnover. The assumption, based on animal studies, was that human urinary excretion of aflatoxin and its metabolites would occur over a period of 24–48 h. In contrast, it was predicted that under chronic exposure conditions aflatoxin–albumin adducts would accumulate to around 30-fold the level of a single exposure, due to the half-life of albumin in man being approximately 20 days (Gan *et al.* 1988). Detailed studies of dietary intake of aflatoxins in relation to biomarkers cast light on some of these assumptions. In The Gambia, for example, Figure 25.4 shows that for the female subject under study, the dietary intake of aflatoxin (ng/day) varied from 152 to 23 940 ng/day (157-fold) over a 7 day period, whilst the urinary metabolite levels varied from 1.73 ng aflatoxin metabolites/mg creatinine to 14.9 ng/mg (8.6-fold) over 4 consecutive days of 24 h urine collection. Over the 8 day period of the investigation, the albumin adduct level varied , <1.5-fold, thus providing a more stable, integrated measure of exposure. It is also notable that the urinary excretion appeared to lag the changes in dietary intake by about 24 h, consistent with a rapid urinary excretion of aflatoxins in exposed people. Thus, a urinary measure, or indeed a measure of dietary intake, could lead to significant misclassification of exposure if only conducted on a single occasion in a situation where day-to-day variation in exposures is so marked. However, when urine levels are averaged over a number of days (Groopman *et al.* 1992b) there is a good correlation with dietary exposure. Thus, ideally for prospective cohort studies, more than one urine collection would be desirable, although rarely possible in large-scale studies. In some situations, e.g. evaluation of intervention strategies, a more rapidly responsive biomarker might be desirable in order to see an effect of the intervention as early as possible in the study. Thus, the choice of biomarker needs to be made on the basis of study design, including sample size and power, logistical considerations and the properties of the individual marker; interpretation of the data generated should take account of these factors.

expressed by the intraclass correlation coefficient. When the intraclass correlation coefficient is 1.0, a person's relative position within the group, independent of exposure, does not change over time. If the intraclass correlation coefficient is 0.0, there is random positioning of the individual's biomarker level relative to the others in the group throughout the time period. The tracking concept is central to interpreting data related to exposure and biomarker levels and requires acquisition of repeated samples from subjects. Unfortunately, data on the temporal patterns of formation and persistence of aflatoxin macromolecular adducts in human samples are very limited. Obviously, chemical-specific biomarkers measured in cross-sectional studies cannot provide information on the predictive value or tracking of an individual's marker level over time. In contrast to the aflatoxin situation, the HBV biomarker tracking has been well characterized and forms the basis for defining chronic infection status (Yim and Lok 2006).

Tracking is important in assessing exposure, and this information is essential in the design of intervention studies to ensure that enough samples are obtained and that collections are made at appropriate intervals. For example, if exposure remains constant and the tracking value for a marker changes over time, it might be assumed that the change in tracking is due to a biological process, such as an alteration in the balance of metabolic pathways responsible for adduct formation. On the other hand, lack of tracking can be attributable to great variance in exposure. Therefore, to determine unequivocally the contributions of intra- and inter-individual variations to biomarker levels, experiments must assess tracking over time.

Case–control and cohort studies

Many published case–control studies have examined the relation of aflatoxin exposure and HCC. Compared with cohort studies, case–control studies are both cost- and time-effective. Unfortunately, case–control studies are initiated at the time of disease diagnosis and it cannot be assumed that the exposure has not appreciably changed over time. The presence of disease, including prior to clinical diagnosis, may affect exposure in cases, par-

ticularly for example if related to diet. This would affect both dietary and biomarker approaches to aflatoxin exposure assessment. Also, such studies involve an assumption that the disease state does not alter metabolism, which if it does occur could affect the exposure-biomarker relationship. There is some evidence to support this risk of reverse causation in a case–control study in China (Hall and Wild 1994). A positive association between dietary estimates of aflatoxin exposure and levels of aflatoxin–albumin biomarker at the individual level was observed in controls but not cases, suggesting that the disease had affected the relationship in the latter group.

Some case–control studies of HCC have nevertheless used estimates of dietary intakes of aflatoxin for exposure assessment. For example, one of the first by Bulatao-Jayme et al. (1982) in the Philippines found that moderate and heavy aflatoxin intake increased HCC risk compared to light intake; however, alcohol was a strong confounding factor and HBV infection was not considered. In a more recent study, Omer et al. (2004) conducted a case–control study of HCC in Sudan and assessed peanut consumption, as a surrogate for aflatoxin exposure, and HBV infection. There was a significant association with both peanut butter consumption and HBV infection, and a more than additive interaction between the two was reported. Kirk et al. (Kirk et al. 2005) similarly used groundnut consumption as a surrogate for aflatoxin consumption in a case–control study in The Gambia and found that the presence of HCC lead to decreases in groundnut consumption; there was no association between self-reported groundnut consumption and HCC or cirrhosis.

Other case–control studies have used biomarkers for aflatoxin exposure assessment. An example was in Taiwan, where aflatoxin–DNA adducts in liver tissue samples were measured (Lunn et al. 1997). The proportion of subjects with a detectable level of DNA adducts was higher for HCC cases than for matched controls [odds ratio (OR) = 5.2]. In the study of Kirk et al. (2004, 2005) the authors examined the p53 codon 249 mutation in the plasma of HCC cases, cirrhosis patients and controls. They found the mutation detectable in 39.8%, 15.3% and 3.5% of the three groups, respectively, with

OR = 20.3 (range 8.19–50.0) for individuals positive for the mutation in plasma. Furthermore, the presence of both the mutation and HBV infection was associated with an OR = 399 (range 48.6–3270). Although a number of negative case–control studies of aflatoxin and HCC have been reported overall (IARC 2002), the evidence pointed to an aetiological role for aflatoxin in human HCC.

Data obtained from cohort studies have the greatest power to determine a true relationship between an exposure and disease outcome, because exposure is assessed in a healthy cohort prior to disease onset. A nested study within the cohort can then be designed to match cases and controls. An advantage of this method is that the controls are truly matched to the cases, since both were recruited at the same time and with the same health status. A major disadvantage, however, is the time needed in follow-up (often years) to accrue the cases. This disadvantage can be partially overcome by enrolling large numbers of people (often tens of thousands) at older ages to ensure case accrual at a reasonable rate.

To date, two major cohort studies with aflatoxin biomarkers have demonstrated the important role of this carcinogen in the aetiology of HCC. The first study, comprising > 18 000 men in Shanghai, examined the interaction of HBV and aflatoxin biomarkers as independent and interactive risk factors for HCC. The nested case–control data revealed a statistically significant increase in the relative risk (RR) of 3.4 for those HCC cases where urinary aflatoxin biomarkers were detected. For HBsAg-positive people the RR = 7, but for individuals with both urinary aflatoxins and positive HBsAg status the RR = 59 (Qian et al. 1994; Ross et al. 1992). These results strongly support a causal relationship between the presence of the chemical- and viral-specific biomarkers and the risk of HCC.

Subsequent cohort studies in Taiwan have substantially confirmed the results from the Shanghai investigation. Wang et al. (1996b) examined HCC cases and controls nested within a cohort and found that in HBV-infected people there was an adjusted OR of 2.8 for detectable compared with non-detectable aflatoxin–albumin adducts and 5.5 for high compared with low levels of aflatoxin metab-olites in urine. In a follow-up study, there was a dose–response relationship between urinary AFM_1 levels and risk of HCC in chronic HBV carriers (Sun et al. 1999). Similar to the Shanghai study, the HCC risk associated with AFB_1 exposure was more striking among the HBV carriers with detectable urinary aflatoxin. Table 25.1 summarizes a number of the cohort studies which have used biomarkers to examine the association between aflatoxin exposure and HCC risk, including interactions with HBV infection.

Clinical trials for reducing aflatoxin exposure and internal dose

Clinical trials and other interventions are designed to translate findings from human and experimental investigations to public health prevention. Both primary (to reduce exposure) and secondary (to alter metabolism and deposition) interventions can use specific biomarkers as endpoints of efficacy. Such biomarkers can also be applied to the pre-selection of exposed individuals for study cohorts, thereby reducing study size requirements. They can also serve as short-term modifiable endpoints (Kensler et al. 1996). In a primary prevention trial the goal is to reduce exposure to aflatoxins in the diet. Interventions can range from attempting to lower mould growth in harvested crops to using trapping agents that block the uptake of ingested aflatoxins. In secondary prevention trials, one goal is to modulate the metabolism of ingested aflatoxin to enhance detoxification processes, thereby reducing internal dose.

The use of aflatoxin biomarkers as efficacy endpoints in primary prevention trials has been recently reported (Turner et al. 2005). This study assessed postharvest measures to restrict aflatoxin contamination of groundnut crops. Six hundred people were monitored and in control villages mean aflatoxin–albumin concentration increased postharvest, from 5.5 pg/mg (95% CI 4.7–6.1) immediately after harvest to 18.7 pg/mg (17.0–20.6) 5 months later. By contrast, mean aflatoxin–albumin concentration in intervention villages after 5 months of groundnut storage was much the same as that immediately postharvest, 7.2 pg/mg (6.2–8.4) vs. 8.0 pg/mg (7.0–9.2).

Table 25.1 Studies of the interaction between aflatoxins and HBV in HCC

Reference	Population	Cohort	Cases	Controls	Biomarker	OR
(Qian *et al.* 1994)	Shanghai, PRC	18 224 males	50	267	Urinary AF biomarker[4]	3.4 (1.1–10.0) AF alone 7.3 (2.2–24) HBsAg alone 59.4 (16.6–212) AF and HBsAg
(Wang *et al.* 1996b)	Taiwan	12 040 males 13 758 females	56	220	Urinary AF metabolites[5]	1.7 (0.3–10.8) AF alone 22.8 (3.6–143.4) HBsAg alone 111.9 (13.8–905) AF and HBsAg
			29 HBsAg +ve	21 HBsAg +ve	Urinary AF metabolites[6]	5.5 (1.3–23.4)
(Chen *et al.* 1996)	Taiwan	6487 4691 males 1796 females	33 (20)	123 (86)[7]	AFB1–albumin adducts	5.5 (1.2–24.5) AF alone 129 (25–659) AF and HBsAg
(Yu *et al.* 1997)	Taiwan	7342 males 4841 HBsAg carriers 2501 non-carriers	43 HBsAg +ve	86 HBsAg +ve	Urinary AFM1	6.0 (1.2–29.0)[1]
(Sun *et al.* 1999)	Qidong County, PRC	145 male HBsAg carriers	22 HBsAg +ve	123 HBsAg +ve	Urinary AFM1[3]	3.3 (1.2–8.7)
(Sun *et al.* 2001)	Taiwan	12 024 males 13 594 females	79 HBsAg +ve	149 HBsAg +ve	Serum AFB1–albumin	2.0 (1.1–3.7)[2]

[1] Highest compared to lowest tertile of AFM1 level; adjusted for educational level, ethnicity, alcohol, cigarette smoking.
[2] Detectable vs. non-detectable; adjusted for sex, age and residence.
[3] Eight monthly urine samples were collected over follow-up and urinary AFM1 analysis was conducted on a pooled sample; AFM1 positive compared to negative.
[4] Presence vs. absence of any aflatoxin biomarker; adjusted for cigarette smoking.
[5] Low vs. high urinary aflatoxin biomarker; adjusted for cigarette smoking and alcohol drinking.
[6] Low vs. high urinary aflatoxin biomarker; adjusted for age, residence, cigarette smoking and alcohol drinking.
[7] Only the numbers of subjects in brackets had samples for analysis of aflatoxin biomarker.

At 5 months, mean adduct concentration in intervention villages was thus < 50% of that in control villages ($p < 0.0001$).

Aflatoxin biomarkers were also used as intermediate endpoints in a Phase IIa chemoprevention trial of oltipraz in Qidong, PRC (Jacobson *et al.* 1997; Kensler *et al.* 1998; Wang *et al.* 1999). This was a placebo-controlled, double-blind study in which participants were randomized to receive placebo or 125 mg oltipraz daily or 500 mg oltipraz weekly. Urinary AFM_1 levels were reduced by 51% compared with the placebo group in persons receiving the 500 mg weekly dose. No significant differences were seen in urinary AFM_1 levels in the 125 mg group

compared with placebo. This effect was thought to be due to inhibition of cytochrome P450 1A2 activity. Median levels of AFB_1-mercapturic acid (a glutathione conjugate derivative) were elevated 2.6-fold in the 125 mg group, but were unchanged in the 500 mg group. Increased AFB_1-mercapturic acid reflects induction of aflatoxin conjugation through the actions of glutathione *S*-transferases. The apparent lack of induction in the 500 mg group probably reflects masking due to diminished AFB,8,9-epoxide formation for conjugation through the inhibition of CYPlA2 seen in this group.

This strategy was extended to chlorophyllin, an anticarcinogen in experimental models when given in large molar excess relative to the carcinogen at or around the time of exposure. One hundred and eighty healthy adults from Qidong were randomly assigned to ingest 100 mg chlorophyllin or a placebo three times a day for 4 months. The primary endpoint was modulation of levels of AFB_1–N^7-Gua in urine samples collected 3 months into the intervention. Chlorophyllin consumption at each meal led to an overall 55% reduction in median urinary levels of this aflatoxin biomarker compared to those taking placebo (Egner et al. 2001). Recently, we tested whether drinking hot water infusions of 3 day-old broccoli sprouts, containing defined concentrations of glucosinolates as a stable precursor of the anticarcinogen sulphoraphane, could alter the disposition of aflatoxin. Two hundred healthy adults drank infusions containing either 400 or < 3 μmol glucoraphanin nightly for 2 weeks. Urinary levels of AFB_1–N^7-Gua were not different between the two intervention arms ($p = 0.68$). However, measurement of urinary levels of dithiocarbamates (sulphoraphane metabolites) indicated striking interindividual differences in bioavailability. Presumably, there were individual differences in the rates of hydrolysis of glucoraphan to sulphoraphane by the intestinal microflora of the study participants. Nonetheless, an inverse association was observed for excretion of dithiocarbamates and AFB_1–N^7-Gua adducts ($r = 0.31$; $p = 0.002$) in individuals receiving broccoli sprout glucosinolates (Kensler et al. 2005).

25.5. SUSCEPTIBILITY

One of the foundations for aflatoxin research in human populations has been the accumulated knowledge concerning the enzymes involved in their metabolism, both in humans and animals (Guengerich et al. 1998). This has permitted a number of case–control studies of the risk of HCC in relation to polymorphisms in cytochrome P450 (CYP), glutathione S-transferase (GST) and other enzymes. However, a frequent limitation of this type of study is that the functional significance of the polymorphism is unclear. In this context, exposure biomarkers have been used to investigate the potential influence of polymorphisms by examining whether there is an association between the

biomarker and polymorphism which could indicate a difference in the way aflatoxin is metabolized at the individual level. For example, Wild et al. (1993) measured serum aflatoxin–albumin in Gambian children in relation to GSTM1 genotype and in Gambian adults in relation to GSTM1, GSTT1, GSTP1 and epoxide hydrolase polymorphisms (Wild et al. 2000) but found no major differences in adduct levels by genotype. Only the GSTM1-null genotype was associated with a modest increase in aflatoxin–albumin levels in adults and this effect was restricted to non-HBV infected individuals. CYP3A4 activity, as judged by urinary cortisol metabolite ratio, was also not associated with albumin adduct level. Similarly, Kensler et al. (1998) found no association between aflatoxin–albumin and GSTM1 genotype in Chinese adults from Qidong County.

The possibility that polymorphisms in DNA repair enzymes could affect the levels of AFB_1–N7-gua adducts has been less extensively studied. Lunn et al. (1999) examined the levels of AFB1–DNA adducts in placental DNA from Taiwanese mothers in relation to polymorphisms in the DNA repair enzyme, XRCC1. The presence of at least one allele of polymorphism 399Gln was associated with a two- to three-fold higher risk of detectable adducts.

In general, these studies illustrate another application of biomarkers in molecular epidemiological studies, viz. to contribute to the biological plausibility of associations between polymorphisms in genes on the causal pathway, in this example aflatoxin metabolism or DNA repair, and disease risk.

25.6. BIOMARKERS TO ELUCIDATE MECHANISMS OF INTERACTION

As detailed above, there is evidence of a multiplicative increase in HCC risk in individuals exposed to both HBV and aflatoxins. Earlier studies in HBV transgenic mice and woodchucks also suggest a synergism between the two risk factors (Bannasch et al. 1995; Sell et al. 1991). An understanding of the molecular mechanisms behind this interaction would be informative in relation to public health measures to reduce HCC incidence. Biomarkers have also been useful in investigating hypotheses in relation to this interaction in the development of HCC.

One possible mechanism of interaction is by chronic HBV infection altering the expression of aflatoxin metabolising enzymes and consequently the amount of mutagenic DNA damage resulting from a given exposure. In this respect, studies in HBV transgenic mouse lineages revealed an induction of CYPs, viz. 1A and 2A5, in association with liver injury consequent to overexpression of the HBV transgenes (Chemin *et al.* 1996, 1999; Kirby *et al.* 1994a). Furthermore, induction of CYP enzymes was been observed in mice and hamsters where liver injury was induced by infection with bacteria and parasites (Chomarat *et al.* 1997; Kirby *et al.* 1994b), suggesting an effect of liver injury *per se* rather than a specific effect of HBV. One study assessed the impact of HBV infection on CYP3A4 activity in aflatoxin-exposed Gambians but reported no association (Wild *et al.* 2000). The potential effects of liver injury on metabolism are not limited to CYP enzymes. For example, an increase in GSTπ was observed in the HBV transgenic mice (Chemin *et al.* 1999) and expression of GSTα class enzymes was significantly decreased in HepG2 cells that were HBV-transfected; transfection of the *HBx* gene into these cells also decreased the amount of GSTα class protein (Jaitovitch-Groisman *et al.* 2000). A study of non-tumourous human liver showed that GST activity is significantly decreased in the presence of HBV DNA (Zhou *et al.* 1997), again suggesting that viral infection may compromise the ability of hepatocytes to detoxify chemical carcinogens.

A more indirect approach to assessing the impact of HBV infection on aflatoxin metabolism has been analogous to that described above for genetic polymorphisms, i.e. to examine the level of aflatoxin biomarkers with respect to viral infection, assuming that this will reflect interindividual differences in metabolism as well as exposure. In West Africa higher aflatoxin–albumin levels have been observed in young children who were HBsAg-positive compared to those who were not (Allen *et al.* 1992; Turner *et al.* 2000; Wild *et al.* 1993). Similar observations have been reported in a study of 200 adolescents from Taiwan (Chen *et al.* 2001). In contrast, this effect of viral infection was not seen in Chinese adults (Wang *et al.* 1996a).

One possibility is that viral infection has more marked effects on aflatoxin metabolism early in life.

An alternative hypothesis regarding the mechanism of interaction between HBV and aflatoxin is that carcinogen exposure may alter viral infection and replication. In ducklings, AFB$_1$ treatment resulted in a significant increase in serum and liver HBV DNA level, and in liver viral RNA and HBV large envelope protein (Barraud *et al.* 1999). This study suggests that AFB1 may lead to enhanced hepadnaviral gene expression. Consistent with this, HepG2 cells transfected with recircularized HBV and treated with AFB1 (10–40 μmol/l) also showed a two- to three-fold increase in HBsAg level 96 h post-treatment (Banerjee *et al.* 2000).

Aflatoxins are potent immunosuppressive agents in animals and yet this has been little examined in human populations. Turner *et al.* (2003) found that Gambian children positive for aflatoxin–albumin adduct had lower salivary IgA levels than children with non-detectable biomarker levels, whilst Jiang *et al.* (2005) reported differences in lymphocyte populations in Ghanain adults in relation to aflatoxin exposure. In principle, an altered immune response to HBV infection as a result of aflatoxin exposure could alter the risk of becoming chronically infected with HBV and the subsequent risk of HCC. The availability of biomarkers permits this hypothesis and other health effects of aflatoxin exposure, e.g. growth impairment in young children (Gong *et al.* 2004) to be examined in field studies.

The above examples do not represent a comprehensive summary of the potential mechanisms of interaction between aflatoxins and HBV. Rather, they are presented to illustrate ways in which biomarkers might be used to investigate mechanisms of action of environmental exposures, thus further contributing to the overall understanding of disease aetiology and underpinning the rationale for prevention strategies.

Figure 25.5 depicts aflatoxin and HBV biomarkers used in mechanistic, population-based studies whilst Figure 25.6 indicates different opportunities for intervention in this mechanistic context.

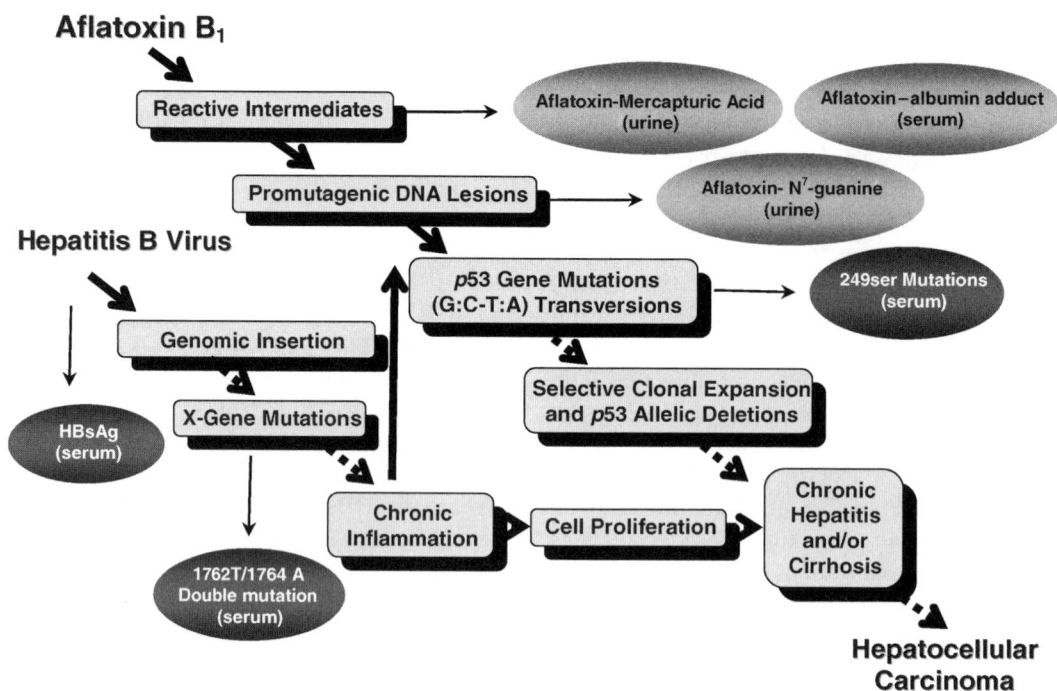

Figure 25.5. Mechanistic-based biomarkers of aflatoxin and HBV targets for intervention in high-risk populations for liver cancer

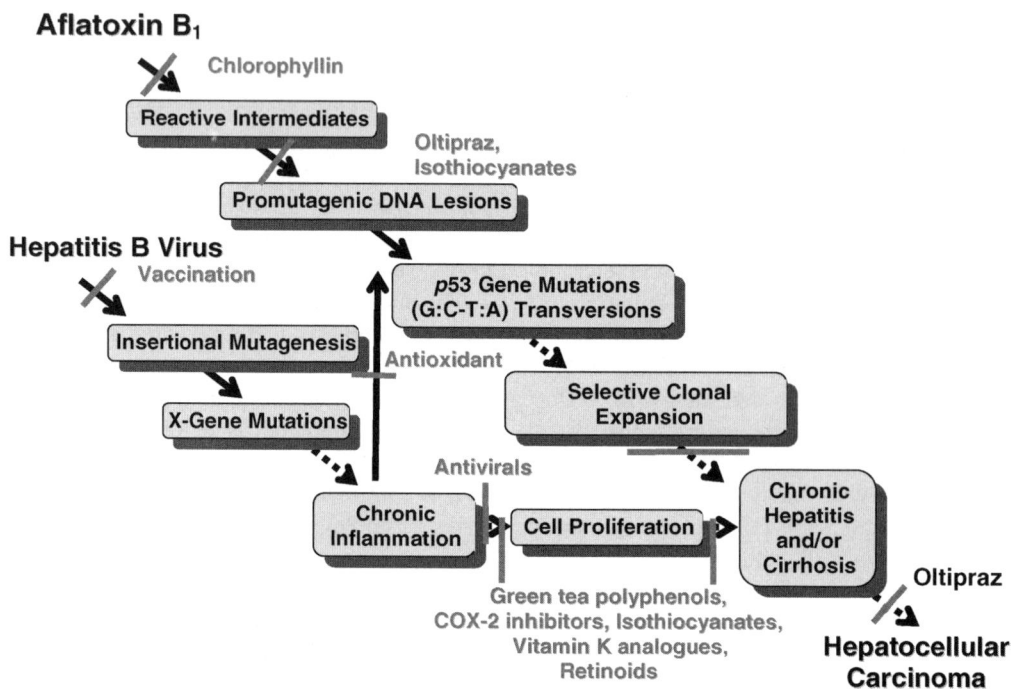

Figure 25.6. Possible intervention strategies for prevantion of aflatoxin-HBV interactions leading to liver cancer

25.7. EARLY DETECTION BIOMARKERS FOR HCC

The development and validation of biomarkers for early detection of disease or for the identification of high-risk individuals is a major translational effort in cancer research. α-Fetoprotein is widely used as a HCC diagnostic marker in high-risk areas because of its ease of use and low cost (Wong *et al.* 2000a). However, this marker suffers from low specificity, owing to its occurrence in diseases other than liver cancer. Moreover, no survival advantage is seen in populations when α-fetoprotein is used in large-scale screening (Chen *et al.* 2003). These inadequacies have contributed to the desire to identify other molecular biomarkers that are possibly more mechanistically associated with HCC development, including hypermethylation of the promoter regions of the *p16*, *p15* and *GSTP1* genes and codon 249 mutations in the *p53* gene (Kirk *et al.* 2005; Shen *et al.* 2002; Wong *et al.* 1999, 2000b). Results from investigations of these genes indicate that these alterations are prevalent in HCC, but there is as of yet limited information on the temporality of these genetic changes prior to clinical diagnosis.

Several studies have now demonstrated that DNA isolated from serum and plasma of cancer patients contains the same genetic aberrations as DNA isolated from an individual's tumour (Jackson *et al.* 2001; Sidransky 2002). The process by which tumour DNA is released into circulating blood is unclear but may result from accelerated necrosis, apoptosis or other processes (Anker *et al.* 1999). While the detection of specific *p53* mutations in liver tumours has provided insight into the aetiology of certain liver cancers, as described above, the application of these specific mutations to the early detection of cancer also offers great promise for prevention (Sidransky and Hollstein 1996). In a seminal study, Kirk *et al.* (2000) reported the detection of codon 249 *p53* mutations in the plasma of liver cancer patients from The Gambia; however, the mutational status of the tumours was not known. These authors also reported a small number of cirrhosis patients having this mutation in plasma and, given the strong relation between cirrhosis and future development of HCC, raised

the possibility of this mutation being an early detection marker. Jackson *et al.* (2001), used short oligonucleotide mass analysis (SOMA), in lieu of DNA sequencing for analysis of specific *p53* mutations in HCC samples. Analysis of 20 plasma and tumour pairs showed 11 tumours containing the specific mutation; six of the paired plasma samples exhibited the same mutation.

The temporality of the detection of this mutation in plasma before and after the clinical diagnosis of HCC was facilitated by the availability of longitudinally collected plasma samples from a cohort of 1638 high-risk individuals in Qidong, PRC, that have been followed since 1992 (Jackson *et al.* 2003). The results showed that in samples collected prior to liver cancer diagnosis, 21.7% of the plasma samples had detectable levels of the codon 249 mutation. The persistence (tracking) of this pre-diagnosis marker in samples collected annually from participants was borderline statistically significant ($p = 0.066$, two-tailed). The codon 249 mutation in *p53* was detected in 44.6% of all plasma samples following the diagnosis of HCC. Collectively these data suggest that nearly one-half of the potential patients with this marker can be detected at least 1 year and in one case 5 years prior to diagnosis.

Using a novel internal standard plasmid, plasma concentrations of *p53* codon 249-mutated DNA were quantified by SOMA in 89 HCC cases, 42 cirrhotic patients, and 131 control subjects without liver disease, all from areas of high aflatoxin exposure regions in The Gambia (Lleonart *et al.* 2005). The hepatocellular carcinoma cases had higher median plasma concentrations of the *p53* mutation (2800 copies/ml; interquartile range, 500–11 000) compared with either cirrhotic (500 copies/ml; interquartile range, 500–2600) or control subjects (500 copies/ml; interquartile range, 500–2000; $p < 0.05$). Levels of $> 10\,000$ copies of *p53* codon 249 mutation/ml plasma were also significantly associated with the diagnosis of HCC (OR, 15; 95% CI 1.6–140) when compared with cirrhotic patients. Potential applications for the quantification of this alteration of DNA in plasma include selection of appropriate high-risk individuals for targeted prophylactic intervention or early therapy.

A double mutation in the HBV genome, an adenine to thymine transversion at nucleotide 1762 and a guanine-to-adenine transition at nucleotide 1764 (*1762T/1764A*), has been found in many HCC cases (Arbuthnot and Kew 2001; Hou *et al.* 1999). Kuang *et al.* (2004) examined, with mass spectrometry, the temporality of the HBV *1762T/1764A* double mutation in plasma and tumours. Initial studies found 52/70 (74.3%) tumours from Qidong, PRC, contained this HBV mutation. Paired plasma samples were available for six of the tumour specimens; four tumours had the HBV *1762T/1764A* mutation, while three of the paired plasma samples were also positive. The potential predictive value of this biomarker was explored using stored plasma samples from a study of 120 residents of Qidong who had been monitored for aflatoxin exposure and HBV infection. After 10 years passive follow-up there were six cases of major liver disease and all had detectable levels of the HBV *1762T/1764A* mutation up to 8 years prior to diagnosis. Finally, 15 liver cancers were selected from a prospective cohort of 1638 high-risk individuals in Qidong and the HBV *1762T/1764A* mutation was detected in 8/15 cases (53.3%) prior to cancer. The tracking of detection of this mutation was statistically significant ($p = 0.022$, two-tailed). We have therefore found that a pre-diagnosis biomarker of specific HBV mutations can be measured in plasma and suggest this marker for use as an intermediate end-point in prevention and intervention trials.

25.8. SUMMARY AND PERSPECTIVES FOR THE FUTURE

Over the past 25 years, the development and application of molecular biomarkers reflecting events from exposure to manifestation of clinical diseases has rapidly expanded our knowledge of the mechanisms of disease pathogenesis. These biomarkers will have increasing potential for early detection, treatment, interventions and prevention. Biomarkers derived from toxicant/carcinogen metabolism include a variety of parent compounds and metabolites in body fluids and excreta, which serve as biomarkers of internal dose.

Molecular epidemiological studies that employ carcinogen–macromolecular adduct measurements are likely to be widely applied in the future and have the potential to generate hypotheses regarding underlying biological mechanisms that subsequently can be tested in the laboratory. Use of advanced techniques, such as rapidly developing metabolomics and proteomics techniques, including NMR–MS and matrix-assisted laser desorption/ionization (MALDI)–MS, and incorporation of validated biomarkers into large-scale studies will help to understand the complex nature of such gene–environment and chemical–biological interactions. In the future, integration of data from these biomarkers, together with other environmental and host susceptibility factors in molecular epidemiological studies of human cancer, will assist in the elucidation of human cancer risk.

The molecular epidemiology investigations of aflatoxin, HBV and HCC probably represent one of the most extensive datasets in the field and this work may serve as a template for future studies of the role of other environmental agents in human diseases with chronic, multi-factorial aetiologies. The development of these biomarkers has been based upon the knowledge of the biochemistry and toxicology of aflatoxins, gleaned from both experimental and human studies. These biomarkers have subsequently been utilized in experimental models to provide data on the modulation of these markers under different situations of disease risk. This systematic approach provides encouragement for design and successful implementation of preventive interventions in exposed human populations.

Acknowledgments

This work was supported in part by Grants R01 CA39416, P01 ES006052 and P30 ES003819 from the US PHS.

References

Aguilar F, Harris CC, Sun T *et al.* 1994. Geographic variation of *p53* mutational profile in non-malignant human liver. *Science* **264**: 1317–1319.

Aguilar F, Hussain SP, Cerutti P. 1993. Aflatoxin-B(1) induces the transversion of G–T in codon 249 of the *p53* tumor-suppressor gene in human hepatocytes. *Proc Natl Acad Sci USA* **90**: 8586–8590.

Allen SJ, Wild CP, Wheeler JG *et al.* 1992. Aflatoxin exposure, malaria and hepatitis-B infection in rural Gambian children. *Trans R Soc Trop Med Hyg* **86**: 426–430.

Anker P, Mulcahy H, Chen XQ *et al.* 1999. Detection of circulating tumour DNA in the blood (plasma/serum) of cancer patients. *Cancer Metast Rev* **18**: 65–73.

Arbuthnot P, Kew M. 2001. Hepatitis B virus and hepatocellular carcinoma. *Int J Exp Pathol* **82**: 77–100.

Banerjee R, Caruccio L, Zhang YJ *et al.* 2000. Effects of carcinogen-induced transcription factors on the activation of hepatitis B virus expression in human hepatoblastoma HepG2 cells and its implication on hepatocellular carcinomas. *Hepatology* **32**: 367–374.

Bannasch P, Khoshkhou NI, Hacker HJ *et al.* 1995. Synergistic hepatocarcinogenic effect of hepadnaviral infection and dietary aflatoxin B-1 in woodchucks. *Cancer Res* **55**: 3318–3330.

Barraud L, Guerret S, Chevallier M *et al.* 1999. Enhanced duck hepatitis B virus gene expression following aflatoxin B-1 exposure. *Hepatol* **29**: 1317–1323.

Bennett RA, Essigmann JM, Wogan GN. 1981. Excretion of an aflatoxin-guanine adduct in the urine of aflatoxin-B1 treated rats. *Cancer Res* **41**: 650–654.

Block TM, Mehta AS, Fimmel CJ *et al.* 2003. Molecular viral oncology of hepatocellular carcinoma. *Oncogene* **22**: 5093–5107.

Bolton MG, Munoz A, Jacobson LP *et al.* 1993. Transient intervention with Oltipraz protects against aflatoxin-induced hepatic tumorigenesis. *Cancer Res* **53**: 3499–3504.

Bosch FX, Munoz N. 1988. *Prospects for Epidemiological Studies on Hepatocellular Cancer as a Model for Assessing Viral and Chemical Interactions.* IARC Scientific Publications No. **89**. IARC: Lyon, France; 427–438.

Bressac B, Kew M, Wands J *et al.* 1991. Selective G-mutation to T-mutation of *p53* gene in hepatocellular carcinoma from Southern Africa. *Nature* **350**: 429–431.

Bulatao-Jayme J, Almero EM, Castro MCA *et al.* 1982. A case–control dietary study of primary liver cancer risk from aflatoxin exposure. *Int J Epidemiol* **11**: 112–119.

Busby WF. 1984. Aflatoxins. In *Chemical Carcinogens*, C. Searle (ed.). American Chemical Society: Washington, DC; 945–1136.

Campbell TC, Caedo JP, Bulataoj J *et al.* 1970. Aflatoxin-M1 in human urine. *Nature* **227**: 403.

Chemin I, Ohgaki H, Chisari FV *et al.* 1999. Altered expression of hepatic carcinogen metabolizing enzymes with liver injury in HBV transgenic mouse lineages expressing various amounts of hepatitis B surface antigen. *Liver* **19**: 81–87.

Chemin I, Takahashi S, Belloc C *et al.* 1996. Differential induction of carcinogen metabolizing enzymes in a transgenic mouse model of fulminant hepatitis. *Hepatology* **24**: 649–656.

Chen CJ, Wang LY, Lu SN *et al.* 1996. Elevated aflatoxin exposure and increased risk of hepatocellular carcinoma. *Hepatology* **24**: 38–42.

Chen JG, Parkin DM, Chen QG *et al.* 2003. Screening for liver cancer: results of a randomized controlled trial in Qidong, China. *J Med Screen* **10**: 204–209.

Chen SY, Chen CJ, Chou SR *et al.* 2001. Association of aflatoxin B-1-albumin adduct levels with hepatitis B surface antigen status among adolescents in Taiwan. *Cancer Epidemiol Biomarkers Prevent* **10**: 1223–1226.

Chomarat P, Sipowicz MA, Diwan BA *et al.* 1997. Distinct time courses of increase in cytochromes P450 1A2, 2A5 and glutathione *S*-transferases during the progressive hepatitis associated with *Helicobacter hepaticus*. *Carcinogenesis* **18**: 2179–2190.

Denissenko MF, Chen JX, Tang MS *et al.* 1997. Cytosine methylation determines hot spots of DNA damage in the human *p53* gene. *Proc Natl Acad Sci USA* **94**: 3893–3898.

Denissenko MF, Koudriakova TB, Smith L *et al.* 1998. The *p53* codon 249 mutational hotspot in hepatocellular carcinoma is not related to selective formation or persistence of aflatoxin B-1 adducts. *Oncogene* **17**: 3007–3014.

Eaton DA, Groopman JD. 1994. *The Toxicology of Aflatoxins: Human Health, Veterinary and Agricultural Significance.* Academic Press: San Diego, CA.

Egner PA, Gange SJ, Dolan PM *et al.* 1995. Levels of aflatoxin–albumin biomarkers in rat plasma are modulated by both long-term and transient interventions with Oltipraz. *Carcinogenesis* **16**: 1769–1773.

Egner PA, Wang JB, Zhu YR *et al.* 2001. Chlorophyllin intervention reduces aflatoxin–DNA adducts in individuals at high risk for liver cancer. *Proc Natl Acad Sci USA* **98**: 14601–14606.

Ellis WO, Smith JP, Simpson BK *et al.* 1991. Aflatoxins in food – occurrence, biosynthesis, effects on organisms, detection, and methods of control. *Crit Rev Food Sci Nutr* **30**: 403–439.

Essigmann JM, Croy RG, Nadzan AM *et al.* 1977. Structural identification of major DNA adduct formed by aflatoxin-B1 *in vitro*. *Proc Natl Acad Sci USA* **74**: 1870–1874.

Foster PL, Eisenstadt E, Miller JH. 1983. Base substitution mutations induced by metabolically activated aflatoxin-B1. *Proc Natl Acad Sci USA Biol Sci* **80**: 2695–2698.

Gan LS, Skipper PL, Peng XC *et al.* 1988. Serum albumin adducts in the molecular epidemiology of aflatoxin carcinogenesis – correlation with aflatoxin-B1 intake

and urinary excretion of aflatoxin M1. *Carcinogenesis*
9: 1323–1325.

Gong YY, Hounsa A, Egal S *et al.* 2004. Postweaning
exposure to aflatoxin results in impaired child growth:
a longitudinal study in Benin, west Africa. *Environ
Health Perspect* **112**: 1334–1338.

Greenblatt MS, Bennett WP, Hollstein M *et al.* 1994.
Mutations in the *p53* tumor-suppressor gene – clues to
cancer etiology and molecular pathogenesis. *Cancer
Res* **54**: 4855–4878.

Groopman JD, Dematos P, Egner PA *et al.* 1992a.
Molecular dosimetry of urinary aflatoxin-N7-gua-
nine and serum aflatoxin albumin adducts predicts
chemoprotection by 1,2-dithiole-3-thione in rats.
Carcinogenesis **13**: 101–106.

Groopman JD, Donahue PR, Zhu JQ *et al.* 1985. Aflatoxin
metabolism in humans – detection of metabolites and
nucleic acid adducts in urine by affinity chromatography. *Proc Natl Acad Sci USA* **82**: 6492–6496.

Groopman JD, Hall AJ, Whittle H *et al.* 1992b. Molecular
dosimetry of aflatoxin–N7–guanine in human urine
obtained in the Gambia, West Africa. *Cancer Epidemiol
Biomarkers Prevent* **1**: 221–227.

Groopman JD, Hasler JA, Trudel LJ *et al.* 1992c.
Molecular dosimetry in rat urine of aflatoxin–N7–
guanine and other aflatoxin metabolites by multiple
monoclonal antibody affinity chromatography and
immunoaffinity high-performance liquid chromatography. *Cancer Res* **52**: 267–274.

Groopman JD, Kensler TW. 1999. The light at the end of
the tunnel for chemical-specific biomarker: daylight or
headlight? *Carcinogenesis* **20**: 1–11.

Groopman JD, Trudel LJ, Donahue PR *et al.* 1984. High-
affinity monoclonal antibodies for aflatoxins and their
application to solid-phase immunoassays. *Proc Natl
Acad Sci USA Biol Sci* **81**: 7728–7731.

Groopman JD, Zhu JQ, Donahue PR *et al.* 1992d. Molecular
dosimetry of urinary aflatoxin–DNA adducts in people living in Guangxi Autonomous Region, People's
Republic Of China. *Cancer Res* **52**: 45–52.

Guengerich FP, Johnson WW, Shimada T *et al.* 1998.
Activation and detoxication of aflatoxin B-1. *Mutat
Res Fund Mol Mech Mutagen* **402**: 121–128.

Hall AJ, Wild CP. 1994. Epidemiology of aflatoxin
related disease. In *The Toxicology of Aflatoxins: Human
Health, Veterinary and Agricultural Significance*, Eaton
DA, Groopman JD (eds). Academic Press: New York;
233–258

Hou JL, Lau GKK, Cheng JJ *et al.* 1999. T-1762/
A(1764) variants of the basal core promoter of hepa-
titis B virus; serological and clinical correlations in
Chinese patients. *Liver* **19**: 411–417.

Hsu IC, Metcalf RA, Sun T *et al.* 1991. Mutational
hotspot in the *p53* gene in human hepatocellular carcinomas. *Nature* **350**: 427–428.

Hulka BS. 1991. Epidemiologic studies using biological markers – issues for epidemiologists. *Cancer
Epidemiol Biomarkers Prevent* **1**: 13–19.

IARC. 2002. *Some Traditional Herbal Medicines, Some
Mycotoxins, Naphthalene and Styrene*. Monographs
on the Evaluation of Carcinogenic Risks to Humans
No. 82. IARC: Lyon, France.

Jackson PE, Kuang SY, Wang JB *et al.* 2003. Prospective
detection of codon 249 mutations in plasma of hepatocellular carcinoma patients. *Carcinogenesis* **24**:
1657–1663.

Jackson PE, Qian GS, Friesen MD *et al.* 2001. Specific
p53 mutations detected in plasma and tumors of hepatocellular carcinoma patients by electrospray ionization mass spectrometry. *Cancer Res* **61**: 33–35.

Jacobson LP, Zhang BC, Zhu YR *et al.* 1997. Oltipraz
chemoprevention trial in Qidong, People's Republic
of China: study design and clinical outcomes. *Cancer
Epidemiol Biomarkers Prevent* **6**: 257–265.

Jaitovitch-Groisman I, Fotouhi-Ardakani N, Schecter RL
et al. 2000. Modulation of glutathione *S*-transferase-α
by hepatitis B virus and the chemopreventive drug
Oltipraz. *J Biol Chem* **275**: 33395–33403.

Jiang Y, Jolly PE, Ellis WO *et al.* 2005. Aflatoxin B1
albumin adduct levels and cellular immune status in
Ghanaians. *Int Immunol* **17**: 807–814.

Kensler TW, Chen JG, Egner PA *et al.* 2005. Effects of
glucosinolate-rich broccoli sprouts on urinary levels
of aflatoxin–DNA adducts and phenanthrene tetraols
in a randomized clinical trial in He Zuo township, Qidong, People's Republic of China. *Cancer
Epidemiol Biomarkers Prevent* **14**: 2605–2613.

Kensler TW, Egner PA, Davidson NE *et al.* 1986.
Modulation of aflatoxin metabolism, aflatoxin–N7–
guanine formation, and hepatic tumorigenesis in rats
fed ethoxyquin – role of induction of glutathione *S*-
transferases. *Cancer Res* **46**: 3924–3931.

Kensler TW, Egner PA, Trush MA *et al.* 1985.
Modification of aflatoxin B-1 binding to DNA *in vivo*
in rats fed phenolic antioxidants, ethoxyquin and a
dithiothione. *Carcinogenesis* **6**: 759–763.

Kensler TW, Egner PA, Wang JB *et al.* 2004.
Chemoprevention of hepatocellular carcinoma in aflatoxin endemic areas. *Gastroenterology* **127**: S310–318.

Kensler TW, Gange SJ, Egner PA *et al.* 1997. Predictive
value of molecular dosimetry: Individual vs. group
effects of oltipraz on aflatoxin–albumin adducts and
risk of liver cancer. *Cancer Epidemiol Biomarkers
Prevent* **6**: 603–610.

Kensler TW, Groopman JD, Wogan GN. 1996. *Use of Carcinogen–DNA and Carcinogen–Protein Adduct Biomarkers for Cohort Selection and as Modifiable End Points in Chemoprevention Trials.* IARC Scientific Publications No 139. IARC: Lyons, France; 237–248.

Kensler TW, He X, Otieno M *et al.* 1998. Oltipraz chemoprevention trial in Qidong, People's Republic of China: modulation of serum aflatoxin albumin adduct biomarkers. *Cancer Epidemiol Biomarkers Prevent* 7: 127–134.

Kensler TW, Qian GS, Chen JG, Groopman JD. 2003. Translational strategies for cancer prevention in liver. *Nat Rev Cancer* 3: 321–329.

Kirby GM, Chemin I, Montesano R *et al.* 1994a. Induction of specific cytochrome P450S involved in aflatoxin B-1 metabolism in hepatitis-B virus transgenic mice. *Mol Carcinog* 11: 74–80.

Kirby GM, Pelkonen P, Vatanasapt V *et al.* 1994b. Association of liver fluke (*Opisthorchis viverrini*) infestation with increased expression of cytochrome-P450 and carcinogen metabolism in male hamster liver. *Mol Carcinog* 11: 81–89.

Kirk GD, Bah E, Montesano R. 2006. Molecular epidemiology of human liver cancer: insights into etiology, pathogenesis and prevention from The Gambia, West Africa. *Carcinogenesis* 27: 2070–2082.

Kirk GD, Camus-Randon AM, Mendy M *et al.* 2000. Ser-249 *p53* mutations in plasma DNA of patients with hepatocellular carcinoma from the Gambia. *J Natl Cancer Inst* 92: 148–153.

Kirk GD, Lesi OA, Mendy M *et al.* 2004. The Gambia Liver Cancer Study: infection with hepatitis B and C and the risk of hepatocellular carcinoma in West Africa. *Hepatology* 39: 211–219.

Kirk GD, Lesi OA, Mendy M *et al.* 2005. 249(ser) *TP53* mutation in plasma DNA, hepatitis B viral infection, and risk of hepatocellular carcinoma. *Oncogene* 24: 5858–5867.

Kuang SY, Jackson PE, Wang JB *et al.* 2004. Specific mutations of hepatitis B virus in plasma predict liver cancer development. *Proc Natl Acad Sci USA* 101: 3575–3580.

Laken SJ, Jackson PE, Kinzler KW *et al.* 1998. Genotyping by mass spectrometric analysis of short DNA fragments. *Nat Biotechnol* 16: 1352–1356.

Lleonart ME, Kirk GD, Villar S *et al.* 2005. Quantitative analysis of plasma TP53 249(Ser)-mutated DNA by electrospray ionization mass spectrometry. *Cancer Epidemiol Biomarkers Prevent* 14: 2956–2962.

Lok AS, Heathcote EJ, Hoofnagle JH. 2001. Management of hepatitis B: 2000 – summary of a workshop. *Gastroenterology* 120: 1828–1853.

Lok ASF, McMahon BJ. 2001. Chronic hepatitis B. *Hepatology* 34: 1225–1241.

Lunn RM, Langlois RG, Hsieh LL *et al.* 1999. *XRCC1* polymorphisms: effects on aflatoxin B-1–DNA adducts and glycophorin A variant frequency. *Cancer Res* 59: 2557–2561.

Lunn RM, Zhang YJ, Wang LY *et al.* 1997. *p53* mutations, chronic hepatitis B virus infection, and aflatoxin exposure in hepatocellular carcinoma in Taiwan. *Cancer Res* 57: 3471–3477.

Omer RE, Kuijsten A, Kadaru AM *et al.* 2004. Population-attributable risk of dietary aflatoxins and hepatitis B virus infection with respect to hepatocellular carcinoma. *Nutrit Cancer Int J* 48: 15–21.

Ozturk M. 1991. *p53* mutation in hepatocellular carcinoma after aflatoxin exposure. *Lancet* 338: 1356–1359.

Poirier MC, Santella RM, Weston A. 2000. Carcinogen macromolecular adducts and their measurement. *Carcinogenesis* 21: 353–359.

Puisieux A, Lim S, Groopman J *et al.* 1991. Selective targeting of *p53* gene mutational hotspots in human cancers by etiologically defined carcinogens. *Cancer Res* 51: 6185–6189.

Qian GS, Ross RK, Yu MC *et al.* 1994. A follow-up study of urinary markers of aflatoxin exposure and liver cancer risk in Shanghai, Peoples Republic of China. *Cancer Epidemiol Biomarkers Prevent* 3: 3–10.

Roebuck BD, Liu YL, Rogers AE *et al.* 1991. Protection against aflatoxin-B1-induced hepatocarcinogenesis in F344 rats by 5-(2-pyrazinyl)-4-methyl-1,2-dithiole-3-thione (oltipraz) – predictive role for short-term molecular dosimetry. *Cancer Res* 51: 5501–5506.

Ross RK, Yuan JM, Yu MC *et al.* 1992. Urinary aflatoxin biomarkers and risk of hepatocellular carcinoma. *Lancet* 339: 943–946.

Sabbioni G, Skipper PL, Buchi G *et al.* 1987. Isolation and characterization of the major serum–albumin adduct formed by aflatoxin B1 *in vivo* in rats. *Carcinogenesis* 8: 819–824.

Santella RM. 1999. Immunological methods for detection of carcinogen–DNA damage in humans. *Cancer Epidemiol Biomarkers Prevent* 8: 733–739.

Scholl PF, Turner PC, Sutcliffe AE *et al.* 2006. Quantitative comparison of aflatoxin B-1 serum albumin adducts in humans by isotope dilution mass spectrometry and ELISA. *Cancer Epidemiol Biomarkers Prevent* 15: 823–826.

Sell S, Hunt JM, Dunsford HA *et al.* 1991. Synergy between hepatitis-B virus expression and chemical hepatocarcinogens in transgenic mice. *Cancer Res* 51: 1278–1285.

Shen LL, Ahuja N, Shen Y *et al.* 2002. DNA methylation and environmental exposures in human hepatocellular carcinoma. *J Natl Cancer Inst* 94: 755–761.

Sidransky D. 2002. Emerging molecular markers of cancer. *Nat Rev Cancer* **2**: 210–219.

Sidransky D, Hollstein M. 1996. Clinical implications of the *p53* gene. *Annu Rev Med* **47**: 285–301.

Sun CA, Wang LY, Chen CJ *et al.* 2001. Genetic polymorphisms of glutathione *S*-transferases M1 and TI associated with susceptibility to aflatoxin-related hepatocarcinogenesis among chronic hepatitis B carriers: a nested case–control study in Taiwan. *Carcinogenesis* **22**: 1289–1294.

Sun ZT, Lu PX, Gail MH *et al.* 1999. Increased risk of hepatocellular carcinoma in male hepatitis B surface antigen carriers with chronic hepatitis who have detectable urinary aflatoxin metabolite M1. *Hepatology* **30**: 379–383.

Tanaka Y, Hanada K, Mizokami M *et al.* 2002. A comparison of the molecular clock of hepatitis C virus in the United States and Japan predicts that hepatocellular carcinoma incidence in the United States will increase over the next two decades. *Proc Natl Acad Sci USA* **99**: 15584–15589.

Turner PC, Mendy M, Whittle H *et al.* 2000. Hepatitis B infection and aflatoxin biomarker levels in Gambian children. *Trop Med Int Health* **5**: 837–841.

Turner PC, Moore SE, Hall AJ *et al.* 2003. Modification of immune function through exposure to dietary aflatoxin in Gambian children. *Environ Health Perspect* **111**: 217–220.

Turner PC, Sylla A, Gong YY *et al.* 2005. Reduction in exposure to carcinogenic aflatoxins by postharvest intervention measures in west Africa: a community-based intervention study. *Lancet* **365**: 1950–1956.

Wang JS, Groopman JD. 1998. Biomarkets for carcinogen exposure: tumor initiation. In *Molecular Biology of the Toxic Response*, Walace K, Puga A (eds). Taylor and Francis: Washington, DC; 145–166

Wang JS, Links JM, Groopman JD. 2001. Molecular epidemiology and biomarkers. In *Genetic Toxicology and Cancer Risk Assessment*, Choi, WN (ed). Marcel Dekker: New York; 269–296

Wang JS, Qian GS, Zarba A *et al.* 1996a. Temporal patterns of aflatoxin–albumin adducts in hepatitis B surface antigen-positive and antigen-negative residents of Daxin, Qidong county, People's Republic of China. *Cancer Epidemiol Biomarkers Prevent* **5**: 253–261.

Wang JS, Shen XN, He X *et al.* 1999. Protective alterations in phase 1 and 2 metabolism of aflatoxin B-1 by oltipraz in residents of Qidong, People's Republic of China. *J Natl Cancer Inst* **91**: 347–354.

Wang LY, Hatch M, Chen CJ *et al.* 1996b. Aflatoxin exposure and risk of hepatocellular carcinoma in Taiwan. *Int J Cancer* **67**: 620–625.

Wang XW, Hussain SP, Huo TI *et al.* 2002. Molecular pathogenesis of human hepatocellular carcinoma. *Toxicology* **181**: 43–47.

Wild CP, Fortuin M, Donato F *et al.* 1993. Aflatoxin, liver enzymes, and hepatitis B virus infection in Gambian children. *Cancer Epidemiol Biomarkers Prevent* **2**: 555–561.

Wild CP, Garner RC, Montesano R *et al.* 1986. Aflatoxin B1 binding to plasma albumin and liver DNA upon chronic administration to rats. *Carcinogenesis* **7**: 853–858.

Wild CP, Hall AJ. 2000. Primary prevention of hepatocellular carcinoma in developing countries. *Mutat Res Revs Mutat Res* **462**: 381–393.

Wild CP, Hudson GJ, Sabbioni G *et al.* 1992. Dietary intake of aflatoxins and the level of albumin-bound aflatoxin in peripheral blood in the Gambia, West Africa. *Cancer Epidemiol Biomarkers Prevent* **1**: 229–234.

Wild CP, Jiang YZ, Allen SJ *et al.* 1990. Aflatoxin–albumin adducts in human sera from different regions of the World. *Carcinogenesis* **11**: 2271–2274.

Wild CP, Turner PC. 2002. The toxicology of aflatoxins as a basis for public health decisions. *Mutagenesis* **17**: 471–481.

Wild CP, Yin F, Turner PC *et al.* 2000. Environmental and genetic determinants of aflatoxin–albumin adducts in the Gambia. *Int J Cancer* **86**: 1–7.

Wogan GN. 1989. Markers of exposure to carcinogens. *Environ Health Perspect* **81**: 9–17.

Wong IH, Lo YM, Lai PB *et al.* 2000a. Relationship of p16 methylation status and serum α-fetoprotein concentration in hepatocellular carcinoma patients. *Clin Chem* **46**: 1420–1422.

Wong IHN, Lo YMD, Zhang J *et al.* 1999. Detection of aberrant p16 methylation in the plasma and serum of liver cancer patients. *Cancer Res* **59**: 71–73.

Wong N, Lai P, Pang E *et al.* 2000b. Genomic aberrations in human hepatocellular carcinomas of differing etiologies. *Clin Cancer Res* **6**: 4000–4009.

Yim HJ, Lok ASF. 2006. Natural history of chronic hepatitis B virus infection: what we knew in 1981 and what we know in 2005. *Hepatology* **43**: S173–181.

Yu MW, Lien JP, Chiu YH *et al.* 1997. Effect of aflatoxin metabolism and DNA adduct formation on hepatocellular carcinoma among chronic hepatitis B carriers in Taiwan. *J Hepatol* **27**: 320–330.

Zhou TL, Evans AA, London WT *et al.* 1997. Glutathione *S*-transferase expression in hepatitis B virus-associated human hepatocellular carcinogenesis. *Cancer Res* **57**: 2749–2753.

26

Complex Exposures – Air Pollution

Steffen Loft*, Elvira Vaclavik Bräuner, Lykke Forchhammer,
Marie Pedersen, Lisbeth E. Knudsen and Peter Møller

University of Copenhagen, Denmark

26.1. INTRODUCTION

Epidemiological studies have consistently associated exposure to ambient air pollution with pulmonary and cardiovascular diseases and cancer (Brunekreef *et al.* 2002; Pope *et al.* 2002). Ambient air contains a complex mixture of toxins, including particulate matter (PM), irritant gases and volatile organic compounds (VOCs). Furthermore, PM represents complex mixtures with enormous variations in size, chemical composition, shape, surface and charge, due to variable sources, atmospheric chemical reactions and meteorological conditions. Nevertheless, for association with health outcomes and regulation, particles are usually only defined by their size as PM_{10} and $PM_{2.5}$, which represent the mass of particles with aerodynamic diameters <10 and 2.5 μm, respectively. Interest is presently focused on the ultrafine particle (UFP) fraction with an aerodynamic diameter <100 nm. These are abundant in numbers but contribute little to particle mass (Sioutas *et al.* 2005). UFPs are considered important with respect to health effects, due to their very high alveolar deposition fraction, large surface area, chemical composition, ability to induce inflammation and potential translocation to systemic circulation (Daigle *et al.* 2003; Donaldson *et al.* 2001, 2002; Kreyling *et al.* 2002; Oberdorster *et al.* 2004; Schins *et al.* 2004; Semmler *et al.* 2004). Vehicle emissions, particularly related to

diesel engines from both heavy-duty and smaller vehicles, are a major source of ambient UFPs, which penetrate indoors (Franck *et al.* 2003; Levy *et al.* 2002). In addition, diesel engines are also major contributors to the ambient levels of the irritant gas NO_2, whereas photochemical atmospheric reactions with organic matter generates the irritant gas ozone, which is consumed by NO in the formation of NO_2. VOCs, such as benzene and polycyclic aromatic hydrocarbons (PAHs), are also emitted from combustion processes in vehicle engines, heating and industry. Distribution of these pollutants varies in time and space, but most of the epidemiologically-based knowledge of associations between air pollution and health effects are based on very crude exposure assessment. This has relied on the assumption that all subjects within a large geographical area have the same exposure, as assessed by few central outdoor air emission monitors. Nevertheless, some studies have attempted to model exposure at residential addresses. Exposure gradients have been achieved by comparing different geographical areas, which requires control for other risk factors, such as smoking, life style, diet and occupation, or by the temporal co-variation in air pollution levels and health outcomes, with possible confounding factors coming from other time-related causes. Accordingly, the complex nature of air pollution makes exposure assessment and establishing a causal relationship with biological effects difficult. Molecular epidemiology may provide strong tools with the application of biomarkers of internal dose, biologically effective dose and early biological effect, as well as markers of individual susceptibility, particularly when applied in combination with assessment of the external dose using personal monitors. Biomarkers with fast response may also be validated in experimental settings with controlled exposure of human subjects. The molecular epidemiology paradigm is depicted in relation to air pollution in Figure 26.1, which serves as an outline of the present chapter.

26.2. PERSONAL MONITORING OF EXTERNAL DOSE

The actual personal exposure is determined not only by ambient levels with large variation in space and time but also by indoor penetration and sources, differences in time-activity patterns as well as anatomical and physiological differences. Personal monitors for PM, gases, VOCs and PAHs are available. Some small portable monitors allow time-resolved monitoring of concentrations of a number of UFPs, whereas most other personal monitors require cumulated collection on some matrix for later analysis, which only allows the measurement of average exposure over time periods. Such equipment has been used to study determinants of personal exposure to $PM_{2.5}$, UFPs, VOCs, NO and PAHs in outdoor and indoor

environments, as well as associations with biomarkers of internal and biologically effective doses (Georgiadis *et al.* 2001; Koistinen *et al.* 2001; Kousa *et al.* 2001; Molnar *et al.* 2006; Sørensen *et al.* 2003b, 2003d, 2005a, 2005b; Vinzents *et al.* 2005). These studies have described exposure in outdoor settings and also shown that indoor sources of air pollutants, including passive smoking, wood smoke, candle burning and cooking are important in relation to personal exposure to air pollutants, in keeping with the fact that people spend around 90% of their time indoors (Jenkins *et al.* 1992). The association between personal exposure and outdoor $PM_{2.5}$ concentration varies notably between studies (Brauer *et al.* 2001; Koistinen *et al.* 2001; Rojas-Bracho *et al.* 2000). Interestingly, some of the stronger associations have been found in elderly study populations (Williams *et al.* 2000).

26.3. BIOMARKERS OF INTERNAL DOSE AND AIR POLLUTANTS

Exposure to carbon monoxide and some VOCs may be monitored in exhaled breath or blood samples, whereas there are no biomarkers of internal dose for the irritant gases (NO_2 and ozone) in air pollution. Identification of biomarkers of internal dose associated with PM is difficult, due to the complex chemical, physical and biological composition and because the causative constituents

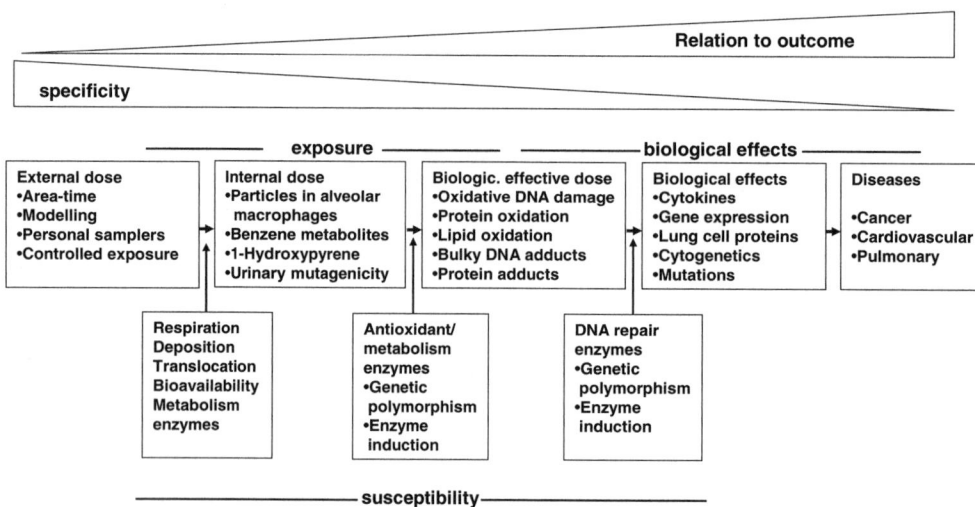

Figure 26.1 The molecular epidemiology paradigm applied to air pollution

related to toxic mechanisms have not yet been fully determined. However, a total burden of PM may be assessed by the carbon load in alveolar macrophages obtained from induced sputum (Kulkarni *et al.* 2005). Urinary mutagenicity has been used to address the total burden of mutagenic material from air pollution (Hansen *et al.* 2004), although diet may also be an important contributor. Another possibility would be to identify relevant tracer or causal compounds in PM, which could be measured in body fluids.

PAH metabolites

Metabolites of PAHs have been proposed as biomarkers of recent exposure to particles (Strickland *et al.* 1999). PAHs are believed to be important compounds among the genotoxic agents present in urban air, where they are primarily associated with the respirable fraction of particles. Most focus has been on 1-hydroxypyrene (1-HP), a urinary excreted metabolite of pyrene mainly found in the volatile fraction of diesel exhaust (Dor *et al.* 1999; Jongeneelen 1997; Keimig *et al.* 1983); diet is also a source. Elevated 1-HP levels have been shown in occupational settings, e.g. in coke-oven workers, road pavers, bus drivers, among smokers and in subjects ingesting charbroiled meat (Buckley *et al.* 1992; Hansen *et al.* 2004; Jongeneelen *et al.* 1990). Similarly, elevated 1-HP excretion has been found among subjects living in areas with very high ambient particle levels, such as in Poland or South Korea (Motykiewicz *et al.* 1998; Yang *et al.* 2003). However, in other studies no difference with respect to 1-HP excretion between subjects from urban and suburban areas (Kyrtopoulos *et al.* 2001; Scherer *et al.* 2000) or between police officers with large differences in personal exposure to benzo(α)pyrene in total suspended particulates was observed (Merlo *et al.* 1998). In a personal monitoring study in Copenhagen, no association between exposure to $PM_{2.5}$ or black smoke and 1-HP excretion was observed (Sørensen *et al.* 2003b). Accordingly, 1-HP appears only to be suited as a biomarker of internal dose for very high levels of exposure. It should also be recognized that the 1-HP excretion may be modified by polymorphisms in genes encoding metabolism enzymes, including *GSTM1*, *GSTT1*

and *CYP1A1* (Alexandrie *et al.* 2000; Nerurkar *et al.* 2000; Yang *et al.* 2003).

Benzene and metabolites

Identification of exposure markers is relatively simple when only a single specific compound, such as benzene, is involved. Benzene can be measured in urine or blood and there are a host of metabolites. Of these, the urinary excretion of *trans-trans-muconic* acid (ttMA) and phenylmercapturic acid (PMA) have been most widely used as biomarkers of internal dose in occupational as well as in ambient air studies (Fustinoni *et al.* 2005; Garte *et al.* 2005; Javelaud *et al.* 1998; Qu *et al.* 2000; Sørensen *et al.* 2003d, 2004). Urinary excretion of PMA and benzene may be the most sensitive biomarkers at low levels of exposure. However, it should be recognized that the excretion of ttMA and PMA is also increased by smoking, due to benzene content in tobacco smoke, and the excretion of PMA may be modified by the genotypes of *GSTM1*, *GSTT1*, *CYP2E1* and *NQO1*, which are involved in the metabolism of benzene (Rossi *et al.* 1999; Sørensen *et al.* 2003d, 2004). Other potential confounding factors include that ingestion of sorbic acid also increases ttMA excretion (Scherer *et al.* 1998).

26.4. BIOMARKERS OF BIOLOGICALLY EFFECTIVE DOSE

Mechanisms of oxidative stress induced by particles and other air pollutants

The mechanisms of PM-induced health effects are believed to involve inflammation and oxidative stress (Figure 26.2) (Donaldson *et al.* 2001, 2002; Knaapen *et al.* 2004; Prahalad *et al.* 2000, 2001; Schins *et al.* 2004; Upadhyay *et al.* 2003). The oxidative stress mediated by PM may arise from mixed sources, involving: (a) direct generation of reactive oxygen species (ROS) from the surface of particles (Fubini *et al.* 2004); (b) soluble compounds such as transition metals or organic compounds; (c) altered function of mitochondria or NADPH-oxidase (Knaapen *et al.* 2004; Li *et al.* 2003; Voelkel *et al.* 2003); and (d) activation of inflammatory cells capable of generating ROS and reactive nitrogen species (Knaapen *et al.* 2004;

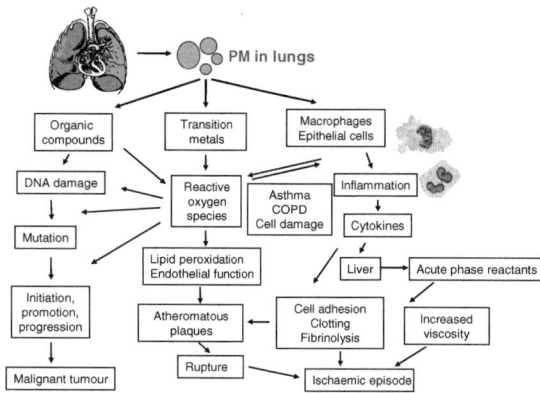

Figure 26.2 Oxidative stress and inflammation as central in the effects of PM on processes involved in cancer, cardiovascular and lung disease

Ma *et al.* 2002; Schins *et al.* 2004). Similarly, ozone and NO_2 are oxidants, whereas benzene metabolism can generate oxidative stress. The organic fraction of urban air particles contains quinone radicals and more may be generated from PAH metabolism with subsequent redox-cycling producing oxidative stress (Bonvallot *et al.* 2001; Dellinger *et al.* 2001; Knaapen *et al.* 2004). Particularly for cancer, bulky DNA adducts from PAH exposure are also thought to be important and particle bound benzo(α)pyrene is proven to be bioavailable for this process (Gerde *et al.* 2001; IARC 1989).

Oxidative stress-induced DNA damage and repair

Free radicals generate a large number of oxidative modifications in DNA, including strand breaks (SB) and base oxidations (Bjelland *et al.* 2003; Cadet *et al.* 2002; Dizdaroglu *et al.* 2002; Dizdaroglu 1992). Among such DNA-damaging products, 8-oxo-7, 8-dihydro-2′-deoxyguanosine (8-oxodG) is probably the most studied oxidation product, due to its relative ease of measurement and pre-mutagenic potential (Kasai 1997). Chemical analysis of oxidative modification in DNA may be hampered by spurious oxidation, although this is less likely in assays based on enzyme nicking of DNA at oxidized bases and determination of SB by methods such as the Comet assay (European Standards Committee on Oxidative DNA Damage (ESCODD) 2003). In DNA, 8-oxodG may be formed by oxidation of

guanine or incorporated during replication or repair as oxidized nucleotides (8-oxodGTP).

Oxidative damage to DNA is repaired by a number of different enzymes, which may change expression level and activity during particle exposure (Risom *et al.* 2005). Oxoguanine glycolase-1 (OGG1) is the base excision repair enzyme that is involved in removal of 8-oxodG (Aburatani *et al.* 1997; Arai *et al.* 2002; Boiteux *et al.* 2000; Klungland *et al.* 1999; Nishimura 2002). The *NUDT1* (formerly known as *MTH1*) gene encodes an 8-oxodGTPase that hydrolyses 8-oxodGTP, and a large number of tumours develop in *NUDT1*-deficient mice (Tsuzuki *et al.* 2001). Other repair pathways for 8-oxodG include nucleotide excision repair processes, mismatch repair and NEIL proteins (Bjelland *et al.* 2003). Due to the dynamics of the repair process, possible changes or differences in the repair rates should be taken into account when interpreting the effects of air pollution on DNA damage levels.

Air pollution exposure and biomarkers of oxidative stress and DNA damage

The knowledge concerning mechanisms of action of air pollutants has prompted the use of markers of oxidative stress and DNA damage, as well as bulky DNA adducts for human biomonitoring in relation to ambient air (Risom *et al.* 2005; Sørensen *et al.* 2003a; Vineis *et al.* 2005). However, most studies have compared groups from different geographical areas or different outdoor and indoor occupations with large exposure gradients, but without personal monitoring. Temporal association between external exposure and biomarkers within the same subject has rarely been addressed. Nevertheless, an acute increase in relevant PAH-related adducts has been shown in cells from induced sputum, originating from the airways, after only 3 h of exposure to environmental tobacco smoke, whereas no response was found in peripheral blood cells (Besaratinia *et al.* 2002).

Recent summaries of cross-sectional studies have rather consistently shown elevated levels of bulky DNA adducts or PAH–protein adducts in populations with substantial PAH exposure in ambient air (Vineis *et al.* 2005). Similarly, urinary excretion of 8-oxodG was elevated in bus drivers from central areas compared with bus drivers from suburban/rural areas of Copenhagen (Loft *et al.* 1999). Association between 8-oxodG in white blood cells and 1-HP excretion has been found among non-occupationally exposed subjects in South Korea (Kim *et al.* 2003). Increased levels of 8-oxodG and DNA SB have been found in the upper airway epithelium in exposed subjects from Mexico city, although it is difficult to know whether this was due to PM, ozone or other exposures (Calderon-Garciduenas *et al.* 1996, 1999; Fortoul *et al.* 2003, 2004; Rojas *et al.* 2000; Valverde *et al.* 1997). Moreover, comparison of rural residents, suburban residents, residents along streets with dense traffic and taxi drivers in Cotonou, Benin, showed clear stepwise differences in SB and oxidized purines in mononuclear blood cells, corresponding to a strong exposure gradient for UFP and benzene (Avogbe *et al.* 2005). In contrast, no difference in SB in lymphocytes was found between subjects from central Athens and rural Greece, although smoking had an effect (Piperakis *et al.* 2000). A similar lack of difference was found between inhabitants of a highly polluted area (Teplice) compared with a low-polluted area in the Czech Republic (Sram *et al.* 2000). PAH–DNA adducts in maternal and infant cord white blood cells were significantly correlated with PM_{10} in Poland (Perera *et al.* 1999), whereas bulky DNA adducts have not been associated with personal $PM_{2.5}$ levels in studies in either Denmark or Greece (Kyrtopoulos *et al.* 2001; Sørensen *et al.* 2003b). It is possible that the variable associations are confounded by dietary and unaccounted indoor exposures, as well as being influenced by variable susceptibility due to genetic polymorphisms, e.g. in GSTs or diets rich in fruits and vegetables. Moreover, most of the employed assays are not well-validated, have laboratory-specific protocols, limited standardization and yield quite different levels between laboratories (Farmer *et al.* 2007; Vineis & Gallo 2007). Thus, measurement of oxidative damage to DNA may suffer from spurious oxidation during sample preparation, whereas postlabelling DNA adducts and two-dimensional thin-layer chromatography may detect more than just bulky adducts.

In light of the relatively modest levels of ambient air pollutants presently found in many Western cities, and to account for indoor sources of exposure, personal monitors of external dose are important for demonstrating associations with biologically effective dose by means of biomarkers with short half-lives, such as DNA base oxidation. By using this approach in a study including 50 subjects measured over the four seasons, associations for personal exposure to $PM_{2.5}$ were demonstrated. In particular, association between the content of vanadium and chromium and the level of 8-oxodG in lymphocytes was shown, although no associations were found in relation to urinary excretion of 8-oxodG (Sørensen *et al.* 2003b, 2005b). UFP may be the most relevant fraction of PM and the availability of new equipment has made personal monitoring possible. This was recently employed in a study on 15 subjects with measurements on 6 different days, of which 5 days included biking in street air and 1 day included biking indoors (Vinzents *et al.* 2005). The individual level of oxidized purines, including 8-oxodG, in mononuclear blood cells was significantly and independently associated with exposure to UFP during biking and while staying indoors; the traffic-related exposure appeared more potent than the indoor UFP.

An association has also been demonstrated between personal exposure to $PM_{2.5}$ and oxidative stress involving protein and lipid in plasma (Sørensen *et al.* 2003c). The protein lesion associated with personal black smoke exposure was 2-amino-adipic semialdehyde, which is an oxidation product of lysine in plasma proteins and reflects changes in damage from a few days up to several weeks of exposure (Young *et al.* 1999, 2002). Similarly, lipid peroxidation in terms of plasma malondialdehyde was significantly related to personal $PM_{2.5}$ exposure in women (Sørensen *et al.* 2003c). These results are in agreement with an earlier study of Copenhagen bus drivers, which found 2-amino-adipic semialdehyde and malondialdehyde levels to be significantly higher in bus drivers from central Copenhagen compared with both bus drivers from rural/suburban areas and Copenhagen postal workers who cycle daily in urban traffic (Autrup *et al.* 1999).

Table 26.1. Use of biomarkers for study of exposure to and effects of air pollution

Biomarker type	Endpoint studied	(Personal) environmental exposure studied	Reference
External dose	Behavioural and environmental determinants	$PM_{2.5}$	(Koistinen et al. 2001)
	Relation to residential indoor, outdoor and workplace concentration	NO_2	(Kousa et al. 2001)
	Relation to bedroom and outdoor concentrations	$PM_{2.5}$, black smoke and NO_2	(Sørensen et al. 2005a)
	Relation to indoor, residential outdoor and background levels	Fine particle trace elements	(Molnar et al. 2006)
	Relation to outdoor concentrations	$PM_{2.5}$	(Brauer et al. 2001; Rojas-Bracho et al. 2000; Williams et al. 2000)
Internal dose	Induced sputum – for total burden of PM	Biomass smoke particles	(Kulkarni et al. 2005)
	Metabolites of PAHs	Particles in urban air	(Strickland et al. 1999)
	1-HP urinary excretion	PAH mixtures, benzene	(Dor et al. 1999; Jongeneelen et al. 1990; Jongeneelen 1997; Kyrtopoulos et al. 2001)
	1-HP urinary excretion	Dietary exposure to PAH	(Buckley et al. 1992)
	1-HP urinary excretion	Ambient air pollution	(Hansen et al. 2004; Motykiewicz et al. 1998; Yang et al. 2003)
	1-HP urinary excretion	Urban and suburban air PAH	(Merlo et al. 1998; Scherer et al. 2000)
	trans-trans-Muconic acid and phenylmercapturic acid	Benzene in urban air	(Fustinoni et al. 2005; Garte et al. 2005; Javelaud et al. 1998; Qu et al. 2000; Sørensen et al. 2003d; Sørensen et al. 2004)
	trans-trans-Muconic acid and phenylmercapturic acid	Smoking (benzene in smoke)	(Rossi et al. 1999)
Biologically effective dose	Bulky DNA adducts	PAH and environmental tobacco smoke	(Georgiadis et al. 2001)
	Oxidative stress in blood	$PM_{2.5}$	(Sørensen et al. 2003c)
	Oxidative damage to DNA	Urban benzene, $PM_{2.5}$, transition metals UFP	(Avogbe et al. 2005; Sørensen et al. 2003b; Sørensen et al. 2003d; Sørensen et al. 2005b; Vinzents et al. 2005)
	PAH-related adducts	Environmental tobacco smoke in pubs	(Besaratinia et al. 2002)
	Bulky DNA adducts	$PM_{2.5}$	(Kyrtopoulos et al. 2001; Sørensen et al. 2003b)
	Urinary excretion of 8-oxo-dG	$PM_{2.5}$, transition metals in $PM_{2.5}$ samples	(Sørensen et al. 2003b; Sørensen et al. 2005b)
	Strand breaks and base oxidations (8-oxo-dG)	UFP	(Vinzents et al. 2005)
	2-Amino-adipic semialdehyde	Black smoke	(Sørensen et al. 2003c)
	Plasma malondialdehyde	$PM_{2.5}$	(Sørensen et al. 2003c)
	2-Amino-adipic semialdehyde and plasma malondialdehyde	Rural, suburban and urban air pollution	(Autrup et al. 1999)

	Urinary excretion of free 8-iso-prostaglandin2alpha	Woodsmoke	(Barregard *et al.* 2006)
Biological effects (inflammation)	Inflammation in airways (IL-5, IL-10, IL-13 and ICAM-1)	NO$_2$, O$_3$, HF	(Bosson *et al.* 2003; Lund *et al.* 2005; Pathmanathan *et al.* 2003; Stenfors *et al.* 2002)
	Redox-sensitive transcription factors and kinases in human airways	Diesel exhaust particles	(Behndig *et al.* 2006; Nordenhall *et al.* 2001; Pourazar *et al.* 2004; Pourazar *et al.* 2005; Stenfors *et al.* 2004)
	Inflammation markers in induced sputum	Diesel exhaust particles and PM$_{2.5}$	(Ghio *et al.* 2004; Nordenhall *et al.* 2000)
	Induced sputum (inflammatory cells, fibronectin, metalloproteinase-p, IL-10)	Occupationally exposed ore miners	(Adelroth *et al.* 2006)
	Exhalation of NO	Ambient PM$_{2.5}$	(Adamkiewicz *et al.* 2004)
	Systemic levels of cytokines (IL-6, TMF-α, C-reactive protein, fibrinogen)	Ambient air pollution and diesel exhaust	(Mills *et al.* 2005; Sørensen *et al.* 2003c)
	Systemic levels of IL-6, serum amyloid A	Wood smoke	(Barregard *et al.* 2006)
	Systemic levels of fibrinogen	PM$_{2.5}$	(Ghio *et al.* 2004)
	Systemic levels of CRP, IL-6	Ambient PM levels	(Dubowsky *et al.* 2006; Peters *et al.* 2001; Seaton *et al.* 1999)
(cell damage)	Plasma Clara cell protein 16	Smoking and other air pollutants	(Berthoin *et al.* 2004; Broeckaert *et al.* 2000)
	Urinary Clara cell protein 16 excretion	O$_3$ and NHCl$_3$	(Bergamaschi *et al.* 2001; Bernard *et al.* 2005; Blomberg *et al.* 2003; Lagerkvist *et al.* 2004)
	Clara cell protein (urine and plasma)	LPS, wood smoke	(Barregard *et al.* 2006; Michel *et al.* 2005)
	Urinary excretion of CC16	PM$_{2.5}$	(Timonen *et al.* 2004)
	Surfactant-associated protein A and B	Ambient air pollution	(Berthoin *et al.* 2004; Hermans *et al.* 2003)
(other)	Micronuclei and altered genome expression in white blood cells	Air pollution	(van Leeuwen *et al.* 2006)
	Cytogenic markers	Occupational exposure and ambient air pollution for children	(Neri *et al.* 2006b; Neri *et al.* 2006a; Norppa *et al.* 2006; Sram *et al.* 2004; Vineis *et al.* 2005)
	Mutations		(Vineis *et al.* 2005)
	Modifying effect of metabolism enzymes on 8-oxodG levels	PM$_{2.5}$	(Sørensen *et al.* 2003b)
Genetic susceptibility	Modifying effect of NQO1 on oxidative DNA damage	Benzene	(Avogbe *et al.* 2005; Nebert *et al.* 2002; Sørensen *et al.* 2003d), (Bauer *et al.* 2003; Kim *et al.* 2004; Wan *et al.* 2002)
	Modifying effect of *OGG1* ser326cys polymorphism on lung cancer	Smoky coal	(Lan *et al.* 2004)
	GST polymorphisms and relation to cardiovascular disease	Ambient air pollution, diesel particles	(Gilliland *et al.* 2004; McCunney 2005; Schwartz *et al.* 2005)

A recent experimental study found increased urinary excretion of free 8-iso-prostaglandin2α, a major F_2-isoprostane and lipid peroxidation product, in nine subjects after 4 h of controlled exposure to relatively high levels of wood smoke (Barregard et al. 2006). These data collectively indicate that inhaled particles can cause protein and lipid oxidation in peripheral blood, which may be important in the pathogenesis of atherosclerosis.

The studies suggest that, even at low exposure concentrations, PM may induce systemic oxidative stress, with effects on DNA as well as other biomolecules, whereas NO_2 shows no such effects (Sørensen et al. 2003b; Sørensen et al. 2003c). At lower levels of exposure there was a small but non-significant relationship between urban background particle concentration and DNA damage in lymphocytes, whereas no association between any measure of exposure and oxidative stress was found. This implies that, compared with background particle concentration, personal exposure is more directly related to the component(s) inducing oxidative stress.

26.5. BIOMARKERS OF BIOLOGICAL EFFECTS

Inflammation

Inflammation is thought to be a central part of the mechanisms of action of PM, NO_2 and ozone with respect to most of the health effects, including pulmonary disease as well as cardiovascular disease and cancer.

The inflammatory process and related oxidative stress may be studied directly in the airways by alveolar lavage, bronchial biopsies, induced sputum or exhaled breath. After controlled exposure to DEP, ozone or NO_2 increased expression of redox-sensitive transcription factors and kinases, cytokines and/or adhesion molecules as well as inflammatory cell infiltration have been found in the bronchial epithelium, whereas the response in the alveolar region was mainly limited to reduced antioxidant concentration (Behndig et al. 2006; Bosson et al. 2003; Lund et al. 2005; Mudway et al. 2004; Pathmanathan et al. 2003; Pourazar et al. 2004, 2005; Stenfors et al. 2002, 2004). Asthma patients may be particularly sensitive in this respect

(Nordenhall et al. 2001). Induced sputum may also be employed to study inflammation after short-term exposure, as shown with controlled exposure to DEP and concentrated $PM_{2.5}$ (Ghio et al. 2004; Nordenhall et al. 2000). Chronic effects can also be studied in induced sputum, as in ore miners with persistently increased levels of inflammatory cells, fibronectin, metalloproteinase-9 and interleukin (IL)-10 in induced sputum compared with controls (Adelroth et al. 2006). Nitric oxide (NO) is produced in inflammatory processes and the exhalation of NO was found to be associated with daily ambient levels of $PM_{2.5}$ in a panel of elderly people from Steubenville, USA (Adamkiewicz et al. 2004). Finally, condensates of exhaled breath may represent the airway lining fluid and the cytokines, oxidative stress products etc. in the condensate can be studied (Liu and Thomas. 2005), although this method has not yet been implemented in relation to air pollution.

Systemic inflammation is thought to be important in the cardiovascular effects of PM. So far, systemic levels of cytokines, such as IL-6 or tumour necrosis factor-α, and the acute phase reactants C-reactive protein and fibrinogen, have not been associated with ambient levels of air pollution or controlled exposures to DEP in healthy subjects (Mills et al. 2005; Sørensen et al. 2003c), whereas a decrease in IL-6 and increased serum amyloid was seen after controlled exposure to wood smoke and increased fibrinogen after exposure to concentrated $PM_{2.5}$ (Barregard et al. 2006; Ghio et al. 2004). In at-risk groups, such as the elderly, particularly those with diabetes, levels of C-reactive protein and/or IL-6 have been found to be associated with daily ambient levels of PM levels (Dubowsky et al. 2006; Peters et al. 2001; Seaton et al. 1999). Secondary to inflammatory responses to exposure to air pollution, the formation and lysis of blood clots may be affected, and this is important for acute coronary events. Indeed, controlled exposure to wood smoke has been found to increase the level of factor VIII and the ratio to the von Willebrand carrying factor (Barregard et al. 2006), whereas the tissue plasminogen activator response important for fibrinolysis was attenuated after exposure to DEP (Mills et al. 2005).

Accordingly, inflammatory responses in the airways have mainly been studied in experimental settings with controlled exposures, and systemic inflammatory responses primarily appear in risk groups at ambient levels or after high exposures in experimental settings.

Cell damage

Clara cell protein 16 (CC16) is expressed as a defence enzyme in Clara cells in the lung and secretion and subsequent plasma levels may be decreased, with chronic damage to lung cells, as induced by smoking or other air pollutants in cross-sectional studies (Berthoin *et al.* 2004; Broeckaert *et al.* 2000). However, short-term damage to the Clara cells may cause an acute increase in the CC16 levels in plasma or urine, reflecting decreased cellular integrity or increased alveolar permeability (Bernard *et al.* 2005; Broeckaert *et al.* 2000). Indeed, controlled exposure to ozone, bacterial lipopolysaccharide and wood smoke has caused increased plasma levels or urinary excretion of CC16 (Barregard *et al.* 2006; Blomberg *et al.* 2003; Michel *et al.* 2005). Effects of ozone have also been shown in studies of ambient air (Bergamaschi *et al.* 2001; Lagerkvist *et al.* 2004). Moreover, urinary CC16 excretion showed co-variation with daily outdoor $PM_{2.5}$ levels in a panel study involving 131 subjects from Helsinki, Ehrfuhrt and Amsterdam (Timonen *et al.* 2004). In addition to CC16, which is eliminated by renal excretion, attention has been drawn to surfactant-associated protein-A and -B, two other pulmonary secretory proteins, which have also shown some association with air pollution in cross-sectional settings (Berthoin *et al.* 2004; Hermans *et al.* 2003).

Gene expression

Noxious insults to any cells should in principle lead to changed expression of genes. Transcriptomics, assessed either genome-wide or as a targeted battery, may have great promise as biomarkers. Changes in gene expression profiles are usually short-term, and this should be relevant for associations with personal exposure monitoring. Indeed, increased expression of inflammation-related genes can be found in human airways after exposure, as described above. However, only toxicological studies have employed a wider transcriptomic approach in the dynamic fashion. Recently, a cross-sectional study showed a higher level of micronuclei as well as an altered pattern of full genome expression in white blood cells in children living in areas with relatively high level of air pollution as compared with children from a low-exposure area (Pedersen *et al.* 2006; van Leeuwen *et al.* 2006). The main biological process that appeared significantly affected was related to nucleosome assembly.

The patterns of resulting proteins, e.g. in plasma, may also prove to be powerful biomarkers, although so far only a few proteins related to inflammation and cell damage have been studied in relation to air pollution (see below).

Cytogenetic markers

Cytogenetic markers in terms of chromosomal aberrations, micronuclei and sister chromatid exchange in lymphocytes stimulated toward division have been studied in a large number of cross-sectional studies comparing different geographical areas or outdoor and indoor occupations (Vineis *et al.* 2005). Many of these, in particular those involving policemen, report increased levels of damage in the exposed population. Moreover, a recent meta-analysis indicates that children living in areas with air pollution or with exposure to other environmental toxins have increased levels of chromosomal aberration and micronuclei (Neri *et al.* 2006b; Neri *et al.* 2006a). New methodological developments suggest that sensitivity can be further enhanced with respect to aberrations by using chromosomal painting (Sram *et al.* 2004). Chromosomal aberrations are of particular interest, because high levels have been shown to be associated with an increased risk of developing cancer in prospective settings, although only at the group level (Norppa *et al.* 2006). However, although these biomarkers are associated with a relevant outcome, they are mainly applicable for cross-sectional studies, because their half-lives are relatively long and it is therefore difficult to follow associations with personal exposure. Combination of cytogenetic biomarkers with biomarkers with early responses may optimize the linkage between exposure and effect.

Mutations

Mutations, particularly in target genes, including oncogenes or tumour suppressor genes, are biomarkers most closely related to outcome. The patterns of such mutations in tumours may also point to mechanisms and sources in relation to air pollution, as reviewed recently (Vineis *et al.* 2005). However, these biomarkers can only be related to cumulative exposures.

26.6. GENETIC SUSCEPTIBILITY AND OXIDATIVE STRESS RELATED TO AIR POLLUTION

Genetic susceptibility related to functional polymorphisms in genes coding for xenobiotic metabolism, antioxidants and DNA repair enzymes are thought to modify the levels of oxidative DNA damage caused by exposure to air pollution, and may need to be considered in the selection of population groups. Only a few studies have addressed this issue with respect to ambient air particles and so far no modifying effects of metabolism enzymes have been found on the association between 8-oxodG levels and personal exposure to $PM_{2.5}$ (Sørensen *et al.* 2003c). For benzene exposure in urban air, the polymorphic NADPH:quinone oxidoreductase (NQO1) catalyses a two-electron reduction of quinones, reducing the risk of their redox cycling, and a modifying effect on oxidative DNA damage would be expected. This has also been shown in two studies in Copenhagen, Denmark, and Cotonou, Benin, where associations were reported between oxidative DNA base damage in terms of 8-oxodG or SB and benzene exposure assessed by urinary excretion of the metabolite PMA in subjects with *NQO1* variant genotypes. In these studies there was no significant association in subjects homozygous for the wild-type (Avogbe *et al.* 2005; Sørensen *et al.* 2003d). This is consistent with the apparent increased susceptibility to benzene toxicity and cancer risk in human subjects associated with variant *NQO1* genotypes and in *NQO$^{-/-}$* mice (Bauer *et al.* 2003; Kim *et al.* 2004; Nebert *et al.* 2002; Wan *et al.* 2002). If urban air particles are present concomitantly with high levels of benzene, it may be difficult to determine exactly which of these is responsible for the effects on DNA damage levels. Indeed, in the study from Cotonou, Benin, there was a strong exposure gradient with respect to both UFP and benzene (Avogbe *et al.* 2005). Nevertheless, NQO1 appeared only to modify the effect of benzene on SB, whereas GSTP1 may have modified the effect of UFP on oxidized purines, including 8-oxodG in DNA, in the present study.

One study has found increased levels of 8-oxodG after biking in ozone-rich ambient air in subjects with the *NQO1* wild-type and *GSTM1*-null genotype, with unchanged levels in other haplotypes. Although ozone is an oxidant, these findings require a complicated mechanistic explanation, if repeatable (Bergamaschi *et al.* 2001).

So far, DNA repair enzymes have not been studied directly in relation to oxidative DNA damage and air pollution. However, the *OGG1* ser326cys polymorphism appeared to modify the risk of lung cancer related to indoor exposure to smoky coal in China, suggesting an important role (Lan *et al.* 2004). Moreover, experimental studies support a role for *OGG1* and its regulation, which could cause differences in DNA damage after short-term exposures to particles and long term exposures, where *OGG1* may be upregulated (Risom *et al.* 2005). Similarly, other defence genes, such as *haem-oxygenase-1*, may be upregulated and modify the effects. This issue needs further investigation in relation to biomarker-based study of DNA damage in relation to air pollution.

Recently, the polymorphisms in GSTs have attracted attention in relation to both cardiovascular- and airway-related health effects of air pollution (McCunney 2005; Schwartz *et al.* 2005), although this has yet to be implemented in biomarker-based studies. Both *GSTM1* and *GSTP1* genotypes were important for the inflammatory response to nasal instillation of DEP (Gilliland *et al.* 2004).

26.7. CONCLUSION

Molecular epidemiology provides strong tools for the study of effects associated with complex exposures, such as air pollution, although there are limitations and it is important to stress the number of confounding factors and experimental difficulties that are involved. Whereas the external dose

Molecular Epidemiology of Chronic Diseases, Edited by C. P. Wild, P. Vineis, and S. Garte